AN INTRODUCTION TO NUMERICAL METHODS AND ANALYSIS

AN INTRODUCTION TO NUMERICAL METHODS AND ANALYSIS

Second Edition

JAMES F. EPPERSON
Mathematical Reviews

Published by John Wiley & Sons, Inc., Hoboken, New Jersey.
Published simultaneously in Canada.

1007561052

For general information on our other products and services please contact our Customer Care Department within the United States at (800) 762-2974, outside the United States at (317) 572-3993 or fax (317) 572-4002.

Wiley also publishes its books in a variety of electronic formats. Some content that appears in print, however, may not be available in electronic formats. For more information about Wiley products, visit our web site at www.wiley.com.

Library of Congress Cataloging-in-Publication Data:

Epperson, James F., author.
  An introduction to numerical methods and analysis / James F. Epperson, Mathematical Reviews. — Second edition.
    pages cm
  Includes bibliographical references and index.
  ISBN 978-1-118-36759-9 (hardback)
  1. Numerical analysis. I. Title.
  QA297.E568 2013
  518—dc23   2013013979

Printed in the United States of America.

10 9 8 7 6 5 4 3 2

To Mom (1920–1986) and Ed (1917–2012)
a story of love, faith, and grace

CONTENTS

PREFACE

Preface to the Second Edition

This third version of the text is officially the Second Edition, because the second version was officially dubbed the Revised Edition. Now that the confusing explanation is out of the way, we can ask the important question: What is new?

- I continue to chase down typographical errors, a process that reminds me of herding cats. I'd like to thank everyone who has sent me information on this, especially Prof. Mark Mills of Central College in Pella, Iowa. I have become resigned to the notion that a typo-free book is the result of a (*slowly* converging) limiting process, and therefore is unlikely to be actually achieved. But I do keep trying.

- The text now assumes that the student is using MATLAB for computations, and many MATLAB routines are discussed and used in examples. I want to emphasize that this book is still a *mathematics* text, not a primer on how to use MATLAB.

- Several biographies were updated as more complete information has become widely available on the Internet, and a few have been added.

- Two sections, one on adaptive quadrature (§5.8.3) and one on adaptive methods for ODEs (§6.9) have been re-written to reflect the decision to rely more on MATLAB.

- Chapter 9 (A Survey of Numerical Methods for Partial Differential Equations) has been extensively re-written, with more examples and graphics.

- New material has been added:

  - Two sections on roots of polynomials. The first (§3.10) introduces the Durand–Kerner algorithm; the second (§8.5) discusses using the companion matrix to find polynomial roots as matrix eigenvalues.

  - A section (§3.12) on very high-order root-finding methods.

  - A section (§4.10) on splines under tension, also known as "taut splines;"

  - Sections on the finite element method for ODEs (§6.10.3) and some PDEs (§9.2);

  - An entire chapter (Chapter 10) on spectral methods[1].

- Several sections have been modified somewhat to reflect advances in computing technology.

- Later in this preface I devote some time to outlining possible chapter and section selections for different kinds of courses using this text.

It might be appropriate for me to describe how I see the material in the book. Basically, I think it breaks down into three categories:

- The fundamentals: All of Chapters 1 and 2, most of Chapters 3 (3.1, 3.2, 3.3, 3.5, 3.8, 3.9), 4 (4.1, 4.2, 4.3, 4.6, 4.7, 4.8, 4.11), and 5 (5.1, 5.2, 5.3, 5.4, 5.7); this is the basic material in numerical methods and analysis and should be accessible to any well-prepared students who have completed a standard calculus sequence.

- Second level: Most of Chapters 6, 7, and 8, plus much of the remaining sections from Chapters 3 (3.4, 3.6, 3.7, 3.10), 4 (4.4, 4.5), and 5 (5.5, 5.6), and some of 6 (6.8) and 7 (7.7); this is the more advanced material and much of it (from Chap. 6) requires a course in ordinary differential equations or (Chaps. 7 and 8) a course in linear algebra. It is still part of the core of numerical methods and analysis, but it requires more background.

- Advanced: Chapters 9 and 10, plus the few remaining sections from Chapters 3, 4, 5, 6, 7, and 8.

- It should go without saying that precisely *what* is considered "second level" or "advanced" is largely a matter of taste.

As always, I would like to thank my employer, *Mathematical Reviews*, and especially the Executive Editor, Graeme Fairweather, for the study leave that gave me the time to prepare (for the most part) this new edition; my editor at John Wiley & Sons, Susanne Steitz-Filler, who does a good job of putting up with me; an anonymous copy-editor at Wiley who saved me from a large number of self-inflicted wounds; and—most of all—my family of spouse Georgia, daughter Elinor, son James, and Border Collie mutts Samantha

---

[1] The material on spectral methods may well not meet with the approval of experts on the subject, as I presented the material in what appears to be a very non-standard way, and I left out a lot of important issues that make spectral methods, especially for time dependent problems, practical. I did it this way because I wanted to write an introduction to the material that would be accessible to students taking a first course in numerical analysis/methods, and also in order to avoid cluttering up the exposition with what I considered to be "side issues." I appreciate that these side issues have to be properly treated to make spectral methods practical, but since this tries to be an elementary text, I wanted to keep the exposition as clean as possible.

and Dylan. James was not yet born when I first began writing this text in 1997, and now he has finished his freshman year of high school; Elinor was in first grade at the beginning and graduated from college during the final editing process for this edition. I'm very proud of them both! And I can never repay the many debts that I owe to my dear spouse.

Online Material

There will almost surely be some online material to supplement this text. At a minimum, there will be

- MATLAB files for computing and/or reading Gaussian quadrature (§5.6) weights and abscissas for $N = 2^m$, $m = 0, 1, 2, \ldots, 10$.

- Similar material for computing and/or reading Clenshaw-Curtis (§10.3) weights and abscissas.

- Color versions of some plots from Chapter 9.

- It is possible that there will be an entire additional section for Chapter 3.

To access the online material, go to

`www.wiley.com/go/epperson2edition`

The webpage should be self-explanatory.

A Note About the Dedication

The previous editions were dedicated to six teachers who had a major influence on the author's mathematics education: Frank Crosby and Ed Croteau of New London High School, New London, CT; Prof. Fred Gehring and Prof. Peter Duren of the University of Michigan Department of Mathematics; and Prof. Richard MacCamy and Prof. George J. Fix of the Department of Mathematics, Carnegie-Mellon University, Pittsburgh, PA. (Prof. Fix served as the author's doctoral advisor.) I still feel an unpayable debt of gratitude to these men, who were outstanding teachers, but I felt it appropriate to express my feelings about my parents for this edition, hence the new dedication to the memory of my mother and step-father.

Course Outlines

One can define several courses from this book, based on the level of preparation of the students and the number of terms the course runs, as well as the level of theoretical detail the instructor wishes to address. Here are some example outlines that might be used.

- A single semester course that does not assume any background in linear algebra or differential equations, and which does not emphasize theoretical analysis of methods:

– Chapter 1 (all sections[2]);
– Chapter 2 (all sections[3]);
– Chapter 3 (Sections 3.1–3.3, 3.8–3.10);
– Chapter 4 (Sections 4.1–4.8);
– Chapter 5 (Sections 5.1–5.7).

- A two-semester course which assumes linear algebra and differential equations for the second semester:

  – Chapter 1 (all sections);
  – Chapter 2 (all sections);
  – Chapter 3 (Sections 3.1–3.3, 3.8–3.10);
  – Chapter 4 (Sections 4.1–4.8);
  – Chapter 5 (Sections 5.1–5.7).
  – Semester break should probably come here.
  – Chapter 6 (6.1–6.6; 6.10 if time/preparation permits)
  – Chapter 7 (7.1–7.6)
  – Chapter 8 (8.1–8.4)
  – Additional material at the instructor's discretion.

- A two-semester course for well-prepared students:

  – Chapter 1 (all sections);
  – Chapter 2 (all sections);
  – Chapter 3 (Sections 3.1–3.10; 3.11 at the discretion of the instructor);
  – Chapter 4 (Sections 4.1–4.11, 4.12.1, 4.12.3; 4.12.2 at the discretion of the instructor);
  – Chapter 5 (Sections 5.1–5.7, 5.8.1; other sections at the discretion of the instructor).
  – Semester break should probably come here.
  – Chapter 6 (6.1–6.8; 6.10 if time/preparation permits; other sections at the discretion of the instructor)
  – Chapter 7 (7.1–7.8; other sections at the discretion of the instructor)
  – Chapter 8 (8.1–8.4)
  – Additional material at the instructor's taste and discretion.

Some sections appear to be left out of all these outlines. Most textbooks are written to include extra material, to facilitate those instructors who would like to expose their students to different material, or as background for independent projects, etc.

I want to encourage anyone—teachers, students, random readers—to contact me with questions, comments, suggestions, or remaining typos. My professional email is still jfe@ams.org

---

[2]§§1.5 and 1.6 are included in order to expose students to the issue of approximation; if an instructor feels that the students in his or her class do not need this exposure, these sections can be skipped in favor of other material from later chapters.
[3]The material on ODEs and tridiagonal systems can be taught to students who have not had a normal ODE or linear algebra course.

Computer Access

Because the author no longer has a traditional academic position, his access to modern software is limited. Most of the examples were done using a very old and limited version of MATLAB from 1994. (Some were done on a Sun workstation, using FORTRAN code, in the late 1990s.) The more involved and newer examples were done using public access computers at the University of Michigan's Duderstadt Center, and the author would like to express his appreciation to this great institution for this.

A Note to the Student

(This is slightly updated from the version in the First Edition.) This book was written to be read. I am under no illusions that this book will compete with the latest popular novel for interest or thrilling narrative. But I have tried very hard to write a book on mathematics that can be read by students. So do not simply buy the book, work the exercises, and sell the book back to the bookstore at the end of the term. Read the text, think about what you have read, and ask your instructor questions about the things that you do not understand.

Numerical methods and analysis is a very different area of mathematics, certainly different from what you have seen in your previous courses. It is not harder, but the differentness of the material makes it seem harder. We worry about different issues than those in other mathematics classes. In a calculus course you are typically asked to compute the derivative or antiderivative of a given function, or to solve some equation for a particular unknown. The task is clearly defined, with a very concrete notion of "the right answer." Here, we are concerned with computing approximations, and this involves a slightly different kind of thinking. We have to understand what we are approximating well enough to construct a reasonable approximation, and we have to be able to think clearly and logically enough to analyze the accuracy and performance of that approximation. One former student has characterized this course material as "rigorously imprecise" or "approximately precise." Both are appropriate descriptions. Rote memorization of procedures is not of use here; it is vital in this course that the student learn the underlying concepts. Numerical mathematics is also *experimental* in nature. A lot can be learned simply by trying something out and seeing how the computation goes.

Preface to the Revised Edition

First, I would like to thank John Wiley for letting me do a Revised Edition of *An Introduction to Numerical Methods and Analysis*, and in particular I would like to thank Susanne Steitz and Laurie Rosatone for making it all possible.

So, what's new about this edition? A number of things. For various reasons, a large number of typographical and similar errors managed to creep into the original edition. These have been aggressively weeded out and fixed in this version. I'd like to thank everyone who emailed me with news of this or that error. In particular, I'd like to acknowledge Marzia Rivi, who translated the first edition into Italian and who emailed me with many typos, Prof. Nicholas Higham of Manchester University, Great Britain, and Mark Mills of Central College in Pella, Iowa. I'm sure there's a place or two where I did something silly like reversing the order of subtraction. If anyone finds any error of any sort, please email me at jfe@ams.org.

I considered adding sections on a couple of new topics, but in the end decided to leave the bulk of the text alone. I spent some time improving the exposition and presentation, but most of the text is the same as the first edition, except for fixing the typos.

I would be remiss if I did not acknowledge the support of my employer, the American Mathematical Society, who granted me a study leave so I could finish this project. Executive Director John Ewing and the Executive Editor of Mathematical Reviews, Kevin Clancey, deserve special mention in this regard. Amy Hendrikson of TeXnology helped with some LaTeX issues, as did my colleague at Mathematical Reviews, Patrick Ion. Another colleague, Maryse Brouwers, an extraordinary grammarian, helped greatly with the final copyediting process.

The original preface has the URL for the text website wrong; just go to www.wiley.com and use their links to find the book. The original preface also has my old professional email. The updated email is jfe@ams.org; anyone with comments on the text is welcome to contact me.

But, as is always the case, it is the author's immediate family who deserve the most credit for support during the writing of a book. So, here goes a big thank you to my wife, Georgia, and my children, Elinor and Jay. Look at it this way, kids: The end result will pay for a few birthdays.

Preface (To the First Edition)

This book is intended for introductory and advanced courses in numerical methods and numerical analysis, for students majoring in mathematics, sciences, and engineering. The book is appropriate for both single-term survey courses or year-long sequences, where students have a basic understanding of at least single-variable calculus and a programming language. (The usual first courses in linear algebra and differential equations are required for the last four chapters.)

To provide maximum teaching flexibility, each chapter and each section begins with the basic, elementary material and gradually builds up to the more advanced material. This same approach is followed with the underlying theory of the methods. Accordingly, one can use the text for a "methods" course that eschews mathematical analysis, simply by not covering the sections that focus on the theoretical material. Or, one can use the text for a survey course by only covering the basic sections, or the extra topics can be covered if you have the luxury of a full year course.

The objective of the text is for students to learn where approximation methods come from, why they work, why they sometimes don't work, and when to use which of many techniques that are available, and to do all this in a style that emphasizes readability and usefulness to the beginning student. While these goals are shared by other texts, it is the development and delivery of the ideas in this text that I think makes it different.

A course in numerical computation—whether it emphasizes the theory or the methods—requires that students think quite differently than in other mathematics courses, yet students are often not experienced in the kind of problem-solving skills and mathematical judgment that a numerical course requires. Many students react to mathematics problems by pigeon-holing them by category, with little thought given to the meaning of the answer. Numerical mathematics demands much more judgment and evaluation in light of the underlying theory, and in the first several weeks of the course it is crucial for students to adapt their way of thinking about and working with these ideas, in order to succeed in the course.

To enable students to attain the appropriate level of mathematical sophistication, this text begins with a review of the important calculus results, and why and where these ideas play an important role in this course. Some of the concepts required for the study of computational mathematics are introduced, and simple approximations using Taylor's theorem are treated in some depth, in order to acquaint students with one of the most common and basic tools in the science of approximation. Computer arithmetic is treated in perhaps more detail than some might think necessary, but it is instructive for many students to see the actual basis for rounding error demonstrated in detail, at least once.

One important element of this text that I have not seen in other texts is the emphasis that is placed on "cause and effect" in numerical mathematics. For example, if we apply the trapezoid rule to (approximately) integrate a function, then the error should go down by a factor of 4 as the mesh decreases by a factor of 2; if this is not what happens, then almost surely there is either an error in the code or the integrand is not sufficiently smooth. While this is obvious to experienced practitioners in the field, it is not obvious to beginning students who are not confident of their mathematical abilities. Many of the exercises and examples are designed to explore this kind of issue.

Two common starting points to the course are root-finding or linear systems, but diving in to the treatment of these ideas often leaves the students confused and wondering what the point of the course is. Instead, this text provides a second chapter designed as a "toolbox" of elementary ideas from across several problem areas; it is one of the important innovations of the text. The goal of the toolbox is to acclimate the students to the culture of numerical methods and analysis, and to show them a variety of simple ideas before proceeding to cover any single topic in depth. It develops some elementary approximations and methods that the students can easily appreciate and understand, and introduces the students, in the context of very simple methods and problems, to the essence of the analytical and coding issues that dominate the course. At the same time, the early development of these tools allows them to be used later in the text in order to derive and explain some algorithms in more detail than is usually the case.

The style of exposition is intended to be more lively and "student friendly" than the average mathematics text. This does not mean that there are no theorems stated and proved correctly, but it does mean that the text is not slavish about it. There is a reason for this: The book is meant to be read by the students. The instructor can render more formal anything in the text that he or she wishes, but if the students do not read the text because they are turned off by an overly dry regimen of definition, theorem, proof, corollary, then all of our effort is for naught. In places, the exposition may seem a bit wordier than necessary, and there is a significant amount of repetition. Both are deliberate. While brevity is indeed better mathematical style, it is not necessarily better pedagogy. Mathematical textbook exposition often suffers from an excess of brevity, with the result that the students cannot follow the arguments as presented in the text. Similarly, repetition aids learning, by reinforcement.

Nonetheless I have tried to make the text mathematically complete. Those who wish to teach a lower-level survey course can skip proofs of many of the more technical results in order to concentrate on the approximations themselves. An effort has been made—not always successfully—to avoid making basic material in one section depend on advanced material from an earlier section.

The topics selected for inclusion are fairly standard, but not encyclopedic. Emerging areas of numerical analysis, such as wavelets, are not (in the author's opinion) appropriate for a first course in the subject. The same reasoning dictated the exclusion of other, more mature areas, such as the finite element method, although that might change in future editions should there be sufficient demand for it. A more detailed treatment of approximation theory, one of the author's favorite topics, was also felt to be poorly suited to a beginning text. It was felt that a better text would be had by doing a good job covering some of the basic ideas, rather than trying to cover everything in the subject.

The text is not specific to any one computing language. Most illustrations of code are made in an informal pseudo-code, while more involved algorithms are shown in a "macro-outline" form, and programming hints and suggestions are scattered throughout the text. The exercises assume that the students have easy access to and working knowledge of software for producing basic Cartesian graphs.

A diskette of programs is *not* provided with the text, a practice that sets this book at odds with many others, but which reflects the author's opinion that students must learn how to write and debug programs that implement the algorithms in order to learn the underlying mathematics. However, since some faculty and some departments structure their courses differently, a collection of program segments in a variety of languages is available on the text web site so that instructors can easily download and then distribute the code to their students. Instructors and students should be aware that these are program *segments*; none of them are intended to be ready-to-run complete programs. Other features of the text web site are discussed below. (*Note:* This material may be removed from the Revised Edition website.)

Exercises run the gamut from simple hand computations that might be characterized as "starter exercises" to challenging derivations and minor proofs to programming exercises designed to test whether or not the students have assimilated the important ideas of each chapter and section. Some of the exercises are taken from application situations, some are more traditionally focused on the mathematical issues for their own sake. Each chapter concludes with a brief section discussing existing software and other references for the topic at hand, and a discussion of material not covered in this text.

Historical notes are scattered throughout the text, with most named mathematicians being accorded at least a paragraph or two of biography when they are first mentioned. This not only indulges my interest in the history of mathematics, but it also serves to engage the interest of the students.

The web site for the text (http://www.wiley.com/epperson) will contain, in addition to the set of code segments mentioned above, a collection of additional exercises for the text, some application modules demonstrating some more involved and more realistic applications of some of the material in the text, and, of course, information about any updates that are going to be made in future editions. Colleagues who wish to submit exercises or make comments about the text are invited to do so by contacting the author at epperson@math.uah.edu.

Notation

Most notation is defined as it appears in the text, but here we include some commonplace items.

$\mathbb{R}$ — The real number line; $\mathbb{R} = (-\infty, \infty)$.

$\mathbb{R}^n$ — The vector space of real vectors of n components.

$\mathbb{R}^{n \times n}$ — The vector space of real $n \times n$ matrices.

$C([a, b])$ — The set of functions f which are defined on the interval $[a, b]$, continuous on all of (a, b), and continuous from the interior of $[a, b]$ at the endpoints.

$C^k([a, b])$ — The set of functions f such that f and its first k derivatives are all in $C([a, b])$.

$C^{p,q}(Q)$ — The set of all functions u that are defined on the two-dimensional domain $Q = \{(x, t) \mid a \leq x \leq b, 0 < t \leq T\}$, and that are p times continuously differentiable in x for all t, and q times continuously differentiable in t for all x.

$\approx$ — Approximately equal. When we say that $A \approx B$, we mean that A and B are approximations to each other. See §1.2.2.

$\equiv$ — Equivalent. When we say that $f(x) = g(x)$, we mean that the two functions agree at the single point x. When we say that $f(x) \equiv g(x)$, we mean that they agree at all points x. The same thing is said by using just the function names, i.e., $f = g$.

$\mathcal{O}$ — On the order of ("big O of"). We say that $A = B + \mathcal{O}(D(h))$ whenever $|A - B| \leq CD(h)$ for some constant C and for all h sufficiently small. See §1.2.3.

u — Machine epsilon. The largest number such that, in computer arithmetic, $1 + \mathbf{u} = 1$. Architecture dependent, of course. See §1.3.

sgn — Sign function. The value of $\text{sgn}(x)$ is 1, -1, or 0, depending on whether or not x is positive, negative, or zero, respectively.

CHAPTER 1

INTRODUCTORY CONCEPTS AND CALCULUS REVIEW

It is best to start this book with a question: What do we mean by "Numerical Methods and Analysis"? What kind of mathematics is this book about?

Generally and broadly speaking, this book covers the mathematics and methodologies that underlie the techniques of *scientific computation*. More prosaically, consider the button on your calculator that computes the sine of the number in the display. Exactly how does the calculator know that correct value? When we speak of using the computer to solve a complicated mathematics or engineering problem, exactly what is involved in making that happen? Are computers "born" with the knowledge of how to solve complicated mathematical and engineering problems? No, of course they are not. Mostly they are *programmed* to do it, and the programs implement algorithms that are based on the kinds of things we talk about in this book .

Textbooks and courses in this area generally follow one of two main themes: Those titled "Numerical Methods" tend to emphasize the implementation of the algorithms, perhaps at the expense of the underlying mathematical theory that explains why the methods work; those titled "Numerical Analysis" tend to emphasize this underlying mathematical theory, perhaps at the expense of some of the implementation issues. The best approach, of course, is to properly mix the study of the algorithms and their implementation ("methods") with the study of the mathematical theory ("analysis") that supports them. This is our goal in this book.

Whenever someone speaks of using a computer to design an airplane, predict the weather, or otherwise solve a complex science or engineering problem, that person is talking about using numerical methods and analysis. The problems and areas of endeavor that use these

An Introduction to Numerical Methods and Analysis, Second Edition. By James F. Epperson
Copyright © 2013 John Wiley & Sons, Inc.

kinds of techniques are continually expanding. For example, *computational mathematics*—another name for the material that we consider here—is now commonly used in the study of financial markets and investment structures, an area of study that does not ordinarily come to mind when we think of "scientific" computation. Similarly, the increasingly frequent use of computer-generated animation in film production is based on a heavy dose of spline approximations, which we introduce in §4.8. And modern weather prediction is based on using numerical methods and analysis to solve the very complicated equations governing fluid flow and heat transfer between and within the atmosphere, oceans, and ground.

There are a number of different ways to break the subject down into component parts. We will discuss the derivation and implementation of the algorithms, and we will also analyze the algorithms, mathematically, in order to learn how best to use them and how best to implement them. In our study of each technique, we will usually be concerned with two issues that often are in competition with each other:

- *Accuracy:* Very few of our computations will yield the exact answer to a problem, so we will have to understand how much error is made, and how to control (or even diminish) that error.

- *Efficiency:* Does the algorithm take an inordinate amount of computer time? This might seem to be an odd question to concern ourselves with—after all, computers are fast, right?—but there are slow ways to do things and fast ways to do things. All else being equal (it rarely is), we prefer the fast ways.

We say that these two issues compete with each other because, generally speaking, the steps that can be taken to make an algorithm more accurate usually make it more costly, that is, less efficient.

There is a third issue of importance, but it does not become as evident as the others (although it is still present) until Chapter 6:

- *Stability*: Does the method produce similar results for similar data? If we change the data by a small amount, do we get vastly different results? If so, we say that the method is unstable, and unstable methods tend to produce unreliable results. It is entirely possible to have an accurate method that is efficiently implemented, yet is horribly unstable; see §6.4.4 for an example of this.

1.1 BASIC TOOLS OF CALCULUS

1.1.1 Taylor's Theorem

Computational mathematics does not require a large amount of background, but it does require a good knowledge of that background. The most important single result in numerical computations, from all of the calculus, is Taylor's Theorem,[1] which we now state:

---

[1]Brook Taylor (1685–1731) was educated at St. John's College of Cambridge University, entering in 1701 and graduating in 1709. He published what we know as Taylor's Theorem in 1715, although it appears that he did not entirely appreciate its larger importance and he certainly did not bother with a formal proof. He was elected a member of the prestigious Royal Society of London in 1712.

Taylor acknowledged that his work was based on that of Newton and Kepler and others, but he did not acknowledge that the same result had been discovered by Johann Bernoulli and published in 1694. (But then Taylor discovered integration by parts first, although Bernoulli claimed the credit.)

Theorem 1.1 (Taylor's Theorem with Remainder) *Let $f(x)$ have $n+1$ continuous derivatives on $[a,b]$ for some $n \geq 0$, and let x, $x_0 \in [a,b]$. Then,*

$$f(x) = p_n(x) + R_n(x)$$

for

$$p_n(x) = \sum_{k=0}^{n} \frac{(x-x_0)^k}{k!} f^{(k)}(x_0), \qquad (1.1)$$

and

$$R_n(x) = \frac{1}{n!} \int_{x_0}^{x} (x-t)^n f^{(n+1)}(t)dt. \qquad (1.2)$$

$\Big)$ *medium-value theorem*

Moreover, there exists a point ξ_x between x and x_0 such that

$$R_n(x) = \frac{(x-x_0)^{n+1}}{(n+1)!} f^{(n+1)}(\xi_x). \qquad (1.3)$$

The point x_0 is usually chosen at the discretion of the user, and is often taken to be 0. Note that the two forms of the remainder are equivalent: the "pointwise" form (1.3) can be derived from the "integral" form (1.2); see Problem 23.

Taylor's Theorem is important because it allows us to represent, exactly, fairly general functions in terms of polynomials with a *known, specified, boundable* error. This allows us to replace, in a computational setting, these same general functions with something that is much simpler—a polynomial—yet at the same time we are able to bound the error that is made. No other tool will be as important to us as Taylor's Theorem, so it is worth spending some time on it here at the outset.

The usual calculus treatment of Taylor's Theorem should leave the student familiar with three particular expansions (for all three of these we have used $x_0 = 0$, which means we really should call them Maclaurin[2] series, but we won't):

$$
\begin{aligned}
e^x &= 1 + x + \frac{1}{2!}x^2 + \frac{1}{3!}x^3 + \ldots + \frac{1}{n!}x^n + \frac{1}{(n+1)!}x^{n+1}e^{\xi_x} \\
&= \sum_{k=0}^{n} \frac{1}{k!}x^k + R_n(x), \\
\sin x &= x - \frac{1}{3!}x^3 + \frac{1}{5!}x^5 + \ldots + \frac{(-1)^n}{(2n+1)!}x^{2n+1} + \frac{(-1)^{n+1}}{(2n+3)!}x^{2n+3}\cos\xi_x \\
&= \sum_{k=0}^{n} \frac{(-1)^k}{(2k+1)!}x^{2k+1} + R_n(x), \quad R_{n+1} \\
\cos x &= 1 - \frac{1}{2!}x^2 + \frac{1}{4!}x^4 + \ldots + \frac{(-1)^n}{(2n)!}x^{2n} + \frac{(-1)^{n+1}}{(2n+2)!}x^{2n+2}\cos\xi_x \\
&= \sum_{k=0}^{n} \frac{(-1)^k}{(2k)!}x^{2k} + R_n(x). \quad R_{2n}
\end{aligned}
$$

---

[2]Colin Maclaurin (1698–1746) was born and lived almost his entire life in Scotland. Educated at Glasgow University, he was professor of mathematics at Aberdeen from 1717 to 1725 and then went to Edinburgh. He worked in a number of areas of mathematics, and is credited with writing one of the first textbooks based on Newton's calculus, *Treatise of fluxions* (1742). The Maclaurin series appears in this book as a special case of Taylor's series.

(Strictly speaking, the indices on the last two remainders should be $2n + 1$ and $2n$, because those are the exponents in the last terms of the expansion, but it is commonplace to present them as we did here.) In fact, Taylor's Theorem provides us with our first and simplest example of an approximation and an error estimate. Consider the problem of approximating the exponential function on the interval $[-1, 1]$. Taylor's Theorem tells us that we can represent e^x using a polynomial with a (known) remainder:

$$e^x = \underbrace{1 + x + \frac{1}{2!}x^2 + \frac{1}{3!}x^3 + \ldots + \frac{1}{n!}x^n}_{p_n(x), \text{ polynomial}} + \underbrace{\frac{1}{(n+1)!}x^{n+1}e^{c_x}}_{R_n(x), \text{ remainder}},$$

where c_x is an unknown point between x and 0. Since we want to consider the most general case, where x can be any point in $[-1, 1]$, we have to consider that c_x can be any point in $[-1, 1]$, as well. For simplicity, let's denote the polynomial by $p_n(x)$, and the remainder by $R_n(x)$, so that the equation above becomes

$$e^x = p_n(x) + R_n(x).$$

Suppose that we want this approximation to be accurate to within 10^{-6} in absolute error, i.e., we want

$$|e^x - p_n(x)| \leq 10^{-6}$$

for *all* x in the interval $[-1, 1]$. Note that if we can make $|R_n(x)| \leq 10^{-6}$ for all $x \in [-1, 1]$, then we will have

$$|e^x - p_n(x)| = |R_n(x)| \leq 10^{-6},$$

so that the error in the approximation will be less than 10^{-6}. The best way to proceed is to create a *simple* upper bound for $|R_n(x)|$, and then use that to determine the number of terms necessary to make this upper bound less than 10^{-6}.

Thus we proceed as follows:

$$
\begin{aligned}
|R_n(x)| &= \frac{|x^{n+1}e^{c_x}|}{(n+1)!}, \\
&= \frac{|x|^{n+1}e^{c_x}}{(n+1)!}, && \text{because } e^z > 0 \text{ for all } z, \\
&\leq \frac{e^{c_x}}{(n+1)!}, && \text{because } |x| \leq 1 \text{ for all } x \in [-1, 1], \\
&\leq \frac{e}{(n+1)!}, && \text{because } e^{c_x} \leq e \text{ for all } x \in [-1, 1].
\end{aligned}
$$

Thus, if we find n such that

$$\frac{1}{(n+1)!}e \leq 10^{-6},$$

then we will have

$$|e^x - p_n(x)| = |R_n(x)| \leq \frac{1}{(n+1)!}e \leq 10^{-6}$$

and we will *know* that the error is less than the desired tolerance, for *all* values of x of interest to us, i.e., all $x \in [-1, 1]$. A little bit of exercise with a calculator shows us that we need to use $n = 9$ to get the desired accuracy. Figure 1.1 shows a plot of the exponential function e^x, the approximation $p_9(x)$, as well as the less accurate $p_2(x)$; since it is impossible to distinguish by eye between the plots for e^x and $p_9(x)$, we have also provided Figure 1.2, which is a plot of the error $e^x - p_9(x)$; note that we begin to lose accuracy as we get away from the interval $[-1, 1]$. This is not surprising. Since the Taylor polynomial is constructed to match f and its first n derivatives at $x = x_0$, it ought to be the case that p_n is a good approximation to f only when x is near x_0.

Here we are in the first section of the text, and we have already constructed our first approximation and proved our first theoretical result. Theoretical result? Where? Why, the error estimate, of course. The material in the previous several lines is a proof of Proposition 1.1.

Proposition 1.1 *Let $p_9(x)$ be the Taylor polynomial of degree 9 for the exponential function. Then, for all $x \in [-1, 1]$, the error in the approximation of $p_9(x)$ to e^x is less than 10^{-6}, i.e.,*

$$|e^x - p_9(x)| \leq 10^{-6}$$

for all $x \in [-1, 1]$.

Although this result is not of major significance to our work—the exponential function is approximated more efficiently by different means—it does illustrate one of the most important aspects of the subject. The result tells us, ahead of time, that we can approximate the exponential function to within 10^{-6} accuracy using a specific polynomial, and this accuracy holds for all x in a specified interval. That is the kind of thing we will be doing throughout the text—constructing approximations to difficult computations that are accurate in some sense, and we *know* how accurate they are.

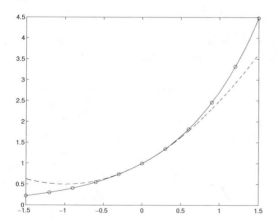

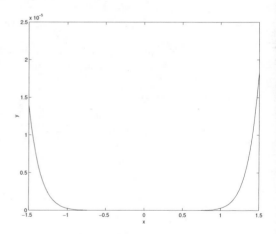

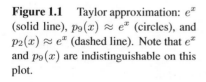

Figure 1.1 Taylor approximation: e^x (solid line), $p_9(x) \approx e^x$ (circles), and $p_2(x) \approx e^x$ (dashed line). Note that e^x and $p_9(x)$ are indistinguishable on this plot.

Figure 1.2 Error in Taylor approximation: $e^x - p_9(x)$.

■ **EXAMPLE 1.1**

Let $f(x) = \sqrt{x + 1}$; then the second-order Taylor polynomial (computed about $x_0 = 0$) is computed at follows:

$$
\begin{aligned}
f(x_0) &= f(0) = 1; \\
f'(x) &= \frac{1}{2}(x + 1)^{-1/2} \Rightarrow f'(x_0) = \frac{1}{2}; \\
f''(x) &= -\frac{1}{2} \times \frac{1}{2}(x + 1)^{-3/2} \Rightarrow f''(x_0) = -\frac{1}{4}; \\
p_2(x) &= f(x_0) + (x - x_0)f'(x_0) + \frac{1}{2}(x - x_0)^2 f''(x_0) = 1 + \frac{1}{2}x - \frac{1}{8}x^2.
\end{aligned}
$$

The error in using p_2 to approximate $\sqrt{x + 1}$ is given by $R_2(x) = \frac{1}{3!}(x - x_0)^3 f'''(\xi_x)$, where ξ_x is between x and x_0. We can simplify the error as follows:

$$
\begin{aligned}
|R_2(x)| &= \left| \frac{1}{3!}(x - x_0)^3 f'''(\xi_x) \right|, \\
&= \frac{1}{6}|x|^3 \left| \frac{3}{2} \times \frac{1}{2} \times \frac{1}{2}(\xi_x + 1)^{-5/2} \right|, \\
&= \frac{1}{16}|x|^3 |\xi_x + 1|^{-5/2}.
\end{aligned}
$$

If we want to consider this for all $x \in [0, 1]$, then, we observe that $x \in [0, 1]$ and ξ_x between x and 0 imply that $\xi_x \in [0, 1]$; therefore

$$
|\xi_x + 1|^{-5/2} \le |0 + 1|^{-5/2} = 1,
$$

so that the upper bound on the error becomes $|R_3(x)| \le 1/16 = 0.0625$, for all $x \in [0, 1]$. If we are only interested in $x \in [0, \frac{1}{2}]$, then the error is much smaller:

$$
\begin{aligned}
|R_2(x)| &= \frac{1}{16}|x|^3 |\xi_x + 1|^{-5/2}, \\
&\le \frac{1}{16}(1/2)^3, \\
&= \frac{1}{128}, \\
&= 0.0078125.
\end{aligned}
$$

■ **EXAMPLE 1.2**

Consider the problem of finding a polynomial approximation to the function $f(x) = \sin \pi x$ that is accurate to within 10^{-4} for all $x \in [-\frac{1}{2}, \frac{1}{2}]$, using $x_0 = 0$. A direct computation with Taylor's Theorem gives us (note that the indexing here is a little different, because we want to take advantage of the fact that the Taylor polynomial here has only odd terms)

$$
p_n(x) = \pi x - \frac{1}{6}\pi^3 x^3 + \frac{1}{120}\pi^5 x^5 + \cdots + (-1)^n \frac{1}{(2n + 1)!}\pi^{2n+1} x^{2n+1},
$$

and

$$R_n(x) = (-1)^{n+1} \frac{1}{(2n+3)!} \pi^{2n+3} x^{2n+3} \cos \xi_x.$$

Thus, the error in the approximation is

$$|f(x) - p_n(x)| = |R_n(x)| = \frac{(\pi x)^{2n+3}}{(2n+3)!} |\cos \xi_x|,$$

where ξ_x is between x and 0; this can be bounded for all $x \in [-\frac{1}{2}, \frac{1}{2}]$ in the following fashion:

$$|R_n(x)| = \frac{|\pi x|^{2n+3}}{(2n+3)!} |\cos \xi_x| \le \frac{(\pi/2)^{2n+3}}{(2n+3)!},$$

and some experimentation with a calculator shows that

$$|R_4(x)| \le 0.3599 \times 10^{-5}, \quad |R_3(x)| \le 0.1604 \times 10^{-3},$$

from which we conclude that $n = 4$ will achieve the desired accuracy, thus we want

$$p_4(x) = \pi x - \frac{1}{6}\pi^3 x^3 + \frac{1}{120}\pi^5 x^5 - \frac{1}{5040}\pi^7 x^7 + \frac{1}{362880}\pi^9 x^9$$

as our approximation. Figure 1.3 shows the error between this polynomial and $f(x)$ over the interval $[-\frac{1}{2}, \frac{1}{2}]$; note that the error is much better than the predicted error, especially in the middle of the interval.

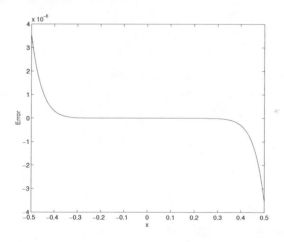

Figure 1.3 Error in Taylor approximation to $f(x) = \sin \pi x$ over $[-\frac{1}{2}, \frac{1}{2}]$.

Although the usual approach is to construct Taylor expansions by directly computing the necessary derivatives of the function f, sometimes a more subtle approach can be used. Consider the problem of computing a Taylor series for the arctangent function, $f(x) = \arctan x$, about the point $x_0 = 0$. It won't take many terms before we get tired of trying to find a general expression for $f^{(n)}$. What do we do?

Recall, from calculus, that

$$\arctan x = \int_0^x \frac{dt}{1 + t^2},$$

so we can get the Taylor series for the arctangent from the Taylor series for $(1 + t^2)^{-1}$ by a simple integration. Now recall the summation formula for the geometric series:

$$\sum_{k=0}^{n} r^k = \frac{1 - r^{n+1}}{1 - r}.$$

If we substitute $r = -t^2$ into this, and re-arrange terms a bit, we get

$$\frac{1}{1 + t^2} = \sum_{k=0}^{n} (-1)^k t^{2k} + \frac{(-t^2)^{n+1}}{1 + t^2}.$$

Thus,

$$
\begin{aligned}
\arctan x &= \int_0^x \frac{dt}{1 + t^2}, \\
&= \int_0^x \left(\sum_{k=0}^{n} (-1)^k t^{2k} + \frac{(-t^2)^{n+1}}{1 + t^2} \right) dt, \\
&= \sum_{k=0}^{n} (-1)^k (2k + 1)^{-1} x^{2k+1} + (-1)^{n+1} \int_0^x \frac{t^{2n+2}}{1 + t^2} dt.
\end{aligned}
$$

This is known as *Gregory's series*[3] for the arctangent and was the basis for one of the early methods for computing π. (See Problems 14 – 16.)

■ **EXAMPLE 1.3**

Let's use the Gregory series to determine the error in a ninth-degree Taylor approximation to the arctangent function. Since $2n + 1 = 9$ implies that $n = 4$, we have

$$R_9(x) = -\int_0^x \frac{t^{10}}{1 + t^2} dt,$$

so that (assuming that $x \geq 0$)

$$|R_9(x)| = \int_0^x \frac{t^{10}}{1 + t^2} dt,$$

and we can bound the error as follows:

$$|R_9(x)| = \int_0^x \frac{t^{10}}{1 + t^2} dt \leq \int_0^x t^{10} dt = \frac{1}{11} x^{11}.$$

So, for all $x \in [-\frac{1}{2}, \frac{1}{2}]$ the remainder is bounded above by

$$|R_9(x)| \leq \frac{1}{11} (1/2)^{11} = 4.44 \times 10^{-5}.$$

---

[3] James Gregory (1638–1675) was born just west of Aberdeen, Scotland, where he first went to university. He later (1664–1668) studied in Italy before returning to Britain. He published a work on optics and several works devoted to finding the areas under curves, including the circle. He knew about what we call Taylor series more than 40 years before Taylor published the theorem, and he might have developed the calculus before Newton did if he had been of a less geometric and more analytical frame of mind.

Finally, we close this section with an illustration of a special kind of Taylor expansion that we will use again and again.

Consider the problem of expanding $f(x+h)$ in a Taylor series, about the point $x_0 = x$. Here h is generally considered to be a small parameter. Direct application of Taylor's Theorem gives us the following:

$$
\begin{aligned}
f(x+h) &= f(x) + ([x+h]-x)f'(x) + \frac{1}{2}([x+h]-x)^2 f''(x) \\
&\quad + \cdots + \frac{1}{n!}([x+h]-x)^n f^{(n)}(x) \\
&\quad + \frac{1}{(n+1)!}([x+h]-x)^{n+1} f^{(n+1)}(\xi), \\
&= f(x) + hf'(x) + \frac{1}{2}h^2 f''(x) + \cdots + \frac{1}{n!}h^n f^{(n)}(x) \\
&\quad + \frac{1}{(n+1)!}h^{n+1} f^{(n+1)}(\xi).
\end{aligned}
$$

This kind of expansion will be useful time and again in our studies.

1.1.2 Mean Value and Extreme Value Theorems

We need to spend some time reviewing other results from calculus that have an impact on numerical methods and analysis. All of these theorems are included in most calculus texts but are usually not emphasized as much as we will need here, or perhaps not in the way that we will need them.

Theorem 1.2 (Mean Value Theorem) *Let f be a given function, continuous on $[a,b]$ and differentiable on (a,b). Then there exists a point $\xi \in [a,b]$ such that*

$$ f'(\xi) = \frac{f(b)-f(a)}{b-a}. \tag{1.4} $$

In the context of calculus, the importance of this result might seem obscure, at best. However, from the point of view of numerical methods and analysis, the Mean Value Theorem (MVT) is probably second in importance only to Taylor's Theorem. Why? Well, consider a slightly reworked form of (1.4):

$$ f(x_1) - f(x_2) = f'(\xi)(x_1 - x_2). $$

Thus, the MVT allows us to replace differences of *function values* with differences of *argument values*, if we scale by the derivative of the function. For example, we can use the MVT to tell us that

$$ |\cos x_1 - \cos x_2| \le |x_1 - x_2|, $$

because the derivative of the cosine is the sine, which is bounded by 1 in absolute value. Note also that the MVT is simply a special case of Taylor's Theorem, for $n=0$.

Similar to the MVT is the Intermediate Value Theorem:

Theorem 1.3 (Intermediate Value Theorem) *Let $f \in C([a,b])$ be given, and assume that W is a value between $f(a)$ and $f(b)$, that is, either $f(a) \le W \le f(b)$, or $f(b) \le W \le f(a)$. Then there exists a point $c \in [a,b]$ such that $f(c) = W$.*

This seems to be a very abstract result; it says that a certain point exists, but doesn't give us much information about its numerical value. (We might ask, why do we care that c exists?) But it is this very theorem that is the basis for our first method for finding the roots of functions, an algorithm called the *bisection method*. (See §3.1.) Moreover, this is the theorem that tells us that a continuous function is one whose graph can be drawn without lifting the pen off the paper, because it says that between any two function values, we have a point on the curve for all possible argument values. Sometimes very abstract results have very concrete consequences.

A related result is the Extreme Value Theorem, which is the basis for the max/min problems that are a staple of most beginning calculus courses.

Theorem 1.4 (Extreme Value Theorem) *Let* $f \in C([a,b])$ *be given; then there exists a point* $m \in [a,b]$ *such that* $f(m) \leq f(x)$ *for all* $x \in [a,b]$, *and a point* $M \in [a,b]$ *such that* $f(M) \geq f(x)$ *for all* $x \in [a,b]$. *Moreover,* f *achieves its maximum and minimum values on* $[a,b]$ *either at the endpoints* a *or* b, *or at a critical point.*

(The student should recall that a *critical point* is a point where the first derivative is either undefined or equal to zero. The theorem thus says that we have one of $M = a$, $M = b$, $f'(M)$ doesn't exist, or $f'(M) = 0$, and similarly for m.)

There are other "mean value theorems," and we need to look at two in particular, as they will come up in our early error analysis.

Theorem 1.5 (Integral Mean Value Theorem) *Let* f *and* g *both be in* $C([a,b])$, *and assume further that* g *does not change sign on* $[a,b]$. *Then there exists a point* $\xi \in [a,b]$ *such that*

$$\int_a^b g(t)f(t)dt = f(\xi) \int_a^b g(t)dt. \tag{1.5}$$

Proof: Since this result is not commonly covered in the calculus sequence, we will go ahead and prove it.

We first assume, without any loss of generality, that $g(t) \geq 0$; the argument changes in an obvious way if g is negative (see Problem 26). Let f_M be the maximum value of the function on the interval, $f_M = \max_{x \in [a,b]} f(x)$, so that $g(t)f(t) \leq g(t)f_M$ for all $t \in (a,b)$; then

$$\int_a^b g(t)f(t)dt \leq \int_a^b g(t)f_M dt = f_M \int_a^b g(t)dt.$$

Similarly, we have

$$\int_a^b g(t)f(t)dt \geq \int_a^b g(t)f_m dt = f_m \int_a^b g(t)dt$$

where $f_m = \min_{x \in [a,b]} f(x)$ is the minimum value of f on the interval. Since g does not change sign on the interval of integration, the only way that we can have

$$\int_a^b g(x)dx = 0$$

is if g is identically zero on the interval, in which case the theorem is trivially true, since both sides of (1.5) would be zero. So we can assume that

$$\int_a^b g(x)dx \neq 0.$$

Now define

$$W = \frac{\int_a^b g(t)f(t)dt}{\int_a^b g(t)dt},$$

so that we have

$$f_m \leq W \leq f_M.$$

By the Extreme Value Theorem, there is a point $M \in [a,b]$ such that $f(M) = f_M$, and similarly there is a point $m \in [a,b]$ such that $f(m) = f_m$. Therefore,

$$f(m) \leq W \leq f(M)$$

and we can apply the Intermediate Value Theorem to establish that there is a point ξ in the interval defined by m and M such that $f(\xi) = W$; but this implies (1.5) (why?) and we are done. •

This result will be useful in Chapter 5 for simplifying some error estimates for numerical integration rules. A related result is the Discrete Average Value Theorem.

Theorem 1.6 (Discrete Average Value Theorem) *Let $f \in C([a,b])$ and consider the sum*

$$S = \sum_{k=1}^n a_k f(x_k),$$

where each point $x_k \in [a,b]$, and the coefficients satisfy

$$a_k \geq 0, \qquad \sum_{k=1}^n a_k = 1.$$

Then there exists a point $\eta \in [a,b]$ such that $f(\eta) = S$, i.e.,

$$f(\eta) = \sum_{k=1}^n a_k f(x_k).$$

Proof: The proof is similar in spirit and technique to the preceding one. We quickly have that

$$S = \sum_{k=1}^n a_k f(x_k) \leq f_M \sum_{k=1}^n a_k = f_M$$

and similarly $S \geq f_m$, where f_M and f_m are defined as in the previous proof. Now define $W = S$ and proceed as before to get that there is a point $\eta \in [a,b]$ such that $f(\eta) = S$. •

All three of the mean value theorems are useful to us in that they allow us to simplify certain expressions that will occur in the process of deriving error estimates for our approximate computations.

Exercises:

1. Show that the third-order Taylor polynomial for $f(x) = (x+1)^{-1}$, about $x_0 = 0$, is

$$p_3(x) = 1 - x + x^2 - x^3.$$

2. What is the third-order Taylor polynomial for $f(x) = \sqrt{x+1}$, about $x_0 = 0$?

3. What is the sixth-order Taylor polynomial for $f(x) = \sqrt{1 + x^2}$, using $x_0 = 0$?

4. Given that

$$R(x) = \frac{|x|^6}{6!} e^{\xi}$$

for $x \in [-1, 1]$, where ξ is between x and 0, find an upper bound for $|R|$, valid for all $x \in [-1, 1]$, that is independent of x and ξ.

5. Repeat the above, but this time require that the upper bound be valid only for all $x \in [-\frac{1}{2}, \frac{1}{2}]$.

6. Given that

$$R(x) = \frac{|x|^4}{4!} \left(\frac{-1}{1 + \xi} \right)$$

for $x \in [-\frac{1}{2}, \frac{1}{2}]$, where ξ is between x and 0, find an upper bound for $|R|$, valid for all $x \in [-\frac{1}{2}, \frac{1}{2}]$, that is independent of x and ξ.

7. Use a Taylor polynomial to find an approximate value for $\sqrt{e}$ that is accurate to within 10^{-3}.

8. What is the fourth-order Taylor polynomial for $f(x) = 1/(x + 1)$, about $x_0 = 0$?

9. What is the fourth-order Taylor polynomial for $f(x) = 1/x$, about $x_0 = 1$?

10. Find the Taylor polynomial of third-order for $\sin x$, using:

    (a) $x_0 = \pi/6$;
    (b) $x_0 = \pi/4$;
    (c) $x_0 = \pi/2$.

11. For each function below, construct the third-order Taylor polynomial approximation, using $x_0 = 0$, and then estimate the error by computing an upper bound on the remainder, over the given interval.

    (a) $f(x) = e^{-x}$, $x \in [0, 1]$;
    (b) $f(x) = \ln(1 + x)$, $x \in [-1, 1]$;
    (c) $f(x) = \sin x$, $x \in [0, \pi]$;
    (d) $f(x) = \ln(1 + x)$, $x \in [-1/2, 1/2]$;
    (e) $f(x) = 1/(x + 1)$, $x \in [-1/2, 1/2]$.

12. Construct a Taylor polynomial approximation that is accurate to within 10^{-3}, over the indicated interval, for each of the following functions, using $x_0 = 0$.

    (a) $f(x) = \sin x$, $x \in [0, \pi]$;
    (b) $f(x) = e^{-x}$, $x \in [0, 1]$;
    (c) $f(x) = \ln(1 + x)$, $x \in [0, 3/4]$;
    (d) $f(x) = 1/(x + 1)$, $x \in [0, 1/2]$;
    (e) $f(x) = \ln(1 + x)$, $x \in [0, 1/2]$.

13. Repeat the above, this time with a desired accuracy of 10^{-6}.

14. Since
$$\frac{\pi}{4} = \arctan 1,$$

    we can estimate π by estimating $\arctan 1$. How many terms are needed in the Gregory series for the arctangent to approximate π to 100 decimal places? 1,000? *Hint:* Use the error term in the Gregory series to predict when the error becomes sufficiently small.

15. Elementary trigonometry can be used to show that

    $$\arctan(1/239) = 4\arctan(1/5) - \arctan(1).$$

    This formula was developed in 1706 by the English astronomer John Machin. Use this to develop a more efficient algorithm for computing π. How many terms are needed to get 100 digits of accuracy with this form? How many terms are needed to get 1,000 digits? *Historical note:* Until 1961 this was the basis for the most commonly used method for computing π to high accuracy.

16. In 1896 a variation on Machin's formula was found:

    $$\arctan(1/239) = \arctan(1) - 6\arctan(1/8) - 2\arctan(1/57),$$

    and this began to be used in 1961 to compute π to high accuracy. How many terms are needed when using this expansion to get 100 digits of π? 1,000 digits?

17. What is the Taylor polynomial of order 3 for $f(x) = x^4 + 1$, using $x_0 = 0$?

18. What is the Taylor polynomial of order 4 for $f(x) = x^4 + 1$, using $x_0 = 0$? Simplify as much as possible.

19. What is the Taylor polynomial of order 2 for $f(x) = x^3 + x$, using $x_0 = 1$?

20. What is the Taylor polynomial of order 3 for $f(x) = x^3 + x$, using $x_0 = 1$? Simplify as much as possible.

21. Let $p(x)$ be an arbitrary polynomial of degree less than or equal to n. What is its Taylor polynomial of degree n, about an arbitrary x_0?

22. The Fresnel integrals are defined as

    $$C(x) = \int_0^x \cos(\pi t^2/2)dt$$

    and

    $$S(x) = \int_0^x \sin(\pi t^2/2)dt.$$

    Use Taylor expansions to find approximations to $C(x)$ and $S(x)$ that are 10^{-4} accurate for all x with $|x| \leq \frac{1}{2}$. *Hint:* Substitute $x = \pi t^2/2$ into the Taylor expansions for the cosine and sine.

23. Use the Integral Mean Value Theorem to show that the "pointwise" form (1.3) of the Taylor remainder (usually called the *Lagrange* form) follows from the "integral" form (1.2) (usually called the *Cauchy* form).

24. For each function in Problem 11, use the Mean Value Theorem to find a value M such that
$$|f(x_1) - f(x_2)| \leq M|x_1 - x_2|$$
is valid for all x_1, x_2 in the interval used in Problem 11.

25. A function is called *monotone* on an interval if its derivative is strictly positive or strictly negative on the interval. Suppose f is continuous and monotone on the interval $[a, b]$, and $f(a)f(b) < 0$; prove that there is exactly one value $\alpha \in [a, b]$ such that $f(\alpha) = 0$.

26. Finish the proof of the Integral Mean Value Theorem (Theorem 1.5) by writing up the argument in the case that g is negative.

27. Prove Theorem 1.6, providing all details.

28. Let $c_k > 0$ be given, $1 \leq k \leq n$, and let $x_k \in [a, b]$, $1 \leq k \leq n$. Then, use the Discrete Average Value Theorem to prove that, for any function $f \in C([a, b])$,

$$\frac{\sum_{k=1}^n c_k f(x_k)}{\sum_{k=1}^n c_k} = f(\xi)$$

for some $\xi \in [a, b]$.

29. Discuss, in your own words, whether or not the following statement is true: "The Taylor polynomial of degree n is the best polynomial approximation of degree n to the given function near the point x_0."

◁ • • • ▷

1.2 ERROR, APPROXIMATE EQUALITY, AND ASYMPTOTIC ORDER NOTATION

We have already talked about the "error" made in a simple Taylor series approximation. Perhaps it is time we got a little more precise.

1.2.1 Error

If A is a quantity we want to compute and A_h is an approximation to that quantity, then the error is the difference between the two:

$$\text{error} = A - A_h;$$

the *absolute error* is simply the absolute value of the error:

$$\text{absolute error} = |A - A_h|; \tag{1.6}$$

and the *relative error* normalizes by the absolute value of the exact value:

$$\text{relative error} = \frac{|A - A_h|}{|A|}, \tag{1.7}$$

where we assume that $A \neq 0$.

Why do we need two different measures of error? Consider the problem of approximating the number

$$x = e^{-16} = 0.1125351747 \times 10^{-6}.$$

Because x is so small, the absolute error in $y = 0$ as an approximation to x is also small. In fact, $|x - y| < 1.2 \times 10^{-7}$, which is decent accuracy in many settings. However, this "approximation" is clearly not a good one.

On the other hand, consider the problem of approximating

$$z = e^{16} = 0.8886110521 \times 10^{7}.$$

Because z is so large, the absolute error in almost any approximation will be large, even though almost all of the digits are matched. For example, if we take $w = 0.8886110517 \times 10^{7}$, then we have $|z - w| = 4 \times 10^{-3}$, hardly very small, even though y matches z to nine decimal digits.

The point is that relative error gives a measure of the number of correct digits in the approximation. Thus,

$$\left| \frac{x - y}{x} \right| = 1,$$

which tells us that not many digits are matched in that example, whereas

$$\left| \frac{z - w}{z} \right| = \frac{4 \times 10^{-3}}{0.8886110521 \times 10^{7}} = 0.4501 \times 10^{-9},$$

which shows that about nine digits are correct. Generally speaking, using a relative error protects us from misjudging the accuracy of an approximation because of scale extremes (very large or very small numbers). As a practical matter, however, we sometimes are not able to obtain an error estimate in the relative sense.

In the definitions (1.6) and (1.7), we have used the subscript h to suggest that, in general, the approximation depends (in part, at least) on a parameter. For the most part, our computations will indeed be constructed this way, usually with either a real parameter h which tends toward zero, or with an integer parameter n which tends toward infinity. So we might want to think in terms of one of the two cases

$$\lim_{h \to 0} A_h = A$$

or

$$\lim_{n \to \infty} A_n = A.$$

In actual applied problems there are, of course, lots of sources of error: simple mistakes, measurement errors, modeling errors, etc. We are concerned here only with the computational errors caused by the need to construct computable approximations. The common terminology is *truncation error* or *approximation error* or *mathematical error*.

1.2.2 Notation: Approximate Equality

If two quantities are approximately equal to each other, we will use the notation "$\approx$" to denote this relationship, as in

$$A \approx B.$$

This is an admittedly vague notion. Is $0.99 \approx 1$? Probably so. Is $0.8 \approx 1$? Maybe not. We will almost always use the $\approx$ symbol in the sense of one of the two contexts outlined previously, of a parameterized set of approximations converging to a limit. Note that the definition of limit means that

$$\lim_{h \to 0} A_h = A \quad \Rightarrow \quad A_h \approx A$$

for all h "sufficiently small" (and similarly for the case of $A_n \to A$ as $n \to \infty$, for n "sufficiently large"). For example, one way to write the definition of the derivative of a function $y = f(x)$ is as follows:

$$\lim_{h \to 0} \frac{f(x+h) - f(x)}{h} = f'(x).$$

We therefore conclude that, for h small enough,

$$\frac{f(x+h) - f(x)}{h} \approx f'(x).$$

Moreover, approximate equality does satisfy the transitive, symmetric, and reflexive properties of what abstract algebra calls an "equivalence relation":

$$A \approx B, \quad B \approx C \Rightarrow A \approx C,$$
$$A \approx B \Rightarrow B \approx A,$$
$$A \approx A.$$

Consequently, we can manipulate approximate equalities much like ordinary equalities (i.e., equations). We can solve them, integrate both sides, etc.

Despite its vagueness, approximate equality is a *very* useful notion to have around in a course devoted to approximations.

1.2.3   Notation: Asymptotic Order

Another notation of use is the so-called "Big O" notation, more formally known as *asymptotic order* notation. Suppose that we have a value y and a family of values $\{y_h\}$, each of which approximates this value,

$$y \approx y_h$$

for small values of h. If we can find a constant $C > 0$, independent of h, such that

$$|y - y_h| \leq C\beta(h) \tag{1.8}$$

for all h sufficiently small, then we say that

$$y = y_h + \mathcal{O}(\beta(h)), \quad \text{as } h \to 0,$$

meaning that $y - y_h$ is "on the order of" $\beta(h)$. Here $\beta(h)$ is a function of the parameter h, and we assume that

$$\lim_{h \to 0} \beta(h) = 0.$$

The utility of this notation is that it allows us to concentrate on the important issue in the approximation—the way that the error $y - y_h$ depends on the parameter h, which is determined by $\beta(h)$—while ignoring the unimportant details such as the precise size of the constant, C. The usage is similar if we have a sequence x_n that approximates a given value x for large values of n. If

$$|x - x_n| \leq C\beta(n) \tag{1.9}$$

for all n sufficiently large, then we say that

$$x = x_n + \mathcal{O}(\beta(n)), \quad \text{as } n \to \infty.$$

The formal definitions are as follows.

Definition 1.1 (Asymptotic Order Notation) *For a given value y, let $\{y_h\}$ be a set of values parameterized by h, which we assume is small, such that $y_h \approx y$ for small h. If there exists a positive function $\beta(h)$, $\beta(h) \to 0$ as $h \to 0$, and a constant $C > 0$, such that for all h sufficiently small,*

$$|y - y_h| \leq C\beta(h),$$

then we say that

$$y = y_h + \mathcal{O}(\beta(h)).$$

Similarly, if $\{y_n\}$ is a set of values parameterized by n, which we assume is large, such that $y_n \approx y$ for large n, and if there exists a positive function $\beta(n)$, $\beta(n) \to 0$ as $n \to \infty$, and a constant $C > 0$, such that for all n sufficiently large,

$$|y - y_n| \leq C\beta(n),$$

then we say that

$$y = y_n + \mathcal{O}(\beta(n)).$$

■ **EXAMPLE 1.4**

Let

$$A = \int_0^\infty e^{-2x} dx,$$
$$A_n = \int_0^n e^{-2x} dx.$$

Simple calculus shows that $A = \frac{1}{2}$ and $A_n = \frac{1}{2} - \frac{1}{2}e^{-2n}$, so that we have $A = A_n + \mathcal{O}(e^{-2n})$. Here $\beta(n) = e^{-2n}$.

■ **EXAMPLE 1.5**

Another example—and many such, in fact—can be generated from Taylor's Theorem. We have that

$$\cos x = 1 - \frac{1}{2}x^2 \cos \xi_x$$

where ξ_x is between x and 0. Since $\cos \xi_x$ is bounded by 1 in absolute value, we easily have that

$$|\cos h - 1| \leq \frac{1}{2}h^2$$

so we can write $\cos h = 1 + \mathcal{O}(h^2)$. Similarly, we can write that $e^h = 1 + h + \mathcal{O}(h^2)$. In both these cases we have $\beta(h) = h^2$.

The following theorem shows how the asymptotic order relationship can be manipulated.

Theorem 1.7 *Let $y = y_h + \mathcal{O}(\beta(h))$ and $z = z_h + \mathcal{O}(\gamma(h))$, with $b\beta(h) > \gamma(h)$ for all h near zero. Then*

$$
\begin{aligned}
y + z &= y_h + z_h + \mathcal{O}(\beta(h) + \gamma(h)), \\
y + z &= y_h + z_h + \mathcal{O}(\beta(h)), \\
Ay &= Ay_h + \mathcal{O}(\beta(h)).
\end{aligned}
$$

In the third equation, A is an arbitrary constant, independent of h.

Proof: We simply show that each relationship above satisfies (1.8) for some constant C. For example,

$$
\begin{aligned}
|(y + z) - (y_h + z_h)| &\leq |y - y_h| + |z - z_h|, \\
&\leq C_1\beta(h) + C_2\gamma(h), \\
&\leq C(\beta(h) + \gamma(h)),
\end{aligned}
$$

where $C = \max(C_1, C_2)$. Thus, $y + z = y_h + z_h + \mathcal{O}(\beta(h) + \gamma(h))$. Moreover, since $b\beta(h) > \gamma(h)$, we also have that

$$
\begin{aligned}
|(y + z) - (y_h + z_h)| &\leq C(\beta(h) + \gamma(h)), \\
&\leq C(\beta(h) + b\beta(h)), \\
&\leq C(1 + b)\beta(h).
\end{aligned}
$$

Also,

$$
\begin{aligned}
|Ay - Ay_h| &= A|y - y_h|, \\
&\leq C_1 A\beta(h), \\
&= C\beta(h),
\end{aligned}
$$

so that $Ay = Ay_h + \mathcal{O}(\beta(h))$. •

A similar result holds for $u_n \approx u$ with $u = u_n + \mathcal{O}(\beta(n))$, $v_n \approx v$ with $v = v_n + \mathcal{O}(\gamma(n))$, and so on.

■ **EXAMPLE 1.6**

We close this section with a simple example that illustrates the utility of the $\mathcal{O}$ notation. Consider the combination of function values

$$
D = -f(x + 2h) + 4f(x + h) - 3f(x),
$$

where f is assumed to be continuous and smooth, and h is a (small) parameter. We can use Taylor's Theorem, together with the definition of the $\mathcal{O}$ notation, to write

$$
f(x + h) = f(x) + hf'(x) + \frac{1}{2}h^2 f''(x) + \mathcal{O}(h^3)
$$

and
$$f(x + 2h) = f(x) + 2hf'(x) + 2h^2 f''(x) + \mathcal{O}(h^3).$$

Therefore,

$$
\begin{aligned}
D &= -f(x + 2h) + 4f(x + h) - 3f(x) \\
&= -(f(x) + 2hf'(x) + 2h^2 f''(x) + \mathcal{O}(h^3)) \\
&\quad + 4(f(x) + hf'(x) + \frac{1}{2}h^2 f''(x) + \mathcal{O}(h^3)) - 3f(x) \\
&= (-1 + 4 - 3)f(x) + (-2h + 4h)f'(x) + (-2h^2 + 2h^2)f''(x) + \mathcal{O}(h^3) \\
&= 2hf'(x) + \mathcal{O}(h^3).
\end{aligned}
$$

If we then solve this for $f'(x)$ we get

$$f'(x) = \frac{-f(x + 2h) + 4f(x + h) - 3f(x)}{2h} + \mathcal{O}(h^2),$$

where we have used the fact that $\mathcal{O}(h^3)/h = \mathcal{O}(h^2)$ (see Problem 10); thus we can use the expression on the right as an approximation to the derivative, and the remainder will be bounded by a constant times h^2. See §§2.2 and 4.5 for more on approximations to the derivative. This particular approximation is derived again, by other means, in §4.5.

Note that we have rather quickly obtained an approximation to the first derivative, along with some notion of the error—it behaves proportionally to h^2—by using the $\mathcal{O}$ notation.

Exercises:

1. Use Taylor's Theorem to show that $e^x = 1 + x + \mathcal{O}(x^2)$ for x sufficiently small.

2. Use Taylor's Theorem to show that $\frac{1 - \cos x}{x} = \frac{1}{2}x + \mathcal{O}(x^3)$ for x sufficiently small.

3. Use Taylor's Theorem to show that

$$\sqrt{1 + x} = 1 + \frac{1}{2}x + \mathcal{O}(x^2)$$

for x sufficiently small.

4. Use Taylor's Theorem to show that

$$(1 + x)^{-1} = 1 - x + x^2 + \mathcal{O}(x^3)$$

for x sufficiently small.

5. Show that
$$\sin x = x + \mathcal{O}(x^3).$$

6. Recall the summation formula

$$1 + r + r^2 + r^3 + \cdots + r^n = \sum_{k=0}^{n} r^k = \frac{1 - r^{n+1}}{1 - r}.$$

Use this to prove that

$$\sum_{k=0}^{n} r^k = \frac{1}{1-r} + \mathcal{O}(r^{n+1}).$$

Hint: What is the *definition* of the $\mathcal{O}$ notation?

7. Use the above result to show that 10 terms ($k = 9$) are all that is needed to compute

$$S = \sum_{k=0}^{\infty} e^{-k}$$

to within 10^{-4} absolute accuracy.

8. Recall the summation formula

$$\sum_{k=1}^{n} k = \frac{n(n+1)}{2}.$$

Use this to show that

$$\sum_{k=1}^{n} k = \frac{1}{2}n^2 + \mathcal{O}(n).$$

9. State and prove the version of Theorem 1.7 that deals with relationships of the form $x = x_n + \mathcal{O}(\beta(n))$.

10. Use the definition of $\mathcal{O}$ to show that if $y = y_h + \mathcal{O}(h^p)$, then $hy = hy_h + \mathcal{O}(h^{p+1})$.

11. Show that if $a_n = \mathcal{O}(n^p)$ and $b_n = \mathcal{O}(n^q)$, then $a_n b_n = \mathcal{O}(n^{p+q})$.

12. Suppose that $y = y_h + \mathcal{O}(\beta(h))$ and $z = z_h + \mathcal{O}(\beta(h))$, for h sufficiently small. Does it follow that $y - z = y_h - z_h$ (for h sufficiently small)?

13. Show that
$$f''(x) = \frac{f(x+h) - 2f(x) + f(x-h)}{h^2} + \mathcal{O}(h^2)$$

for all h sufficiently small. *Hint:* Expand $f(x \pm h)$ out to the fourth-order terms.

14. Explain, in your own words, why it is necessary that the constant C in (1.8) be independent of h.

1.3 A PRIMER ON COMPUTER ARITHMETIC

We need to spend some time reviewing how the computer actually does arithmetic. The reason for this is simple: Computer arithmetic is generally inexact, and while the errors that are made are very small, they can accumulate under some circumstances and actually dominate the calculation. Thus, we need to understand computer arithmetic well enough to anticipate and deal with this phenomenon.

Most computer languages use what is called *floating-point arithmetic*. Although the details differ from machine to machine, the basic idea is the same. Every number is represented using a (fixed, finite) number of binary digits, usually called *bits*. A typical implementation would represent the number in the form

$$x = \sigma \times f \times \beta^{t-p}.$$

Here σ is the sign of the number (± 1), denoted by a single bit; f is the mantissa or fraction; β is the base of the internal number system, usually binary ($\beta = 2$) or hexadecimal ($\beta = 16$), although other systems have sometimes been used; t is the (shifted) exponent, i.e., the value that is actually stored; and p is the shift required to recover the actual exponent. (Shifting in the exponent is done to avoid the need for a sign bit in the exponent itself.) The number would be stored by storing only the values of σ, f, and t. The standard way to represent the computer word containing a floating-point number is as follows:

σ	t	f

To keep things simple, we will use a base 2 representation here for our floating point examples.

The total number of bits devoted to the number would be fixed by the computer architecture, and the fraction and exponent would each be allowed a certain number of bits. For example, a common situation in older, "mainframe" architectures would allow 32 bits for the entire word,[4] assigned as follows: 24 bits for the fraction, 7 bits for the exponent, and a single bit for the sign.

Note that this imposes limits on the numbers that can be represented. For example, a 7 bit exponent means that

$$0 \leq t \leq 127.$$

In order to allow for a nearly equal range of positive and negative exponents, a shift p is employed, and in this case should be taken to be $p = 63$, so that

$$-63 \leq t - p \leq 64.$$

Attempts to create larger exponents result in what is called an *overflow*. Attempts to create smaller exponents result in an *underflow*[5]. The fraction is also limited in size by the number of bits available:

$$0 \leq f \leq \sum_{k=1}^{24} 2^{-k} = 1 - 2^{-24}.$$

In practice, most architectures assume that the fraction is *normalized* to be between β^{-1} and 1; any leading zeroes would be dropped and the exponent adjusted accordingly.[6] Thus,

---

[4]A *word* is the largest unit of computer storage. Usually a word consists of two or more *bytes* which themselves consist of a certain number of bits, typically 8.

[5]Most modern computers adhere to the so-called IEEE standard for arithmetic, which uses a type of *extended-floating-point* number system. In addition to ordinary numbers, the IEEE standard allows for results Inf (infinity) and NaN (not a number), and includes rules for manipulating with ordinary floating-point numbers and these special values. For example, x/Inf yields 0 as a result, while x/0 yields plus or minus Inf, depending on the sign of x. Most manipulations with NaN return NaN as their result. In older arithmetic schemes, if an overflow or divide-by-zero occurred, program execution usually terminated.

[6]Some architectures take advantage of this assumption to avoid actually storing that leading bit—all the basic arithmetic algorithms are written to assume an extra leading 1—and thus they are able to get 25 bits of information into 24 bits of storage space.

we actually have

$$\frac{1}{2} \le f \le \sum_{k=1}^{24} 2^{-k} = 1 - 2^{-24}.$$

The errors in computer arithmetic come about because the floating-point numbers are allowed only a fixed number of bits, and not every number can be exactly represented in a fixed number of bits. The common name for the error that is caused by this finite representation is *rounding error*.

Let's consider the simple addition of two numbers,

$$x = 0.1, \quad y = 0.00003$$

assuming that they are represented in the scheme outlined above. The exact answer, of course, is

$$z = x + y = 0.10003.$$

We note first that neither of these numbers can be represented exactly in our scheme. The best that we can do is[7]

$$
\begin{aligned}
\tilde{x} &= 0.00011001\ 10011001\ 10011001\ 100_2 \\
&= 0.0999999940395...
\end{aligned}
$$

and

$$
\begin{aligned}
\tilde{y} &= 0.00000000\ 00000001\ 11110111\ 01010001\ 0000010_2 \\
&= 0.0000299999992421....\ .
\end{aligned}
$$

Thus we have (admittedly, small) errors being made even before the addition occurs.

In our floating-point scheme these two numbers would be stored as

$$
\begin{aligned}
\tilde{x} &= 0.11001100\ 11001100\ 11001100_2 \times 2^{60-63}, \\
\tilde{y} &= 0.11111011\ 10101000\ 10000010_2 \times 2^{48-63}.
\end{aligned}
$$

Because the two numbers are of somewhat different sizes, a normalization is required in order to get equal exponents. One way to do this—the precise details would depend on the particular computer architecture—would give us

$$
\begin{aligned}
\tilde{x} &= 1100\ 11001100\ 11001100\ 11000000\ 00000000_2 \times 2^{-39}, \\
\tilde{y} &= 0000\ 00000000\ 11111011\ 10101000\ 10000010_2 \times 2^{-39},
\end{aligned}
$$

and we can now add the mantissas to get the sum, which is first written as

$$\tilde{w} = 1100\ 11001101\ 11001000\ 01101000\ 10000010_2 \times 2^{-39}.$$

Note that the fraction here is too long to be stored with only 24 bits. What is done with the extra bits? Depending on the machine, they are either thrown away, regardless of their size (chopping), or the result would be rounded up or down, depending on size (rounding). Rounding is more accurate, of course, but chopping is faster. If we round, then we have

$$
\begin{aligned}
\tilde{z} &= 11001100\ 11011100\ 10000111_2 \times 2^{-27} \\
&= 0.11001100\ 11011100\ 10000111_2 \times 2^{60-63} \\
&= 0.100029997527599...
\end{aligned}
$$

---

[7]Note that we have used the subscript "2" to indicate that the number should be interpreted as a base 2 fraction.

and the error is
$$|z - \tilde{z}| = 0.24724... \times 10^{-8}.$$

Let's next assume that our machine uses chopping, in which case we end up with

$$
\begin{aligned}
\tilde{z} &= 11001100\ 11011100\ 10000110_2 \times 2^{-27} \\
&= 0.11001100\ 11011100\ 10000110_2 \times 2^{60-63} \\
&= 0.1000299900770...
\end{aligned}
$$

and a final error of
$$|z - \tilde{z}| = 0.992298... \times 10^{-8}.$$

Note that the chopping error is indeed larger than the error when we used rounding. Similar errors would, of course, occur with the other arithmetic operations.

The difference here—whether we chop or round—is indeed very small, and we might be tempted to ignore it as being too small to worry about. In fact, this is usually the case. But it is possible for the effects of different rounding errors to combine in such a way as to dominate and ruin a calculation. We can illustrate this point with simple decimal arithmetic as long as we insist on using only a small number of digits.

Consider an eight-digit approximation to $a = e^{-(1/100)^2} = 0.99990001$ and a similar approximation to $b = e^{-(1/1000)^2} = 0.99999900$. By construction, both of these numbers are accurate to eight decimal digits. What about their difference $c = a - b = -0.00009899$? How many accurate digits do we have here? The answer is: only *four*. Because we were subtracting two nearly equal numbers, we lost a great deal of accuracy. This phenomenon is called *subtractive cancellation*. If we had started with more accurate approximations, then the difference would contain more accurate digits; try this by looking at the 16-digit values

$$
\begin{aligned}
a &= 0.9999000049998333, \\
b &= 0.9999990000005000, \\
c &= -0.0000989950006667.
\end{aligned}
$$

The result c is now accurate to 12 digits.

To see how subtractive cancellation can destroy almost all the accuracy in a calculation, consider the quantity

$$D = (f(x_1) - f(x_2)) - (f(x_2) - f(x_3))$$

where $f(x) = e^{-x^2}$ and $x_1 = 999/10,000, x_2 = 1/10, x_3 = 1,001/10,000$. Then the calculation, as organized above and done in eight-digit arithmetic, yields $D = -0.1 \times 10^{-7}$. But when we do it in 16-digit arithmetic, we get $D = -0.194049768 \times 10^{-7}$. The eight-digit calculation had *no* accurate digits. Subtractive cancellation is therefore something to avoid as much as possible, and to be aware of when it is unavoidable.

Sometimes the problem with rounding error can be eliminated by increasing the precision of the computation. Traditionally, floating-point arithmetic systems used a single word for each number (*single-precision*) by default, and a second word could be used (*double-precision*) by properly specifying the type of data format to be used. Most languages now use double-word arithmetic by default[8]. Sometimes the entire extra word is used to extend

---

[8] MATLAB allows the user to declare variables to be single-precision via the `single` command. Operations with single-precision variables are done in single-precision.

the length of the fraction; sometimes the length of the exponent is extended as well. *If changing the precision of a calculation dramatically changes the results, then it is almost certain that the computation is being seriously affected by rounding errors.*

Another example of subtractive cancellation occurs with the evaluation of the function

$$f(x) = \frac{e^x - 1}{x}$$

for values of x near zero. L'Hôpital's Rule[9] can be easily used to show that

$$\lim_{x \to 0} f(x) = 1,$$

and Taylor's Theorem can be used to show that

$$f(x) = 1 + \frac{1}{2}x + \mathcal{O}(x^2)$$

for small x, but the evaluation of f for x near 0 will exhibit subtractive cancellation that is amplified by the division by x (since x is small precisely when the subtractive cancellation is worst). Table 1.1 shows the results of computing f using single-precision and double-precision floating-point arithmetic. Note that the error in the single-precision results increases dramatically for $x \leq 2^{-12} = 1/4096$, which is not that small a number. A second threshold of inaccuracy is reached at around $x = 2^{-24}$. Note that the use of double-precision arithmetic defers the onset and lessens the severity of the error, but does not eliminate it entirely, as the last few rows show. (Even though the limiting value of f is 1, the error in the computed value for $x = 2^{-30}$ is still non-zero; $f(2^{-30}) - 1 = 0.4657 \times 10^{-9}$. Although this error would be acceptable in single-precision arithmetic, it is not acceptable for double-precision arithmetic.) How do we fix this?

One approach would be to use Taylor's Theorem:

$$
\begin{aligned}
f(x) &= \frac{\left(1 + x + \frac{1}{2}x^2 + \frac{1}{6}x^3 + \dots \frac{1}{n!}x^n + \frac{1}{(n+1)!}x^{n+1}e^{c_x}\right) - 1}{x}, \\
&= 1 + \frac{1}{2}x + \frac{1}{6}x^2 + \dots + \frac{1}{n!}x^{n-1} + \frac{1}{(n+1)!}x^n e^{c_x},
\end{aligned}
$$

where c_x is between x and 0; the value of n would depend on our required accuracy. We would thus define f, for computational purposes, as

$$f(x) = \begin{cases} 1 + \frac{1}{2}x + \frac{1}{6}x^2 + \dots + \frac{1}{n!}x^{n-1}, & |x| \text{ close to } 0, \\ x^{-1}(e^x - 1), & \text{otherwise.} \end{cases}$$

If we wanted more accuracy, we would use more terms in the Taylor expansion.

We close this section with a definition. In a floating-point computer system, there will exist many nonzero numbers x such that

$$1 + x = 1$$

---

[9]Guillaume François Antoine, Marquis de L'Hôpital (1661–1704) was not trained in mathematics, but took it up after resigning from the military due to poor eyesight. He studied with Johann Bernoulli in the 1690s, and in 1692 published what is generally regarded to be the first calculus textbook, *Analyse des infiniment petits pour l'intelligence des lignes courbes*. What we know as "L'Hôpital's rule" first appears in this book, and is probably actually due to Bernoulli.

Table 1.1   Illustration of subtractive cancellation, using $f(x) = \frac{(e^x - 1)}{x}$.

k	$x_k = 2^{-k}$	single-precision	double-precision
1	0.50000000000000E+00	0.12974424362183E+01	0.12974425414003D+01
2	0.25000000000000E+00	0.11361017227173E+01	0.11361016667510D+01
3	0.12500000000000E+00	0.10651874542236E+01	0.10651876245346D+01
4	0.62500000000000E-01	0.10319118499756E+01	0.10319113426858D+01
5	0.31250000000000E-01	0.10157890319824E+01	0.10157890399713D+01
6	0.15625000000000E-01	0.10078506469727E+01	0.10078533495479D+01
7	0.78125000000000E-02	0.10039215087891E+01	0.10039164424253D+01
8	0.39062500000000E-02	0.10019531250000E+01	0.10019556706170D+01
9	0.19531250000000E-02	0.10009765625000E+01	0.10009771985934D+01
10	0.97656250000000E-03	0.10004882812500E+01	0.10004884402344D+01
11	0.48828125000000E-03	0.10002441406250E+01	0.10002441803663D+01
12	0.24414062500000E-03	0.10000000000000E+01	0.10001220802469D+01
13	0.12207031250000E-03	0.10000000000000E+01	0.10000610376392D+01
14	0.61035156250000E-04	0.10000000000000E+01	0.10000305182002D+01
15	0.30517578125000E-04	0.10000000000000E+01	0.10000152589419D+01
16	0.15258789062500E-04	0.10000000000000E+01	0.10000076294382D+01
17	0.76293945312500E-05	0.10000000000000E+01	0.10000038146973D+01
18	0.38146972656250E-05	0.10000000000000E+01	0.10000019073486D+01
19	0.19073486328125E-05	0.10000000000000E+01	0.10000009536743D+01
20	0.95367431640625E-06	0.10000000000000E+01	0.10000004768372D+01
21	0.47683715820312E-06	0.10000000000000E+01	0.10000002384186D+01
22	0.23841857910156E-06	0.10000000000000E+01	0.10000001192093D+01
23	0.11920928955078E-06	0.10000000000000E+01	0.10000000596046D+01
24	0.59604644775391E-07	0.00000000000000E+00	0.10000000298023D+01
25	0.29802322387695E-07	0.00000000000000E+00	0.10000000149012D+01
26	0.14901161193848E-07	0.00000000000000E+00	0.10000000000000D+01
27	0.74505805969238E-08	0.00000000000000E+00	0.10000000000000D+01
28	0.37252902984619E-08	0.00000000000000E+00	0.10000000000000D+01
29	0.18626451492310E-08	0.00000000000000E+00	0.10000000000000D+01
30	0.93132257461548E-09	0.00000000000000E+00	0.10000000000000D+01

within computer precision. For instance, in the 32-bit implementation outlined at the beginning of the section, it is clear that this will hold for any $x < 2^{-24}$, since $1 + 2^{-24}$ in binary will require 25 bits of storage, and we only have 24 to work with. We define the *machine rounding unit*, or *machine epsilon*, **u**, to be the largest such number:

Definition 1.2 (Machine Epsilon) *The machine epsilon (alternatively, the machine rounding unit),* **u**, *is the largest floating-point number x such that $x + 1$ cannot be distinguished from 1 on the computer:*

$$\mathbf{u} = \max\{x \mid 1 + x = 1, \text{in computer arithmetic}\}.$$

It is possible to compute the machine epsilon based on knowledge of the floating-point number system on the computer in question.

■ **EXAMPLE 1.7**

Suppose that we want to compute the machine epsilon for a simple three-digit decimal computer which rounds (i.e., a machine that does decimal arithmetic, keeps only three digits, and rounds its results). Thus, we can write any number x that is stored on the computer as $x = d_1 d_2 d_3 \times 10^t$, where t is the exponent and $d_1 d_2 d_3$ represent the decimal digits of the fraction. Now, based on the definition of machine epsilon, we know that $x_1 = 1.00 \times 10^{-2}$ is too large, because

$$1 + x_1 = 1.00 + 0.01 = 1.01 \neq 1.00.$$

On the other hand, we know that $x_2 = 1.00 \times 10^{-3}$ is certainly small enough (perhaps too small) because we have

$$1 + x_2 = 1.00 + 0.001 = 1.001 \Rightarrow 1.00 = 1.00.$$

Thus, the computer cannot distinguish between 1 and $1 + x_2$. But **u** is the *largest* such number, so we have to look a bit further. It's not a matter of using the right formula; it is really a matter of experimentation and trial and error. We have

$$1 + 0.002 = 1.002 \rightarrow 1.00 = 1,$$

so we know that $x_3 = 2.00 \times 10^{-3}$ is small enough; the same argument would apply to $x_4 = 3.00 \times 10^{-3}$ and $x_5 = 4.00 \times 10^{-3}$. But

$$1 + 0.005 = 1.005 \rightarrow 1.01 \neq 1$$

(because the machine rounds its computations), thus $x_6 = 5.00 \times 10^{-3}$ is too large. But it is just barely too large, as any smaller number would have resulted in the sum being rounded down to 1.00. Thus we want the next smaller number *within the floating-point number system*, i.e., we want $\mathbf{u} = 4.99 \times 10^{-3}$. This gives us

$$1.00 + 0.00499 = 1.00499 \rightarrow 1.00 = 1,$$

and it is clear that any larger number that we can represent in our floating-point system would result in the sum being rounded up to $1.01 \neq 1$. The student ought to be able to show, that if the machine chops instead of rounds, then $\mathbf{u} = 9.99 \times 10^{-3}$.

There are a number of interesting consequences of the existence of the machine epsilon. For example, consider the numbers

$$a = 1, b = \mathbf{u}, c = \mathbf{u}.$$

If we want to add up these numbers, we quickly learn that the way we organize the calculation matters. For example,

$$(a + b) + c = (1 + \mathbf{u}) + \mathbf{u} = 1 + \mathbf{u} = 1,$$

whereas

$$a + (b + c) = 1 + 2\mathbf{u} \neq 1.$$

From a technical point of view, this means that the associative law of arithmetic (for addition) does not hold in floating-point arithmetic systems. In other words,

the order in which we do operations sometimes matters.

It is possible to estimate the machine epsilon by constructing a loop that adds increasingly small numbers to 1, and only terminates this when the result cannot be distinguished from 1.

The basic result on computer arithmetic is the following.

Theorem 1.8 (Computer Arithmetic Error) *Let $*$ denote any of the basic binary operations (addition, subtraction, multiplication, or division), and let $fl(x)$ denote the floating point value of x. Then there exists a constant $C > 0$ such that, for all x and y,*

$$|x * y - fl(x * y)| \leq C\mathbf{u}|x * y|.$$

Thus, the computer value of $x * y$ is relatively accurate to within $\mathcal{O}(\mathbf{u})$.

The point of this section is that we should now be aware that computer arithmetic is not the 100% reliable thing we might have thought it was. However, the errors that crop up do tend to appear only when extremely large or small numbers are involved (exceptions do occur, so we have to be careful), and are themselves very small. Moreover, rounding errors tend to cancel themselves out over the long run. Rounding error and related effects are things we have to watch out for and be aware of, but they are *usually* dominated by the mathematical error that is made in constructing the approximations. Exceptions to this rule of thumb do exist, however, as we shall see in §2.2 and §4.12.1.

Exercises:

1. In each problem below, A is the exact value, and A_h is an approximation to A. Find the absolute error and the relative error.

   (a) $A = \pi$, $A_h = 22/7$;
   (b) $A = e$, $A_h = 2.71828$;
   (c) $A = \frac{1}{6}$, $A_h = 0.1667$;
   (d) $A = \frac{1}{6}$, $A_h = 0.1666$.

2. Perform the indicated computations in each of three ways: (i) Exactly; (ii) Using three-digit decimal arithmetic, with chopping; (iii) Using three-digit decimal arithmetic, with rounding. For both approximations, compute the absolute error and the relative error.

(a) $\frac{1}{6} + \frac{1}{10}$;

(b) $\frac{1}{6} \times \frac{1}{10}$;

(c) $\frac{1}{9} + \left(\frac{1}{7} + \frac{1}{6}\right)$;

(d) $\left(\frac{1}{7} + \frac{1}{6}\right) + \frac{1}{9}$.

3. For each function below explain why a naive construction will be susceptible to significant rounding error (for x near certain values), and explain how to avoid this error.

(a) $f(x) = (\sqrt{x+9} - 3)x^{-1}$;

(b) $f(x) = x^{-1}(1 - \cos x)$;

(c) $f(x) = (1 - x)^{-1}(\ln x - \sin \pi x)$;

(d) $f(x) = (\cos(\pi + x) - \cos \pi)x^{-1}$;

(e) $f(x) = (e^{1+x} - e^{1-x})(2x)^{-1}$.

4. For $f(x) = (e^x - 1)/x$, how many terms in a Taylor expansion are needed to get single-precision accuracy (seven-decimal digits) for all $x \in [0, \frac{1}{2}]$? How many terms are needed for double-precision accuracy (14 decimal digits) over this same range?

5. Using single-precision arithmetic only, carry out each of the following computations, using first the form on the left side of the equals sign, then using the form on the right side, and compare the two results. Comment on what you get in light of the material in §1.3.

(a) $(x + \epsilon)^3 - 1 = x^3 + 3x^2\epsilon + 3x\epsilon^2 + \epsilon^3 - 1$, $x = 1.0$, $\epsilon = 0.000001$;

(b) $-b + \sqrt{b^2 - 2c} = 2c(-b - \sqrt{b^2 - 2c})^{-1}$, $b = 1,000$, $c = \pi$.

6. Consider the sum

$$S = \sum_{k=0}^{m} e^{-14(1 - e^{-0.05k})},$$

where $m = 2 \times 10^5$. Again using only single-precision arithmetic, compute this two ways: First, by summing in the order indicated in the formula; second, by summing *backwards*, that is, starting with the $k = 200,000$ term and ending with the $k = 0$ term. Compare your results and comment on them.

7. Using the computer of your choice, find three values a, b, and c, such that

$$(a + b) + c \neq a + (b + c).$$

Repeat using your pocket calculator.

8. Assume that we are using three-digit decimal arithmetic. For $\epsilon = 0.0001$, $a_1 = 5$, compute

$$a_2 = a_0 + \left(\frac{1}{\epsilon}\right) a_1$$

for a_0 equal to each of 1, 2, and 3. Comment.

9. Let $\epsilon \leq \mathbf{u}$. Explain, in your own words, why the computation

$$a_2 = a_0 + \left(\frac{1}{\epsilon}\right) a_1$$

is potentially rife with rounding error. (Assume that a_0 and a_1 are of comparable size.) *Hint:* See Problem 8.

10. Using the computer and language of your choice, write a program to estimate the machine epsilon.

11. We can compute e^{-x} using Taylor polynomials in two ways, either using

$$e^{-x} \approx 1 - x + \frac{1}{2}x^2 - \frac{1}{6}x^3 + \dots$$

or using

$$e^{-x} \approx \frac{1}{1 + x + \frac{1}{2}x^2 + \frac{1}{6}x^3 + \dots}.$$

Discuss, in your own words, which approach is more accurate. In particular, which one is more (or less) susceptible to rounding error?

12. What is the machine epsilon for a computer that uses binary arithmetic, 24 bits for the fraction, and rounds? What if it chops?

13. What is the machine epsilon for a computer that uses *octal* (base 8) arithmetic, assuming that it retains eight octal digits in the fraction?

◁ ● ● ▷

1.4 A WORD ON COMPUTER LANGUAGES AND SOFTWARE

In the early 1970s (when the author was an undergraduate student) the standard computer language for scientific computation was FORTRAN, with Algol being perhaps the second choice. BASIC, in many incarnations, was also a possibility. By the late 1980s, Pascal had entered the fray and FORTRAN was considered passé in some quarters, especially since various easy-to-use integrated environment packages for Pascal were being marketed to personal computer users. By the 1990s, Pascal was fading away, but we now had C, or C++, or even Java. And FORTRAN, despite the predictions of many and the desires of many more, was still with us. In addition, the increased power of personal computers meant that software packages such as MATLAB or MathCAD or Maple or Mathematica might be used to do scientific computation. (MATLAB, in particular, has had a tremendous influence on scientific computation.)

In short, if you don't like the present state of affairs with regard to computing languages, wait around a little while—it will change.

However, it is impossible to ignore the dominant place taken in recent years by MATLAB, which has been very successful in becoming the scientific programming language of choice for many applications. The ease with which it solves complicated problems and provides easy-to-use graphical tools is almost unsurpassed. Earlier editions of this book spoke of being "language neutral," but, starting with this Second Edition, the assumption will be that the students are coding in MATLAB. References to specific MATLAB commands will be made in a computer-font typeface, thus: `rand`, which is the command for generating random numbers, vectors, or matrices.

If the instructor or some students wish to program in a different language, this book is not so wedded to MATLAB as to make that impossible. The only significant consequence will be that some exercises may have to be skipped.

Most examples are given in a generic pseudo-code which is heavily based on MATLAB, but occasionally, raw MATLAB code will be given. The author is still of the opinion that students should be comfortable in as many different languages as possible—even though most scientific programming *today* might be done in MATLAB, it is still the case that there is a lot of computer code (called *legacy code*) that is still being used and that was written in FORTRAN or Pascal or Algol or C.

The more involved algorithms will be presented in more of an "outline" style, rather than in a line-by-line of code style.

Finally, it needs to be said: This book is not intended as a text in how to use MATLAB, but as a text in the mathematics of numerical methods.

There exist a number of sources for good mathematical software. Traditionally, two of the best sources were the IMSL and NAG libraries, collections of FORTRAN routines for a wide variety of computational tasks. More specialized packages have also been developed, notably QUADPACK (numerical integration), LINPACK (linear algebra), EISPACK (eigenvalue methods), LAPACK (an updated package that combines and actually replaces LINPACK and EISPACK), and others. A repository of public-domain mathematics software is maintained at NETLIB.[10] MATLAB has created a number of specialized "toolboxes" for numerical computation, a list of which can be found at `http://www.mathworks.com/products/`. Many authors now put their own codes up on websites for public access.

There are also some "freeware" versions of MATLAB-like software which can be installed on individual PCs.

Despite the wide availability of general-purpose mathematical software, it is still important for students to learn how to write and (most important) debug their own codes. For this reason many of the exercises in this book involve computer programming.

The text assumes that students are familiar with the use of elementary packages (such as MATLAB or Maple or Mathematica) for producing simple plots of functions. Whenever the exercises call for graphs or plots to be produced, it is assumed that such modern technology will be used. Students should also feel free (with their instructor's permission, of course) to use Maple or Mathematica to simplify some of the more involved manipulations in the exercises.

1.5   SIMPLE APPROXIMATIONS

We have already used Taylor series to construct a simple approximation to the exponential function. Here we present a similar but slightly more involved example to reinforce the basic ideas and also to illustrate the use of the asymptotic order notation introduced above.

---

[10]On the World Wide Web at `http://www.netlib.org`.

The *error function* occurs often in probability theory and other areas of applied mathematics (the solution of heat conduction problems, for example). It is defined by an integral:

$$\mathrm{erf}(x) = \frac{2}{\sqrt{\pi}} \int_0^x e^{-t^2} dt.$$

It is not possible to evaluate this integral by means of the Fundamental Theorem of Calculus; there is no elementary anti-derivative for e^{-t^2}. In Chapter 5 we will derive techniques that can be applied to directly approximate the integral. Here we will use Taylor's Theorem to approximate the integrand as a polynomial, and exactly integrate that polynomial.

This is a fundamental idea in numerical methods: When confronted with a computation that cannot be done exactly, we often replace the relevant function with something simpler which approximates it, and carry out the computation exactly *on the simple approximation.*

We might be tempted to compute the Taylor approximation to e^{-t^2} by appealing directly to the formula (1.1), but this will lead to a lot of unnecessary work. We know that

$$
\begin{aligned}
e^x &= 1 + x + \frac{1}{2!}x^2 + \frac{1}{3!}x^3 + \ldots + \frac{1}{k!}x^k + \frac{x^{k+1}}{(k+1)!}e^c \\
&= p_k(x) + R_k(x),
\end{aligned}
$$

where

$$p_k(x) = \sum_{i=0}^{k} \frac{x^i}{i!}$$

and

$$R_k(x) = \frac{x^{k+1}}{(k+1)!}e^c.$$

Therefore, we can get the expansion and error for e^{-t^2} by a simple substitution:

$$e^{-t^2} = p_k(-t^2) + R_k(-t^2).$$

Thus, we have

$$
\begin{aligned}
\mathrm{erf}(x) &= \frac{2}{\sqrt{\pi}} \int_0^x p_k(-t^2) dt + \frac{2}{\sqrt{\pi}} \int_0^x R_k(-t^2) dt \\
&= \frac{2}{\sqrt{\pi}} \left(\sum_{i=0}^{k} \frac{(-1)^i x^{2i+1}}{(2i+1)i!} \right) + \frac{2}{\sqrt{\pi}} \int_0^x R_k(-t^2) dt.
\end{aligned}
$$

For simplicity, define the polynomial that approximates the error function by q_k:

$$
\begin{aligned}
q_k(x) &= \frac{2}{\sqrt{\pi}} \left(\sum_{i=0}^{k} \frac{(-1)^i x^{2i+1}}{(2i+1)i!} \right) \\
&= \frac{2}{\sqrt{\pi}} \left(x - \frac{1}{3}x^3 + \frac{1}{10}x^5 - \frac{1}{42}x^7 + \ldots + \frac{(-1)^k x^{2k+1}}{(2k+1)k!} \right)
\end{aligned}
$$

so that we have

$$\mathrm{erf}(x) - q_k(x) = \frac{2}{\sqrt{\pi}} \int_0^x R_k(-t^2) dt.$$

We want to simplify this error and produce an error bound, so we will also set

$$E_k(x) = \frac{2}{\sqrt{\pi}} \int_0^x R_k(-t^2)dt.$$

Thus,

$$\text{erf}(x) - q_k(x) = E_k(x). \tag{1.10}$$

To simplify and bound the error, we note that

$$
\begin{aligned}
E_k(x) &= \frac{2}{\sqrt{\pi}} \int_0^x R_k(-t^2)dt, \\
&= \frac{2}{\sqrt{\pi}} \int_0^x \frac{(-t^2)^{k+1}}{(k+1)!} e^c dt, \\
&= \frac{2(-1)^{k+1}}{(k+1)!\sqrt{\pi}} \int_0^x t^{2k+2} e^c dt,
\end{aligned}
$$

where c depends on t, and $-t^2 \le c \le 0$. Now, since both functions in the integrand above are positive, we can apply the Integral Mean Value Theorem to get

$$\int_0^x t^{2k+2} e^c dt = e^\xi \int_0^x t^{2k+2} dt = e^\xi \frac{x^{2k+3}}{2k+3}$$

for some ξ between 0 and $-x^2$. Thus,

$$E_k(x) = \frac{2(-1)^{k+1}x^{2k+3}}{(2k+3)(k+1)!\sqrt{\pi}} e^\xi. \tag{1.11}$$

This error statement, although applicable only to a very elementary problem, nonetheless contains all of the common features that will appear in the error estimates for more realistic algorithms applied to more complicated problems. Thus, it is worthwhile discussing it before we proceed to simplify it further.

Let us write the error in a more structured form:

$$E_k(x) = \left(\frac{2}{\sqrt{\pi}} \right) \left(\frac{(-1)^{k+1}x^{2k+3}}{(2k+3)(k+1)!} \right) e^\xi = C\delta_k(x)M,$$

where

$$C = \frac{2}{\sqrt{\pi}}, \quad \delta_k(x) = \frac{(-1)^{k+1}x^{2k+3}}{(2k+3)(k+1)!}, \quad M = e^\xi.$$

This divides the error into three distinct parts:

1. A raw numerical constant, C;

2. An expression depending on the computational parameters (k, x) of the problem; this is $\delta_k(x)$;

3. A factor that depends on the function or its derivatives, evaluated at an indeterminate, unknown point; this is M.

We are most interested in the second part of the error. The raw numerical constant is usually of less interest, since we cannot do anything to make it smaller. The function-dependent part of the error is of some concern to us in more complicated approximations; here we simply note that we are not able to actually compute that value in most cases, so we resort to upper bounds.

It is the parameter-dependent part of the error that determines convergence and accuracy and how fast we achieve either of them, and that is, at least partly, under our control. In fact, we might use the asymptotic order notation to write (1.10) as

$$\mathrm{erf}(x) = q_k(x) + \mathcal{O}(\delta_k(x)).$$

Alternatively, we could use the approximate equality notation to write

$$\mathrm{erf}(x) \approx q_k(x),$$

where it is understood (perhaps only implicitly) that this approximation is valid only for k large or for x small. We might want to simplify the error a little more by removing one of the variables from $\delta_k(x)$. For example, if we were only interested in values of x between 0 and 2, then we could easily establish that

$$|\delta_k(x)| \le \frac{2^{2k+3}}{(2k+3)(k+1)!}$$

for all $x \in [0, 2]$. Moreover, the $2k + 3$ factor in the denominator is not going to affect the error bound nearly as much as the factorial, so we can further simplify to get

$$|\delta_k(x)| \le \frac{8}{5}\frac{2^{2k}}{(k+1)!} = \frac{8}{5}\frac{4^k}{(k+1)!}$$

for all $k \ge 1$. Thus, we can write

$$\mathrm{erf}(x) = q_k(x) + \mathcal{O}\left(\frac{4^k}{(k+1)!}\right) \quad 0 \le x \le 2.$$

This extra simplification has indeed increased our estimate of the error, but only slightly, and not in a manner that ignores the most important factors in the convergence of the approximation—the factorial and the power. The benefit we get from this slightly increased error estimate is the ability to quickly and easily gauge the accuracy of an approximation using a specified number of terms. For example, we find that the error in a 20-term approximation is on the order of 2.2×10^{-8}, whereas a 10-term approximation is accurate only to within about 0.026.

Exercises:

1. Consider the error (1.11) in approximating the error function. If we restrict ourselves to $k \le 3$, then over what range of values of x is the approximation accurate to within 10^{-3}?

2. If we are interested only in $x \in [0, \frac{1}{2}]$, then how many terms—equivalently, what degree of polynomial—do we need in the error function approximation to get an accuracy of 10^{-4}?

3. Repeat the above for $x \in [0, 1]$.

4. Assume that $x \in [0, 1]$ and write the error in the approximation to the error function using the asymptotic order notation.

5. For

$$f(x) = \int_0^x t^{-1} \sin t \, dt, \quad x \in [-\pi/4, \pi/4],$$

construct a Taylor approximation that is accurate to within 10^{-4} over the indicated interval.

6. Repeat the above for

$$f(x) = \int_0^x e^{-t^2} dt, \quad x \in \left[-\frac{1}{2}, \frac{1}{2}\right].$$

7. Construct a Taylor approximation for

$$f(x) = \int_0^x t^{-p} e^{-t^2} dt, \quad 0 \le p < 1, \quad x \in \left[0, \frac{1}{2}\right]$$

that is accurate to within 10^{-3} for *all* values of p in the indicated range.

8. Does it make a difference in Problem 7 if we restrict p to $p \in [0, \frac{1}{4}]$?

9. What is the error in the Taylor polynomial of degree 5 for $f(x) = 1/x$, using $x_0 = 3/4$, for $x \in [\frac{1}{2}, 1]$?

10. How many terms must be taken in the above to get an error of less than 10^{-2}? 10^{-4}?

11. What is the error in a Taylor polynomial of degree 4 for $f(x) = \sqrt{x}$ using $x_0 = 9/16$, for all $x \in [1/4, 1]$?

12. Consider the rational function

$$r(x) = \frac{1 + \frac{1}{2}x}{1 - \frac{1}{2}x};$$

carry out the indicated division to write this as

$$r(x) = p(x) + R(x),$$

where $p(x)$ is a polynomial, and $R_1(x)$ is a remainder term in the form $R_1 = x^2 R(x)$, where R is a proper rational function, i.e., one with the degree of the numerator strictly less than the degree of the denominator. Can you relate $p(x)$ to a Taylor expansion for the exponential function? Bound the error $|e^x - r(x)|$, assuming that $x \le 0$, if you can.

13. Repeat the analysis in the previous problem for the rational function

$$r(x) = \frac{1 + \frac{1}{2}x + \frac{1}{12}x^2}{1 - \frac{1}{2}x + \frac{1}{12}x^2}.$$

Can you get a better error in this case?

14. Finally, consider the rational function[11]

$$r(x) = \frac{1 + \frac{1}{3}x}{1 - \frac{2}{3}x + \frac{1}{6}x^2}.$$

By dividing out the rational function into a polynomial $p(x)$ plus a remainder, bound the error $|e^x - r(x)|$ for all $x \leq 0$. Try to get as high an accuracy as you can in terms of powers of x.

$$\triangleleft \bullet \bullet \bullet \triangleright$$

1.6   APPLICATION: APPROXIMATING THE NATURAL LOGARITHM

In this section we will put together many of the basic ideas from previous sections to construct a reasonable approximation to the natural logarithm function. The primary tool will be Taylor's Theorem, and our goal will be to produce an approximation to the logarithm that is accurate to within $\epsilon = 10^{-16}$.

We first observe that if we can assume that $\tau = \ln 2$ is known to arbitrary precision, then we really only need to construct a logarithm approximation that is valid over a short interval. This follows because of the way that the computer stores numbers. Since any z is stored as $z = f \cdot 2^\beta$ for $\frac{1}{2} \leq f \leq 1$ and some β, we have

$$\ln z = \ln f + \beta \ln 2.$$

Since $f \in [\frac{1}{2}, 1]$, we will get the best results by choosing the center of the expansion to be $x_0 = \frac{3}{4}$. This is a little unusual, but not unprecedented. The Taylor expansion for the logarithm is then (you ought to verify this)

$$
\begin{aligned}
\ln x \;=\; & \ln x_0 + \frac{x - x_0}{x_0} - \frac{1}{2}\left(\frac{x - x_0}{x_0}\right)^2 + \ldots + (-1)^{n-1}\frac{1}{n}\left(\frac{x - x_0}{x_0}\right)^n \\
& + \; (-1)^n \int_{x_0}^{x} (x - t)^n t^{-n-1} dt.
\end{aligned}
\tag{1.12}
$$

Since the remainder here is a little more complicated than in the usual case, let's look at it carefully. We have

$$R_n(x) = (-1)^n \int_{x_0}^{x} (x - t)^n t^{-n-1} dt$$

so that

$$|R_n(x)| = \left|\int_{x_0}^{x} (x - t)^n t^{-n-1} dt\right|.$$

We now take upper bounds using the elementary fact that

$$\left|\int_{a}^{b} f(x)dx\right| \leq |b - a| \max_{x \in [a,b]} |f(x)|.$$

----
[11]The functions in these exercises are all examples of what are known as *Padé approximations*.

Thus,

$$|R_n(x)| \leq |x - x_0| \max_t \left| \frac{(x-t)^n}{t^{n+1}} \right|.$$

Now, we know that $x_0 = \frac{3}{4}$, that $x \in [\frac{1}{2}, 1]$, and that t is between x and x_0. It follows, then, that

$$|x - x_0| \leq \frac{1}{4}$$

and that

$$\left| \frac{(x-t)^n}{t^{n+1}} \right| = |t|^{-1} \left| \frac{x}{t} - 1 \right|^n \leq 2 \left| \frac{x-t}{t} \right|^n.$$

At this point we have to proceed carefully. We want to bound the function

$$g(t) = \frac{x-t}{t}$$

in absolute value. What do we know about t? Since t is between x and x_0, we have either $x_0 \leq t \leq x$ or $x \leq t \leq x_0$. We have $g'(t) = -x/t^2$ so there are no critical points (since $x \neq 0$). Therefore, the Extreme Value Theorem tells us that the maximum and minimum of g occurs at the endpoints, that is, for $t = x$ and for $t = x_0$. Since $g(x) = 0$, we clearly have $|g(t)| \leq |g(x_0)|$. Thus, substituting into the previous inequality, we have

$$\left| \frac{(x-t)^n}{t^{n+1}} \right| \leq 2 \left| \frac{x-x_0}{x_0} \right|^n \leq 2 \left| \frac{\frac{1}{4}}{\frac{3}{4}} \right|^n.$$

Therefore,

$$|R_n(x)| \leq \frac{1}{2} \left(\frac{1}{3} \right)^n,$$

from which we conclude that $n = 33$ is sufficient to guarantee that $|R_n(x)| \leq 10^{-16}$ for all $x \in [\frac{1}{2}, 1]$. It thus requires a 33-degree polynomial to approximate the logarithm in this fashion, along with an accurate representation for $\ln 2$ and $\ln \frac{3}{4}$.

Problem 5 asks you to implement this as a subprogram and check it against the intrinsic natural logarithm function on your computer.

Can this be improved? Yes, it is possible to construct an equally accurate logarithm approximation that uses fewer computations. Problem 6 asks you to look into this by using a clever combination of logarithm expansions that results in faster convergence.

Exercises:

1. Write each of the following in the form $x = f \times 2^\beta$ for some $f \in [\frac{1}{2}, 1]$.

   (a) $x = 13$;

   (b) $x = 25$;

   (c) $x = \frac{1}{3}$;

   (d) $x = \frac{1}{10}$.

2. For each value in the previous problem, compute the logarithm approximation using the degree 4 Taylor polynomial from (1.12). What is the error compared to the logarithm on your calculator?

3. Repeat the above for the degree 6 Taylor approximation.

4. Repeat the above for the degree 10 Taylor approximation.

5. Implement (as a computer program) the logarithm approximation constructed in this section. Compare it to the intrinsic logarithm function over the interval $[\frac{1}{2}, 1]$. What is the maximum observed error?

6. Let's consider how we might improve on our logarithm approximation from this section.

   (a) Compute the Taylor expansions, with remainder, for $\ln(1 + x)$ and $\ln(1 - x)$ (use the integral form of the remainder).

   (b) Combine the two to get the Taylor expansion for

   $$f(x) = \ln\left(\frac{1 - x}{1 + x}\right).$$

   What is the remainder in this expansion?

   (c) Given $z \in [\frac{1}{2}, 1]$ show how to compute x such that $z = (1 - x)(1 + x)^{-1}$. What interval contains x?

   (d) Use the answer to part (c) to construct an approximation to $\ln z$ that is accurate to within 10^{-16}.

7. Use the logarithm expansion from the previous problem, but limited to the degree 4 case, to compute approximations to the logarithm of each value in the first problem of this section.

8. Repeat the above, using the degree 10 approximation.

9. Implement (as a computer program) the logarithm approximation constructed in Problem 6. Compare it to the intrinsic logarithm function over the interval $[\frac{1}{2}, 1]$. What is the maximum observed error?

10. Try to use the ideas from this section to construct an approximation to the reciprocal function, $f(x) = x^{-1}$, that is accurate to within 10^{-16} over the interval $[\frac{1}{2}, 1]$.

1.7 A BRIEF HISTORY OF COMPUTING

The development of numerical methods and analysis is closely tied to the development of modern computing equipment, so it makes sense to include this brief essay on the history of computing, drawn from a variety of sources, including the author's own experiences as a student and faculty member.[12] A standard source—written by a numerical analyst—is *The Computer: From Pascal to von Neumann* [13].

The history of computing—by which we mean "the history of the development of machine computation"—is probably older than most students think. Devices like the

---

[12]Besides, it enables the author to indulge his avocational love of history.

abacus have been around for centuries. The first mechanical calculator, sometimes known as the Pascaline, was developed and constructed by the French mathematician Blaise Pascal (1623–1662) in 1645. The machine worked via a set of wheels and gears, and could do the basic arithmetic operations. Slide rules were also developed in the 1600s based on the emerging science of logarithms, and remained the most common personal computing device until the early 1970s. In the 1820s and 1830s, Charles Babbage designed a "Difference Engine" and a more complex "Analytical Engine," both of which were supposed to aid in numerical calculations. Neither was completed by Babbage, although modern examples have been built using Babbage's plans[13]. At this time, and for several decades further on, the word "computer" meant a *person* whose job it was to perform extensive calculations *by hand*.

Babbage's Analytical Engine used punched cards for input, a technology first introduced in the early 19th Century for controlling the pattern being created by automatic looms, and then later refined by Hermann Hollerith (1860–1929) to process data for the 1890 U.S. Census. Eventually, in the period 1950-1975, the punched card became the mainstay input medium for computer operations. In the wake of his success with the 1890 Census, Hollerith founded a firm, Tabulating Machine Company, in 1896, which in 1924 was re-named the International Business Machines Corporation[14].

Meanwhile, developments on a different front were being made by physicists, first in Britain, and then in the United States. A planimeter is a device that can be used to compute the area of an irregular figure that it is used to trace. The first such device was probably built by the German engineer, J. H. Hermann, in 1814. The emminent physicist, James Clerk Maxwell, invented one in 1855. James Thomson, a Scottish engineer and brother to the physicist Sir William Thomson (later Lord Kelvin), produced one based on Maxwell's design in the early 1860s. Thomson did not do much with his device until 1876, when it was realized that this device could be used to compute some integrals (which, after all, are just areas defined by curves) needed for a project of Sir William's[15]. This was the beginning of the development of so-called *analog computers*, which work by mimicking the process under study[16]. Sir William wrote and published a paper[17] describing how a mechanical planimeter-based device could be constructed to "solve" a general second order linear differential equation, but the technology did not (yet) exist to build such a machine.

In 1897, the American physicists A.A. Michaelson (the first American to win a Nobel Prize in Physics) and S. W. Stratton built a "harmonic analyzer" for computing Fourier coefficients that was able to surmount many of the difficulties that Thomson's efforts had encountered. Their machine was able to handle Fourier series of as many as 20 terms

---

[13]Pehr Georg Scheutz of Sweden built an example of the Difference Engine, with Babbage's blessing and assistance, which earned Scheutz a knighthood from the King of Sweden. Babbage appears to have had personality issues that made it difficult for him to carry any of his ideas through to completion, as he would get a project partially done, conceive of a better way to do things, get that project partially done, and so on, in an almost unending sequence. See [13], p. 24.

[14]And, yes, the classic "IBM punch card" got its precise dimensions from the dollar bill of the 1890s—doing so allowed Hollerith to use the same machines to make his punch cards that the government used to print and cut currency.

[15]Specifically, Sir William Thomson needed to compute some Fourier coefficients to determine the periodicity of the tides.

[16]In his undergraduate days, the author took a course in electronic analog computing. Using plug boards, the students "programmed" the machine by constructing a circuit whose voltage would follow exactly the desired differential equation. When power was applied to the circuit, a graph of the solution—the output voltage—would be produced. It was a fun course.

[17]"Mechanical Integration of the General Linear Differential Equation of any Order with Variable Coefficients," *Proceedings of the Royal Society of London*, 1876. Available online.

(Thomson was only able to treat two Fourier terms), on the basis of which they obtained funding to develop and build a second machine that could handle 80 Fourier terms.

This led to the development, in the 1930s and 1940s, of several versions of an electrically driven mechanical machine (commonly known as a "differential analyzer") for solving differential equations. The primary effort in the United States was led by Vannevar Bush of MIT. For a brief time circa 1930 or so, these were the state-of-the-art in machine computation. At about the same time, a number of efforts based on telephone-relay technology were developed, notably under the direction of Howard Aiken at Harvard. Also, John Atanasoff developed an electronic computing machine at Iowa State during the 1930s; this machine was dedicated to solving systems of linear equations and was not really a programmable device.

In the United States, the direct impetus for the development of more-capable computing devices came largely from the U.S. Army's Ballistics Research Laboratory at Aberdeen Proving Ground, which needed better facilities in order to produce so-called "firing tables" for use with artillery; the effort was led by John Mauchly and J.P. Eckert of the Moore School of Engineering at the University of Pennsylvania. In Britain, a team led by Alan Turing (who contributed a great deal to the theory of machine computation) developed a computing machine known as the "bombe" for decoding intercepted German communications that had been encoded by the so-called ENIGMA device. A German effort, known as the Z3, was destroyed in a bombing raid on Berlin in December of 1943.

The American effort, dubbed ENIAC (Electronic Numerical Integrator and Computer), is widely considered to be the first true programmable computing machine. By modern standards it was enormously difficult to use and very limited—it was "programmed" by essentially re-wiring it—but it was in service for about 10 years. The finished machine consisted of over 17,000 vacuum tubes, 70,000 resistors, 10,000 capacitors, 1,500 relays, and 6,000 switches. In size, it filled three walls of a large room, being 100 feet long, 10 feet high, and 3 feet deep. It operated at a clock rate of about 100,000 cycles per second (100 kHz). It could multiply 10-digit numbers at a rate of about 357 per second.

It is at roughly this point that John von Neumann enters the picture. Hungarian-born and incredibly brilliant, von Neumann was working on a number of scientific projects for the United States government in support of the war effort, most notably the Manhattan Project (the effort to build the atomic bomb), which needed to perform extensive shock wave computations in order to make the "implosion trigger" work. A chance meeting between Herman Goldstine (working on the development of ENIAC) and von Neumann in a Philadelphia train station led to von Neumann's involvement with the ENIAC, and eventually to his seminal unpublished paper on computer design, "First Draft of a Report on the EDVAC" [12]. (Although this document was not formally published until 1993, it was widely circulated and is considered as the basis of modern computer architecture[18].)

After the war, von Neumann led an effort at the Institute for Advanced Study (IAS) to develop an improved machine, largely for the purpose of doing the necessary computations for the development of the hydrogen bomb, although some of the earliest numerical weather prediction tests were run on this machine, which was in service from 1951 until mid-1958. The story of this machine and its development is the subject of the book, *Turing's Cathedral* [10]. The IAS machine could perform about 1400 multiplications per second. (ENIAC's multiplication rate was about 357 per second.)

---

[18]As is pointed out in the online version, a significant number of typos crept in to the preparation of the 1993 publication; these are corrected in the online document.

The invention of the transistor in 1947 led to the development of a new generation of machines based on this new, faster, more reliable, technology. Companies such as Sperry Rand, Honeywell, and (eventually) IBM began marketing commercial computers in the 1950s and 1960s. The development of the integrated-circuit microprocessor in the early 1970s eventually led to today's personal computers and ubiquitous digital technology.

Parallel with large computers is the development of electronic calculators. In the early 1960s (and before, of course), calculators were large, bulky, electro-mechanical devices largely used for office accounting. (The author's father had one in his medical office, which the author liked to play with as a toddler.) The author first experienced more advanced technology at the end of his senior year in high school, when several electronic devices—crude by modern standards—were bought by the school. They could do the basic operations plus compute trig and exponential functions, but were not programmable. (They may have had some small memory.) During the early 1970s, Hewlett-Packard and Texas Instruments began selling hand-held calculators with heretofore unheard of capabilities. The author distinctly remembers the first time he saw one, which occurred in the fall of 1972, when a physics professor at the University of Michigan pulled one out to do a particularly ugly computation he had just derived as part of a lecture on electromagnetism. The students—who barely had time to get out their slide rules before he had the answer—were very impressed. Within a very few years, scientific calculators were ubiquitous among students. They have continued to evolve, and now routinely include graphics and some symbolic computation capabilities.

When the author was an undergraduate student in the early 1970s, a typical university computer system consisted of a single large "mainframe" computer which students and faculty usually accessed through a central "computer center." Programs were written on punched cards, which were read in by specialized devices. Some access through teletypewriter terminals existed. By the time the author was in graduate school in the late 1970s, punched card access was becoming obsolete, and teletypewriters were being replaced by video terminals. In contrast, modern college and university libraries usually have scores of desktop computers connected to a network of servers. Each of those individual desktop units is faster and more powerful than the mainframes on which the author first learned to program in the early 1970s.

As new devices such as tablets, E-readers and smart phones continue to be developed and introduced, it is safe to assume that the digital age will continue to evolve around us. No doubt some future student will happen upon an old copy of this book, and marvel at the archaic tools we were working with in the early 21st Century!

1.8   LITERATURE REVIEW

There are a lot of textbooks in numerical analysis and numerical methods. Some, like [8], [9], and [16], are considered classics. A list, by no means exhaustive, of numerical analysis or numerical methods texts is given in the References.

All of these books give decent treatments of the basic topics. Some are more mathematical than the others; some are designed for less well-prepared students. The books [2, 4, 19] are intended for a graduate student audience; [3, 5, 6, 7, 11, 15, 17, 18] are intended for an undergraduate audience. The presentation in this text has been heavily influenced by [4] and the earlier editions of [6], as well as by the author's experiences teaching at the University of Georgia and the University of Alabama in Huntsville.

An interesting and light-hearted collection of projects for a numerical methods course can be found in [14].

There exist more specialized books that treat only specific topics, such as the root-finding problem, or numerical integration, and so on. These will be discussed in the appropriate chapters.

REFERENCES

1. Acton, Forman S.; *Numerical Methods That Work*, Mathematical Association of America, Washington, DC, 1990 (originally published by Harper & Row, 1970).

2. Allen, Myron B., and Isaacson, Eli L.; *Numerical Analysis for Applied Science*, John Wiley & Sons, Inc., New York, 1998.

3. Asaithambi, A.S.; *Numerical Analysis, Theory and Practice*, Saunders College Publishing, Fort Worth, TX, 1995.

4. Atkinson, Kendall; *An Introduction to Numerical Analysis*, John Wiley & Sons, Inc., New York, 1989 (2nd edition).

5. Atkinson, Kendall; *Elementary Numerical Analysis*, John Wiley & Sons, Inc., New York, 1993 (2nd edition).

6. Burden, Richard, and Faires, J. Douglas; *Numerical Analysis*, Brooks/Cole, Pacific Grove, CA, 2012 (9th Edition).

7. Cheney, Ward, and Kincaid, David; *Numerical Mathematics and Computing*, Brooks/Cole Publishing, Pacific Grove, CA, 2012 (7th edition).

8. Conte, S.D., and de Boor, Carl; *Elementary Numerical Analysis*, McGraw-Hill, New York, 1980 (3rd edition).

9. Dahlquist, Germund, and Bjorck, Ake; *Numerical Methods*, Prentice-Hall, Englewood Cliffs, NJ, 1974.

10. Dyson, George; *Turing's Cathedral*, Vintage, New York, 2012.

11. Faires, J. Douglas and Burden, Richard; *Numerical Methods*, Brooks/Cole, Pacific Grove, CA, 2012 (4th edition).

12. Godfrey, Michael (ed.), von Neumann, John; "First Draft of a Report on the EDVAC," *IEEE Annals on the History of Computing*, vol. 27, no. 4, pp. 27-75, 1993. Available online, with commentary, at http://qss.stanford.edu/ godfrey/vonNeumann/vnedvac.pdf.

13. Goldstine, Herman H., *The Computer: From Pascal to von Neumann*, Princeton University Press, Princeton, NJ, 1980.

14. Grandine, Thomas; *The Numerical Methods Programming Projects Book*, Oxford Science Publications, Oxford, UK, 1970.

15. Gregory, John, and Redmond, Don; *Introduction to Numerical Analysis*, Jones & Bartlett Publishers, Boston, 1994.

16. Isaacson, Eugene, and Keller, Herbert B.; *Analysis of Numerical Methods*, Dover Publications, New York, 1994; originally published by John Wiley & Sons, Inc., New York, 1966.

17. Kahaner, David, Moler, Cleve, and Nash, Stephen; *Numerical Methods and Software*, Prentice-Hall, Englewood Cliffs, NJ, 1989.

18. Maron, Melvin J., and Lopez, Robert J.; *Numerical Analysis, a Practical Approach*, Wadsworth, Belmont, CA, 1991 (3rd edition).

19. Stoer, J., and Bulirsch, R.; *Introduction to Numerical Analysis*, Springer-Verlag, New York, 1980.

CHAPTER 2

A SURVEY OF SIMPLE METHODS AND TOOLS

In this chapter we want to offer a few very simple examples of approximations, algorithms, and error estimates, partly because they will be useful in developing or implementing some of the methods presented in later chapters, and partly to ease our way into the subject material. What we want to do is to acclimate the student into the broad area of numerical computations without restricting ourselves to one small corner of the subject. In addition, we show how some of the simplest approximation techniques (difference methods for derivative approximation, linear interpolation, solution of tridiagonal systems) can be used as the basis for computational schemes in more involved settings (Euler's method for the initial value problem, the trapezoid rule for numerical integration, the approximate solution of two-point boundary value problems). The goal here is to introduce the reader to the basic ideas of numerical method and analysis by looking at simple techniques across a broad spectrum of problem areas.

2.1 HORNER'S RULE AND NESTED MULTIPLICATION

In Chapter 1 we devoted some time to the construction of polynomial approximations to given functions. It might be good if we discussed the best way to evaluate those approximations efficiently; hence this section.

The most efficient way to evaluate a polynomial is by *nested multiplication*. If we have

$$p_n(x) = a_0 + a_1 x + a_2 x^2 + \cdots + a_n x^n,$$

An Introduction to Numerical Methods and Analysis, Second Edition. By James F. Epperson
Copyright © 2013 John Wiley & Sons, Inc.

then we factor out each power of x as far as it will go, thus getting

$$p_n(x) = a_0 + x(a_1 + x(a_2 + \cdots + x(a_{n-1} + a_n x) \cdots)).$$

Computation with the second form of the polynomial will take $n + 1$ multiplications and n additions. Computation with the first form will take the same amount of work, *plus* the cost of forming the powers $x^2, x^3, \ldots, x^n$.

For example, we could write

$$q(x) = 1 + x + 3x^2 - 6x^3$$

as

$$q(x) = 1 + x(1 + x(3 - 6x)).$$

An algorithmic form of nested multiplication can be written very simply, as the following pseudo-code illustrates:

Algorithm 2.1 *Horner's Rule for Polynomial Evaluation (pseudo-code).*

```
!-------------------------------------------------------
!
!     Assumes that the polynomial coefficients
!     are stored in the array a(j),j=0..n.
!     px = value of polynomial upon completion
!         of the code.
!
      px = a(n)
      for k = n-1 downto 0
            px = a(k) + px*x
      endfor
```

This is known as *Horner's rule*.[1] It can be easily modified to give the first derivative as well. Returning to our earlier notation and examples, we have

$$p'_n(x) = a_1 + 2a_2 x + 3a_3 x^2 + \cdots + na_n x^{n-1},$$

so that

$$p'_n(x) = a_1 + x(2a_2 + x(3a_3 + \cdots + x((n-1)a_{n-1} + na_n x)\ldots)).$$

In the specific case of q, we have

$$q'(x) = 1 + (3 \times 2)x + (-6 \times 3)x^2,$$

which we write as

$$q'(x) = 1 + x(6 - 18x).$$

---

[1] William George Horner (1786–1837) was born in Bristol, England, and spent much of his life as a schoolmaster in Bristol or, after 1809, in Bath. His paper on solving algebraic equations, which supposedly contains the algorithm we know as Horner's rule, was published in 1819, but it has been alleged that Horner's rule is, in fact, not covered there. He eventually published it in a separate paper in 1830, although it is now known that the Italian Ruffini had previously published a very similar idea.

Pseudo-code for the derivative is a modest change to the original algorithm.

Algorithm 2.2 *Horner's Rule for Polynomial Derivative Evaluation (pseudo-code).*

```
!-----------------------------------------------------
!
!      Assumes that the polynomial coefficients
!      are stored in the array a(j),j=0..n.
!      dp = value of derivative upon completion
!
       dp = n*a(n)
       for k=n-1 down to 1
               dp = k*a(k) + dp*x
       endfor
```

The student should check that both programs perform as is asserted here.

If the intermediate values in the computation of $p(x)$ are saved, then the subsequent computation of the derivative can be done more cheaply. Suppose that we define the values $b_k, k = 1, 2, \ldots, n$, according to

$$b_k = xb_{k-1} + a_{n-k}, \quad b_0 = a_n,$$

so that $b_n = p(x)$. Now define the values c_k according to

$$c_k = xc_{k-1} + b_k, \quad c_0 = b_0.$$

Then (note that the b_k and c_k are functions of x)

$$b_0' = 0, \quad b_k' = b_{k-1} + xb_{k-1}',$$

and, in particular,

$$b_1' = b_0.$$

Therefore,

$$c_1 = xc_0 + b_1 = xb_0 + b_1 = xb_1' + b_1 = b_2',$$

$$c_2 = xc_1 + b_2 = xb_2' + b_2 = b_3',$$

and so on. In general, as can be established by an inductive proof, $c_k = b_{k+1}'$. Therefore, since $b_n = p(x)$, it follows that $c_{n-1} = p'(x)$.

So a more efficient algorithm might be the following.

Algorithm 2.3 *A More Efficient Implementation of Horner's Rule (pseudo-code).*

```
!
!      Assumes that the polynomial coefficients
!      are stored in the array a(j),j=0..n.
!      b(n) = value of polynomial upon completion
!           of the code.
!
       b(0) = a(n)
       for k = 1 to n
             b(k) = x*b(k-1) + a(n-k)
       endfor
!
!      c(n-1) = value of derivative upon completion
!           of the code
!
       c(0) = b(0)
       for k = 1 to n-1
             c(k) = x*c(k-1) + b(k)
       endfor
```

Programming Hint: Be careful when writing Horner's rule for a polynomial which has some coefficients that are negative. It is best to simply store the coefficients as negative numbers rather than try to (correctly) distribute the negative signs throughout the nested multiplication. Thus, we write

$$p(x) = 1 - 2x - x^2 + x^3$$

as

$$p(x) = 1 + x((-2) + x((-1) + x)).$$

To do otherwise is to invite a lot of headache and potential error.

Exercises:

1. Write each of the following polynomials in nested form:

   (a) $x^3 + 3x + 2$;

   (b) $x^6 + 2x^4 + 4x^2 + 1$;

   (c) $5x^6 + x^5 + 3x^4 + 3x^3 + x^2 + 1$;

   (d) $x^2 + 5x + 6$.

2. Write each of the following polynomials in nested form, but this time take advantage of the fact that they involve only even powers of x to minimize the computations.

   (a) $1 + x^2 + \frac{1}{2}x^4 + \frac{1}{6}x^6$;

   (b) $1 - \frac{1}{2}x^2 + \frac{1}{24}x^4$.

3. Write each of the following polynomials in nested form:

(a) $1 - x + x^2 - x^3$;

(b) $1 - x^2 + \frac{1}{2}x^4 - \frac{1}{6}x^6$;

(c) $1 - x + \frac{1}{2}x^2 - \frac{1}{3}x^3 - \frac{1}{4}x^5$.

4. Write a computer code that takes a polynomial, defined by its coefficients, and evaluates that polynomial and its first derivative using Horner's rule. Test this code by applying it to each of the polynomials in Problem 1.

5. Repeat the above, using the polynomials in Problem 2 as the test set.

6. Repeat the above, using the polynomials in Problem 3 as the test set.

7. Consider the polynomial

$$p(x) = 1 + (x - 1) + \frac{1}{6}(x - 1)(x - 2) + \frac{1}{7}(x - 1)(x - 2)(x - 4).$$

This can be written in "nested" form by factoring out each binomial term as far as it will go, thus:

$$p(x) = 1 + (x - 1)\left(1 + (x - 2)\left(\frac{1}{6} + \frac{1}{7}(x - 4)\right)\right).$$

Write each of the following polynomials in this kind of nested form:

(a) $p(x) = 1 + \frac{1}{3}x - \frac{1}{60}x(x - 3)$;

(b) $p(x) = -1 + \frac{6}{7}(x - 1/2) - \frac{5}{21}(x - 1/2)(x - 4) + \frac{1}{7}(x - 1/2)(x - 4)(x - 2)$;

(c) $p(x) = 3 + \frac{1}{5}(x - 8) - \frac{1}{60}(x - 8)(x - 3)$.

8. Write a computer code that computes polynomial values using the type of nested form used in the previous problem, and test it on each of the polynomials in that problem.

9. Write a computer code to do Horner's rule on a polynomial defined by its coefficients. Test it out by using the polynomials in the previous problems. Verify that the same values are obtained when Horner's rule is used as for a naive evaluation.

10. Write out the Taylor polynomial of degree 5 for approximating the exponential function, using $x_0 = 0$, and the Horner form. Repeat for the degree 5 Taylor approximation to the sine function. (Be sure to take advantage of the fact that the Taylor expansion to the sine uses only odd powers.)

11. For each function in Problem 11 of §1.1, write the polynomial approximation in Horner form, and use this as the basis for a computer program that approximates the function. Compare the accuracy you actually achieve (based on the built-in intrinsic functions on your computer) to that which was theoretically established. Be sure to check that the required accuracy is achieved over the entire interval in question.

12. Repeat the above, but this time compare the accuracy of the *derivative* approximation constructed by taking the derivative of the approximating polynomial. Be sure to use the derivative form of Horner's rule to evaluate the polynomial.

◁ ● ● ● ▷

2.2   DIFFERENCE APPROXIMATIONS TO THE DERIVATIVE

One of the simplest uses of Taylor's Theorem as a means of constructing approximations involves the use of difference quotients to approximate the derivative of a known function f. Intuitively, this is obvious from the definition of the derivative:

$$f'(x) = \lim_{h \to 0} \frac{f(x+h) - f(x)}{h} \Rightarrow f'(x) \approx \frac{f(x+h) - f(x)}{h}.$$

The challenge for us is to make this vague statement more precise (just how accurate is this approximation, in terms of the parameter h?) and to see if anything better (more accurate) can be found.

We determine the accuracy of the approximation by a very simple computation involving Taylor's Theorem:

$$f'(x) - \frac{f(x+h) - f(x)}{h} \;\; = \;\; f'(x) - \frac{hf'(x) + \frac{1}{2}h^2 f''(\xi_{x,h})}{h} = -\frac{1}{2}hf''(\xi_{x,h}),$$

so that we have

$$f'(x) - \frac{f(x+h) - f(x)}{h} = -\frac{1}{2}hf''(\xi_{x,h}) = \mathcal{O}(h). \tag{2.1}$$

Thus the error is roughly proportional to the parameter h.

Can we do better? Yes, and the improvement does not take a lot of work. Consider the two Taylor expansions:

$$f(x+h) = f(x) + hf'(x) + \frac{1}{2}h^2 f''(x) + \frac{1}{6}h^3 f'''(\xi_1) \tag{2.2}$$

and

$$f(x-h) = f(x) - hf'(x) + \frac{1}{2}h^2 f''(x) - \frac{1}{6}h^3 f'''(\xi_2). \tag{2.3}$$

Now subtract these to get

$$f(x+h) - f(x-h) = 2hf'(x) + \frac{1}{6}h^3 f'''(\xi_1) + \frac{1}{6}h^3 f'''(\xi_2).$$

We can solve this for $f'(x)$ to get

$$f'(x) = \frac{f(x+h) - f(x-h)}{2h} - \frac{1}{6}h^2 \frac{f'''(\xi_1) + f'''(\xi_2)}{2} \tag{2.4}$$

or, using the Discrete Average Value Theorem,

$$f'(x) = \frac{f(x+h) - f(x-h)}{2h} - \frac{1}{6}h^2 f'''(\xi_{x,h}), \tag{2.5}$$

where $\xi_{x,h}$ depends on both x and h. It is worthwhile to compare the two approximations and error estimates. We note first that the two approximations cost about the same to compute: Both require two function evaluations and only a handful of other arithmetic operations. Since the function is arbitrary, its cost of computation might be quite large, so we view that as the most significant part of the cost. However, the error estimates are quite different. The first approximation has an error that depends on f'' and is proportional to h;

the second approximation depends on f''' and is proportional to h^2. Clearly, we will want to use h small to get the best accuracy, and thus the second estimate will tend to be better, since $h^2 < h$ will tend to be more important than the different sizes of f'' and f'''.

(**Note:** The approximation $f'(x) \approx \frac{f(x+h)-f(x)}{h}$ is usually called the *forward difference approximation*, while $f'(x) \approx \frac{f(x+h)-f(x-h)}{2h}$ is the *central difference approximation*. The reader is encouraged to think about why these names are appropriate.)

■ **EXAMPLE 2.1**

Consider, for example, the task of approximating the derivative of $f(x) = e^x$ at $x = 1$. The exact value, of course, is $f'(x) = e^x \Rightarrow f'(1) = e$. Using $h = 1/8$ and the one-sided difference (2.1) we get

$$f'(1) \approx \frac{e^{1.125} - e}{0.125} = 2.895480164,$$

while the same value of h with the centered difference (2.5) yields

$$f'(1) \approx \frac{e^{1.125} - e^{0.875}}{0.25} = 2.72536622.$$

The error in the first approximation is $-0.177...$, but the error in the second approximation is only $-7.084... \times 10^{-3}$.

To further illustrate these differences in accuracy, let's continue computing with the same example, but take more and smaller values of h. Let

$$D_1(h) = \frac{f(1+h) - f(1)}{h}$$

and

$$D_2(h) = \frac{f(1+h) - f(1-h)}{2h},$$

with the corresponding errors

$$E_1(h) = f'(1) - D_1(h)$$

and

$$E_2(h) = f'(1) - D_2(h).$$

Then Table 2.1 gives the results of the computation using a sequence of decreasing values of h.

Several comments might be in order here. Note that the error E_2 went down by about a factor of 4 each time we cut h in half, whereas the error E_1 only went down by a factor of 2. This, of course, follows directly from the error estimates for both methods. The importance here is that we learn that the error in the second approximation goes down at a faster rate than the error in the first one.

However, if we continue the sequence of computations, something disturbing happens. Look at Table 2.2, in which we keep going with smaller and smaller values of h for $D_2(h)$. The last column gives the ratio of errors; the error estimate asserts that this should be nearly 4, yet in the last two entries it is clearly not even close to 4, and in fact the error actually begins to increase. What is going on here?

Table 2.1   Example of derivative approximation to $f(x) = e^x$ at $x = 1$.

h^{-1}	$D_1(h)$	$E_1(h) = f'(1) - D_1(h)$	$D_2(h)$	$E_2(h) = f'(1) - D_2(h)$
2	3.526814461	$-0.8085327148e + 00$	2.832967758	$-0.1146860123e + 00$
4	3.088244438	$-0.3699626923e + 00$	2.746685505	$-0.2840375900e - 01$
8	2.895481110	$-0.1771993637e + 00$	2.725366592	$-0.7084846497e - 02$
16	2.805027008	$-0.8674526215e - 01$	2.720052719	$-0.1770973206e - 02$
32	2.761199951	$-0.4291820526e - 01$	2.718723297	$-0.4415512085e - 03$
64	2.739639282	$-0.2135753632e - 01$	2.718391418	$-0.1096725464e - 03$
128	2.728942871	$-0.1066112518e - 01$	2.718307495	$-0.2574920654e - 04$

Table 2.2   Illustration of rounding error in derivative approximations, using $f(x) = e^x$, $x = 1$.

h^{-1}	$D_2(h)$	$E_2(h) = f'(1) - D_2(h)$	$E_2(h)/E_2(h/2)$
2	2.832967758	$-0.1146860123e + 00$	N/A
4	2.746685505	$-0.2840375900e - 01$	4.038
8	2.725366592	$-0.7084846497e - 02$	4.009
16	2.720052719	$-0.1770973206e - 02$	4.001
32	2.718723297	$-0.4415512085e - 03$	4.011
64	2.718391418	$-0.1096725464e - 03$	4.026
128	2.718307495	$-0.2574920654e - 04$	4.259
256	2.718292236	$-0.1049041748e - 04$	2.455
512	2.718261719	$0.2002716064e - 04$	-0.524

(*Note:* The use of "N/A" in a table entry in anywhere in this book means that the particular data element does not exist and should simply be ignored.)

Our error estimates, made above, completely ignored the issue of rounding error. This is not unusual; most of our error estimates will be made in this fashion, concentrating on the mathematical error instead of the rounding error. But we already know that the rounding error can intrude on a computation, and this is our first example of that.

We can illustrate what is going on here by a little bit of abstraction. Let $\tilde{f}(x)$ denote the function computation as actually done on the computer, that is, the one polluted with rounding error. Define $\epsilon(x) = f(x) - \tilde{f}(x)$ as the error between the function as computed in infinite precision and as actually computed on the machine. This error will be small, but it won't be zero. The approximate derivative that we compute is constructed with $\tilde{f}$, not f, so we define

$$\tilde{D}_2(h) = \frac{\tilde{f}(x + h) - \tilde{f}(x - h)}{2h}$$

and we now want to bound the error $f'(x) - \tilde{D}_2(h)$. Note that this is the error between the quantity we want to compute $[f'(1)]$ and the quantity we can actually compute $[\tilde{D}_2(h)]$. We have

$$f'(x) - \tilde{D}_2(h) = f'(x) - \frac{\tilde{f}(x + h) - \tilde{f}(x - h)}{2h},$$

which we write as

$$
\begin{aligned}
f'(x) - \tilde{D}_2(h) &= & f'(x) - \frac{f(x+h)-f(x-h)}{2h} &+ & \frac{f(x+h)-f(x-h)}{2h} - \frac{\tilde{f}(x+h)-\tilde{f}(x-h)}{2h} \\
&= & -\frac{1}{6}h^2 f'''(\xi_{x,h}) &+ & \frac{f(x+h)-f(x-h)-\tilde{f}(x+h)+\tilde{f}(x-h)}{2h} \\
&= & \underbrace{-\frac{1}{6}h^2 f'''(\xi_{x,h})}_{\text{error due to approximation}} &+ & \underbrace{\frac{\epsilon(x+h)-\epsilon(x-h)}{2h}}_{\text{error due to rounding}}.
\end{aligned}
$$

Generally, the numerator in the rounding-error fraction will *not* go to zero as $h \to 0$, since we will always have some (small) amount of rounding error present. (Bear in mind that the error due to rounding can be assumed to include the error in doing the subtraction $f(x+h) - f(x-h)$, which, as we saw in §1.3, will be prone to large amounts of error as h gets smaller.) But since we are dividing the rounding error by $2h$, we do expect that this term will begin to grow as h gets small, and this in fact is what is observed in Table 2.2.

Figure 2.1 shows a plot of $\log_{10}|f'(1) - D_2(h)|$ versus $1/h$ for h ranging from $\frac{1}{2}$ all the way to $1/2048$. Note that for h much smaller than 0.01 (approximately), the plot ceases to show a decrease in the error; instead the error oscillates wildly about, creating a "cloud" rather than a smooth curve. Note also that this cloud is trending slightly upwards, meaning that the error is acually beginning to *increase*. Smaller values of h would only have continued to confirm this trend.

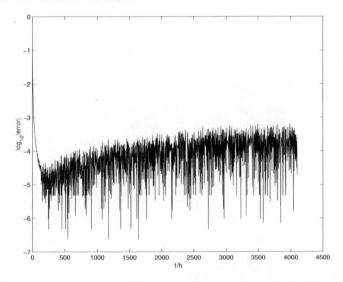

Figure 2.1   Illustration of rounding error in approximating derivatives.

To confirm that rounding error is the culprit, we can repeat the calculation at higher precision. The data for Tables 2.1 and 2.2 were produced using computer code in single precision (about six or seven decimal digits of accuracy). If we repeat the computations for Table 2.2 in double precision (about 15 decimal digits), we get the results shown in Table 2.3. Note that we can go all the way down to $h = 2^{-15}$ without rounding error seriously affecting the computation, although it appears from the last two entries that, even at this level of precision, the effects of rounding error are beginning to become apparent. Figure 2.2 shows the same plot as in Figure 2.1, except this time we used the double-precision data. Note the absence of the cloud effect and the steady decrease of the error.

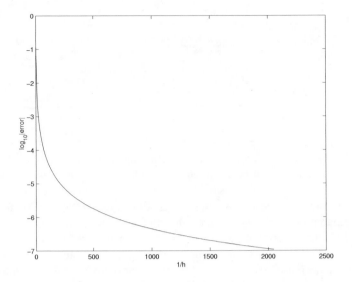

Figure 2.2   Double-precision version of Figure 2.1.

Table 2.3   Illustration of rounding error in derivative approximations, using $f(x) = e^x$, $x = 1$ (double-precision).

h^{-1}	$D_2(h)$	$E_2(h) = f'(1) - D_2(h)$	$E_2(h)/E_2(h/2)$
2	2.832967800	$-0.1146859712e + 00$	N/A
4	2.746685882	$-0.2840405324e - 01$	4.038
8	2.725366220	$-0.7084391345e - 02$	4.009
16	2.720051889	$-0.1770060412e - 02$	4.002
32	2.718724279	$-0.4424502865e - 03$	4.001
64	2.718392437	$-0.1106085209e - 03$	4.000
128	2.718309480	$-0.2765187706e - 04$	4.000
256	2.718288741	$-0.6912953448e - 05$	4.000
512	2.718283557	$-0.1728237360e - 05$	4.000
1024	2.718282261	$-0.4320591627e - 06$	4.000
2048	2.718281936	$-0.1080143859e - 06$	4.000
4096	2.718281855	$-0.2700387380e - 07$	4.000
8192	2.718281835	$-0.6751245785e - 08$	4.000
16384	2.718281830	$-0.1688998275e - 08$	3.997
32768	2.718281829	$-0.4193436709e - 09$	4.028

More involved combinations of $f(x \pm h)$ can be used to construct more accurate approximations to f' and higher-order derivatives. Problem 12 asks you to derive an approximation to the second derivative that will be used in §2.7 to construct approximate solutions to certain differential equation problems.

It might seem silly to use these methods to compute approximate derivative values for the exponential function—after all, we can compute that derivative rather easily by basic calculus—but we need to keep in mind that we won't always have a neat and tidy formula for f that allows us to compute f' directly. The function f might be the result of a lengthy computation for which a simple formula is not really available. In addition, we can use formulas such as those developed in this section to compute (approximate) derivative values for functions defined only as a table of values. Finally, we can use the formulas derived in this section to replace the derivative in other equations in order to produce approximation schemes for other problems. The next section on Euler's method is just one example of this.

Exercises:

1. Use the methods of this section to show that

$$f'(x) = \frac{f(x) - f(x-h)}{h} + \mathcal{O}(h).$$

2. Compute, by hand, approximations to $f'(1)$ for each of the following functions, using $h = 1/16$ and each of the derivative approximations contained in (2.1) and (2.5).

    (a) $f(x) = \sqrt{x+1}$;
    (b) $f(x) = \arctan x$;
    (c) $f(x) = \sin \pi x$;
    (d) $f(x) = e^{-x}$;
    (e) $f(x) = \ln x$.

3. Write a computer program that uses the same derivative approximations as in the previous problem to approximate the first derivative at $x = 1$ for each of the following functions, using $h^{-1} = 4, 8, 16, 32$. Verify that the predicted theoretical accuracy is obtained—in other words, show that your results are consistent with the analysis in this section.

    (a) $f(x) = \sqrt{x+1}$;
    (b) $f(x) = \arctan x$;
    (c) $f(x) = \sin \pi x$.
    (d) $f(x) = e^{-x}$;
    (e) $f(x) = \ln x$.

4. Use the approximations from this section to fill in approximations to the missing values in Table 2.4.

5. Use the error estimate (2.5) for the centered difference approximation to the first derivative to prove that this approximation will be *exact* for any quadratic polynomial.

6. Find coefficients A, B, and C so that

    (a) $f'(x) = Af(x) + Bf(x+h) + Cf(x+2h) + \mathcal{O}(h^2)$;
    (b) $f'(x) = Af(x) + Bf(x-h) + Cf(x-2h) + \mathcal{O}(h^2)$.

    Hint: Use Taylor's Theorem.

Table 2.4   Table for Problem 4.

x	$f(x)$	$f'(x)$
1.00	1.0000000000	
1.10	0.9513507699	
1.20	0.9181687424	
1.30	0.8974706963	
1.40	0.8872638175	
1.50	0.8862269255	
1.60	0.8935153493	
1.70	0.9086387329	
1.80	0.9313837710	
1.90	0.9617658319	
2.00	1.0000000000	

7. Fill in the data from Problem 4 using methods that are $\mathcal{O}(h^2)$ at each point. *Hint:* See the previous problem.

8. Use Taylor's Theorem to show that the approximation

$$f'(x) \approx \frac{8f(x+h) - 8f(x-h) - f(x+2h) + f(x-2h)}{12h}$$

is $\mathcal{O}(h^4)$.

9. Use the derivative approximation from Problem 8 to approximate $f'(1)$ for the same functions as in Problem 3. Verify that the expected rate of decrease is observed for the error.

10. Use the derivative approximation from Problem 8 to fill in as much as possible of the table in Problem 4.

11. Let $f(x) = \arctan x$. Use the derivative approximation from Problem 8 to approximate $f'(\frac{1}{4}\pi)$ using $h^{-1} = 2, 4, 8, \ldots$. Try to take h small enough that the rounding error effect begins to dominate the mathematical error. For what value of h does this begin to occur? (You may have to restrict yourself to working in single precision.)

12. Use Taylor expansions for $f(x \pm h)$ to derive an $\mathcal{O}(h^2)$ accurate approximation to f'' using $f(x)$ and $f(x \pm h)$. Provide all the details of the error estimate. *Hint:* Go out as far as the fourth derivative term, and then add the two expansions.

13. Let $h > 0$ and $\eta > 0$ be given, where $\eta = \theta h$, for $0 < \theta < 1$. Let f be some smooth function. Use Taylor expansions for $f(x + h)$ and $f(x - \eta)$ in terms of f and its derivatives at x in order to construct an approximation to $f'(x)$ that depends on $f(x + h)$, $f(x)$, and $f(x - \eta)$, and which is $\mathcal{O}(h^2)$ accurate. Check your work by verifying that for $\theta = 1 \Rightarrow \eta = h$ you get the same results as those in the text.

14. Write a computer program to test the approximation to the second derivative from Problem 12 by applying it to estimate $f''(1)$ for each of the following functions, using $h^{-1} = 4, 8, 16, 32$. Verify that the predicted theoretical accuracy is obtained.

(a) $f(x) = e^{-x}$;

(b) $f(x) = \cos \pi x$;

(c) $f(x) = \sqrt{1+x}$;

(d) $f(x) = \ln x$.

15. Define the following function:

$$f(x) = \ln(e^{\sqrt{x^2+1}} \sin(\pi x) + \tan \pi x).$$

Compute values of f' over the range $[0, \frac{1}{4}]$ using two methods:

(a) Using the centered difference formula from (2.5);

(b) Using ordinary calculus to find the formula for f'.

Comment.

16. Let $f(x) = e^x$, and consider the problem of approximating $f'(1)$, as in the text. Let $D_1(h)$ be the difference approximation in (2.1). Using the appropriate values in Table 2.1, compute the new approximations

$$\Delta_1(h) = 2D_1(h) - D_1(2h);$$

and compare these values to the exact derivative value. Are they more or less accurate than the corresponding values of D_1? Try to deduce, from your calculations, how the error depends on h.

17. Repeat the above idea for $f(x) = \arctan(x)$, $x = 1$ (but this time you will have to compute the original $D_1(h)$ values).

18. By keeping more terms in the Taylor expansion for $f(x + h)$, show that the error in the derivative approximation (2.1) can be written as

$$f'(x) - \left(\frac{f(x+h) - f(x)}{h} \right) = -\frac{1}{2}hf''(x) - \frac{1}{6}h^2 f'''(x) - \dots . \qquad (2.6)$$

Use this to construct a derivative approximation involving $f(x)$, $f(x + h)$, and $f(x + 2h)$ that is $\mathcal{O}(h^2)$ accurate. *Hint:* Use (2.6) to write down the error in the approximation

$$f'(x) \approx \frac{f(x+2h) - f(x)}{2h}$$

and combine the two error expansions so that the terms that are $\mathcal{O}(h)$ are eliminated.

19. Apply the method derived above to the list of functions in Problem 3, and confirm that the method is as accurate in practice as is claimed.

20. Let $f(x) = e^x$, and consider the problem of approximating $f'(1)$, as in the text. Let $D_2(h)$ be the difference approximation in (2.5). Using the appropriate values in Table 2.1, compute the new approximations

$$\Delta_2(h) = (4D_2(h) - D_2(2h))/3;$$

and compare these values to the exact derivative value. Are they more or less accurate that the corresponding values of D_2? Try to deduce, from your calculations, how the error depends on h.

21. Repeat the above idea for $f(x) = \arctan(x)$, $x = 1$ (but this time you will have to compute the original $D_2(h)$ values).

22. The same ideas as in Problem 18 can be applied to the centered difference approximation (2.5). Show that in this case the error satisfies

$$f'(x) - \frac{f(x+h) - f(x-h)}{2h} = -\frac{1}{6}h^2 f'''(x) - \frac{1}{120}h^4 f'''''(x) - \dots . \quad (2.7)$$

Use this to construct a derivative approximation involving $f(x \pm h)$, and $f(x \pm 2h)$ that is $\mathcal{O}(h^4)$ accurate.

23. Apply the method derived above to the list of functions in Problem 3, and confirm that the method is as accurate in practice as is claimed.

◁ ● ● ▷

2.3   APPLICATION: EULER'S METHOD FOR INITIAL VALUE PROBLEMS

One immediate application of difference methods for approximating derivatives is the approximate solution of initial value problems for ordinary differential equations. The usual general form of such a problem is

$$y' = f(t, y), \quad y(t_0) = y_0, \quad (2.8)$$

where f is a known function of t and y, and t_0 and y_0 are given values. The object in solving this problem is to find y as a function of t; in the usual sophomore-level course in ordinary differential equations, the student learns a number of techniques for analytically solving (2.8), based on assuming any one of a number of special forms for f. Here we will use one of our derivative approximations to construct a method for approximately solving (2.8).

We use (2.1) to replace the derivative in (2.8):

$$\frac{y(t+h) - y(t)}{h} = f(t, y(t)) + \frac{1}{2}hy''(t_h),$$

which can be simplified slightly to become

$$y(t+h) = y(t) + hf(t, y(t)) + \frac{1}{2}h^2 y''(t_h). \quad (2.9)$$

This suggests the following numerical method:

1. Define a sequence of t values (called a *grid*) according to $t_n = t_0 + nh$, where h is a set parameter (called the *mesh spacing* or *grid size*; we will encounter this kind of thing often in later topics).

2. Compute the values y_n from y_0 and the t grid values, according to

$$y_{n+1} = y_n + hf(t_n, y_n). \tag{2.10}$$

Note that this follows from (2.9) by dropping the error term and adjusting the notation slightly.

Equation (2.10) defines what is known as *Euler's method*[2] for solving (approximately) initial value problems for ordinary differential equations. Figure 2.3 shows what is happening, geometrically.

An example might be in order at this point.

■ **EXAMPLE 2.2**

Consider the very simple initial value problem

$$y' = -y + \sin t, \quad y(0) = 1.$$

This has exact solution $y(t) = \frac{3}{2}e^{-t} + \frac{1}{2}(\sin t - \cos t)$, found by using the kinds of methods taught in the usual ODE course. If we apply Euler's method to this, using $h = \frac{1}{4}$, we get the following results.

Step 1: We have $h = \frac{1}{4}$, so $t_1 = h = \frac{1}{4}$ and y_0 is given as 1. Then,

$$y_1 = y_0 + hf(t_0, y_0) = 1 + \frac{1}{4}(-1 + \sin 0) = \frac{3}{4}.$$

Thus, $y(1/4) \approx 0.75$, and the error in this approximation is $e_1 = y(1/4) - y_1 = 0.8074469434 - 0.75 = 0.0574469434$.

Step 2: We have $t_2 = 2h = \frac{1}{2}$ and $y_1 = 0.75$ from the Step 1. Then,

$$y_2 = y_1 + hf(t_1, y_1) = 3/4 + \frac{1}{4}\left(-3/4 + \sin\frac{1}{4}\right) = 0.6243509898.$$

Thus, $y(1/2) \approx 0.6243509898$, and the error in this approximation is $e_2 = y(1/2) - y_2 = 0.7107174779 - 0.6243509898 = 0.0863664881$.

---

[2]Leonhard Euler (1707–1783) was one of the two greatest mathematicians of the post-Newton age, the other being Carl Friedrich Gauss. Euler was born in Basel, Switzerland, and educated at the University of Basel, at first with an eye toward following in his father's career as a Lutheran minister. With the assistance of his tutor and mentor Johann Bernoulli, however, he was able to convince his father to let him pursue a career in mathematics. In 1727 Euler joined the St. Petersburg Academy of Sciences in Russia, where he remained until 1741, at which time he joined the Berlin Academy of Sciences at the invitation of the Prussian king, Frederick the Great. After some disputes with the monarch, Euler left Berlin in 1766 and returned to St. Petersburg.

Euler's contributions to mathematics are almost unmatched in their breadth. He published an enormous amount of material, in a wide variety of areas, including infinite series, special functions (a field of study that he practically invented), number theory, complex variables, and hydrodynamics. His name is attached to countless results in mathematics, from Euler's formula relating the trigonometric functions to complex exponentials, to the Euler–Cauchy differential equations, to Euler's formula relating the number of sides, edges, and vertices in a polyhedron. His influence on notation is still felt today, as it was Euler who introduced e, π, and $i = \sqrt{-1}$ into the literature as standard symbols, in addition to the use of Σ for denoting summations, and cos and sin for the cosine and sine of an angle. Euler's collected works, published between 1911 and 1975, encompass *72* volumes!

The method for numerically solving differential equations that bears his name was apparently first presented in the period 1768–1769, in the two-volume work known as *Institutiones calculi integralis*. The theoretical basis for the convergence of the method was laid down by Augustin Louis Cauchy in the mid-1800s and by Rudolf Lipschitz in the late 1800s.

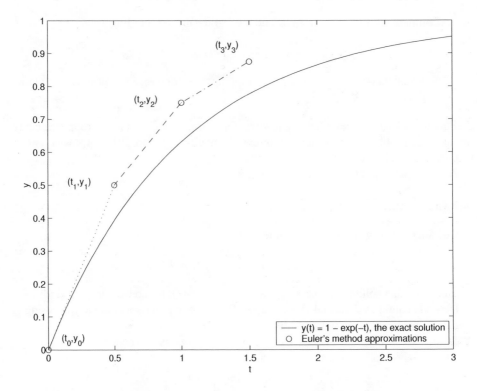

Figure 2.3   Geometric derivation of Euler's method.

If we instead use $h = 1/8$ and continue the computation out to $t = 1$, then we get Table 2.5.

If we cut the mesh size in half, again, to $h = 1/16$, then we get the results in Table 2.6. Note that for $h = 1/8$, the maximum error is given by 4.425×10^{-2}, whereas for $h = 1/16$ it is given by 2.140×10^{-2}. This suggests (but does not prove) that Euler's method is $\mathcal{O}(h)$ accurate, something we will prove in Chapter 6, where we undertake a more wide-ranging study of numerical methods for ordinary differential equations. This is adequate but not outstanding accuracy; we would prefer a method that was $\mathcal{O}(h^p)$ accurate for $p \geq 2$.

Figure 2.4 shows the exact solution (solid line), the approximate solution computed with $h = 1/8$ (denoted by the asterisks), and the approximate solution computed with $h = 1/16$ (denoted by the plus signs). Note that the plus signs (those values computed with a smaller mesh[3]) do appear to be more accurate.

Writing a computer code for Euler's method is not difficult. If we assume that h, the mesh size, is given, along with N, the number of steps to take, then the code will look something like the code given in Algorithm 2.4.

---

[3] In several sections of this book we will have reason to talk about "mesh spacing" or "grid sizes." Different folks have different terminological preferences here, but the fact is that numerical analysts use the terms *mesh size* and *grid size* almost interchangeably.

Table 2.5   Euler's method applied to $y' + y = \sin t$, $h = 1/8$.

t_k	y_k	Error $= y(t_k) - y_k$
0.00000	$0.1000000000e + 01$	$0.0000000000e + 00$
0.12500	$0.8750000000e + 00$	$0.1498388695e - 01$
0.25000	$0.7812093417e + 00$	$0.2623760171e - 01$
0.37500	$0.7144836689e + 00$	$0.3433270290e - 01$
0.50000	$0.6709572764e + 00$	$0.3976020153e - 01$
0.62500	$0.6470158092e + 00$	$0.4294341032e - 01$
0.75000	$0.6392759921e + 00$	$0.4424878254e - 01$
0.87500	$0.6445713381e + 00$	$0.4399501342e - 01$
1.00000	$0.6599428586e + 00$	$0.4246064258e - 01$

Table 2.6   Euler's method applied to $y' + y = \sin t$, $h = 1/16$.

t_k	y_k	Error $= y(t_k) - y_k$
0.00000	$0.1000000000e + 01$	$0.0000000000e + 00$
0.06250	$0.9375000000e + 00$	$0.3825497791e - 02$
0.12500	$0.8828099574e + 00$	$0.7173929590e - 02$
0.18750	$0.8354265059e + 00$	$0.1008216323e - 01$
0.25000	$0.7948625553e + 00$	$0.1258438808e - 01$
0.31250	$0.7606463930e + 00$	$0.1471233364e - 01$
0.37500	$0.7323209006e + 00$	$0.1649547113e - 01$
0.43750	$0.7094428774e + 00$	$0.1796119911e - 01$
0.50000	$0.6915824637e + 00$	$0.1913501427e - 01$
0.56250	$0.6783226558e + 00$	$0.2004066841e - 01$
0.62500	$0.6692589069e + 00$	$0.2070031255e - 01$
0.68750	$0.6639988048e + 00$	$0.2113462907e - 01$
0.75000	$0.6621618220e + 00$	$0.2136295267e - 01$
0.81250	$0.6633791306e + 00$	$0.2140338100e - 01$
0.87500	$0.6672934759e + 00$	$0.2127287562e - 01$
0.93750	$0.6735591026e + 00$	$0.2098735402e - 01$
1.00000	$0.6818417279e + 00$	$0.2056177329e - 01$

Algorithm 2.4 *Euler's Method (pseudo-code).*

```
input h, N, y0, t0
external f
y = y0
t = t0
for k=1 to N do
     yn = y + h*f(t,y)
     y = yn
     t = t + h
endfor
```

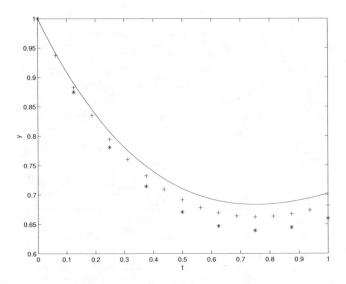

Figure 2.4   Exact and approximate solutions to $y' + y = \sin t$; asterisks denote the approximate solution for $h = 1/8$, and plus signs denote the approximate solution for $h = 1/16$.

Programming Hints: (1) Note that the pseudo-code assumes that there is some externally written subprogram or function that computes the values of $f(t, y)$. (2) When writing a code for the exercises, be very sure that you compare y_n to the exact solution at $t = t_n$. It is very easy to accidentally write the program so that you are comparing y_n to $y(t_{n+1})$.

Exercises:

1. Use Euler's method with $h = 0.25$ to compute approximate solution values for the initial value problem
$$y' = \sin(t + y), \quad y(0) = 1.$$
   You should get $y_4 = 1.851566895$ (be sure that your calculator is set in radians).

2. Repeat the above with $h = 0.20$. What value do you now get for $y_5 \approx y(1)$?

3. Repeat the above with $h = 0.125$. What value do you now get for $y_8 \approx y(1)$?

4. Use Euler's method with $h = 0.25$ to compute approximate solution values for
$$y' = e^{t-y}, \quad y(0) = -1.$$
   What approximate value do you get for $y(1) = 0.7353256638$?

5. Repeat the above with $h = 0.20$. What value do you now get for $y_5 \approx y(1)$?

6. Repeat the above with $h = 0.125$. What value do you now get for $y_8 \approx y(1)$?

7. Use Euler's method with $h = 0.0625$ to compute approximate solution values over the interval $0 \le t \le 1$ for the initial value problem
$$y' = t - y, \quad y(0) = 2,$$

which has exact solution $y(t) = 3e^{-t} + t - 1$. Plot your approximate solution as a function of t, and plot the error as a function of t.

8. Repeat the above for the equation

$$y' = e^{t-y}, \quad y(0) = -1,$$

which has exact solution $y = \ln(e^t - 1 + e^{-1})$.

9. Repeat the above for the equation

$$y' + y = \sin t, \quad y(0) = -1,$$

which has exact solution $y = (\cos t + \sin t - e^{-t})/2$.

10. Use Euler's method to compute approximate solutions to each of the initial value problems below, using $h^{-1} = 2, 4, 8, 16$. Compute the maximum error over the interval $[0, 1]$ for each value of h. Plot your approximate solutions for the $h = 1/16$ case. *Hint:* Verify that your code works by using it to reproduce the results given for the examples in the text.

(a) $y' + 4y = 1$, $y(0) = 1$; $y(t) = \frac{1}{4}(3e^{-4t} + 1)$;

(b) $y' = -y \ln y$, $y(0) = 3$; $y(t) = e^{(\ln 3)e^{-t}}$;

(c) $y' = t - y$, $y(0) = 2$; $y(t) = 3e^{-t} + t - 1$.

11. Consider the approximate values in Tables 2.5 and 2.6. Let y_k^8 denote the approximate values for $h = 1/8$, and y_k^{16} denote the approximate values for $h = 1/16$. Note that

$$y_k^8 \approx y(k/8)$$

and

$$y_{2k}^{16} \approx y(2k/16) = y(k/8) \approx y_k^8;$$

thus y_k^8 and y_{2k}^{16} are both approximations to the same value. Compute the set of new approximations

$$u_k = 2y_{2k}^{16} - y_k^8$$

and compare these to the corresponding exact solution values. Are they better or worse as an approximation?

12. Apply the basic idea from the previous problem to the approximation of solutions to

$$y' = e^{t-y}, \quad y(0) = -1,$$

which has exact solution $y = \ln(e^t - 1 + e^{-1})$.

13. Assume that the function f satisfies

$$|f(t, y) - f(t, z)| \le K|y - z|$$

for some constant K. Use this and (2.9)–(2.10) to show that the error $|y(t_{n+1}) - y_{n+1}|$ satisfies the recursion

$$|y(t_{n+1}) - y_{n+1}| \le (1 + Kh)|y(t_n) - y_n| + \frac{1}{2}h^2 Y_2$$

where

$$Y_2 = \max_t |y''(t)|.$$

◁ • • ▷

2.4   LINEAR INTERPOLATION

Before hand calculators became a commonplace tool, high school algebra students were taught how to estimate intermediate values for tables of log and trig functions, etc., by a process known as *interpolation*. Even though this particular application is no longer as important as it once was, interpolation is still an important tool in computational mathematics. For example, almost all graphs produced by computers are actually the result of something called *piecewise linear interpolation* in which the machine draws a very large number of very small straight lines to represent the curve.

Given a set of data points x_k, typically called *nodes*, we say that the function p interpolates the function f at these nodes if $p(x_k) = f(x_k)$, for all k. Sometimes we only have one function, p, and a set of function values y_k; in this case we say that p interpolates the data if $p(x_k) = y_k$, for all k. In the most common kind of interpolation, the function p (the *interpolant*) is a polynomial. Usually, we are most interested in the extent to which $p \approx f$ or the extent to which p represents the data.

The interpolation problem gets a full treatment in Chapter 4; here we want to consider only the special case of *linear* interpolation, by which we mean using a straight line to approximate a given function. Since it only takes two points to determine a straight line, we assume that we are given two nodes x_0 and x_1, and a function, f. Since we want to find the equation of a straight line that passes through the two points $(x_0, f(x_0))$ and $(x_1, f(x_1))$, it is not difficult to show that

$$p_1(x) = \frac{x_1 - x}{x_1 - x_0} f(x_0) + \frac{x - x_0}{x_1 - x_0} f(x_1) \qquad (2.11)$$

is what we are looking for. The student should confirm that the interpolatory conditions are satisfied, i.e., that $p_1(x_i) = f(x_i), i = 0, 1$.

Figure 2.5 shows an example of a linear interpolant to the exponential function using the two points $x_0 = 0$ and $x_1 = 1$. Clearly, the approximation is not extremely accurate, but if the nodes were closer together then it might be more accurate, at least between the nodes.

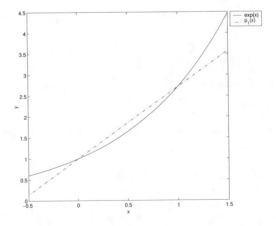

Figure 2.5   Linear interpolation to the exponential function on $[0, 1]$.

Just how accurate is linear interpolation? We can investigate this using not much more than Rolle's Theorem.[4] Define

$$E(x) = f(x) - p_1(x), \quad w(x) = (x - x_0)(x - x_1), \quad G(x) = E(x) - \frac{w(x)}{w(t)}E(t),$$

where t is some fixed value in (x_0, x_1) (i.e., $x_0 < t < x_1$). (The point t can actually be anywhere, but this restriction is usually the case.) Note that

$$G(x_0) = 0, \quad G(x_1) = 0, \quad G(t) = 0.$$

Then Rolle's Theorem states that there exists a point η_0 between x_0 and t, such that $G'(\eta_0) = 0$, and similarly, a point η_1 between x_1 and t such that $G'(\eta_1) = 0$. But now we can apply Rolle's Theorem to G', and assert that there exists a point ξ between η_0 and η_1, such that $G''(\xi) = 0$. But

$$G''(x) = f''(x) - \frac{2}{w(t)}E(t),$$

so that

$$G''(\xi) = 0 \Rightarrow f''(\xi) - \frac{2}{w(t)}E(t) = 0;$$

and we finally have

$$f(t) - p_1(t) = \frac{1}{2}(t - x_0)(t - x_1)f''(\xi), \tag{2.12}$$

for any point $t \in [x_0, x_1]$. The example plot given in Figure 2.5 shows clearly that the error in the approximation will grow rapidly outside the interval $[x_0, x_1]$, so we will confine our efforts at getting an upper bound to the error on that interval. Taking absolute values, and assuming the worst case for the second derivative term (since we can't do anything else with it),

$$|f(x) - p_1(x)| \leq \frac{1}{2}|(x - x_0)(x - x_1)| \max_{x_0 \leq t \leq x_1} |f''(t)|,$$

$$\leq \frac{1}{2}\left(\max_{x_0 \leq t \leq x_1} |(t - x_0)(t - x_1)|\right)\left(\max_{x_0 \leq t \leq x_1} |f''(t)|\right),$$

so the upper bound on the error depends on our obtaining the maximum of the function

$$g(x) = |(x - x_0)(x - x_1)| = (x_1 - x)(x - x_0).$$

Since $g(x_0) = g(x_1) = 0$, the Extreme Value Theorem says that the maximum value of g on the interval $[x_0, x_1]$ will be found at a critical point. Thus we apply some ordinary calculus to get

$$g'(x) = x_1 - x - x + x_0 = (x_0 + x_1) - 2x,$$

which implies that the critical point is

$$x_c = \frac{1}{2}(x_0 + x_1),$$

---

[4]The student should recall that Rolle's Theorem is a special case of the Mean Value Theorem in which $f(a) = f(b)$, thus, there exists ξ such that $f'(\xi) = 0$. Michel Rolle (1652–1719), born in France, was largely self-taught, and supported himself as an assistant to several attorneys, then as a scribe when he moved to Paris in 1675. Rolle's Theorem first appeared in 1692.

i.e., the midpoint of the interval. The maximum value of the function is then given by

$$g(x_c) = \frac{1}{4}(x_1 - x_0)^2,$$

giving us, finally, that our error is bounded according to

$$|f(x) - p_1(x)| \leq \frac{1}{8}(x_1 - x_0)^2 \left(\max_{x_0 \leq x \leq x_1} |f''(x)| \right). \tag{2.13}$$

We can summarize this as a theorem.

Theorem 2.1 (Linear Interpolation Error) *Let* $f \in C^2([x_0, x_1])$ *and let* $p_1(x)$ *be the linear polynomial that interpolates* f *at* x_0 *and* x_1. *Then, for all* $x \in [x_0, x_1]$,

$$|f(x) - p_1(x)| \leq \frac{1}{2}|(x - x_0)(x - x_1)| \max_{x_0 \leq x \leq x_1} |f''(x)| \leq \frac{1}{8}(x_1 - x_0)^2 \max_{x_0 \leq x \leq x_1} |f''(x)|.$$

Let's now use this to bound the error in an old-time application, estimating intermediate values from tables of data.

■ **EXAMPLE 2.3**

Recall the error function

$$\text{erf}(x) = \frac{2}{\sqrt{\pi}} \int_0^x e^{-t^2} dt.$$

Table 2.7 gives values of the $\text{erf}(x)$ in increments of 0.1 over the interval $[0, 1]$.

Table 2.7   Table of $\text{erf}(x)$ values.

x	$\text{erf}(x)$
0.0	0.00000000000000
0.1	0.11246291601828
0.2	0.22270258921048
0.3	0.32862675945913
0.4	0.42839235504667
0.5	0.52049987781305
0.6	0.60385609084793
0.7	0.67780119383742
0.8	0.74210096470766
0.9	0.79690821242283
1.0	0.84270079294971

Suppose that we need to know $\text{erf}(0.14)$; then we can interpolate using the values at $x_0 = 0.1$ and $x_1 = 0.2$. The polynomial becomes

$$p_1(x) = \frac{0.2 - x}{0.1}(0.11246291601828) + \frac{x - 0.1}{0.1}(0.22270258921048)$$

so that

$$\begin{aligned} p_1(0.14) &= \frac{0.06}{0.1}(0.11246291601828) + \frac{0.04}{0.1}(0.22270258921048) \\ &= 0.15655878529516 \approx \text{erf}(0.14). \end{aligned}$$

We can bound the error in this approximation using Theorem 2.1. We have

$$\mathrm{erf}'(x) = \frac{2}{\sqrt{\pi}}e^{-x^2} \Rightarrow \mathrm{erf}''(x) = -\frac{4x}{\sqrt{\pi}}e^{-x^2};$$

a crude upper bound on the interval $[0.1, 0.2]$ is

$$|\mathrm{erf}''(x)| \leq \frac{4 \times 0.2}{\sqrt{\pi}}e^{-0.1^2} = 0.4468606427,$$

thus the error is bounded above by

$$|\mathrm{erf}(0.14) - p_1(0.14)| \leq \frac{1}{8}(0.1)^2 \times 0.4468606427 = 5.585758034 \times 10^{-4}.$$

Similarly, we get that

$$\mathrm{erf}(0.72) \approx 0.69066114801147$$

with an error that is bounded by $1.38254972 \times 10^{-3}$.

A common and related technique for approximating a curve is to use *piecewise linear interpolation*, in which we break up the interval of interest into subintervals and then use a different linear interpolant on each subinterval. The effect is to approximate the curve by a set of connected straight lines.

■ **EXAMPLE 2.4**

Consider the problem of constructing a piecewise linear approximation to $f(x) = \log_2(x)$ using the nodes $\frac{1}{4}, \frac{1}{2}, 1$. We construct the separate linear polynomials over each pair of adjacent nodes:

$$Q_1(x) = \left(\frac{\frac{1}{2} - x}{\frac{1}{2} - \frac{1}{4}}\right)\log_2\left(\frac{1}{4}\right) + \left(\frac{x - \frac{1}{4}}{\frac{1}{2} - \frac{1}{4}}\right)\log_2\left(\frac{1}{2}\right) = 4x - 3$$

$$Q_2(x) = \left(\frac{1 - x}{1 - \frac{1}{2}}\right)\log_2\left(\frac{1}{2}\right) + \left(\frac{x - \frac{1}{2}}{1 - \frac{1}{2}}\right)\log_2(1) = 2x - 2.$$

Thus, the piecewise polynomial function is defined as

$$q(x) = \begin{cases} 4x - 3, & \frac{1}{4} \leq x \leq \frac{1}{2}; \\ 2x - 2, & \frac{1}{2} \leq x \leq 1. \end{cases}$$

Figure 2.6 shows a plot of $\log_2(x)$ and the piecewise polynomial approximation.

The error is estimated by looking at the error in each individual polynomial approximation. We have

$$|\log_2(x) - Q_1(x)| \leq \frac{1}{8}\left(\frac{1}{2} - \frac{1}{4}\right)^2 \max_{t \in [\frac{1}{4}, \frac{1}{2}]} |\log_2(e)t^{-2}| = 0.1803368801...$$

and this holds for all $x \in [\frac{1}{4}, \frac{1}{2}]$; furthermore,

$$|\log_2(x) - Q_2(x)| \leq \frac{1}{8}\left(1 - \frac{1}{2}\right)^2 \max_{t \in [\frac{1}{2}, 1]} |\log_2(e)t^{-2}| = 0.1803368801...$$

and this holds for all $x \in [\frac{1}{2}, 1]$. Thus,

$$|\log_2(x) - q(x)| \leq 0.1803368801...$$

for all $x \in [\frac{1}{4}, 1]$. (Usually, the error in each piece will *not* be the same. That is an artifact of the example.

The more general case of piecewise polynomial interpolation with higher-degree polynomials will be taken up in §4.7.

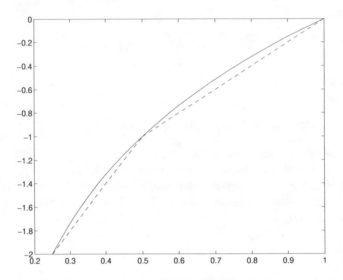

Figure 2.6   Piecewise linear approximation to $y = \log_2(x)$.

In addition, linear interpolation is a powerful tool that we will return to again and again in order to construct approximations for more involved computations, and the accuracy of those approximations will depend heavily on the accuracy of the underlying linear interpolation. In fact, we will use the estimate proved in this section to establish the accuracy and convergence properties of the secant method for finding roots of equations (§3.11.3), the trapezoid rule for numerically approximating integrals (§2.5), and several other methods for more sophisticated problems.

Exercises:

1. Use linear interpolation to find approximations to the following values of the error function, using Table 2.7 in the text. For each case, give an upper bound on the error in the approximation.

   (a) $\mathrm{erf}(0.56)$;

   (b) $\mathrm{erf}(0.07)$;

   (c) $\mathrm{erf}(0.34)$;

   (d) $\mathrm{erf}(0.12)$;

   (e) $\mathrm{erf}(0.89)$.

2. The gamma function, denoted by $\Gamma(x)$, occurs in a number of applications, most notably probability theory and the solution of certain differential equations. It is

basically the generalization of the factorial function to non-integer values, in that $\Gamma(n+1) = n!$. Table 2.8 gives values of $\Gamma(x)$ for x between 1 and 2. Use linear interpolation to approximate values of $\Gamma(x)$ as given below.

(a) $\Gamma(1.930) = 0.9723969178$;

(b) $\Gamma(1.290) = 0.8990415863$;

(c) $\Gamma(1.005) = 0.9971385354$;

(d) $\Gamma(1.635) = 0.8979334930$.

Table 2.8   Table of $\Gamma(x)$ values.

x	$\Gamma(x)$
1.00	1.0000000000
1.10	0.9513507699
1.20	0.9181687424
1.30	0.8974706963
1.40	0.8872638175
1.50	0.8862269255
1.60	0.8935153493
1.70	0.9086387329
1.80	0.9313837710
1.90	0.9617658319
2.00	1.0000000000

3. Theorem 2.1 requires an upper bound on the second derivative of the function being interpolated, and this is not always available as a practical matter. However, if a table of values is available, we can use a difference approximation to the second derivative to *estimate* the upper bound on the derivative, and hence the error. Recall, from Problem 12 of §2.2, that

$$f''(x) \approx \frac{f(x-h) - 2f(x) + f(x+h)}{h^2}.$$

Assume that the function values are given at the equally spaced grid points $x_k = a + kh$ for some grid spacing, h. Using the estimate

$$\max_{x_k \leq x \leq x_{k+1}} |f''(x)| \approx \max \left\{ \left| \frac{f(x_{k-1}) - 2f(x_k) + f(x_{k+1})}{h^2} \right|, \left| \frac{f(x_k) - 2f(x_{k+1}) + f(x_{k+2})}{h^2} \right| \right\}$$

to approximate the derivative upper bound, estimate the error made in using linear interpolation to approximate each of the following values of $\Gamma(x)$, based on Table 2.8.

(a) $\Gamma(1.290) = 0.8990415863$;

(b) $\Gamma(1.579) = 0.8913230181$;

(c) $\Gamma(1.456) = 0.8856168100$;

(d) $\Gamma(1.314) = 0.8954464400$;

(e) $\Gamma(1.713) = 0.9111663772$.

4. Construct a linear interpolating polynomial to the function $f(x) = x^{-1}$ using $x_0 = \frac{1}{2}$ and $x_1 = 1$ as the nodes. What is the upper bound on the error over the interval $[\frac{1}{2}, 1]$, according to the error estimate?

5. Repeat the above for $f(x) = \sqrt{x}$, using the interval $[\frac{1}{4}, 1]$.

6. Repeat the above for $f(x) = x^{1/3}$, using the interval $[\frac{1}{8}, 1]$.

7. If we want to apply linear interpolation to the sine function and obtain an accuracy of 10^{-6}, how close together do the entries in the table have to be? What if we change the error criterion to 10^{-3}?

8. Repeat the above for $f(x) = \tan x$, for $x \in [-\pi/4, \pi/4]$.

9. If we try to approximate the logarithm by using a table of logarithm entries, together with linear interpolation, to construct the approximation to $\ln(1+x)$ over the interval $[-\frac{1}{2}, 0]$, how many points are required to get an approximation that is accurate to within 10^{-14}?

10. Construct a piecewise linear approximation to $f(x) = \sqrt{x}$ over the interval $[\frac{1}{4}, 1]$ using the nodes $\frac{1}{4}, \frac{9}{16}, 1$. What is the maximum error in this approximation?

11. Repeat the above for $f(x) = x^{1/3}$ over the interval $[\frac{1}{8}, 1]$, using the nodes $\frac{1}{8}, \frac{27}{64}, 1$.

2.5 APPLICATION—THE TRAPEZOID RULE

One of the most important applications of linear interpolation is the construction of the trapezoid rule for approximating definite integrals. Define the integration of interest as $I(f)$, thus:

$$I(f) = \int_a^b f(x)dx.$$

Let $p_1(x)$ be the linear polynomial that interpolates f at $x = a$ and $x = b$:

$$p_1(x) = \frac{x-a}{b-a}f(b) + \frac{b-x}{b-a}f(a).$$

Then the basic trapezoid rule is defined by exactly integrating $p_1(x)$:

$$T_1(f) = I(p_1) = \int_a^b p_1(x)dx = \frac{1}{2}(b-a)(f(b) + f(a)). \tag{2.14}$$

Note here that we are constructing an approximation by replacing the "exact function" (f) by a simpler function that approximates it (p_1), and doing the desired calculation (integration) *exactly* on the simpler function.

Figure 2.7 illustrates this approximation. The integrand is $\frac{1}{2} + \sin \pi x$ over the interval $[\frac{1}{2}, 1]$.

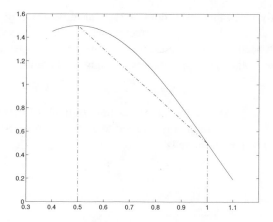

Figure 2.7   Illustration of the basic trapezoid rule.

How accurate is this rule? Clearly the error is $I(f) - T_1(f)$, and the illustration suggests that this might be large. We can analyze the error because the construction of T_1 as the exact integral of p_1 allows us to use the interpolation error theory from §2.4:

$$
\begin{aligned}
I(f) - T_1(f) &= I(f) - I(p_1), \\
&= \int_a^b f(x)dx - \int_a^b p_1(x)dx, \\
&= \int_a^b \left(f(x) - p_1(x) \right) dx, \\
&= \frac{1}{2} \int_a^b (x-a)(x-b)f''(\xi_x)dx,
\end{aligned}
$$

where $\xi_x \in [a,b]$ depends on x. Since $(x-a)(x-b)$ does not change sign on $[a,b]$, we can now apply the Integral Mean Value Theorem to get an error estimate:

$$
\begin{aligned}
\int_a^b (x-a)(x-b)f''(\xi_x)dx &= f''(\eta) \int_a^b (x-a)(x-b)dx, \\
&= -\frac{1}{6}(b-a)^3 f''(\eta),
\end{aligned}
$$

so that we have

$$
I(f) - T_1(f) = -\frac{1}{12}(b-a)^3 f''(\eta), \tag{2.15}
$$

where $\eta \in [a,b]$. Formally, then, we have proved the following theorem:

Theorem 2.2 (Trapezoid Rule Error Estimate, Single Subinterval) *Let $f \in C^2([a,b])$ and let p_1 interpolate f at a and b. Define $T_1(f) = I(p_1)$. Then there exists $\eta \in [a,b]$ such that*

$$
I(f) - T_1(f) = -\frac{1}{12}(b-a)^3 f''(\eta). \tag{2.16}
$$

Although perfectly valid, this result does not appear to be especially useful, since the error will only be small if the length of the integration interval, $b - a$, is small; but it is only a first step. What we need to do now is use more points in the approximation. To this end we subdivide the interval $[a, b]$ into n subintervals using meshpoints $x_i, 0 \leq i \leq n$:

$$a = x_0 < x_1 < x_2 \ldots < x_{n-1} < x_n = b. \tag{2.17}$$

Over each subinterval $[x_{i-1}, x_i]$ we apply the basic trapezoid rule (2.14) to get the *composite trapezoid rule*, which is more commonly referred to as "the n-subinterval trapezoid rule" (or, just the *trapezoid rule* when there is no possibility of confusion):

$$
\begin{aligned}
I(f) &= \sum_{i=1}^{n} \int_{x_{i-1}}^{x_i} f(x)dx, \\
&\approx \sum_{i=1}^{n} \frac{1}{2}(x_i - x_{i-1})(f(x_{i-1}) + f(x_i)), \\
&= T_n(f).
\end{aligned}
$$

If we use a *uniform* grid, in which the meshpoints are equally spaced so that $x_i - x_{i-1} = h$, for all i, then this simplifies a great deal:

$$T_n(f) = \frac{h}{2}\left(f(x_0) + 2f(x_1) + 2f(x_2) + \ldots + 2f(x_{n-1}) + f(x_n)\right). \tag{2.18}$$

Figure 2.8 shows this applied to the same integral of $f(x) = \frac{1}{2} + \sin \pi x$ as was used in Figure 2.7. Note that even with very few additional points in the grid, the apparent error is substantially reduced from the previous case. What we have done here, essentially, is exactly integrate a piecewise linear interpolation of f.

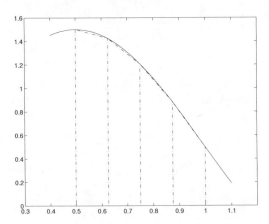

Figure 2.8   Illustration of the trapezoid rule.

The accuracy of the n-subinterval rule can be easily established from the previous result.

Theorem 2.3 (Trapezoid Rule Error Estimate, Uniform Grid) *Let $f \in C^2([a, b])$ and let $T_n(f)$ be the n-subinterval trapezoid rule approximation to $I(f)$, using a uniform grid. Then there exists $\xi_h \in [a, b]$, depending on h, such that*

$$I(f) - T_n(f) = -\frac{b-a}{12}h^2 f''(\xi_h). \tag{2.19}$$

Proof: We simply apply the previous error result to each subinterval:

$$I(f) - T_n(f) = \sum_{i=1}^{n} \left(\int_{x_{i-1}}^{x_i} f(x)dx - \frac{1}{2}(x_i + x_{i-1})(f(x_i) + f(x_{i-1})) \right),$$

$$= -\sum_{i=1}^{n} \frac{1}{12}h^3 f''(\xi_{i,h}), \xi_{i,h} \in [x_{i-1}, x_i],$$

$$= -\frac{1}{12}h^3 \sum_{i=1}^{n} f''(\xi_{i,h}),$$

$$= -\frac{1}{12}h^2 \frac{b-a}{n} \sum_{i=1}^{n} f''(\xi_{i,h}),$$

$$= -\frac{b-a}{12}h^2 \left(\frac{1}{n} \sum_{i=1}^{n} f''(\xi_{i,h}) \right).$$

We can now apply the Discrete Average Value Theorem to simplify the remaining sum: There exists $\xi_h \in [a, b]$ such that

$$\frac{1}{n} \sum_{i=1}^{n} f''(\xi_{i,h}) = f''(\xi_h), \tag{2.20}$$

where we note that $\xi_h \in [a, b]$ does depend on the value of h. This completes the proof of the theorem. •

This theorem tells us quite a bit about the n-subinterval trapezoid rule. First, it tells us that the numerical approximation will converge to the exact value, as long as $f \in C^2([a, b])$:

$$\lim_{n \to \infty} |I(f) - T_n(f)| = \lim_{h \to 0} \frac{b-a}{12}h^2 |f''(\xi_h)| \leq \left(\lim_{h \to 0} \frac{b-a}{12}h^2 \right) \left(\max_{x \in [a,b]} |f''| \right) = 0,$$

so that

$$\lim_{n \to \infty} T_n(f) = I(f).$$

Second, it tells us how fast this convergence occurs, since the error goes like h^2. Perhaps the best way to see this is to note that doubling the number of subintervals results in the error going down by roughly a factor of 4. We have to say "roughly" because the value of ξ_h does depend on h and so will shift about a bit as we change the mesh size.

It is worth emphasizing at this point that the value of the error estimate is not that we are able to *a priori* predict the error—although we sometimes can use the error theory to *estimate* the error—but rather that the error estimate tells us how rapidly the error will decrease as a function of the mesh size, h. This will be useful in Chapter 5 when we seek to *improve* the trapezoid rule.

An example will illustrate.

■ **EXAMPLE 2.5**

Let $f(x) = e^x$ and $[a, b] = [0, 1]$ so that $I(f) = e - 1 = 1.71828...$. Then the trapezoid rule using a single subinterval is given by

$$T_1(f) = \frac{1}{2} \times 1 \times (e^0 + e^1) = (1 + e)/2 = 1.859140914,$$

whereas the trapezoid rule using two subintervals is given by

$$T_2(f) = \frac{1}{2} \times \frac{1}{2} \times (e^0 + 2e^{1/2} + e^1) = \frac{1}{4}(1 + 2\sqrt{e} + e) = 1.753931093.$$

We note that the second value is considerably more accurate, as expected.

If we now continue the computation using a sequence of equally spaced grids with $h^{-1} = 2, 4, 8, ..., 2048$, then we get the results shown in Table 2.9. Note that the last column, which gives the ratio of the previous error to the current error, does show that the error decreases by roughly a factor of 4 as we double the number of points.

Table 2.9 Trapezoid example, $f(x) = e^x$, $[a, b] = [0, 1]$.

n	$T_n(f)$	$I(f) - T_n(f)$	Error Ratio
2	1.753931092	$-0.356493E - 01$	N/A
4	1.727221905	$-0.894008E - 02$	3.9876
8	1.720518592	$-0.223676E - 02$	3.9969
16	1.718841129	$-0.559300E - 03$	3.9992
32	1.718421660	$-0.139832E - 03$	3.9998
64	1.718316787	$-0.349584E - 04$	4.0000
128	1.718290568	$-0.873962E - 05$	4.0000
256	1.718284013	$-0.218491E - 05$	4.0000
512	1.718282375	$-0.546227E - 06$	4.0000
1024	1.718281965	$-0.136557E - 06$	4.0000
2048	1.718281863	$-0.341392E - 07$	4.0000

Programming Hint: We should note that the device of doubling the number of subintervals and observing that the error goes down by roughly a factor of 4 is a useful device for debugging trapezoid rule programs: If your program does not produce errors that behave this way, then there are only three possible conclusions to draw: (1) Your test function is not C^2 on the entire interval of integration; (2) Your code has an error; or, (3) Your example is one of a special class of functions (periodic functions, integrated over an integer multiple of the period) for which the trapezoid rule is "super-accurate," in which case the error starts out very small and does not always decrease in a regular fashion; see §5.8.1. You should not accept results that do not display the correct rate of error decrease unless you can demonstrate that either the first or third of these possibilities holds for your example.

■ **EXAMPLE 2.6**

The error theory can sometimes be used to predict how small the mesh must be to achieve a specified desired error. For example, consider the integral

$$I(f) = \int_0^1 e^{-x^2} dx.$$

How small does h have to be to guarantee that $|I(f) - T_n(f)| \leq 10^{-3}$?

This is a relatively simple exercise. We know that

$$|I(f) - T_n(f)| = \frac{1}{12}h^2|f''(\xi)| \leq \frac{1}{12}h^2 \max_{x \in [0,1]} |f''(x)|.$$

Ordinary calculus shows us that

$$f''(x) = e^{-x^2}\left(4x^2 - 2\right)$$

so that, for all $x \in [0,1]$,

$$|f''(x)| = |e^{-x^2}\left(4x^2 - 2\right)| \leq e^0\left(4 + 2\right) = 6.$$

Therefore,

$$|I(f) - T_n(f)| \leq 10^{-3} \Leftarrow \frac{1}{12}h^2(6) \leq 10^{-3},$$

which implies that we need $h \leq \sqrt{0.002} = 0.044721....$ The difficult part of this computation is in getting an upper bound on the second derivative factor in the error estimate.

The assumption of a uniform grid is useful but not required. If the mesh is not uniform, then the result of Theorem 2.3 is changed slightly:

Theorem 2.4 (Trapezoid Rule Error Estimate, Non-uniform Grid) *Let $f \in C^2([a,b])$ and let $T_n(f)$ be the n-subinterval trapezoid rule approximation to $I(f)$, using the non-uniform grid defined by*

$$a = x_0 < x_1 < x_2 < \ldots < x_{n-1} < x_n = b,$$

with $h_i = x_i - x_{i-1}$ and $h = \max_i h_i$. Then

$$|I(f) - T_n(f)| \leq \frac{b-a}{12}h^2 \max_{x \in [a,b]} |f''|. \tag{2.21}$$

The student should think about how to prove this result.

We close this section with a piece of pseudo-code that shows how to implement the trapezoid rule. Note that this code explicitly makes use of the assumed uniformity of the mesh spacing.

Algorithm 2.5 *Simple Trapezoid Rule (pseudo-code).*

```
input a, b, n; external f:
    sum = 0.5*(f(a) + f(b))
    h = (b - a)/n
    for i=1 to n-1 do
        x = a + i*h
        sum = sum + f(x)
    endfor
    trapezoid = h*sum
end code
```

Programming Hints: (1) When writing a program for *any* integration method, test it first on something for which you know the exact value of the integral, and verify that your approximations are converging to the exact value at the proper *rate* as h decreases. (This assumes that the integrand has enough continuous derivatives, of course.) (2) If your code works on one example but not on others, it is a good bet that the problem is with the coding of the other examples. Make sure that your "exact values" are indeed correct, for instance.

We close this section with a brief discussion of the *stability* of the trapezoid rule. Consider the integral

$$I(f) = \int_a^b f(x)dx$$

and its trapezoid rule approximation, $T_n(f)$. What happens if we slightly perturb f to produce a new function, say, $g(x) = f(x) + \epsilon(x)$ where $\epsilon(x)$ is always "small." How much change do we produce in the trapezoid rule values? In other words, can we bound the difference $|T_n(f) - T_n(g)|$ in terms of $|f(x) - g(x)| = |\epsilon(x)|$? We have

$$T_n(f) = h \sum_{i=0}^{n} {}'' f(x_i), \quad T_n(g) = h \sum_{i=0}^{n} {}'' g(x_i),$$

where the double prime on the summation symbols means that the first and last terms are multiplied by $\frac{1}{2}$. Thus,

$$T_n(f) - T_n(g) = h \sum_{i=0}^{n} {}'' \left(f(x_i) - g(x_i) \right) = h \sum_{i=0}^{n} {}'' \epsilon(x_i),$$

so that

$$|T_n(f) - T_n(g)| \le \left(\max_{x \in [a,b]} |\epsilon(x)| \right) h \sum_{i=0}^{n} {}'' 1 = (b-a) \max_{x \in [a,b]} |\epsilon(x)|.$$

Therefore, if $\max_{x \in [a,b]} |\epsilon(x)|$ is "small," so is $|T_n(f) - T_n(g)|$, and in fact if $\max_{x \in [a,b]} |\epsilon(x)| \to 0$, then $T_n(g) \to T_n(f)$; thus, we conclude that the trapezoid rule is a stable numerical method. Almost all methods for numerically approximating integrals are stable, in fact. Note that this is in marked contrast to the situation for numerical differentiation.

Modern versions of MATLAB have a function, `trapz`, which performs the trapezoid rule. Students should not use this to do any of the exercises without their instructor's knowledge *and* permission.

Chapter 5 contains more material on the approximation of definite integrals.

Exercises:

1. Use the trapezoid rule with $h = \frac{1}{4}$ to approximate the integral

$$I = \int_0^1 x^3 dx = \frac{1}{4}.$$

You should find that $T_4 = 17/64$. How small does h have to be for the error to be less than 10^{-3}? 10^{-6}?

2. Use the trapezoid rule and $h = \pi/4$ to approximate the integral

$$I = \int_0^{\pi/2} \sin x dx = 1.$$

How small does h have to be for the error to be less than 10^{-3}? 10^{-6}?

3. Repeat the above with $h = \pi/6$.

4. Apply the trapezoid rule with $h = \frac{1}{8}$, to approximate the integral

$$I = \int_0^1 \frac{1}{\sqrt{1 + x^4}} dx = 0.92703733865069.$$

How small does h have to be for the error to be less than 10^{-3}? 10^{-6}?

5. Apply the trapezoid rule with $h = \frac{1}{8}$, to approximate the integral

$$I = \int_0^1 x(1 - x^2) dx = \frac{1}{4}.$$

Feel free to use a computer program or a calculator, as you wish. How small does h have to be to get that the error is less than 10^{-3}? 10^{-6}?

6. Apply the trapezoid rule with $h = \frac{1}{8}$ to approximate the integral

$$I = \int_0^1 \ln(1 + x) dx = 2\ln 2 - 1.$$

How small does h have to be for the error to be less than 10^{-3}? 10^{-6}?

7. Apply the trapezoid rule with $h = \frac{1}{8}$ to approximate the integral

$$I = \int_0^1 \frac{1}{1 + x^3} dx = \frac{1}{3}\ln 2 + \frac{1}{9}\sqrt{3}\pi.$$

How small does h have to be for the error to be less than 10^{-3}? 10^{-6}?

8. Apply the trapezoid rule with $h = \frac{1}{8}$ to approximate the integral

$$I = \int_1^2 e^{-x^2} dx = 0.1352572580.$$

How small does h have to be for the error to be less than 10^{-3}? 10^{-6}?

9. Let I_8 denote the value you obtained in the previous problem. Repeat the computation, this time using $h = \frac{1}{4}$, and call this approximate value I_4. Then compute

$$I_R = (4I_8 - I_4)/3$$

and compare this to the exact value of the integral.

10. Repeat the above for the integral

$$I = \int_0^1 \frac{1}{1 + x^3} dx = \frac{1}{3}\ln 2 + \frac{1}{9}\sqrt{3}\pi.$$

11. For each integral below, write a program to do the trapezoid rule using the sequence of mesh sizes $h = \frac{1}{2}(b - a), \frac{1}{4}(b - a), \frac{1}{8}(b - a), \ldots, \frac{1}{128}(b - a)$, where $b - a$ is the length of the given interval. Verify that the expected rate of decrease of the error is obtained.

(a) $f(x) = x^2 e^{-x}$, $[0, 2]$, $I(f) = 2 - 10e^{-2} = 0.646647168$;

(b) $f(x) = 1/(1 + 25x^2)$, $[0, 1]$, $I(f) = \frac{1}{5} \arctan(5)$;

(c) $f(x) = \sqrt{1 - x^2}$, $[-1, 1]$, $I(f) = \pi/2$;

(d) $f(x) = \ln x$, $[1, 3]$, $I(f) = 3 \ln 3 - 2 = 1.295836867$;

(e) $f(x) = x^{5/2}$, $[0, 1]$, $I(f) = 2/7$;

(f) $f(x) = e^{-x} \sin 4x$, $[0, \pi]$, $I(f) = \frac{4}{17}(1 - e^{-\pi}) = 0.2251261368$.

12. For each integral in Problem 11, how small does h have to be, according to our error theory, to achieve an accuracy of at least 10^{-3}? 10^{-6}?

13. Apply the trapezoid rule to the integral

$$I = \int_0^1 \sqrt{x}\, dx = \frac{2}{3}$$

using a sequence of uniform grids with $h = \frac{1}{2}, \frac{1}{4}, \ldots$. Do we get the expected rate of convergence? Explain.

14. The length of a curve $y = g(x)$, for x between a and b, is given by the integral

$$L(g) = \int_a^b \sqrt{1 + [g'(x)]^2}\, dx.$$

Use the trapezoid rule with $h = \pi/4$ and $h = \pi/16$ to find the length of one "arch" of the sine curve.

15. Use the trapezoid rule to find the length of the logarithm curve between $a = 1$ and $b = e$, using $n = 4$ and $n = 16$.

16. What should h be to guarantee an accuracy of 10^{-8} when using the trapezoid rule for each of the following integrals.

(a)
$$I(f) = \int_0^1 e^{-x^2}\, dx;$$

(b)
$$I(f) = \int_1^3 \ln x\, dx;$$

(c)
$$I(f) = \int_{-5}^5 \frac{1}{1 + x^2}\, dx;$$

(d)
$$I(f) = \int_0^1 \cos(\pi x/2) dx.$$

17. Since the natural logarithm is defined as an integral,

$$\ln x = \int_1^x \frac{1}{t} dt, \tag{2.22}$$

it is possible to use the trapezoid rule (or any other numerical integration rule) to construct approximations to $\ln x$.

(a) Show that using the trapezoid rule on the integral (2.22) results in the series approximation (for $x \in [1, 2]$)

$$\ln x \approx \frac{x^2 - 1}{2nx} + \sum_{k=1}^{n-1} \frac{x - 1}{n + k(x - 1)}.$$

Hint: What are a and b in the integral that defines the logarithm?

(b) How many terms are needed in this approximation to get an error of less than 10^{-8} for all $x \in [1, 2]$? How many terms are needed for an error of less than 10^{-15} over the same interval?

(c) Implement this series for a predicted accuracy of 10^{-8} and compare it to the intrinsic natural logarithm function on your computer, over the interval $[1, 2]$. Is the expected accuracy achieved?

(d) If we were only interested in the interval $[1, 3/2]$, how many terms would be needed for the accuracy specified in part (b)?

(e) Is it possible to reduce the computation of $\ln x$ for all $x > 0$ to the computation of $\ln z$ for $z \in [1, 3/2]$? Explain.

18. How small must h be to compute the error function,

$$\text{erf}(x) = \frac{2}{\sqrt{\pi}} \int_0^x e^{-t^2} dt,$$

using the trapezoid rule, to within 10^{-8} accuracy for all $x \in [0, 1]$?

19. Use the data in Table 2.8 to compute

$$I = \int_1^2 \Gamma(x) dx$$

using $h = 0.2$ and $h = 0.1$.

20. Use the data in Table 2.7 to compute

$$I = \int_0^1 \text{erf}(x) dx$$

using $h = 0.2$ and $h = 0.1$.

21. Show that the trapezoid rule is *exact* for all linear polynomials.

22. Prove Theorem 2.4.

23. Extend the discussion on stability to include changes in the interval of integration instead of changes in the function. State and prove a theorem that bounds the change in the trapezoid rule approximation to

$$I(f) = \int_a^b f(x) dx$$

when the upper limit of integration changes from b to $b + \epsilon$, but f remains the same.

24. Consider a function $\tilde{f}(x)$ that is the floating-point representation of $f(x)$; thus, $f - \tilde{f}$ is the rounding error in computing f. If we assume that $|f(x) - \tilde{f}(x)| \le \epsilon$ for all x, show that

$$|T_n(f) - T_n(\tilde{f})| \le \epsilon(b - a).$$

What does this say about the effects of rounding error on the trapezoid rule?

2.6   SOLUTION OF TRIDIAGONAL LINEAR SYSTEMS

The solution of linear systems of equations is one of the most important areas of computational mathematics. A complete treatment is often given in a separate course, and there exist a number of texts that are devoted entirely to the subject of numerical linear algebra. We present some of that material in Chapter 7; in this section we discuss one particular problem of special interest—the solution of tridiagonal systems of linear equations.

Recall that a system of linear equations can be written in the form

$$
\begin{aligned}
a_{11}x_1 + a_{12}x_2 + \cdots + a_{1n}x_n &= f_1 \\
a_{21}x_1 + a_{22}x_2 + \cdots + a_{2n}x_n &= f_2 \\
&\vdots \\
a_{n1}x_1 + a_{n2}x_2 + \cdots + a_{nn}x_n &= f_n
\end{aligned}
$$

where the a_{ij} and f_i are knowns, and the x_i are the unknowns we wish to find. This can be put into matrix–vector form as

$$Ax = f, \tag{2.23}$$

where A is the $n \times n$ matrix having a_{ij} as its entries, and x and f are real-valued vectors with components x_i and f_i, respectively.

A common special case of a linear system occurs when A is *tridiagonal*, that is, there are only three "diagonals" in A that contain non-zero elements: the main diagonal, the first diagonal above the main diagonal (the super-diagonal), and the first diagonal below the main diagonal (the sub-diagonal). Thus A has the special form

$$
A = \begin{bmatrix}
a_{11} & a_{12} & 0 & \cdots & & 0 \\
a_{21} & a_{22} & a_{23} & \ddots & & 0 \\
0 & a_{32} & a_{33} & \ddots & & 0 \\
\vdots & \ddots & \ddots & \ddots & & a_{n-1,n} \\
0 & \cdots & 0 & & a_{n,n-1} & a_{nn}
\end{bmatrix}.
$$

For example,

$$
A = \begin{bmatrix}
4 & 2 & 0 & 0 & 0 \\
1 & 5 & 6 & 0 & 0 \\
0 & 3 & 9 & 5 & 0 \\
0 & 0 & 0 & 1 & 2 \\
0 & 0 & 0 & 2 & 4
\end{bmatrix}
$$

is tridiagonal. This makes the solution of the system, under certain assumptions, quite easy. Before proceeding with the algorithm, let's make a notational simplification. Instead of writing the system in the traditional matrix notation and indexing style, we will use l_i, d_i, and u_i to denote the lower-diagonal, diagonal, and upper-diagonal elements:

$$\begin{cases} l_i = a_{i,i-1}, & 2 \leq i \leq n, \\ d_i = a_{ii}, & 1 \leq i \leq n, \\ u_i = a_{i,i+1}, & 1 \leq i \leq n-1, \end{cases}$$

where we adopt the convention that $l_1 = 0$ and $u_n = 0$. Under this notation, the augmented matrix corresponding to the system is

$$[A \,|\, f] = \begin{bmatrix} d_1 & u_1 & 0 & \cdots & 0 & \bigm| & f_1 \\ l_2 & d_2 & u_2 & \ddots & 0 & \bigm| & f_2 \\ 0 & l_3 & d_3 & \ddots & 0 & \bigm| & \vdots \\ \vdots & \ddots & \ddots & \ddots & u_{n-1} & \bigm| & f_{n-1} \\ 0 & \cdots & 0 & l_n & d_n & \bigm| & f_n \end{bmatrix}, \tag{2.24}$$

and we can store the entire problem using just the four vectors for l, d, u, and f, instead of the entire $n \times n$ matrix, which is mostly zeroes anyway. (A common and convenient shorthand notation is to write

$$A = \text{tridiag}(l_i, d_i, u_i)$$

for the tridiagonal matrix A.)

Recall from linear algebra that the standard means of solving a linear system (called *Gaussian elimination*) is to eliminate all the components of A below the main diagonal, in other words, we reduce A to *triangular* form. In our case, this is easy because we only have to eliminate a single term each time, since there is only a single element below the main diagonal in each column. Thus, we would multiply the first equation by l_2/d_1 and subtract this from the second equation, to get

$$[A \,|\, f] \sim \begin{bmatrix} d_1 & u_1 & 0 & \cdots & 0 & \bigm| & f_1 \\ 0 & d_2 - u_1(l_2/d_1) & u_2 & \ddots & 0 & \bigm| & f_2 - f_1(l_2/d_1) \\ 0 & l_3 & d_3 & \ddots & 0 & \bigm| & \vdots \\ \vdots & \ddots & \ddots & \ddots & u_{n-1} & \bigm| & f_{n-1} \\ 0 & \cdots & 0 & l_n & d_n & \bigm| & f_n \end{bmatrix}$$

and then continue on with each successive row. Note that carrying out this step requires that $d_1 \neq 0$, and continuing to the next step is going to require that $d_2 - u_1(l_2/d_1) \neq 0$. Assuming the corresponding statement to be the case on each remaining row, we can reduce (2.24) to the equivalent system

$$[T \,|\, g] = \begin{bmatrix} \delta_1 & u_1 & 0 & \cdots & 0 & \bigm| & g_1 \\ 0 & \delta_2 & u_2 & \ddots & 0 & \bigm| & g_2 \\ 0 & 0 & \delta_3 & \ddots & 0 & \bigm| & \vdots \\ \vdots & \ddots & \ddots & \ddots & u_{n-1} & \bigm| & g_{n-1} \\ 0 & \cdots & 0 & 0 & \delta_n & \bigm| & g_n \end{bmatrix},$$

where
$$\delta_1 = d_1, \ \delta_2 = d_2 - u_1(l_2/\delta_1), \ \delta_3 = d_3 - u_2(l_3/\delta_2),$$
and, generally,
$$\delta_k = d_k - u_{k-1}(l_k/\delta_{k-1}), \ 2 \le k \le n.$$
Similarly,
$$g_1 = f_1, \ g_2 = f_2 - g_1(l_2/\delta_1), \ g_3 = f_3 - g_2(l_3/\delta_2),$$
so that the general form is
$$g_k = f_k - g_{k-1}(l_k/\delta_{k-1}), \ 2 \le k \le n.$$

The matrix $[T \mid g]$ is *row equivalent* to the original augmented matrix $[A \mid f]$, meaning that we can progress from one to the other using elementary row operations; thus, the two augmented matrices represent systems with exactly the same solution sets. Moreover, the solution is now easy to obtain, since we can solve the last equation, $\delta_n x_n = g_n$, to get $x_n = g_n/\delta_n$, and then use this value in the previous equation to get x_{n-1}, and so on, to get each solution component. Again, carrying out this stage of the computation requires the assumption that each $\delta_k \ne 0, 1 \le k \le n$.

The first stage of the computation (reducing the tridiagonal matrix A to the *triangular* matrix T) is generally called the *elimination step*, and the second stage is generally called the *backward solution* (or *backsolve*) step. A pseudo-code algorithm for this process is given below. Note that we do not store the entire matrix, but only the three vectors needed to define the elements in the non-zero diagonals; in addition, we did not use different variable names for the d_i, δ_i, and so on, but rather, *overwrote* the original variable with the new values. This saves storage when working with large problems.

Algorithm 2.6 *Pseudo-code for the Solution of Tridiagonal Systems*

```
!
!    Elimination stage
!
        for i=2 to n
            d(i) = d(i) - u(i-1)*l(i)/d(i-1)
            f(i) = f(i) - f(i-1)*l(i)/d(i-1)
        endfor
!
!    Backsolve stage (bottom row is a special case)
!
        x(n) = f(n)/d(n)
        for i=n-1 downto 1
            x(i) = (f(i) - u(i)*x(i+1))/d(i)
        endfor
```

Programming Hints: Always test your codes on simple examples for which you know the solution, and for which you can easily do the computation manually. In the case of the tridiagonal system, start with a small $(3 \times 3, 4 \times 4)$ system; if your computed solution is not almost exactly the same as the true solution, then your code has an error in it someplace. Do the algorithm by hand and then print out what the computer has for the final values of d, u, etc., so that you can see where your code deviates from the correct computation. When you don't know the exact solution, compute the *residual*, defined by $r = b - Ax$. If the computed solution is correct, the residual should be almost exactly zero in each of the components. (*Note:* If the example matrix is "nearly singular"—a concept that will be more precisely defined in Chapter 7—then it is possible that the rounding error will be quite large, even for small 3×3 examples. To avoid this potential problem, make sure that your examples have diagonal elements that are large compared to the absolute sum of the off-diagonal elements.)

At this point, we look at several examples.

■ **EXAMPLE 2.7**

Consider the system

$$\begin{bmatrix} 4 & 1 & 0 & 0 \\ 1 & 4 & 1 & 0 \\ 0 & 1 & 4 & 1 \\ 0 & 0 & 1 & 4 \end{bmatrix} \begin{bmatrix} x_1 \\ x_2 \\ x_3 \\ x_4 \end{bmatrix} = \begin{bmatrix} 6 \\ 12 \\ 18 \\ 19 \end{bmatrix}. \tag{2.25}$$

If we do the elimination stage of the algorithm, we get the new system (written now as a single augmented matrix):

$$[T \,|\, g] = \begin{bmatrix} 4 & 1 & 0 & 0 & | & 6 \\ 0 & \frac{15}{4} & 1 & 0 & | & \frac{21}{2} \\ 0 & 0 & \frac{56}{15} & 1 & | & \frac{76}{5} \\ 0 & 0 & 0 & \frac{209}{56} & | & \frac{209}{14} \end{bmatrix},$$

from which we get the solution

$$x_1 = 1, \quad x_2 = 2, \quad x_3 = 3, \quad x_4 = 4.$$

On the other hand, if we look at the system

$$\begin{bmatrix} 1 & 4 & 0 & 0 \\ 1 & 4 & 1 & 0 \\ 0 & 1 & 4 & 1 \\ 0 & 0 & 1 & 4 \end{bmatrix} \begin{bmatrix} x_1 \\ x_2 \\ x_3 \\ x_4 \end{bmatrix} = \begin{bmatrix} 9 \\ 12 \\ 18 \\ 19 \end{bmatrix}, \tag{2.26}$$

then the elimination stage produces the following augmented matrix after a single pass through the first loop:

$$[T \,|\, g] = \begin{bmatrix} 1 & 4 & 0 & 0 & | & 9 \\ 0 & 0 & 1 & 0 & | & 3 \\ 0 & 1 & 4 & 1 & | & 18 \\ 0 & 0 & 1 & 4 & | & 19 \end{bmatrix}.$$

Clearly, we cannot continue the process, for we would have to divide by zero in the next step. We should point out, however, that this does *not* mean that the system in question does not have a solution; it simply means that the algorithm fails to work on this example. The reader should check that

$$x_1 = 1, \quad x_2 = 2, \quad x_3 = 3, \quad x_4 = 4$$

is indeed the solution to the above system.

A complete discussion of this problem and similar issues in numerical linear algebra is deferred until Chapter 7; for now we content ourselves with stating and proving a common condition that is sufficient to guarantee that the tridiagonal solution algorithm presented here will work. We require a definition, first.

Definition 2.1 (Diagonal Dominance for Tridiagonal Matrices) *A tridiagonal matrix is called* diagonally dominant *if*

$$d_i > |l_i| + |u_i| > 0, \quad 1 \le i \le n. \tag{2.27}$$

For example, the matrix

$$A_1 = \begin{bmatrix} 6 & 1 & 0 & 0 \\ 2 & 6 & 3 & 0 \\ 0 & 6 & 9 & 0 \\ 0 & 0 & 3 & 4 \end{bmatrix}$$

is diagonally dominant, but

$$A_2 = \begin{bmatrix} 6 & 1 & 0 & 0 \\ 2 & 5 & 3 & 0 \\ 0 & 6 & 9 & 0 \\ 0 & 0 & 3 & 4 \end{bmatrix}$$

is not. (The inequality does not hold in the second row.)

If the matrix is diagonally dominant, then we can easily show that we do not need to worry about the diagonal elements becoming zero during the elimination process.

Theorem 2.5 *If the tridiagonal matrix A is diagonally dominant, then Algorithm 2.6 will succeed in producing the correct solution to the original linear system, within the limitations of rounding error.*

Proof: The diagonal dominance condition directly shows that $d_1 = \delta_1 \ne 0$, so all that remains is to show that each $\delta_k = d_k - u_{k-1} l_k / \delta_{k-1} \ne 0$, for $2 \le k \le n$. Assume momentarily that $l_2 \ne 0$. We have

$$
\begin{aligned}
\delta_2 &= d_2 - u_1 l_2 / d_1, \\
&\ge d_2 - |u_1 l_2 / d_1|, \\
&\ge |u_2| + |l_2| - |l_2|\theta_1, \\
&\ge (|u_2| + |l_2|)(1 - \theta_1),
\end{aligned}
$$

for $\theta_1 = |u_1|/|d_1| < 1$. Therefore, $\delta_2 > 0$ since $l_2 \ne 0$. If, on the other hand, $l_2 = 0$, then we have $\delta_2 = d_2 - 0 = d_2 > 0$. We can repeat the same argument for each index, and that completes the proof. •

Note that in our example (2.25) the matrix was indeed diagonally dominant, whereas in the next one (2.26), it was not; condition (2.27) fails in the first row.

Note also that we did not discuss any error issues in this section. The reason is actually very simple—there is no approximation being made, so there is no mathematical error to worry about. In the absence of rounding error, this algorithm will produce the exact solution of the tridiagonal linear system, *if* it is able to run to completion, i.e., without divisions by zero.

MATLAB of course has many simple commands to solve linear systems problems. We will work with many of these in Chapter 7, and will do a little bit in one of the exercises here.

Exercises:

1. Use Algorithm 2.6 to compute the solution to the following system of equations:

$$
\begin{bmatrix}
4 & 2 & 0 & 0 \\
1 & 4 & 1 & 0 \\
0 & 1 & 4 & 1 \\
0 & 0 & 2 & 4
\end{bmatrix}
\begin{bmatrix}
x_1 \\ x_2 \\ x_3 \\ x_4
\end{bmatrix}
=
\begin{bmatrix}
\pi/9 \\ \sqrt{3}/2 \\ \sqrt{3}/2 \\ -\pi/9
\end{bmatrix}.
$$

2. Write a computer code to solve the previous problem.

3. Use the Algorithm 2.6 to solve the following system of equations:

$$
\begin{bmatrix}
6 & 1 & 0 & 0 \\
2 & 4 & 1 & 0 \\
0 & 1 & 4 & 2 \\
0 & 0 & 1 & 6
\end{bmatrix}
\begin{bmatrix}
x_1 \\ x_2 \\ x_3 \\ x_4
\end{bmatrix}
=
\begin{bmatrix}
8 \\ 13 \\ 22 \\ 27
\end{bmatrix}.
$$

   You should get the solution $x = (1, 2, 3, 4)^T$.

4. Write a computer code to solve the previous problem.

5. The diagonal dominance condition is an example of a *sufficient* but not *necessary* condition. That is, Algorithm 2.6 will often work for systems that are *not* diagonally dominant. Show that the following system is not diagonally dominant, but then use Algorithm 2.6 to compute the solution to it:

$$
\begin{bmatrix}
1 & \frac{1}{2} & 0 & 0 \\
\frac{1}{2} & \frac{1}{3} & \frac{1}{4} & 0 \\
0 & \frac{1}{4} & \frac{1}{5} & \frac{1}{6} \\
0 & 0 & \frac{1}{6} & \frac{1}{7}
\end{bmatrix}
\begin{bmatrix}
x_1 \\ x_2 \\ x_3 \\ x_4
\end{bmatrix}
=
\begin{bmatrix}
2 \\ 23/12 \\ 53/30 \\ 15/14
\end{bmatrix}.
$$

   You should get the solution $x = (1, 2, 3, 4)^T$.

6. Use Algorithm 2.6 to compute the solution to the following system of equations:

$$
\begin{bmatrix}
1 & \frac{1}{2} & 0 & 0 \\
\frac{1}{2} & \frac{1}{3} & \frac{1}{4} & 0 \\
0 & \frac{1}{4} & \frac{1}{5} & \frac{1}{6} \\
0 & 0 & \frac{1}{6} & \frac{1}{7}
\end{bmatrix}
\begin{bmatrix}
x_1 \\ x_2 \\ x_3 \\ x_4
\end{bmatrix}
=
\begin{bmatrix}
2 \\ 2 \\ 53/30 \\ 15/14
\end{bmatrix}.
$$

   Note that this is a very small change from the previous problem, since the only difference is that f_2 has changed by only $1/12$. How much has the answer changed?

7. Write a computer code to do the previous two problems.

8. Verify that the following system is diagonally dominant, and use Algorithm 2.6 to find the solution.

$$
\begin{bmatrix}
\frac{1}{2} & \frac{10}{21} & 0 & 0 \\
\frac{1}{4} & \frac{1}{3} & \frac{1}{13} & 0 \\
0 & \frac{1}{5} & \frac{1}{4} & \frac{1}{21} \\
0 & 0 & \frac{1}{6} & \frac{1}{5}
\end{bmatrix}
\begin{bmatrix}
x_1 \\ x_2 \\ x_3 \\ x_4
\end{bmatrix}
=
\begin{bmatrix}
61/42 \\ 179/156 \\ 563/420 \\ 13/10
\end{bmatrix}.
$$

9. Use Algorithm 2.6 to find the solution to this system:

$$
\begin{bmatrix}
\frac{1}{2} & \frac{10}{21} & 0 & 0 \\
\frac{1}{4} & \frac{1}{3} & \frac{1}{13} & 0 \\
0 & \frac{1}{5} & \frac{1}{4} & \frac{1}{21} \\
0 & 0 & \frac{1}{6} & \frac{1}{5}
\end{bmatrix}
\begin{bmatrix}
x_1 \\ x_2 \\ x_3 \\ x_4
\end{bmatrix}
=
\begin{bmatrix}
61/42 \\ 180/156 \\ 563/420 \\ 13/10
\end{bmatrix}.
$$

Note that the right side here is different from that in the previous problem by only a small amount in the f_2 component. Comment on your results here as compared to those in the previous problem.

10. Write a computer code to do the previous two problems.

11. Write a code that carries out Algorithm 2.6, and test it on the following system of equations:

$$ Tx = b, $$

where T is 10×10 with

$$
t_{ij} =
\begin{cases}
1, & |i - j| = 1; \\
j + 1, & i = j; \\
0, & \text{otherwise.}
\end{cases}
$$

and $b_i = 1$ for all i. Check your results by computing the residual $r = b - Tx$. What is the largest component (in absolute value) of r? (You could also check your results by using MATLAB's backslash operator to solve the system.)

12. Extend Algorithm 2.6 to a *pentadiagonal matrix*, that is, one with five non-zero diagonals. Write a program to carry out this solution algorithm, and apply it to the system

$$
\begin{bmatrix}
4 & 2 & 1 & 0 \\
1 & 4 & 1 & 1 \\
1 & 1 & 4 & 1 \\
0 & 1 & 2 & 4
\end{bmatrix}
\begin{bmatrix}
x_1 \\ x_2 \\ x_3 \\ x_4
\end{bmatrix}
=
\begin{bmatrix}
1 \\ 1 \\ 1 \\ 1
\end{bmatrix}.
$$

Check your results by again computing the residual vector, or by using MATLAB's backslash operation.

13. Consider the family of tridiagonal problems defined by the matrix $K_n \in \mathbb{R}^{n \times n}$, with

$$ K_n = \text{tridiag}(-1, 2, -1) $$

and a randomly defined right-hand-side vector. (Use rand to generate the random vectors.) Solve the system over the range $4 \leq n \leq 100$; use the flops command to estimate the number of operations required for each case, and plot the result as a function of n.

◁ ● ● ● ▷

2.7 APPLICATION: SIMPLE TWO-POINT BOUNDARY VALUE PROBLEMS

Consider the problem of finding a function $u(x)$ such that

$$-u'' + u = f(x), \quad x \in [0,1], \tag{2.28}$$
$$u(0) = u(1) = 0. \tag{2.29}$$

Here, f is a known function and we seek u that satisfies the differential equation and the conditions at $x = 0$ and $x = 1$. This is an example of a *two-point boundary value problem*, something we treat in more detail in §6.10, but we have enough tools now to compute solutions, although we cannot analyze the error completely. Problems of this sort occur in many areas of applied mathematics and engineering, most notably perhaps in structural mechanics.

Divide the interval $[0,1]$ into n equal subintervals $[x_{k-1}, x_k]$, according to

$$0 = x_0 < x_1 < x_2 < \cdots < x_{n-1} < x_n = 1,$$

with $x_k - x_{k-1} = h$ (therefore $x_k = kh$), and let $U_h(x)$ denote the approximation to $u(x)$. We can use Taylor expansions similar to (2.2) and (2.3) (just take more terms) to derive an approximation to the second derivative, by adding them (instead of subtracting them). After a little manipulation (see Problem 12 in §2.2) we find that

$$u''(x) - \frac{u(x-h) - 2u(x) + u(x+h)}{h^2} = \frac{1}{12}h^2 u^{(4)}(\eta_x), \tag{2.30}$$

for some $\eta_x \in [x-h, x+h]$. Therefore, the differential equation $-u'' + u = f(x)$ implies that

$$\frac{-u(x-h) + 2u(x) - u(x+h)}{h^2} + u(x) = f(x) + \frac{1}{12}h^2 u^{(4)}(\eta_x).$$

To define our approximation, we drop the remainder term and replace u with U_h, which yields

$$\frac{-U_h(x-h) + 2U_h(x) - U_h(x+h)}{h^2} + U_h(x) = f(x). \tag{2.31}$$

This holds for all $x \in [0,1]$. To get a practical computational problem, we impose (2.31) only on our grid points, i.e., we seek the $n-1$ values $U_k = U_h(x_k)$, $1 \le k \le n-1$, which satisfy (recall that the boundary conditions will force $U_0 = U_n = 0$)

$$-U_{k-1} + (2 + h^2)U_k - U_{k+1} = h^2 f(x_k), \quad 1 \le k \le n-1.$$

This is a tridiagonal system of linear equations. Written out in matrix–vector form, we have

$$\begin{bmatrix} 2+h^2 & -1 & 0 & \cdots & \cdots & 0 \\ -1 & 2+h^2 & -1 & 0 & \cdots & 0 \\ 0 & -1 & 2+h^2 & -1 & 0 & \vdots \\ \vdots & & \ddots & \ddots & \ddots & 0 \\ \vdots & & & \ddots & \ddots & -1 \\ 0 & \cdots & \cdots & 0 & -1 & 2+h^2 \end{bmatrix} \begin{bmatrix} U_1 \\ U_2 \\ \vdots \\ \vdots \\ U_{n-2} \\ U_{n-1} \end{bmatrix} = \begin{bmatrix} h^2 f(x_1) \\ h^2 f(x_2) \\ \vdots \\ \vdots \\ h^2 f(x_{n-2}) \\ h^2 f(x_{n-1}) \end{bmatrix}.$$

Moreover, it is diagonally dominant, so we can apply Algorithm 2.6 to produce solutions.

■ EXAMPLE 2.8

If we take the specific case of $h = \frac{1}{4}$ and $f(x) = 1$, then we have a 3×3 system to solve, that is,

$$
\begin{bmatrix}
2.0625 & -1 & 0 \\
-1 & 2.0625 & -1 \\
0 & -1 & 2.0625
\end{bmatrix}
\begin{bmatrix}
U_1 \\
U_2 \\
U_3
\end{bmatrix}
=
\begin{bmatrix}
0.0625 \\
0.0625 \\
0.0625
\end{bmatrix}
$$

The solution is easily found, using Algorithm 2.6, to be

$$U_1 = 0.08492201039861, \quad U_2 = 0.11265164644714, \quad U_3 = 0.08492201039861.$$

Thus we have $u(1/4) \approx 0.08492201039861$, $u(1/2) \approx 0.11265164644714$, and $u(3/4) \approx 0.08492201039861$.

Table 2.10 shows the errors we get between $u(x_k)$ and U_k as h decreases from $1/4$ to $1/1024$, for this same choice of $f(x)$. The exact solution is given by

$$u(x) = 1 - \frac{e-1}{e^2-1}e^x - \frac{e^2-e}{e^2-1}e^{-x}.$$

Note that the error is decreasing by roughly a factor of 4 each time we double the number of points in the grid. Based on what we saw with the difference approximations to the derivatives, as well as the trapezoid rule, we suspect this means that the method is $\mathcal{O}(h^2)$ accurate, something we will be able to prove in §6.10.1, assuming that the solution is smooth enough. Figure 2.9 shows a plot of the approximate solution and exact solution for the $h = \frac{1}{8}$ case. Note that the approximate solution

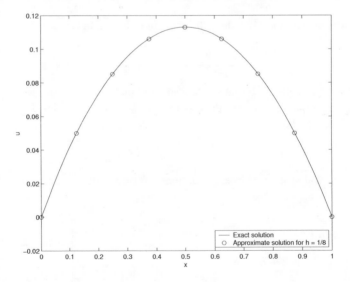

Figure 2.9   Approximate solution to two-point BVP $-u'' + u = 1$, for $h = 1/8$.

is virtually exact, even for this coarse mesh. (This of course will not always be the case.)

Table 2.10   Two-point boundary value problem results.

h^{-1}	$\max |u(x_k) - U_k|$
4	$0.529469582786e - 03$
8	$0.133142281084e - 03$
16	$0.333344381367e - 04$
32	$0.833667068484e - 05$
64	$0.208435910093e - 05$
128	$0.521101741688e - 06$
256	$0.130276180538e - 06$
512	$0.325690907577e - 07$
1024	$0.814227994750e - 08$

Programming Hint: Although our grid on $[0, 1]$ has n subintervals, note that the matrix system is $(n-1) \times (n-1)$, *not* $n \times n$. This is because the values at both $x = 0$ and $x = 1$ are known and so are not part of the system.

Exercises:

1. Solve, by hand, the two-point BVP (2.28)–(2.29) when $f(x) = x$, using $h = \frac{1}{4}$. Write out the linear system explicitly prior to solution. You should get the following 3×3 system:

$$\begin{bmatrix} 2.0625 & -1 & 0 \\ -1 & 2.0625 & -1 \\ 0 & -1 & 2.0625 \end{bmatrix} \begin{bmatrix} U_1 \\ U_2 \\ U_3 \end{bmatrix} = \begin{bmatrix} 0.015625 \\ 0.03125 \\ 0.046875 \end{bmatrix}$$

2. Repeat the above, this time using $h = \frac{1}{5}$. What is the system now?

3. Repeat it again, this time using $h = \frac{1}{8}$. What is the system now? (For this problem, you probably will want to use a computer code to actually solve the system.)

4. Write a program that solves the two-point BVP (2.28)–(2.29), where f is as given below.

    (a) $f(x) = 4e^{-x} - 4xe^{-x}$, $u(x) = x(1-x)e^{-x}$;
    (b) $f(x) = (\pi^2 + 1) \sin \pi x$, $u(x) = \sin \pi x$;
    (c) $f(x) = \pi(\pi \sin \pi x + 2 \cos \pi x)e^{-x}$, $u(x) = e^{-x} \sin \pi x$;
    (d) $f(x) = 3 - \frac{1}{x} - (x^2 - x - 2) \log x$, $u(x) = x(1-x) \log x$.

    The exact solutions are as given. Using $h^{-1} = 4, 8, 16, \ldots$, do we get the same kind of accuracy as in Table 2.10? Explain why or why not.

5. Try to apply the ideas of this section to approximating the solution of the two-point BVP

$$\begin{aligned} -u'' + u' + u &= 1, \quad x \in [0, 1] \\ u(0) = 0, u(1) &= 0. \end{aligned}$$

Can we get a tridiagonal system to which Algorithm 2.6 can be applied? *Hint:* Consider some of the approximations from §2.2; use the most accurate ones that can be easily used.

6. Solve the two-point BVP problem

$$-u'' + 64u' + u = 1, \quad x \in [0, 1]$$
$$u(0) = 0, u(1) = 0,$$

using a range of mesh sizes, starting with $h = 1/4, 1/8$, and going as far as $h = 1/256$. Comment on your results.

7. Generalize the solution of the two-point BVP to the case where $u(0) = g_0 \neq 0$ and $u(1) = g_1 \neq 0$. Apply this to the solution of the problem

$$-u'' + u = 0, \quad x \in [0, 1],$$
$$u(0) = 2, u(1) = 1,$$

which has the exact solution

$$u(x) = \left(\frac{e - 2}{e^2 - 1} \right) e^x + \left(\frac{2e - 1}{e^2 - 1} \right) e^{-x}.$$

Solve this for a range of values of the mesh. Do we get the expected $\mathcal{O}(h^2)$ accuracy?

8. Consider the problem of determining the deflection of a thin beam, supported at both ends, due to a uniform load being placed along the beam. In one simple model, the deflection $u(x)$ as a function of position x along the beam satisfies the BVP

$$-u'' + pu = qx(L - x), \quad 0 < x < L;$$
$$u(0) = u(L) = 0.$$

Here p is a constant that depends on the material properties of the beam, L is the length of the beam, and q depends on the material properties of the beam, as well as on the size of the load placed on the beam. For a 6-foot-long beam, with $p = 7 \times 10^{-6}$ and $q = 4 \times 10^{-7}$, what is the maximum deflection of the beam? Use a fine enough grid that you can be confident of the accuracy of your results. Note that this problem is slightly more general than our example (2.28)–(2.29); you will have to adapt our method to this more general case.

9. Repeat the above problem, but this time use a 3-foot-long beam. How much does the maximum deflection change, and is it larger or smaller?

10. Repeat the beam problem again, but this time use a 12-foot-long beam.

11. Try to apply the ideas of this section to the solution of the *nonlinear* BVP defined by

$$-u'' + u^2 = 1, \quad 0 < x < 1;$$
$$u(0) = u(1) = 0.$$

Write out the systems of equations for the specific case of $h = \frac{1}{4}$. What goes wrong? Why can't we proceed with the approximate solution?

◁ • • ▷

CHAPTER 3

ROOT-FINDING

A problem that most students should be familiar with from ordinary algebra is that of finding the root of an equation $f(x) = 0$, i.e., the value of the argument that makes f zero. More precisely, if the function is defined as $y = f(x)$, we seek the value α such that

$$f(\alpha) = 0.$$

The precise terminology is that α is a *zero* of the function f, or a *root* of the equation $f(x) = 0$.[1] Note that we have not yet specified what kind of function f is. The obvious case is when f is an ordinary real-valued function of a single real variable x, but we can also consider the problem when f is a vector-valued function of a vector-valued variable, in which case the expression above is a system of equations; this more complicated case is discussed in Chapter 7.

In this chapter we consider only the simple case where f is a scalar real-valued function of a single real-valued variable. We will discuss three basic methods for finding the point α: the bisection method, Newton's method, and the secant method. We then consider a broad class of ideas coming under the heading of *fixed-point theory*, which will enable us to broaden and extend our understanding of iterations in general, whether applied to root-finding problems or not. Finally, we will discuss some variants of Newton's method and other advanced topics.

---

[1] The distinction is often muddled in practice, however. For example, it is commonplace to speak of the "roots of a polynomial," when one should in fact speak only of the "zeroes of a polynomial." In this book, we have used the terms interchangeably, as denoting where a function is equal to zero.

An Introduction to Numerical Methods and Analysis, Second Edition. By James F. Epperson
Copyright © 2013 John Wiley & Sons, Inc.

3.1 THE BISECTION METHOD

Bisection is a marvelously simple idea that is based on little more than the continuity of the function f. Suppose we know that $f(a)f(b) < 0$. This means that f is negative at one point and positive at the other. If we assume that f is continuous, then it follows (by the Intermediate Value Theorem) that there must be some value between a and b at which f is zero. In other words, we know that there is a root α between a and b. (*Note:* There may be more than one root in the interval.)

Now let's try to use these ideas to find α. Let c be the midpoint of the interval $[a, b]$, i.e.,

$$c = \frac{1}{2}(a + b)$$

and consider the product $f(a)f(c)$. There are three possibilities:

1. $f(a)f(c) < 0$; this means that a root (there might be more than one) is between a and c, i.e., $\alpha \in [a, c]$.

2. $f(a)f(c) = 0$; if we assume that we already know $f(a) \neq 0$, this means that $f(c) = 0$, thus $\alpha = c$ and we have found a root.

3. $f(a)f(c) > 0$; this means that a root must lie in the other half of the interval, i.e., $\alpha \in [c, b]$.

At first glance, this is helpful only if we get the second case and land right on top of a root, and this does not seem very likely. However, a second look reveals that if (1) or (3) hold, we now have a root localized to an interval ($[a, c]$ or $[c, b]$) that is *half* the length of the original interval $[a, b]$. If we now repeat the process, the interval of uncertainty is again decreased in half, and so on, until we have the root localized to within any tolerance we desire.

■ **EXAMPLE 3.1**

If $f(x) = 2 - e^x$, and we take the original interval to be $[a, b] = [0, 1]$, then the first several steps of the computation are as follows:

$$
\begin{aligned}
f(a) &= 1, f(b) = -0.7183 \Rightarrow c = [0 + 1]/2 = 1/2; f(c) = 0.3513 > 0 \\
&\Rightarrow [a, b] \leftarrow [1/2, 1]; \\
f(a) &= 0.3513, f(b) = -0.7183 \Rightarrow c = [1/2 + 1]/2 = 3/4; f(c) = -0.1170 < 0 \\
&\Rightarrow [a, b] \leftarrow [1/2, 3/4]; \\
f(a) &= 0.3513, f(b) = -0.1170 \Rightarrow c = [1/2 + 3/4]/2 = 5/8; f(c) = 0.1318 > 0 \\
&\Rightarrow [a, b] \leftarrow [5/8, 3/4].
\end{aligned}
$$

Thus, we have reduced the "interval of uncertainty" from $[0, 1]$, which has length 1, to $[5/8, 3/4]$, which has length $\frac{1}{8} = 0.125$. If we were to continue the process we would eventually have the root localized to within an interval of length as small as we want, since each step cuts the interval of uncertainty in half. Figure 3.1 shows a graphical view of this example, with the marks on the horizontal axis showing how the interval around the root changes with each iteration.

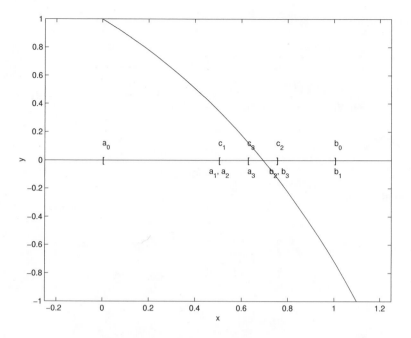

Figure 3.1 Bisection applied to $y = 2 - e^x$.

A formal statement is given in Algorithm 3.1.

Algorithm 3.1 *Bisection Method (Outline Form).*

1. *Given an initial interval $[a_0, b_0] = [a, b]$, set $k = 0$ and proceed as follows:*

2. *Compute $c_{k+1} = a_k + \frac{1}{2}(b_k - a_k)$;*

3. *If $f(c_{k+1})f(a_k) < 0$, set $a_{k+1} = a_k$, $b_{k+1} = c_{k+1}$;*

4. *If $f(c_{k+1})f(b_k) < 0$, set $b_{k+1} = b_k$, $a_{k+1} = c_{k+1}$;*

5. *Update k and go to Step 2.*

Each step is decreasing an upper bound on the (absolute) error by a factor of 2, a fact that we can establish rigorously.

Theorem 3.1 (Bisection Convergence and Error) *Let $[a_0, b_0] = [a, b]$ be the initial interval, with $f(a)f(b) < 0$. Define the approximate root as $x_n = c_n = (b_{n-1} + a_{n-1})/2$. Then there exists a root $\alpha \in [a, b]$ such that*

$$|\alpha - x_n| \le \left(\frac{1}{2}\right)^n (b - a). \tag{3.1}$$

Moreover, to achieve an accuracy of

$$|\alpha - x_n| \le \epsilon,$$

it suffices to take

$$n \geq \frac{\log(b-a) - \log \epsilon}{\log 2}.$$ (3.2)

Proof: We first observe, trivially, that

$$b_n - a_n = \frac{1}{2}(b_{n-1} - a_{n-1}),$$

which implies that

$$b_n - a_n = \left(\frac{1}{2}\right)^n (b_0 - a_0).$$

Therefore,

$$|\alpha - x_n| \leq \frac{1}{2}(b_{n-1} - a_{n-1})$$

implies that

$$
\begin{aligned}
|\alpha - x_n| &\leq \frac{1}{2}(b_{n-1} - a_{n-1}), \\
&= \frac{1}{2}\left(\frac{1}{2}\right)^{n-1}(b_0 - a_0), \\
&= \left(\frac{1}{2}\right)^n (b_0 - a_0),
\end{aligned}
$$

and we have proved the estimate. To establish the bound on the number of iterations, n, we simply observe that

$$\left(\frac{1}{2}\right)^n (b-a) \leq \epsilon,$$ (3.3)

together with (3.1), implies

$$|\alpha - x_n| \leq \epsilon$$

and we solve (3.3) for n to get (3.2). •

The observant reader will have noticed that Algorithm 3.1 is written without any stopping criterion. This is because we don't really need one, because Theorem 3.1 tells us precisely how many steps we need to take to achieve a desired accuracy.

The bisection process can be easily reduced to a pseudocode, as follows:

Algorithm 3.2 *Bisection Method (Pseudo-code).*

```
input a,b,eps
external f
fa = f(a)
fb = f(b)
if f(a)*f(b) > 0 then stop
n  = fix((log(b-a) - log(eps))/log(2)) + 1
for i=1 to n do
     c  = a + 0.5*(b - a)
     fc = f(c)
     if fa*fc < 0 then
          b = c
          fb = fc
     else
          if fa*fc > 0 then
               a = c
               fa = fc
          else
               alpha = c
               return
          endif
     endif
endfor
```

Programming Hints: Note that we have used the formula

$$c = a + 0.5(b - a) \tag{3.4}$$

to compute c, rather than the more obvious choice,

$$c = (a + b)/2. \tag{3.5}$$

The reason is that for very large values of a and b, (3.5) can lead to a computational overflow, whereas (3.4) will not. Note also that we were very careful in the algorithm to re-use the function values from iteration to iteration. A careless implementation of bisection will have two function calls each time through the loop, but it is possible to get by with only one function call. This is important because the function evaluation is outside our control and might be the most expensive part of the computation. An example of a function that could be quite expensive to compute is given in the discussion of "shooting methods" for nonlinear boundary value problems, in §6.10.2. Finally, although we will not usually bother to show output statements in any of the algorithms, it is a good idea to have the code print out the entire sequence of iterates, along with the corresponding function values, rather than just the final answer. This provides useful information for debugging and validating the code.

Table 3.1 shows the result of running this algorithm on the function $f(x) = 2 - e^x$, for which $\alpha = \log 2 = 0.6931471806$.

Bisection is an example of what is known as a *global method*, that is, it always converges, no matter how far you start from the actual root (assuming, of course, that you have the

Table 3.1   Bisection method applied to $f(x) = 2 - e^x$.

n	x_n	$b_{n-1} - a_{n-1}$	a_n	b_n
1	0.500000000000	1.000000000000	0.500000000000	1.000000000000
2	0.750000000000	0.500000000000	0.500000000000	0.750000000000
3	0.625000000000	0.250000000000	0.625000000000	0.750000000000
4	0.687500000000	0.125000000000	0.687500000000	0.750000000000
5	0.718750000000	0.062500000000	0.687500000000	0.718750000000
6	0.703125000000	0.031250000000	0.687500000000	0.703125000000
7	0.695312500000	0.015625000000	0.687500000000	0.695312500000
8	0.691406250000	0.007812500000	0.691406250000	0.695312500000
9	0.693359375000	0.003906250000	0.691406250000	0.693359375000
10	0.692382812500	0.001953125000	0.692382812500	0.693359375000
11	0.692871093750	0.000976562500	0.692871093750	0.693359375000
12	0.693115234375	0.000488281250	0.693115234375	0.693359375000
13	0.693237304688	0.000244140625	0.693115234375	0.693237304688
14	0.693176269531	0.000122070312	0.693115234375	0.693176269531
15	0.693145751953	0.000061035156	0.693145751953	0.693176269531
16	0.693161010742	0.000030517578	0.693145751953	0.693161010742
17	0.693153381348	0.000015258789	0.693145751953	0.693153381348
18	0.693149566650	0.000007629395	0.693145751953	0.693149566650
19	0.693147659302	0.000003814697	0.693145751953	0.693147659302
20	0.693146705627	0.000001907349	0.693146705627	0.693147659302

root "bracketed," i.e., $f(a)f(b) < 0$. Most of the other methods that we will study are only *locally* convergent, meaning that we must start the iteration with a "good enough" approximation to the root. One disadvantage of bisection is that it cannot be used to find roots when the function is tangent to the axis and does not pass through the axis—think about using bisection to find the root of $f(x) = x^2$, for example. Another disadvantage is that it converges slowly compared to other methods—note that Theorem 3.1 implies that a decrease in the initial error by a factor of $2^{10} \approx 1,000$ requires 10 iterations—however, it can be used in conjunction with a more rapidly converging local method to make a rather effective root-finding algorithm. (See §3.11.5.)

Exercises:

1. Do three iterations (by hand) of the bisection method, applied to $f(x) = x^3 - 2$, using $a = 0$ and $b = 2$.

2. For each of the functions listed below, do a calculation by hand (i.e., with a calculator) to find the root to an accuracy of 0.1. This process will take at most five iterations for all of these, and fewer for several of them.

   (a) $f(x) = x - e^{-x^2}, [a, b] = [0, 1]$;
   (b) $f(x) = \ln x + x, [a, b] = [\frac{1}{10}, 1]$;
   (c) $f(x) = x^3 - 3, [a, b] = [0, 3]$;
   (d) $f(x) = x^6 - x - 1, [a, b] = [0, 2]$;

(e) $f(x) = 3 - 2^x$, $[a,b] = [0,2]$.

3. Write a program that uses the bisection method to find the root of a given function on a given interval, and apply this program to find the roots of the functions below on the indicated intervals. Use the relationship (3.2) to determine *a priori* the number of steps necessary for the root to be accurate to within 10^{-6}.

   (a) $f(x) = x^3 - 2$, $[a,b] = [0,2]$;

   (b) $f(x) = e^x - 2$, $[a,b] = [0,1]$;

   (c) $f(x) = x - e^{-x}$, $[a,b] = [0,1]$;

   (d) $f(x) = x^6 - x - 1$, $[a,b] = [0,2]$;

   (e) $f(x) = x^3 - 2x - 5$, $[a,b] = [0,3]$;

   (f) $f(x) = 1 - 2xe^{-x/2}$, $[a,b] = [0,2]$;

   (g) $f(x) = 5 - x^{-1}$, $[a,b] = [0.1, 0.25]$;

   (h) $f(x) = x^2 - \sin x$, $[a,b] = [0,\pi]$.

4. Use Algorithm 3.2 to solve the nonlinear equation $x = \cos x$. Choose your own initial interval by some judicious experimentation with a calculator.

5. Use Algorithm 3.2 to solve the nonlinear equation $x = e^{-x}$. Again, choose your own initial interval by some judicious experimentation with a calculator.

6. If you borrow L dollars at an annual interest rate of r (in decimal form, so 5% is written as 0.05), for a period of m years, then the size of the monthly payment, M, is given by the *annuity equation*

$$L = \frac{12M}{r} \left[1 - (1 + r/12)^{-12m} \right].$$

The author needs to borrow $150,000 to buy the new house that he wants, and he can only afford to pay $600 per month. Assuming a 30-year mortgage, use the bisection method to determine what interest rate he can afford to pay. (Should the author perhaps find some rich relatives to help him out here?)

7. What is the interest rate that the author can afford if he only has to borrow $100,000?

8. Consider the problem of modeling the position of the liquid–solid boundary in a substance that is melting due to the application of heat at one end. In a simplified model,[2] if the initial position of the interface is taken to be $x = 0$, then the interface moves according to

$$x = 2\beta\sqrt{t}$$

where β satisfies the nonlinear equation

$$\frac{(T_M - T_0)k}{\lambda\sqrt{\pi}} e^{-\beta^2/k} = \beta \operatorname{erf}(\beta/\sqrt{k}).$$

---

[2]See L.I. Rubinstein, *The Stefan Problem*, American Mathematical Society, Providence, RI, 1971.

Here T_M is the melting temperature (absolute scale), $T_0 < T_M$ is the applied temperature, k and λ are parameters dependent on the material properties of the substance involved, and $\text{erf}(z)$ is the *error function*, defined by

$$\text{erf}(z) = \frac{2}{\sqrt{\pi}} \int_0^z e^{-t^2}\,dt.$$

MATLAB has an intrinsic error function, `erf`. If you are not using MATLAB, the error function can be accurately approximated by

$$E(z) = 1 - (a_1\xi + a_2\xi^2 + a_3\xi^3)e^{-z^2},$$

where $\xi = 1/(1 + pz)$ and

$$p = 0.47047,\ a_1 = 0.3480242,\ a_2 = -0.0958798,\ a_3 = 0.747856.$$

(a) Show that finding β is equivalent to finding the root α of the one-parameter family of functions defined by

$$f(z) = \theta e^{-z^2} - z\,\text{erf}(z). \tag{3.6}$$

What is θ? How is α related to β?

(b) Find the value of α corresponding to $\theta = 0.001, 0.1, 10, 1000$. Use the bisection method, and either MATLAB or $E(z)$ to approximate the error function.

9. A variation on the bisection method is known as *regula-falsi*, or, *the method of false position*. Given an interval $[a, b]$, with $f(a)f(b) < 0$, the new point c is defined by finding where the straight line connecting $(a, f(a))$ and $(b, f(b))$ crosses the axis. Show that this yields

$$c = b - f(b)(b - a)/(f(b) - f(a)).$$

10. Do three iterations (by hand) of regula-falsi (see Problem 9), applied to $f(x) = x^3 - 2$, using $a = 0$ and $b = 2$. Compare to your results for Problem 1.

11. Modify Algorithm 3.2 to perform regula-falsi (see Problem 9), and use the new method to find the same roots as in Problem 3. Stop the program when the difference between consecutive iterates is less than 10^{-6}, i.e., when $|x_{k+1} - x_k| \le 10^{-6}$.

12. Repeat Problem 8(b), using your regula-falsi program.

13. The bisection method will always cut the interval of uncertainty in half, but regula-falsi might cut the interval by less, or might cut it by more. Do both bisection and regula-falsi on the function $f(x) = e^{-4x} - \frac{1}{10}$, using the initial interval $[0, 5]$. Which one gets to the root the fastest?

14. Apply both bisection and regula-falsi to the following functions on the indicated intervals. Comment on your results in the light of how the methods are supposed to behave.

(a) $f(x) = 1/(x^2 + 1)$, $[a, b] = [0, 5]$;
(b) $f(x) = 1/(x - 1)$, $[a, b] = [0, 3]$.

◁ ● ● ▷

3.2 NEWTON'S METHOD: DERIVATION AND EXAMPLES

Newton's method[3] is the classic algorithm for finding roots of functions. It is often introduced in the calculus sequence as an application of the derivative of a function. Historically, it appears to have been first used by Newton in 1669, although the ideas were known to others beforehand.[4] In fact, the ancient Babylonians had a method for approximating square roots that is essentially Newton's method.[5]

There are two good derivations of Newton's method, one that is geometric and one that is analytic. We will discuss both, beginning with the geometric.

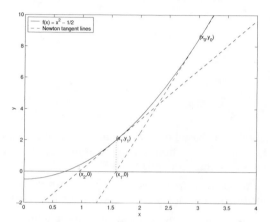

Figure 3.2 Newton's method for $y = x^2 - 1/2$.

Consider Figure 3.2. We wish to find a root, α, of $y = f(x)$, given an "initial guess" of x_0. How do we improve upon this initial guess to get a better and better approximation? The fundamental idea in Newton's method is to use the tangent line approximation to the function f at the point $(x_0, f(x_0))$. The point-slope formula for the equation of the straight line gives us

$$\frac{y - y_0}{x - x_0} = f'(x_0);$$

---

[3]Isaac Newton (1642–1727) is generally regarded as one of the greatest mathematicians of all time. He entered Trinity College, Cambridge, in 1661 and graduated with a B.A. degree in 1665. In 1668, he received a master's degree, and was appointed Lucasian Professor of Mathematics, one of the most prestigious positions in English academia at the time. In his later years, Newton served in Parliament and was Warden of the Mint. In 1703, he was elected president of the Royal Society of London, of which he had been a member since 1672. Two years later he was knighted by Queen Anne.

Newton is given co-credit, along with the German Wilhelm Gottfried von Leibniz, for the discovery and development of the calculus, work that Newton did in the period 1664–1666, but did not publish until years later, thus laying the groundwork for an ugly argument with Leibniz over who should get credit for the discovery. In 1687, at the urging of the astronomer Edmund Halley, Newton published his ground-breaking compilation of mathematics and science, *Principia Mathematica*, which is apparently the first place that the root-finding method that bears his name appears, although he probably had used it as early as 1669.

[4]See the book, *A History of Numerical Analysis from the 16th Through the 19th Century* (pp. 64ff), by Herman Goldstine, Springer-Verlag, New York, 1977, for a discussion of this and other historical aspects of numerical methods. The method is sometimes called Newton–Raphson, in honor of Joseph Raphson, who published the idea before Newton did.

[5]Carl Boyer, *A History of Mathematics*, John Wiley & Sons, Inc., New York, (1st edition), 1968, p. 449.

thus, we have a straight line with equation

$$y = \ell_0(x) = f(x_0) + f'(x_0)(x - x_0).$$

To find where this crosses the axis, we set $y = 0$ and solve for x:

$$x = x_0 - \frac{f(x_0)}{f'(x_0)}.$$

Call this new approximate value x_1, and note that (at least in the case of Figure 3.2) it is much closer to the root α. Now continue the process with another straight line to get

$$x_2 = x_1 - \frac{f(x_1)}{f'(x_1)};$$

or, generally,

$$x_{n+1} = x_n - \frac{f(x_n)}{f'(x_n)}. \tag{3.7}$$

This is Newton's method. It is based on a very simple idea, one that is fundamental and repeated time and again in the derivation of numerical methods (in fact, we have already seen it in the derivation of the trapezoid rule in §2.5):

> *Replace a general function by a simpler function, and do the required computation* exactly *on the simpler function.*

Thus, in this instance, we replaced the general function f with the simple function (a straight line) ℓ_0 and found the *exact* root of ℓ_0.

The second derivation of Newton's method depends on our analytical workhorse, Taylor's Theorem. Given a value $x_n \approx \alpha$, we expand f in a Taylor series about x_n:

$$f(x) = f(x_n) + (x - x_n)f'(x_n) + \frac{1}{2}(x - x_n)^2 f''(\xi_n),$$

where ξ_n is between x and x_n. To get a useful algorithm out of this, we set $f(x) = 0$ and solve for x in terms of $f(x_n), f'(x_n)$, and the remainder:

$$x = x_n - \frac{f(x_n)}{f'(x_n)} - \frac{1}{2}(x - x_n)^2 \frac{f''(\xi_n)}{f'(x_n)}, \tag{3.8}$$

and now define x_{n+1} to be the quantity that results from (3.8) when we drop the remainder term; thus,

$$x_{n+1} = x_n - \frac{f(x_n)}{f'(x_n)}$$

which defines the next point. This, of course, is Newton's method, again. This derivation makes use of another fundamental and oft-repeated idea in numerical methods and analysis:

> *Given an expression in terms of something simple plus a remainder, generate a numerical approximation by dropping the remainder term (and adjusting the notation, perhaps).*

Thus, in (3.8) we dropped the remainder term to define x_{n+1}.

The geometric derivation suggests that Newton's method ought to work very well, although the formula (3.7) does hint that there might be problems if $f'(x_n) \approx 0$ were ever true. Before proceeding with any analysis, let's look at an example.

■ **EXAMPLE 3.2**

Consider applying Newton's method to the function $f(x) = 2 - e^x$, which is the same as we used for bisection in §3.1. We choose $x_0 = 0$ and then compute as follows:

$$
\begin{aligned}
x_1 &= x_0 - \frac{2 - e^{x_0}}{-e^{x_0}} = -\frac{2-1}{-1} = 1; \\
x_2 &= x_1 - \frac{2 - e^{x_1}}{-e^{x_1}} = 1 - \frac{2-e}{-e} = 0.7357588823; \\
x_3 &= x_2 - \frac{2 - e^{x_2}}{-e^{x_2}} = 0.7357588823 - \frac{2 - e^{0.7357588823}}{-e^{0.7357588823}} = 0.6940422999.
\end{aligned}
$$

Table 3.2 gives the results of continuing the computation for several more steps. Note that the convergence here was much more rapid than for bisection—in six iterations bisection had achieved only about a digit and a half of accuracy, while Newton has over *thirteen* accurate digits.

Note, by the way, that the number of correct digits in the Newton calculation is doubling, roughly, every iteration. If this were typical behavior, it would indeed suggest that Newton's method is very fast. Note also the use of the quantity $\log_{10}(b_n - x_n)$; this is an upper bound on the base-10 logarithm of the error, and thus it gives an estimate of the number of correct digits in the approximation, which is a useful measure of the accuracy of the approximation. (This estimate is only valid if the error is less than 1.)

Table 3.2   Newton's method for $f(x) = 2 - e^x$.

n	x_n	$\alpha - x_n$	$\log_{10}(\alpha - x_n)$
0	0.000000000000	0.693147180560	-0.1592
1	1.000000000000	0.306852819440	-0.5131
2	0.735758882343	0.042611701783	-1.3705
3	0.694042299919	0.000895119359	-3.0481
4	0.693147581060	0.000000400500	-6.3974
5	0.693147180560	0.000000000000	-13.0961

Newton's method is *not* a global method, however. There are examples for which convergence will be poor or even for which convergence doesn't occur. Usually, this can be cured by obtaining a better initial guess, but sometimes we have to take x_0 *very* close to α in order to obtain convergence.

Consider Figure 3.3; this is the graph of $f(x) = \frac{4}{3}e^{2-\frac{x}{2}}(1 + x^{-1}\log x)$. Note that there is a single root in the vicinity of $x = 0.5$. But notice that the shape of the graph of f will cause problems for Newton's method unless the initial point is carefully chosen. If $x_0 \in [0.8, 1.2]$ (approximately), then the tangent line generated by Newton's method will predict that $x_1 < 0$, which is outside the domain of definition of f. If $x_0 > 1.2$ (approximately), then Newton's method will generate a sequence of values $x_{n+1} > x_n$, growing without bound as $n \to \infty$. (Note that, in a sense, Newton's method is working in this latter case, as $\lim_{x \to \infty} f(x) = 0$, i.e., there is a root at infinity. But this is not the root we want to find.)

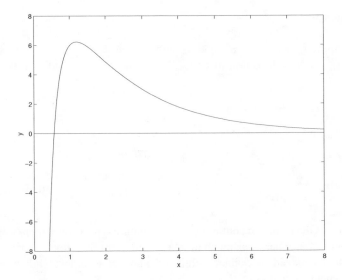

Figure 3.3   A function for which Newton's method will not work well, unless x_0 is carefully chosen.

A simpler example can be found by considering the function

$$f(x) = \frac{20x - 1}{19x},$$

which has a single root at $\alpha = 0.05$. The graph of f is given in Figure 3.4, and from this we can see that convergence will not occur except for a narrow range of values near the root, since Newton's method will predict $x_1 < 0$ for x_0 much larger than α. Table 3.3 shows the value of x_1 produced for each of a sequence of values of x_0. Note that we do not get $x_1 > 0$ until $x_0 = \frac{1}{16}$, which is very close to the actual root.

Table 3.3   Newton's method (first step, only) applied to $f(x) = \frac{20x-1}{19x}$.

x_0	x_1
1.00000000	−18.00000000
0.50000000	−4.00000000
0.25000000	−0.75000000
0.12500000	−0.06250000
0.06250000	0.04687500

It is even possible to construct a function f such that, for some initial guesses, Newton's method will cycle indefinitely. Consider $f(x) = \arctan(x)$. This has a single root at $x = 0$, of course. However, for $x_0 = 1.39174520027...$, it is not hard to show that $x_1 = -1.39174520027...$ and $x_2 = 1.39174520027... = x_0$. In other words, Newton's method will just hop back and forth between these two values. (The student might want to consider how to compute this special value of x. See Problem 9.)

So, where are we with the convergence of Newton's method? The one example that we did above showed rapid convergence, but then we quickly followed that up with two

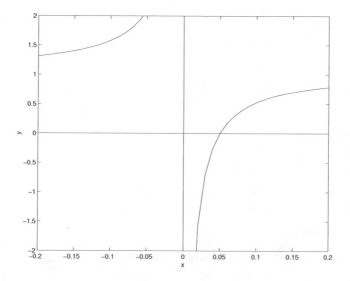

Figure 3.4   A second function for which Newton's method will not work well, unless x_0 is very close to α.

examples of what would clearly be less than marvelous performance. Under what conditions can we expect Newton's method to converge? Is there any sort of theory that we can rely on to give us confidence in this method? Note that in the absence of any such theory, we cannot use Newton's method with any expectation or confidence that the results it reports will have any meaning.

The answer is actually very simple. We will defer a detailed discussion of it until §3.6, but the conclusion can be stated fairly succinctly:

> If f, f', and f'' are all continuous near the root, and if f' does not equal zero at the root, then Newton's method will converge whenever the initial guess is sufficiently close to the root. Moreover, this convergence will be very rapid, with the number of correct digits roughly doubling each iteration.

In this context, "sufficiently close" implies that if we keep taking x_0 closer and closer to the root, we will eventually find an x_0 such that the iteration converges.

Exercises:

1. Write down the iteration for Newton's method as applied to the function $f(x) = x^2 - 2$. Simplify the computation as much as possible. What has been accomplished if we find the root of this function?

2. Write down the iteration for Newton's method as applied to the function $f(x) = x^3 - 2$. Simplify the computation as much as possible. What has been accomplished if we find the root of this function?

3. Generalize the preceeding two exercises by writing down the iteration for Newton's method as applied to $f(x) = x^n - a$.

4. Write down the iteration for Newton's method as applied to the function $f(x) = a - x^{-1}$. Simplify the resulting computation as much as possible. What has been accomplished if we find the root of this function?

5. Do three iterations of Newton's method (by hand) for each of the following functions:

   (a) $f(x) = x^6 - x - 1$, $x_0 = 1$;

   (b) $f(x) = x + \tan x$, $x_0 = 3$;

   (c) $f(x) = 1 - 2xe^{-x/2}$, $x_0 = 0$;

   (d) $f(x) = 5 - x^{-1}$, $x_0 = \frac{1}{2}$;

   (e) $f(x) = x^2 - \sin x$, $x_0 = \frac{1}{2}$;

   (f) $f(x) = x^3 - 2x - 5$, $x_0 = 2$;

   (g) $f(x) = e^x - 2$, $x_0 = 1$;

   (h) $f(x) = x^3 - 2$, $x_0 = 1$;

   (i) $f(x) = x - e^{-x}$, $x_0 = 1$;

   (j) $f(x) = 2 - x^{-1}\ln x$, $x_0 = \frac{1}{3}$.

6. Do three iterations of Newton's method for $f(x) = 3 - e^x$, using $x_0 = 1$. Repeat, using $x_0 = 2, 4, 8, 16$. Comment on your results.

7. Draw the graph of a single function f that satisfies all of the following:

   (a) f is defined and differentiable for all x;

   (b) There is a unique root $\alpha > 0$;

   (c) Newton's method will converge for any $x_0 > \alpha$;

   (d) Newton's method will diverge for all $x_0 < 0$.

   Note that the requirement here is to find a single function that satisfies all of these conditions.

8. Draw the graph of a single function f that satisfies all of the following:

   (a) f is defined and differentiable for all x;

   (b) There is a single root $\alpha \in [a, b]$, for some interval $[a, b]$;

   (c) There is a second root, $\beta < a$;

   (d) Newton's method will converge to α for all $x_0 \in [a, b]$;

   (e) Newton's method will converge to β for all $x_0 \leq \beta$.

   Note that the requirement here, again, is to find a single function that satisfies all of these conditions.

9. Write down the iteration for Newton's method for finding the root of the arctangent function. From this formulate an equation that must be satisfied by the value $x = \beta$, in order to have the Newton iteration cycle back and forth between β and $-\beta$. *Hint:* If $x_n = \beta$, and Newton's method is supposed to give us $x_{n+1} = -\beta$, what is an equation satisfied by β?

10. Compute the value of β from the previous problem. *Hint:* Use bisection to solve the equation for β.

11. Use Newton's method on the computer of your choice to compute the root $\alpha = 0$ of the arctangent function. Use the value of β from the previous problem as your x_0

and comment on your results. Repeat using $x_0 = \beta/2$ and $x_0 = \beta - \epsilon$, where ϵ is $\mathcal{O}(\mathbf{u})$.

$$\triangleleft \bullet \bullet \bullet \triangleright$$

3.3 HOW TO STOP NEWTON'S METHOD

In the bisection method we were able to compute, ahead of time, the number of iterations needed to achieve a given desired accuracy. Although we have not completed our study of Newton's method, it is the case that we will not be able to do this for Newton's method. (More correctly, the computation that might be used is too difficult to be practical; see §3.6.) So the question arises: How do we stop the iteration?

Ideally, we would want to stop when the error, $\alpha - x_n$ is sufficiently small. Of course, we cannot use this since we don't know the value of α! However, as long as f' is not zero near the root, we can relate this error to a computable quantity. The Mean Value Theorem tells us that

$$f(\alpha) - f(x_n) = f'(c_n)(\alpha - x_n),$$

where c_n is some value between α and x_n. Therefore, we can solve for $\alpha - x_n$ and write

$$
\begin{aligned}
\alpha - x_n &= \frac{f(\alpha) - f(x_n)}{f'(c_n)}, \\
&= -\frac{f(x_n)}{f'(c_n)}, \\
&= (x_{n+1} - x_n)\frac{f'(x_n)}{f'(c_n)}.
\end{aligned}
$$

Note that in the last step we used the Newton iteration to replace $-f(x_n)$ with $(x_{n+1} - x_n)f'(x_n)$. We have, then, that

$$\frac{\alpha - x_n}{x_{n+1} - x_n} = C_n \tag{3.9}$$

for $C_n = \frac{f'(x_n)}{f'(c_n)}$. Thus, the error $\alpha - x_n$ is a simple multiple of the *computable*[6] quantity $x_{n+1} - x_n$. Morever, if we assume that convergence is occurring, and that $f'(\alpha) \neq 0$, then the constant of proportionality satisfies

$$\lim_{n \to \infty} |C_n| = 1. \tag{3.10}$$

(The student should consider why this is true.) It follows, then, that

$$\lim_{n \to \infty} \frac{|\alpha - x_n|}{|x_{n+1} - x_n|} = 1.$$

As a consequence, we can use the computable quantity $|x_{n+1} - x_n|$ to measure convergence. Usually, one stops the iteration when a modest multiple of this is small; for example,

$$5|x_{n+1} - x_n| \leq \epsilon$$

---

[6]This is another important rule in numerical methods and analysis: Use computable *estimates* of quantities that cannot be easily computed.

where $\epsilon > 0$ is a user-defined tolerance.[7]

A word of warning is in order, however. The results in this section are based on some very casual analysis, and although it can all be made precise and rigorous, the fact remains that it is possible to have $|x_{n+1} - x_n|$ small and yet have x_{n+1} not very close to α. This can happen, for example, if $f'(x_n)$ is very large compared to $f(x_n)$. For this reason, it is commonplace to add a term to the error check to make sure that the function value itself is small; that is, we stop when

$$|f(x_n)| + |x_n - x_{n-1}| \leq \epsilon/5.$$

Exercises:

1. Under the assumption that $f'(\alpha) \neq 0$ and $x_n \to \alpha$, prove (3.10); be sure to provide all the details. *Hint:* Expand f and f' in Taylor series about $x = \alpha$.

2. We could also stop the iteration when $|f(x_n)|$ was sufficiently small. Use the Mean Value Theorem plus the fact that $f(\alpha) = 0$ to show that, if f' is continuous and non-zero near α, then there are constants c_1 and c_2 such that

$$c_1|f(x_n)| \leq |\alpha - x_n| \leq c_2|f(x_n)|.$$

Comment on this result.

3. Write a computer program that uses Newton's method to find the root of a given function, and apply this program to find the root of the following functions, using x_0 as given. Stop the iteration when the error as estimated by $|x_{n+1} - x_n|$ is less than 10^{-6}. Compare to your results for bisection.

   (a) $f(x) = 1 - 2xe^{-x/2}$, $x_0 = 0$;
   (b) $f(x) = 5 - x^{-1}$, $x_0 = \frac{1}{4}$;
   (c) $f(x) = x^3 - 2x - 5$, $x_0 = 2$;
   (d) $f(x) = e^x - 2$, $x_0 = 1$;
   (e) $f(x) = x - e^{-x}$, $x_0 = 1$;
   (f) $f(x) = x^6 - x - 1$, $x_0 = 1$;
   (g) $f(x) = x^2 - \sin x$, $x_0 = \frac{1}{2}$;
   (h) $f(x) = x^3 - 2$, $x_0 = 1$;
   (i) $f(x) = x + \tan x$, $x_0 = 3$;
   (j) $f(x) = 2 - x^{-1} \ln x$, $x_0 = \frac{1}{3}$.

4. Figure 3.5 shows the geometry of a planetary orbit[8] around the sun. The position of the sun is given by S, the position of the planet is given by P. Let x denote the angle

---

[7]I would like to acknowledge some personal comments from Prof. Nick Higham of the School of Mathematics at the University of Manchester in the United Kingdom, who pointed out a minor error in the first edition of this section, and also suggested how to repair my analysis. In this connection (as well as many others), I would like to recommend Professor Higham's book, *Accuracy and Stability of Numerical Algorithms*, SIAM Publishing, Philadelphia, 2nd Edition, 2002

[8]For the background material for this and the next problem, the author is indebted to the interesting calculus text by Alexander J. Hahn, *Basic Calculus: From Archimedes to Newton to its Role in Science*, published in 1998 by Springer-Verlag, New York.

defined by P_0OA, measured in radians. The dotted line is a circle concentric to the ellipse and having a radius equal to the major axis of the ellipse. Let T be the total period of the planet, and let t be the time required for the planet to go from A to P. Then Kepler's equation from orbital mechanics, relating x and t, is

$$x - \epsilon \sin x = \frac{2\pi t}{T}.$$

Here ϵ is the eccentricity of the elliptical orbit (the extent to which it deviates from a circle). For an orbit of eccentricity $\epsilon = 0.01$ (roughly equivalent to that of the Earth), what is the value of x corresponding to $t = T/4$? What is the value of x corresponding to $t = T/8$? Use Newton's method to solve the required equation.

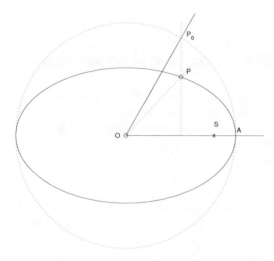

Figure 3.5   Orbital geometry for Problems 4 and 5

5. Consider now a highly eccentric orbit, such as that of a comet, for which $\epsilon = 0.9$ might be appropriate. What is the value of x corresponding to $t = T/4$? What is the value of x corresponding to $t = T/8$?

6. Consider the problem of putting water pipes far enough underground to avoid frozen pipes should the external temperature suddenly drop. Let T_0 be the initial temperature of the ground, and assume that the external air temperature suddenly drops to a new value $T < T_0$. Then a simple model of how the underground temperature responds to the change in external temperature tells us that

$$\frac{u(x,t) - T}{T_0 - T} = \text{erf}\left(\frac{x}{2\sqrt{at}}\right).$$

Here $u(x,t)$ is the temperature at a depth of x feet t seconds after the temperature change, and a is the thermal conductivity of the soil. Suppose that $a = 1.25 \times 10^{-6}\,\text{ft}^2/\text{sec}$. How deep must the pipe be buried to guarantee that the temperature does not reach 0° C for 30 days after a temperature shift of 40° C? If your computing environment does not have an intrinsic error function, use the approximation presented in Problem 8 of §3.1.

7. In the previous problem, produce a plot of temperature u at depths of 6 and 12 feet, as a function of time (days), in response to a temperature shift of $40°$ C for the following initial temperatures:

   (a) $T_0 = 40°$ C;

   (b) $T_0 = 50°$ C;

   (c) $T_0 = 60°$ C.

8. Use Newton's method to solve Problem 6 from §3.1.

9. Repeat the above, assuming that the mortgage is only 15 years in length.

3.4 APPLICATION: DIVISION USING NEWTON'S METHOD

In this section we will investigate an old algorithm that was actually used in the early days of computing to approximate the division operation.[9] The purpose is to illustrate the use of Newton's method and the analysis of the resulting iteration, in particular to demonstrate how we can predict the performance of the iteration for a wide range of parameter values.

Consider the function

$$f(x) = a - \frac{1}{x}.$$

Clearly, $f(\alpha) = 0 \Rightarrow \alpha = a^{-1}$, so we can "divide" by a by finding the root of f and then multiplying. If we use Newton's method to find the root, we get the iteration

$$x_{n+1} = x_n - \left(\frac{a - x_n^{-1}}{x_n^{-2}} \right) = x_n(2 - ax_n).$$

Thus, we can approximate the reciprocal of a without having to do division.

The questions of most interest to us are: When does this iteration converge, and how fast? What initial guesses x_0 will work for us?

We begin by using the information about the way the computer stores numbers. We have

$$a = b \times 2^{t-p}$$

for $b \in [\frac{1}{2}, 1]$, so it suffices to be able to compute the reciprocal for numbers between $\frac{1}{2}$ and 1. We assume, then, without loss of generality, that $a \in [\frac{1}{2}, 1]$.

To analyze the behavior of the iteration, we use the *residual*, defined by

$$r_n = 1 - ax_n.$$

We then have that the error satisfies $\alpha - x_n = e_n = r_n/a = 1/a - x_n$ and, moreover,

$$r_{n+1} = 1 - ax_{n+1} = 1 - a[x_n(2 - ax_n)] = 1 - a[x_n(1 + r_n)] = 1 - (1 - r_n)(1 + r_n) = r_n^2,$$

from which it follows that

$$r_n = r_0^{2^n}.$$

---

[9] Much of this presentation is taken from Kendall Atkinson's textbook, *An Introduction to Numerical Analysis*, John Wiley & Sons, Inc. New York, 1989 (2nd edition).

If $|r_0| < 1$, then we will get $r_n \to 0$, and very rapidly. In fact, we have that the absolute error satisfies

$$|e_{n+1}| = |ae_n^2| = ae_n^2$$

and the relative error satisfies (recall that $\alpha = 1/a$)

$$\left| \frac{e_{n+1}}{\alpha} \right| = \left(\frac{e_n}{\alpha} \right)^2$$

so that we have

$$\left| \frac{e_n}{\alpha} \right| = \left(\frac{e_0}{\alpha} \right)^{2^n} = (r_0)^{2^n}.$$

Since the error e_n goes to zero if and only if the residual r_n does, we have convergence so long as $|r_0| < 1$, which is equivalent to saying that

$$0 < x_0 < \frac{2}{a} = 2\alpha.$$

How do we get an x_0 that satisifes this?

Recall linear interpolation from Chapter 2. We can construct the linear interpolate to the function $y = x^{-1}$ and use this to *estimate* a^{-1}, and then use this value as x_0. The linear interpolate using the endpoints of $[\frac{1}{2}, 1]$ as the nodes is easily found to be

$$p_1(x) = 3 - 2x,$$

so that

$$x_0 = p_1(a) = 3 - 2a.$$

The linear interpolation error formula (2.13) shows us that

$$\left| \frac{1}{a} - p_1(a) \right| \leq \frac{1}{2}; \qquad (3.11)$$

hence, $|x_0 - \alpha| \leq \frac{1}{2}$, which is equivalent to saying that $|r_0| = a|e_0| = a/2 \leq \frac{1}{2}$. Thus, we will get convergence using this for our initial value. In terms of the relative error, we have

$$\left| \frac{e_0}{\alpha} \right| \leq \frac{\frac{1}{2}}{\alpha} \leq \frac{1}{2}.$$

A mere six iterations (about 21 total operations) yield relative accuracy on the order of 10^{-20}.

The estimate (3.11) is actually very conservative, because of the way that the bound on the second derivative of $1/x$ was treated. Because of the simplicity of the functions involved, we can actually apply some ordinary calculus to determine that the initial error is *much* smaller than predicted above.

Define the error in the initial guess as

$$g(x) = \frac{1}{x} - (3 - 2x);$$

we want to find the maximum of this (in absolute value) over the interval $[\frac{1}{2}, 1]$. Taking the derivative gives us

$$g'(x) = \frac{-1}{x^2} + 2,$$

which implies that critical points exist at $x = \pm \frac{1}{2}\sqrt{2}$. The critical point at $x = -\frac{1}{2}\sqrt{2}$ is outside our interval of interest, so we ignore it and compute

$$g(\sqrt{2}/2) = \frac{2}{\sqrt{2}} - 3 + \sqrt{2} = 2\sqrt{2} - 3 = -0.1716... .$$

In addition, we have that $g(1/2) = g(1) = 0$. Therefore, the *smallest* value of $g(x)$ over $[\frac{1}{2}, 1]$ is (to four digits) -0.1716. But it therefore follows that the *largest* value of $|g(x)|$ over the same interval is $0.1716 < 0.172$. Using this as the upper bound on e_0 shows that we need only five iterations to get relative accuracy on the order of 10^{-20}.

■ **EXAMPLE 3.3**

Let's try this out on $a = 0.8$, so that we are trying to approximate $(0.8)^{-1} = 1.25$. The initial guess is found from the linear interpolate as

$$x_0 = 3 - 2 \times 0.8 = 1.4.$$

Newton's method then gives us

$$
\begin{aligned}
x_1 &= x_0(2 - (0.8)x_0) = (1.4)(2 - (0.8)(1.4)) = 1.232; \\
x_2 &= x_1(2 - (0.8)x_1) = (1.232)(2 - (0.8)(1.232)) = 1.2497408; \\
x_3 &= x_2(2 - (0.8)x_2) = (1.2497408)(2 - (0.8)(1.2497408)) = 1.249999946; \\
x_4 &= x_3(2 - (0.8)x_3) = (1.249999946)(2 - (0.8)(1.249999946)) = 1.25.
\end{aligned}
$$

Thus, we converge in only four iterations (a total of about 14 operations).

■ **EXAMPLE 3.4**

Suppose that we have a *quadratic* approximation to $1/t$ that we use to generate x_0. For example, consider the quadratic

$$p_2(t) = \frac{1}{3}\left(8t^2 - 18t + 13\right).$$

Note that $p_2(\frac{1}{2}) = 2$, $p_2(3/4) = 4/3$, and $p_2(1) = 1$; therefore we see that p_2 interpolates to $1/t$ for $t = \frac{1}{2}$, $t = \frac{3}{4}$, and $t = 1$. (We will see in Chapter 4 how to construct p_2.) Since we don't have an error theory (yet) for this type of interpolation, how can we estimate the error between $p_2(a)$ and $1/a$?

Again, we can use ordinary calculus. Define the error as

$$E(t) = t^{-1} - p_2(t),$$

so that

$$E'(t) = -t^{-2} - \frac{1}{3}\left(16t - 18\right).$$

Then $E'(t) = 0$ if and only if

$$0 = -t^{-2} - \frac{1}{3}\left(16t - 18\right)$$

or

$$16t^3 - 18t^2 + 3 = 0.$$

This doesn't factor, but we can use a computer algebra package to plot it and see that it has roots near $t = 1$, $t = 1/2$, and $t = -\frac{1}{2}$. We can then use Newton's method or bisection to find those roots, getting (to three places)

$$t_1 = 0.886, \quad t_2 = 0.595, \quad t_3 = -0.356$$

as our three critical points. Since t_3 is not in the interval $[1/2, 1]$, we are not interested in it anymore. We then evaluate $E(t)$ at the remaining critical points plus the endpoints, getting

$$E(t_1) = 0.018012294, \quad E(t_2) = -0.026728214, \quad E(1/2) = 0, \quad E(1) = 0.$$

Therefore (note the use of the Extreme Value Theorem here),

$$|E(t)| \leq 0.026728214$$

is the upper bound on the error for $t \in [1/2, 1]$. Hence,

$$e_0 = |1/a - p_2(a)| \leq 0.026728214.$$

This is substantially smaller than we got with the linear polynomial. We get around 10^{-25} accuracy in only four iterations, and the total number of operations is now 16.

The analysis we have done here shows us how to reliably choose x_0 to ensure convergence for all values of a that are of interest, for this particular application. However, for most cases we will need more powerful tools, which we will derive in the next section.

Exercises:

1. Test the method derived in this section by using it to approximate $1/0.75 = 1.333...$, with x_0 as suggested in the text. Don't write a computer program, just use a hand calculator.

2. Repeat the above for $a = 2/3$.

3. Repeat the above for $a = 0.6$.

4. Based on the material in this section, write each of the following numbers in the form

$$a = b \times 2^k,$$

where $b \in [\frac{1}{2}, 1]$, and then use Newton's method to find the reciprocal of b, and hence of a.

   (a) $a = 7$;

   (b) $a = \pi$;

   (c) $a = 6$;

   (d) $a = 5$;

   (e) $a = 3$.

Be sure to use the initial value as generated in this section and only do as many iterations as are necessary for 10^{-16} accuracy. Compare your values to the intrinsic reciprocal function on your computer or calculator. This can be done on a computer or on a hand calculator.

5. Test the method derived in this section by writing a computer program to implement it. Test the program by having it compute $1/0.75 = 1.333...$, with x_0 as suggested in the text, as well as $1/(2/3) = 1.5$.

6. How close an initial guess is needed for the method to converge to within a relative error of 10^{-14} in only *three* iterations?

7. Consider the quadratic polynomial

$$q_2(x) = 4.328427157 - 6.058874583x + 2.745166048x^2.$$

What is the error when we use this to generate the initial guess? How many steps are required to get 10^{-20} accuracy? How many operations?

8. Repeat Problem 7, this time with the polynomial

$$q_3(x) = 5.742640783 - 12.11948252x + 11.14228488x^2 - 3.767968453x^3.$$

9. Modify your program from Problem 5 to use each of p_2, q_2, and q_3 to compute $1/a$ for $a = 0.75$ and $a = 0.6$. Compare what you get to the values produced by ordinary division on your computer. (Remember to use Horner's rule to evaluate the polynomials!)

10. Modify the numerical method to directly compute the ratio b/a, rather than just the reciprocal of a. Is the error analysis affected? (*Note:* Your algorithm should approximate the ratio b/a without using any divisions at all.)

11. Test your modification by using it to compute the value of the ratio $4/5 = 0.8$.

3.5 THE NEWTON ERROR FORMULA

The derivation of Newton's method from Taylor's Theorem strongly suggested, but did not prove, that the error at each step of the iteration is related to the square of the previous error. We can, in fact, make that a precise statement, and it becomes a key tool in deriving a precise theory of convergence for Newton's method.

Theorem 3.2 (Newton Error Formula) *Let $f \in C^2(I)$ be given, for some interval $I \subset \mathbb{R}$, with $f(\alpha) = 0$ for some $\alpha \in I$. For a given $x_n \in I$, define*

$$x_{n+1} = x_n - \frac{f(x_n)}{f'(x_n)}.$$

Then there exists a point ξ_n between α and x_n such that

$$(\alpha - x_{n+1}) = -\frac{1}{2}(\alpha - x_n)^2 \frac{f''(\xi_n)}{f'(x_n)}. \tag{3.12}$$

Proof: We begin by expanding f in a Taylor series about $x = x_n$:

$$f(x) = f(x_n) + (x - x_n)f'(x_n) + \frac{1}{2}(x - x_n)^2 f''(\xi_n);$$

here ξ_n is between x and α. Now set $x = \alpha$ to get

$$0 = f(x_n) + (\alpha - x_n)f'(x_n) + \frac{1}{2}(\alpha - x_n)^2 f''(\xi_n).$$

Divide both sides by $f'(x_n)$ and re-arrange, which yields

$$(x_n - \alpha) - \frac{f(x_n)}{f'(x_n)} = \frac{1}{2}(\alpha - x_n)^2 \frac{f''(\xi_n)}{f'(x_n)}.$$

The left side simplifies to

$$(x_n - \alpha) - \frac{f(x_n)}{f'(x_n)} = \left(x_n - \frac{f(x_n)}{f'(x_n)} \right) - \alpha = x_{n+1} - \alpha,$$

from which (3.12) follows by a simple substitution, and we are done. •

Note that this result shows that the error at one step goes like the *square* of the error at the previous step. Thus, once the error becomes small enough, it begins to decrease rapidly. In fact, if we assume that convergence is occurring (with $f'(\alpha) \neq 0$), so that

$$\lim_{n \to \infty} x_n = \alpha,$$

then we can say that $f'(x_n) \approx f'(\alpha)$ and $f''(\xi_n) \approx f''(\alpha)$, so that

$$(\alpha - x_{n+1}) \approx C(\alpha - x_n)^2, \tag{3.13}$$

where

$$C = -\frac{1}{2}\frac{f''(\alpha)}{f'(\alpha)}.$$

The approximate equality (3.13) suggests, to some extent, how close x_0 has to be to α for convergence to occur. We will address this more precisely in §3.6.

In addition, if convergence is occurring, then we can easily show that

$$\lim_{n \to \infty} \frac{\alpha - x_{n+1}}{(\alpha - x_n)^2} = -\frac{f''(\alpha)}{2f'(\alpha)}. \tag{3.14}$$

This follows from the squeeze principle of calculus (together with the continuity of f' and f''), which shows that the convergence of the iteration ($x_n \to \alpha$) forces $\xi_n \to \alpha$.

Next, we will explain why it is often said that a converging Newton iteration will double the number of accurate decimal digits at each step.

■ **EXAMPLE 3.5**

Let x_n be a sequence of values generated by applying Newton's method to a smooth function f, and assume that the sequence is converging to the root α. Then it follows from (3.12) that

$$|\alpha - x_{n+1}| = \frac{1}{2}(\alpha - x_n)^2 \frac{|f''(\xi_n)|}{|f'(x_n)|},$$

so that, taking common logarithms of both sides, we have

$$
\begin{aligned}
\log_{10} |\alpha - x_{n+1}| &= 2 \log_{10} |\alpha - x_n| + \log_{10} \left(\frac{|f''(\xi_n)|}{2|f'(x_n)|} \right) \\
&= 2 \log_{10} |\alpha - x_n| + b_n.
\end{aligned}
\tag{3.15}
$$

As discussed previously, the quantity

$$
d_n = \log_{10} |\alpha - x_n|
$$

can be interpreted as the number of correct decimal digits in the approximation (as long as the error is less than 1), and (3.15) shows that (ignoring the additive b_n term) this doubles each iteration.

We close this section with a definition that allows us to characterize how rapidly a sequence of values is converging to a limit. Bear in mind that, although we are indeed applying this to the root-finding problem, the definition can be used for any sequence of numbers that are approaching a limit.

Definition 3.1 (Order of Convergence for Sequences) *Let x_n be a sequence of values converging to α, such that*

$$
\lim_{n \to \infty} \frac{\alpha - x_{n+1}}{(\alpha - x_n)^p} = C
\tag{3.16}
$$

for some non-zero (finite) constant C, and some p. Then p is called the order of convergence *for the sequence.*

The requirement that C be non-zero and finite actually forces p to be a single, unique value. If p is too large, then the denominator shrinks too fast and the ratio grows without bound to $\pm\infty$; if p is too small, then the numerator vanishes too quickly and the ratio shrinks to zero as $n \to \infty$. One special case that is of interest to us is when $p = 2$, as in Newton's method; this we call *quadratic convergence*. The case $p = 1$ is called *linear convergence*, and requires that $|C| < 1$ (see Problem 10).

Another interesting special case is when the sequence satisfies

$$
\lim_{n \to \infty} \frac{\alpha - x_{n+1}}{\alpha - x_n} = 0
\tag{3.17}
$$

but

$$
\lim_{n \to \infty} \frac{\alpha - x_{n+1}}{(\alpha - x_n)^2} = \pm\infty;
\tag{3.18}
$$

i.e., the sequence does not converge quadratically. In this case, the sequence is called *superlinearly convergent*, because (3.17) suggests that it converges faster than linear, but (3.18) means that it is not quadratic.

Generally speaking, having p larger means that convergence is more rapid, since (3.16) implies

$$
\alpha - x_{n+1} \approx C(\alpha - x_n)^p.
\tag{3.19}
$$

The Newton error estimate (3.12) strongly suggests that Newton's method is quadratically convergent, something we will prove to be the case in §3.6. We can illustrate its use in analyzing a particular example here; a more thorough discussion (for another example) may be found in §3.7.

■ **EXAMPLE 3.6**

Consider the problem of finding the root of $f(x) = e^{-ax} - x$. We want to be able to compute this root for a in the range $0 < a \leq 1$, and right now our only tools are bisection or Newton's method. If we apply the Newton error formula to this problem, we get

$$\alpha - x_{n+1} = -\frac{1}{2}(\alpha - x_n)^2 \frac{a^2 e^{-a\xi_n}}{-ae^{-ax_n} - 1} = \frac{1}{2}(\alpha - x_n)^2 \frac{a^2 e^{-a\xi_n}}{ae^{-ax_n} + 1},$$

where ξ_n is between x_n and α. Note that this implies (since the right side is positive) that $\alpha - x_{n+1} \geq 0$; thus, for $n \geq 1$, we have $x_n \leq \xi_n \leq \alpha$. Moreover, for any $x_0 > 0$, the Newton iteration itself always stays positive: $x_n > 0$. This enables us to write

$$\left| \frac{a^2 e^{-a\xi_n}}{ae^{-ax_n} + 1} \right| \leq a \left| \frac{ae^{-ax_n}}{ae^{-ax_n} + 1} \right| \leq a \leq 1.$$

Thus,

$$|\alpha - x_{n+1}| \leq \frac{1}{2}(\alpha - x_n)^2.$$

If the initial error is less than 1, then we can quickly show (just by going through the recursion) that we have

$$|\alpha - x_5| \leq 5 \times 10^{-10}.$$

Hence, five iterations are enough to compute our root, no matter what the value of a (as long as $a \in (0, 1]$). How do we find an initial guess such that $|\alpha - x_0| \leq 1$? We note that $f(0) = 1$ and $f(2) < 0$ for any $a > 0$; hence, the root must lie in the interval $[0, 2]$. Therefore, $x_0 = 1$ will always satisfy $|\alpha - x_0| \leq 1$.

Note that we have performed this analysis without doing any numerical computations, and that it applies for all $a \in (0, 1]$. Even though we don't know the value of the root, nor the value of the one parameter in the problem, we are able to determine an initial guess such that the iteration will converge in five steps. We won't always be able to do this complete an analysis of a problem, but this example does illustrate the issues that are involved, and some of the tools that are required.

Exercises:

1. If f is such that $|f''(x)| \leq 3$ for all x and $|f'(x)| \geq 1$ for all x, and if the initial error in Newton's method is less than $\frac{1}{2}$, what is an upper bound on the error at each of the first three steps? *Hint:* Use the Newton error formula from this section.

2. If f is now such that $|f''(x)| \leq 4$ for all x but $|f'(x)| \geq 2$ for all x, and if the initial error in Newton's method is less than $\frac{1}{3}$, what is an upper bound on the error at each of the first three steps?

3. Consider the left-hand column of data in Table 3.4. Supposedly, this comes from applying Newton's method to a smooth function whose derivative does not vanish at the root. Use the limit result (3.14) to determine whether or not the program is working. *Hint:* Use $\alpha \approx x_7$.

4. Repeat the previous exercise, using the right-hand column of data in Table 3.4.

Table 3.4   Data for Problems 3 and 4.

n	x_n (Prob. 3)	x_n (Prob. 4)
0	10.0	0.000000000000
1	5.25	0.626665778573
2	3.1011904761905	0.889216318667
3	2.3567372726442	0.970768039664
4	2.2391572227372	0.992585155212
5	2.2360701085329	0.998139415613
6	2.2360679775008	0.999534421170
7	2.2360679774998	0.999883578197
8	N/A	0.999970892852
9	N/A	0.999992723105
10	N/A	0.999998180783

5. Apply Newton's method to find the root of $f(x) = 2 - e^x$, for which the exact solution is $\alpha = \ln 2$. Perform the following experiments:

   (a) Compute the ratio

   $$R_n = \frac{\alpha - x_{n+1}}{(\alpha - x_n)^2}$$

   and observe whether or not it converges to the correct value as $n \to \infty$. (Hint: See (3.14).)

   (b) Compute the modified ratio

   $$R_n(p) = \frac{\alpha - x_{n+1}}{(\alpha - x_n)^p}$$

   for various $p \neq 2$, but near 2. What happens? Comment, in light of the definition of order of convergence.

6. Consider applying Newton's method to find the root of the function $f(x) = 4x - \cos x$. Assume that we want accuracy to within 10^{-8}. Use the Newton error estimate (3.12) to show that the iteration will converge for all $x_0 \in [-2, 2]$. How many iterations will be needed for the iteration to converge? Compare with the corresponding results for bisection, using an initial interval of $[-2, 2]$.

7. Verify by actual computation that your results in the previous problem were correct; i.e., apply Newton's method to $f(x) = 4x - \cos x$. Do you converge to the specified accuracy in the correct number of iterations, and does this convergence appear to be occurring for all choices of x_0?

8. Consider now the function $f(x) = 7x - \cos(2\pi x)$. Show that a root exists on the interval $[0, 1]$, and then use the Newton error estimate to determine how close x_0 has to be to the root to guarantee convergence.

9. Investigate your results in the previous problem by applying Newton's method to $f(x) = 7x - \cos(2\pi x)$, using several choices of x_0 within the interval $[0, 1]$. Comment on your results in light of the theory of the method. *Note:* This can be done by a very modest modification of your existing Newton program.

10. Show that if x_n is a sequence converging linearly to a value α, then the constant C in (3.16) must satisfy $|C| < 1$. *Hint:* Assume that $|C| > 1$ and prove a contradiction to the convergence assumption.

11. Explain, in your own words, why the assumptions $f'(x_n) \approx f'(\alpha)$ and $f''(\xi_n) \approx f''(\alpha)$ are valid if we have $x_n \to \alpha$.

$$\lhd \bullet \bullet \bullet \rhd$$

3.6 NEWTON'S METHOD: THEORY AND CONVERGENCE

The Newton error estimate (3.12) gives us a very precise idea of how the error changes from iteration to iteration, but it is not specific enough by itself to imply convergence or establish any kind of accuracy in practice, save for specific examples like those in the exercises for §3.5. To get a general theory that is widely applicable to more general problems, we need to do a little more work, and that is the purpose of this section.

First we prove a theorem that makes very stringent assumptions on the behavior of the function f. We then show that those assumptions are not entirely necessary in order to prove *local convergence*, that is, convergence under the assumption of a sufficiently close initial guess.

Theorem 3.3 *Assume that f is defined and twice continuously differentiable for all x, with $f(\alpha) = 0$ for some α. Define the ratio[10]*

$$M = \frac{\max_{x \in \mathbb{R}} |f''(x)|}{2 \min_{x \in \mathbb{R}} |f'(x)|} \tag{3.20}$$

and assume that $M < \infty$. Then, for any x_0 such that

$$M|\alpha - x_0| < 1, \tag{3.21}$$

the Newton iteration converges. Moreover,

$$|\alpha - x_n| \leq M^{-1} \left(M|\alpha - x_0| \right)^{2^n}.$$

Proof: The proof is based almost completely on the Newton error formula (3.12). We have

$$\begin{aligned} |\alpha - x_{n+1}| &= \frac{1}{2}(\alpha - x_n)^2 \frac{|f''(\xi_n)|}{|f'(x_n)|}, \\ &\leq (\alpha - x_n)^2 M. \end{aligned}$$

To simplify the notation, let $e_n = |\alpha - x_n|$. Then we have

$$e_{n+1} \leq e_n^2 M,$$

so that

$$\begin{aligned} e_1 &\leq e_0^2 M \\ e_2 &\leq e_1^2 M \leq (e_0^2 M)^2 M = e_0^4 M^3 = M^{-1}(Me_0)^4 \\ e_3 &\leq e_2^2 M \leq (e_1^2 M)^2 M = e_0^8 M^7 = M^{-1}(Me_0)^8 \end{aligned}$$

---

[10]Recall that $\mathbb{R} = (-\infty, \infty)$.

and an inductive argument easily shows that

$$e_n \leq M^{-1} \left(M e_0 \right)^{2^n}. \tag{3.22}$$

It is clear that the right side goes to zero as $n \to \infty$ if and only if $M e_0 = M|\alpha - x_0| < 1$, which completes the proof. •

A couple of observations might be in order here. First, note that the convergence, when it occurs, is *very* rapid, since the exponent in (3.22) grows exponentially. For example, if $M e_0 = 0.5$, we have

$$e_5 \leq M^{-1} (0.5)^{32} = M^{-1} \times 2.328... \times 10^{-10}.$$

Second, it is not very practical to use the condition (3.21) to predict *a priori* what value of x_0 to use. The point of the theorem is not to help us choose x_0, but to explain to us how it is that the iteration converges, to show us what issues are important in obtaining convergence, and to assure us that if we can find a value of x_0 that is close enough to the root, then we will get convergence (and rapid convergence).

But recall that this theorem assumed that the function f was defined and twice continuously differentiable everywhere, in addition to the assumption of the bound (3.20). These are very restrictive assumptions and we would like to remove them, if possible. The key issue is to understand that assuming continuity of f and its first two derivatives *near* the root is actually enough to achieve essentially the same result as in Theorem 3.3. Establishing this, precisely, is a somewhat involved task, however.

Theorem 3.4 *Let $f \in C^2(I)$, where $\alpha \in I \subset \mathbb{R}$ is a root and I is an open interval. Assume that $f'(\alpha) \neq 0$, and let the values x_n be defined by applying Newton's method to f. Then, for x_0 sufficiently close to α, we have that*

$$\lim_{n \to \infty} x_n = \alpha \tag{3.23}$$

and

$$\lim_{n \to \infty} \frac{\alpha - x_{n+1}}{(\alpha - x_n)^2} = -\frac{f''(\alpha)}{2 f'(\alpha)}. \tag{3.24}$$

Proof: Since $f'(\alpha) \neq 0$ and f' is continuous, we can find a closed interval around α (perhaps a very small interval) such that $f'(x) \neq 0$ throughout this interval. Call this interval J, and note that we can assume, without loss of generality, that $J \subset I$ and $J = \{x \mid \alpha - \epsilon \leq x \leq \alpha + \epsilon\}$, where $\epsilon > 0$ is small enough for J to be contained in I and to satisfy $f'(x) \neq 0$ in J.

Now, since J is closed, and in addition f' does not vanish on J and f'' is continuous on J, the ratio

$$M = \frac{\max_{x \in J} |f''(x)|}{2 \min_{x \in J} |f'(x)|}$$

is bounded. (This follows from the Extreme Value Theorem.) Take $x_0 \in J$ and use Newton's method to define x_1:

$$x_1 = x_0 - \frac{f(x_0)}{f'(x_0)}.$$

We need to take x_0 sufficiently close to α that $x_1 \in J$. The Newton error formula (3.12) tells us that

$$\alpha - x_1 = -\frac{1}{2}(\alpha - x_0)^2 \frac{f''(\xi_0)}{f'(x_0)},$$

where ξ_0 is between α and x_0, and therefore $\xi_0 \in J$. Thus,

$$|\alpha - x_1| = \frac{1}{2}(\alpha - x_0)^2 \frac{|f''(\xi_0)|}{|f'(x_0)|} \leq (\alpha - x_0)^2 \frac{\max_{x \in J}|f''(x)|}{2\min_{x \in J}|f'(x)|} = (\alpha - x_0)^2 M.$$

If we choose x_0 so that $M|\alpha - x_0| < 1$, then we have

$$|\alpha - x_1| \leq (\alpha - x_0)^2 M = |\alpha - x_0|\,(|\alpha - x_0|M) < |\alpha - x_0|,$$

which forces $x_1 \in J$ (since it is closer to α than x_0 is). The same argument can now be used to show that x_2 (defined from x_1 by Newton's method) is also in J and, recursively, the entire sequence of Newton iterates is in J. We can now essentially follow the same argument as used in Theorem 3.3.[11] We have

$$
\begin{aligned}
|\alpha - x_{n+1}| &= \frac{1}{2}(\alpha - x_n)^2 \frac{|f''(\xi_n)|}{|f'(x_n)|} \\
&\leq (\alpha - x_n)^2 M
\end{aligned}
$$

where the fact that all the iterates are in J is crucial because it allows us to bound the derivative ratio on the right side. From this we quickly get that, for $e_n = |\alpha - x_n|$, again,

$$e_n \leq M^{-1}\left(Me_0\right)^{2^n}.$$

so that for $Me_0 = M|\alpha - x_0| < 1$ (which defines "sufficiently close to α") convergence occurs and (3.23) is proven. To get (3.24), we have

$$\alpha - x_{n+1} = -\frac{1}{2}(\alpha - x_n)^2 \frac{f''(\xi_n)}{f'(x_n)}$$

from which we get

$$\frac{\alpha - x_{n+1}}{(\alpha - x_n)^2} = -\frac{1}{2}\frac{f''(\xi_n)}{f'(x_n)}.$$

Now take the limit as $n \to \infty$ of both sides. Since $x_n \to \alpha$ and ξ_n is between α and x_n, it follows (from the "squeeze theorem" of calculus) that $\xi_n \to \alpha$ as well. Continuity then shows that

$$
\begin{aligned}
\lim_{n\to\infty} \frac{\alpha - x_{n+1}}{(\alpha - x_n)^2} &= -\lim_{n\to\infty}\frac{1}{2}\frac{f''(\xi_n)}{f'(x_n)}, \\
&= -\frac{f''(\lim_{n\to\infty}\xi_n)}{2f'(\lim_{n\to\infty}x_n)}, \\
&= -\frac{f''(\alpha)}{2f'(\alpha)},
\end{aligned}
$$

---

[11] Note that all of the hard work with the technical issues of continuity and intervals was so that we could get to this point and use an already verified argument, i.e., the argument from Theorem 3.3.

which completes the proof. •

Exercises:

1. Consider a function f that satisfies the properties:

   (a) There exists a unique root $\alpha \in [0, 1]$;

   (b) For all real x we have $f'(x) \geq 2$ and $0 \leq f''(x) \leq 3$.

   Show that for $x_0 = \frac{1}{2}$, Newton's method will converge to within 10^{-6} of the actual root in four iterations. How long would bisection take to achieve this accuracy?

2. Consider a function f that satisfies the properties:

   (a) There exists a unique root $\alpha \in [2, 3]$;

   (b) For all real x we have $f'(x) \geq 3$ and $0 \leq f''(x) \leq 5$.

   Using $x_0 = 5/2$, will Newton's method converge, and if so, how many iterations are required to get 10^{-4} accuracy?

3. Repeat the above, this time aiming for 10^{-20} accuracy.

4. Consider a function f that satisfies the following properties:

   (a) There exists a unique root $\alpha \in [-1, 3]$;

   (b) For all real x we have $f'(x) \geq 4$ and $-6 \leq f''(x) \leq 3$.

   Using $x_0 = 1$, will Newton's method converge, and if so, how many iterations are required to get 10^{-4} accuracy?

5. Repeat the above, this time aiming for 10^{-20} accuracy.

6. Consider a function that satisfies the following properties:

   (a) f is defined and twice continuously differentiable for all x;

   (b) f has a unique root $\alpha \in [-1, 1]$;

   (c) $|f'(x)| \geq 2$ for all x;

   (d) $|f''(x)| \leq 5$ for all x.

   Can we conclude that Newton's method will converge for all $x_0 \in [-1, 1]$? If so, how many iterations are required to get 10^{-6} accuracy? If not, how many bisection steps must we complete to get the initial interval small enough so that Newton's method will converge? (For *any* choice of x_0.) How many total function evaluations (bisection plus Newton) are required to get 10^{-6} accuracy?

7. Repeat the above for a function satisfying the following properties:

   (a) f is defined and twice continuously differentiable for all x;

   (b) f has a unique root $\alpha \in [-1, 2]$;

   (c) $f'(x) \leq -3$ for all x;

   (d) $|f''(x)| \leq 4$ for all x.

8. Consider the function $f(x) = x - a\sin x - b$, where a and b are positive parameters, with $a < 1$. Will the initial guess $x_0 = b$ always lead to convergence? If not, what additional condition on a or the initial guess needs to be made?

9. Write a program using Newton's method to find the root of $f(x) = 1 - e^{-x^2}$; the exact value is $\alpha = 0$. Compute the ratio

$$R_n = \frac{\alpha - x_{n+1}}{(\alpha - x_n)^2}$$

as the iteration progresses and comment (carefully!) on your results, in light of the material in this section.

10. A *monotone function* is one whose derivative never changes sign; the function is either always increasing or always decreasing. Show that a monotone function can have at most one root.

11. If f is a smooth monotone function with a root $x = \alpha$, will Newton's method always converge to this root, for any choice of x_0? (Provide either a counter-example or a valid proof.)

◁ ● ● ▷

3.7 APPLICATION: COMPUTATION OF THE SQUARE ROOT

(Much of this presentation is based on an exercise in Kendall Atkinson's text, *An Introduction to Numerical Analysis*, published by John Wiley & Sons, Inc.)

An interesting application of Newton's method is to the computation of $\sqrt{a}$. Consider now the function $f(x) = x^2 - a$. Clearly, the roots of this function are $\alpha = \pm\sqrt{a}$, so if we can find the (positive) root of f, we have computed $\sqrt{a}$.

Recall that the floating-point representation of a (assuming a binary base) is

$$a = \tilde{b} \times 2^{t-p},$$

where $\tilde{b} \in [\frac{1}{2}, 1]$ is the fraction, and $t - p$ is the binary exponent. If $t - p$ is odd, we shift one factor of 2 from the fraction to the exponent; this allows us to write

$$a = b \times 2^{2k},$$

where $b \in [\frac{1}{4}, 1]$. Thus,

$$\sqrt{a} = \sqrt{b} \times 2^k$$

and all we have to do is compute the square root of a number in the interval $[\frac{1}{4}, 1]$. Can we find an initial guess such that Newton's method will always converge for b on this interval? If so, how rapidly will it converge?

The Newton error formula (3.12) applied to $f(x) = x^2 - b$ tells us that

$$\sqrt{b} - x_{n+1} = -(\sqrt{b} - x_n)^2 \left(\frac{2}{4x_n}\right) = -(\sqrt{b} - x_n)^2 \left(\frac{1}{2x_n}\right), \qquad (3.25)$$

so that the relative error satisfies

$$\left| \frac{\sqrt{b} - x_{n+1}}{\sqrt{b}} \right| = \left(\frac{\sqrt{b} - x_n}{\sqrt{b}} \right)^2 \left| \frac{\sqrt{b}}{2x_n} \right|. \tag{3.26}$$

Now, Newton's method applied to f is

$$x_{n+1} = x_n - \frac{x_n^2 - b}{2x_n} = \frac{1}{2} \left(x_n + \frac{b}{x_n} \right),$$

so that if $x_0 > 0$, then $x_n > 0$ for all n. But then (3.25) implies that

$$\sqrt{b} - x_{n+1} \leq 0$$

for all $n \geq 0$. Thus, $\alpha = \sqrt{b} \leq x_n$ for all $n \geq 1$. (We could choose x_0 less than $\sqrt{b}$, but all subsequent iterates would be greater than $\sqrt{b}$.) It therefore follows, from (3.26), that since $|\sqrt{b}/x_n| \leq 1$,

$$\left| \frac{\sqrt{b} - x_{n+1}}{\sqrt{b}} \right| \leq \frac{1}{2} \left(\frac{\sqrt{b} - x_n}{\sqrt{b}} \right)^2.$$

We can solve the recursion quickly to get

$$\left| \frac{\sqrt{b} - x_n}{\sqrt{b}} \right| \leq 2 \left(\frac{\sqrt{b} - x_0}{2\sqrt{b}} \right)^{2^n}. \tag{3.27}$$

We now turn our attention to finding an initial guess. The fact that we are going to be applying the Newton iteration to finding $\sqrt{b}$ means that we need only look at the interval $[\frac{1}{4}, 1]$. We could just take $x_0 = \frac{5}{8}$, which is the midpoint of the interval and implies that $|\alpha - x_0| \leq 3/8$, but we can do better by using *linear interpolation* to the square root function. We take our nodes to be $\frac{1}{4}$ and 1 and apply the linear interpolation formula (2.11) to $g(x) = \sqrt{x}$. We then have that

$$p_1(x) = \left(\frac{x - 1/4}{3/4} \right) \sqrt{1} + \left(\frac{1 - x}{3/4} \right) \sqrt{1/4} = \frac{2x + 1}{3},$$

and we take $x_0 = p_1(b)$; therefore, $x_0 = \frac{1}{3}(2b + 1)$. Moreover, we can use the error estimate (2.13) to show that

$$\begin{aligned} |\sqrt{b} - x_0| &\leq \frac{1}{8} \left(1 - \frac{1}{4} \right)^2 \max_{\frac{1}{4} \leq t \leq 1} \left| \frac{1}{4} t^{-3/2} \right|, \\ &= \frac{9}{64}, \end{aligned}$$

so that the relative error in $\sqrt{b}$ (which equals the relative error in $\sqrt{a}$) becomes

$$\left| \frac{\sqrt{b} - x_n}{\sqrt{b}} \right| \leq 2 \left(\frac{9}{128\sqrt{b}} \right)^{2^n} \leq 2 \left(\frac{9}{64} \right)^{2^n}.$$

Since $\frac{9}{64} < 0.15$, this will decrease very rapidly. In fact, we have that for $n = 4$, the relative error is less than 4.7×10^{-14}. Thus, four steps of Newton's method—about 15 total operations, counting the cost of forming x_0—suffice to compute the square root of a number to a very high (relative) accuracy.

Exercises:

1. Based on the material in this section, write each of the following numbers in the form

$$a = b \times 2^k,$$

   where $b \in [\frac{1}{4}, 1]$ and k is even, and then use Newton's method to find the square root of b and hence of a.

   (a) $a = \pi$;

   (b) $a = 5$;

   (c) $a = 7$;

   (d) $a = 3$;

   (e) $a = 6$.

   Be sure to use the initial value as generated in this section and only do as many iterations as are necessary for 10^{-16} accuracy. Compare your values to the intrinsic square root function on your computer or calculator.

2. In the example discussed in this section, how many iterations are required for the *absolute* error to be less than 2^{-48}?

3. How accurate is the result in only three iterations, based on the initial guess used here?

4. What does the initial error have to be for the relative error to be less than 10^{-14} after only *two* iterations? Hint: See 3.27.

5. Extend the derivation and analysis of this section to the problem of computing *cube* roots by solving for the roots of $f(x) = x^3 - a$. Be sure to cover all the major points that were covered in the text. Is using linear interpolation to get x_0 accurate enough to guarantee convergence of the iteration?

6. Consider using the polynomial

$$p_2(x) = \frac{9}{35} + \frac{22}{21}x - \frac{32}{105}x^2$$

   to generate the initial guess. What is the maximum error in $|\sqrt{x} - p_2(x)|$ over the interval $[\frac{1}{4}, 1]$? How many iterations (and hence, how many operations) are needed to get (relative) accuracy of 10^{-16}? (Feel free to compute the maximum value experimentally, but you must justify the accuracy of your value in this case with some kind of argument.)

7. Repeat the above using

$$q_2(x) = 0.2645796916 + 1.0302824818x - 0.2983646911x^2.$$

8. Repeat again, using

$$q_1(x) = 0.647941993x + 0.3667102689.$$

9. Rewrite your program using p_2, q_2, and q_1 to generate the initial guess, applying it to each of the values in Problem 1. Compare your results to the square root function on your system.

10. If we use a piecewise linear interpolation to construct x_0, using the nodes $\frac{1}{4}$, $\frac{9}{16}$, and 1, what is the initial error, and how many iterations are now needed to get 10^{-16} accuracy? How many operations are involved in this computation?

3.8 THE SECANT METHOD: DERIVATION AND EXAMPLES

An obvious drawback of Newton's method is that it requires us to have a formula for the derivative of f. For classroom examples this is not an issue, but in the "real world" it can be. Suppose, for example, that f is not defined by a simple and tidy formula such as $f(x) = 2 - e^x$, but is instead "defined" by a separate subprogram that uses about 2,000 lines of involved computer code. Even if we could theoretically write a formula for f from which f' could be constructed, is this a practical task to set for ourselves?

One obvious way to deal with this problem is to use an approximation to the derivative in the Newton formula. For example, in §2.2 we saw that

$$f'(x) \approx \frac{f(x+h) - f(x)}{h},$$

so we could use this in (3.7) to get the new iteration

$$x_{n+1} = x_n - f(x_n)\left[\frac{h}{f(x_n + h) - f(x_n)}\right].$$

We will, in fact, analyze the convergence of this method in §3.11.2. In this section, we want to pursue a similar, but slightly different idea.

Newton's method is derived, geometrically, by drawing a tangent line from the current approximate root down to the axis. The derivative is required because we use the tangent line. If, instead, we used a *secant line* (i.e., one passing through two points on the curve instead of just one), then no derivative would be required.

Let x_0 and x_1 be given—thus we have two initial guesses—and consider Figure 3.6. Construct the line that passes through $(x_0, f(x_0))$ and $(x_1, f(x_1))$ and use its root to define the next iterate, x_2. The line is defined by the formula

$$\frac{y - f(x_1)}{x - x_1} = \frac{f(x_1) - f(x_0)}{x_1 - x_0},$$

so that the next iterate is given by (setting $y = 0$, and solving for $x = x_2$ in the equation above):

$$x_2 = x_1 - f(x_1)\left[\frac{x_1 - x_0}{f(x_1) - f(x_0)}\right].$$

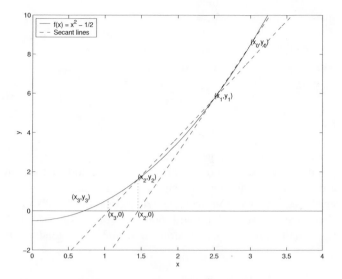

Figure 3.6   Illustration of secant method.

More generally, then, we have

$$x_{n+1} = x_n - f(x_n) \left[\frac{x_n - x_{n-1}}{f(x_n) - f(x_{n-1})} \right].$$ (3.28)

This is the *secant method*. Note that it is consistent with Newton's method, where we use the approximation

$$f'(x_n) \approx \frac{f(x_n) - f(x_{n-1})}{x_n - x_{n-1}}.$$

The error in this derivative approximation is proportional to $x_n - x_{n-1}$. Thus, if we assume that the iteration is converging (so that $x_n - x_{n-1} \to 0$), then the secant method becomes more and more like Newton's method. Hence, we expect rapid convergence for x_0 near α.

The secant method has a number of advantages over Newton's method. Not only does it not require the derivative, but it can be coded in such a way (see below) as to require only a single function evaluation per iteration. Newton requires two: one for the function and one for the derivative. Thus the secant method is about half as costly per step as Newton's method.

But the important question is: How well does it perform?

■ **EXAMPLE 3.7**

If we apply the secant method to the same example that we have used heretofore, $f(x) = 2 - e^x$, using $x_0 = 0$ and $x_1 = 1$, we get the following results for the first few iterations:

$$
\begin{aligned}
x_2 &= x_1 - f(x_1)(x_1 - x_0)/(f(x_1) - f(x_0)) \\
&= 1 - (2 - e)(1 - 0)/(1 - e) = 0.5819767068 \\
x_3 &= x_2 - f(x_2)(x_2 - x_1)/(f(x_2) - f(x_1)) \\
&= 0.5819767068 - (2 - e^{0.5819767068})(0.5819767068 - 1)/(e - e^{0.5819767068}) \\
&= 0.6766927037.
\end{aligned}
$$

As we continue the iteration, we get the results shown in Table 3.5. Note that we converged to almost the same root as for Newton's method, but this time in six iterations as opposed to five for Newton. However, recall that Newton is more costly per iteration: Newton achieved its accuracy with a total of 10 calls to the function f and the derivative f'; secant did it with only six calls, all to f. Secant (for this example) was actually *more* efficient in terms of the number of total function calls.

Table 3.5 The secant method applied to $f(x) = 2 - e^x$.

n	x_n	$\alpha - x_n$
0	0.000000000000	0.693147180560
1	1.000000000000	0.306852819440
2	0.581976706869	0.111170473691
3	0.676692703760	0.016454476800
4	0.694081399681	0.000934219121
5	0.693139474645	0.000007705915
6	0.693147176961	0.000000003599
7	0.693147180560	0.000000000000

The secant method and Newton's method suffers from many of the same ills, however, since the underlying geometric idea is pretty much the same. The fundamental convergence theory is almost exactly the same, in essence: If the initial guesses are both sufficiently close to the root, then the method will converge. A fairly complete discussion of the convergence of the secant method in given in §3.11.3. At this point we simply note, without proof, that an error formula for the secant method can be derived, which says that

$$
\alpha - x_{n+1} = -\frac{1}{2}(\alpha - x_n)(\alpha - x_{n-1})\frac{f''(\xi_n)}{f'(\eta_n)}, \tag{3.29}
$$

where $\min\{\alpha, x_n, x_{n-1}\} \le \xi_n, \eta_n \le \max\{\alpha, x_n, x_{n-1}\}$. Note that this error formula is very much akin to the one for Newton's method. Note also that it shows that

$$
|\alpha - x_{n+1}| \approx C|\alpha - x_n||\alpha - x_{n-1}|
$$

for $x_n \approx \alpha$, $x_{n+1} \approx \alpha$, and $x_{n-1} \approx \alpha$. Thus, the error goes like the product of the two previous errors, and it follows easily that the secant method is superlinear when it converges.

In addition, we can easily establish that a relationship similar to (3.9) holds for the secant iterates as well as the Newton iterates (the definition of the constant C_n is slightly different); thus, we can use the difference of consecutive approximations as a stopping criterion.

Algorithm 3.3 *Secant Method*

```
input x0, x1, tol, n
external f
        f0 = f(x0)
        f1 = f(x1)
        for i=1 to n do
                x = x1 - f1*(x1 - x0)/(f1 - f0)
                fx = f(x)
                x0 = x1
                x1 = x
                f0 = f1
                f1 = fx
                if abs(x1 - x0) < tol then
                        root = x1
                        stop
                endif
        endfor
```

We close this section with an informal statement of the convergence result for the secant method.

> If f, f', and f'' are all continuous near the root, and if f' does not equal zero at the root, then the secant method will converge whenever the initial guess is sufficiently close to the root. Moreover, this convergence will be superlinear, in the sense that

$$\lim_{n \to \infty} \frac{\alpha - x_{n+1}}{\alpha - x_n} = 0.$$

Note that, fundamentally, this is almost the same as for Newton's method.

Exercises:

1. Do three steps of the secant method for $f(x) = x^3 - 2$, using $x_0 = 0$ and $x_1 = 1$.

2. Repeat the above using $x_0 = 1$, $x_1 = 0$. Comment.

3. Apply the secant method to the same functions as in Problem 3 of §3.1, using x_0, x_1 equal to the endpoints of the given interval. Stop the iteration when the error as estimated by $|x_n - x_{n-1}|$ is less than 10^{-6}. Compare to your results for Newton and bisection in the earlier exercises.

4. For the secant method, prove that

$$\alpha - x_{n+1} = C_n(x_{n+1} - x_n),$$

where $C_n \to 1$ as $n \to \infty$, so long as the iteration converges. Hint: Follow what we did in §3.3 for Newton's method.

5. Assume (3.29) and prove that if the secant method converges, then it is superlinear.

6. Assume (3.29) and consider a function f such that:

   (a) There is a unique root on the interval $[0, 4]$;
   (b) $|f''(x)| \leq 2$ for all $x \in [0, 4]$;
   (c) $f'(x) \geq 5$ for all $x \in [0, 4]$.

   Can we prove that the secant iteration will converge for any $x_0, x_1 \in [0, 4]$? If so, how many iterations are required to get an error that is less than 10^{-8}? If convergence is not guaranteed for all $x \in [0, 4]$, how many steps of bisection are needed before convergence *will* be guaranteed for secant?

7. Repeat the above problem under the following assumptions:

   (a) There is a unique root on the interval $[0, 9]$;
   (b) $|f''(x)| \leq 6$ for all $x \in [0, 9]$;
   (c) $f'(x) \geq 2$ for all $x \in 0, 9]$.

   Can we prove that the secant iteration will converge for any $x_0, x_1 \in [0, 9]$? If so, how many iterations are required to get an error that is less than 10^{-8}? If convergence is not guaranteed for all $x \in [0, 9]$, how many steps of bisection are needed before convergence *will* be guaranteed for secant?

8. Repeat Problem 8 of §3.1, but this time find α for the set of θ values defined by

   $$\theta_k = 10^{k/4},$$

   for k ranging from -24 all the way to 24. Construct a plot of α versus $\log_{10} \theta$.

9. Comment, in your own words, on the differences between the secant method and regula-falsi (see Problem 9 of §3.1).

◁ ● ● ▷

3.9   FIXED-POINT ITERATION

So far, we have looked at three methods for approximating the roots of a given function $f(x)$: Bisection, Newton's method, and the secant method. The first two can be considered as instances of simple iteration, in which we recursively substitute values into a function or process to obtain the next value. (The secant method can be put into the framework as well, but only imperfectly, since each secant value is determined by the *two* previous values.)

In this section, we will study simple iteration for its own sake, generally divorced from any considerations or connections to root-finding problems. Our goal, however, is to use the added understanding of simple iteration to enhance our understanding of and ability to solve root-finding problems.

Consider Newton's method as applied to $f(x) = x^2 - a$:

$$x_{n+1} = \frac{1}{2} \left(x_n + \frac{a}{x_n} \right). \tag{3.30}$$

As $n \to \infty$, we know that $x_n \to \alpha = \sqrt{a}$. (In this case, convergence occurs for any $x_0 > 0$.) We can write (3.30) more abstractly as

$$x_{n+1} = g(x_n) \tag{3.31}$$

for $g(x) = \frac{1}{2}(x + ax^{-1})$. Note that

$$f(\alpha) = 0 \iff \alpha = g(\alpha).$$

The fact that $\alpha = g(\alpha)$ is interesting. This defines α, which we already know as the root of f (the point where the graph of $y = f(x)$ crosses the x-axis), to be a point where the graph of the new function $y = g(x)$ crosses the line $y = x$. Because $\alpha = g(\alpha)$ shows that $g(\alpha)$ "stays" at α, this kind of point is called a *fixed point* of the function g, and an iteration of the form (3.31) is called a *fixed-point iteration* for g.

We have already seen that we can establish a connection between a root-finding problem and a fixed-point problem using Newton's method. But this is clearly not the only way to do so. For example, each of the following iterations can be derived from the equation $x^2 - a = 0$; thus, for each one, $\alpha = g(\alpha) \iff f(\alpha) = 0 \iff \alpha = \sqrt{a}$. To verify, simply let $x_n = x_{n+1} = x$, and solve for x.

1. $x_{n+1} = x_n + \frac{1}{2}(x_n^2 - a)$; $g(x) = x + \frac{1}{2}(x^2 - a)$;

2. $x_{n+1} = a/x_n$; $g(x) = a/x$;

3. $x_{n+1} = a + x_n - x_n^2$; $g(x) = a + x - x^2$.

So, for a given function g, a number of questions can be raised:

1. Under what conditions does a fixed point exist?

2. Under what conditions does the iteration (3.31) converge?

3. If the iteration converges, how fast does it converge?

In the remainder of this section, we will outline the theory and practice of fixed-point iteration, mostly from the perspective of applying it to the root-finding problem. Our objective is to use the fixed-point theory to inform us about certain aspects of root-finding problems.

Students should be advised not to confuse the notion of a root of a function with that of a fixed point for the same function. Generally, for a given function f, the roots and fixed points (if any exist) are not the same. Figure 3.7 might be instructive in this regard; it shows the graph of $y = f(x) = x^2 - \frac{1}{2}$ and $y = x$. The roots of f are $\alpha = \pm\sqrt{1/2}$, but the fixed point of f is where it intersects the graph of $y = x$, and this occurs at $x = \frac{1}{2}(1 + \sqrt{3}) \approx 1.366$. When we use a fixed-point iteration to find a root of a function f, the root is a fixed point of a *different* function, g, not f. The fixed point of f, if one even exists, is usually not related at all to the roots of f. Further clarification in this regard might come from comparing Figures 3.8 and 3.9. Figure 3.8 shows two iterations of Newton's method, graphically, as done on $f(x) = \frac{1}{2}x^2 - \frac{1}{2}$. The dashed lines are the tangent lines being used to find the next iterate. Figure 3.9 shows fixed-point iteration being performed on $g(x) = \frac{1}{2}e^{-x}$. Here the dashed lines show how x_0 is mapped onto $y = x$, and then projected onto the curve $y = g(x)$ to get the next iterate. Geometrically, we are looking for different points on the curve.

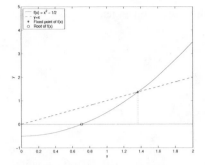

Figure 3.7 Fixed point versus root; the root is where the curve crosses the x-axis, the fixed point is where the curve crosses the line $y = x$.

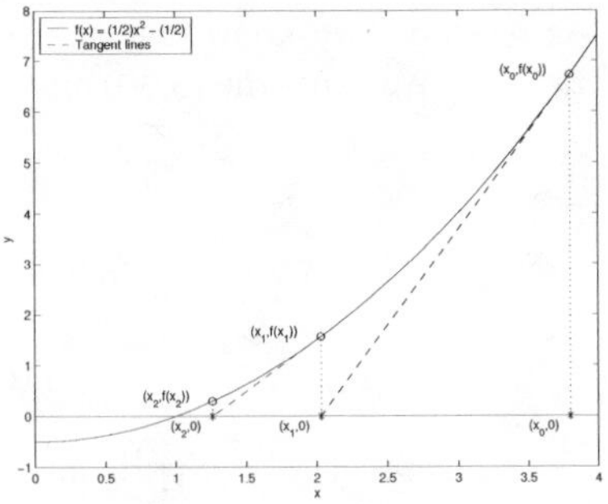

Figure 3.8 Newton's method.

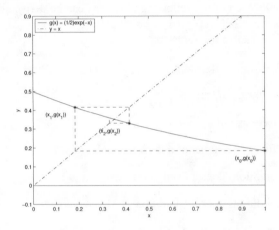

Figure 3.9 Fixed-point iteration.

Before proceeding with the formal development of the important results, let's do some informal investigation of a fixed-point iteration. We have

$$\alpha = g(\alpha)$$

and

$$x_{n+1} = g(x_n), \tag{3.32}$$

so that

$$\alpha - x_{n+1} = g(\alpha) - g(x_n) = g'(\xi_n)(\alpha - x_n), \tag{3.33}$$

where, in order to use the Mean Value Theorem, we have of course assumed that g was differentiable. If $|g'(x)| < 1$ near the fixed point, then we ought to be able to show that

$$|\alpha - x_n| < c^n |\alpha - x_0|$$

for some positive value $c < 1$, which would of course imply convergence. Once we have convergence established, a limit result quickly follows from (3.33), which establishes that the iteration converges linearly, since (3.33) implies that

$$\frac{\alpha - x_{n+1}}{\alpha - x_n} = g'(\xi_n) \to g'(\alpha).$$

We can investigate the iteration (3.32) and the results we have rather casually derived by experimenting with a function g whose derivative is less than 1.

■ **EXAMPLE 3.8**

Let's take $g(x) = \frac{1}{2}e^{-x}$ as our example. Will the iteration

$$x_{n+1} = \frac{1}{2}e^{-x_n}$$

converge?

Let's take $x_0 = 0$; this is a somewhat arbitrary choice, but let's see where it leads us. We have

$$\begin{aligned}
x_1 &= g(x_0) = \frac{1}{2}e^0 = \frac{1}{2}, \\
x_2 &= g(x_1) = \frac{1}{2}e^{-1/2} = 0.3032653299, \\
x_3 &= g(x_2) = \frac{1}{2}e^{-0.3032653299} = 0.369201575, \\
x_4 &= g(x_3) = \frac{1}{2}e^{-0.369201575} = 0.3456430253.
\end{aligned}$$

If we continue, we get the values in Table 3.6, along with the results for two other values of x_0. The computation was stopped at $n = 15$; clearly, all three iterations are converging to the fixed point $\alpha \approx 0.3517$, but it is worth noting that none of the sequences is converging to its limit as fast as either Newton's method or the secant method. Figure 3.10 shows the graph of $y = g(x)$ and $y = x$ to illustrate where the fixed point is located.

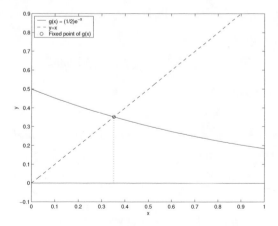

Figure 3.10   Fixed point illustration.

Table 3.6   Fixed-point iteration example, $g(x) = \frac{1}{2}e^{-x}$.

n	x_n, when $x_0 = 0$	x_n, when $x_0 = 5$	x_n, when $x_0 = -5$
0	0.	5.0000000000000	−5.0000000000000
1	0.50000000000000	3.3689734995427e-03	74.206579551288
2	0.30326532985632	0.49831834756204	2.9611607506452e−33
3	0.36920157498737	0.30377574578843	0.50000000000000
4	0.34564302521408	0.36901317670616	0.30326532985632
5	0.35388254815500	0.34570814990043	0.36920157498737
6	0.35097870435331	0.35385950241548	0.34564302521408
7	0.35199937290228	0.35098679301031	0.35388254815500
8	0.35164028150092	0.35199652571160	0.35097870435331
9	0.35176657517651	0.35164128268928	0.35199937290228
10	0.35172215208802	0.35176622299208	0.35164028150092
11	0.35173777701935	0.35172227595910	0.35176657517651
12	0.35173228118368	0.35173773344922	0.35172215208802
13	0.35173421425181	0.35173229650870	0.35173777701935
14	0.35173353432626	0.35173420886147	0.35173228118368
15	0.35173377347896	0.35173353622223	0.35173421425181

The important results in this section can be covered in three theorems, which we will now state and prove. Although this makes for a lengthy and perhaps abstract theoretical interlude, it is necessary to understand the theory underlying fixed-point iterations in order to understand (and appreciate) what fixed-point iteration gives us in terms of practical algorithms for root-finding.

The first theorem summarizes the conditions under which a given function g will have a fixed point, the conditions under which a fixed-point iteration will converge for any x_0 on a given interval, and also provides us with error estimates for this convergence.

Theorem 3.5 (Fixed-point Existence and Iteration Convergence Theory) *Let $g \in C([a, b])$ with $a \le g(x) \le b$ for all $x \in [a, b]$; then:*

1. g has at least one fixed point $\alpha \in [a, b]$;

2. If there exists a value $\gamma < 1$ such that

$$|g(x) - g(y)| \le \gamma |x - y| \qquad (3.34)$$

for all x and y in $[a, b]$, then:

(a) α is unique;

(b) The iteration $x_{n+1} = g(x_n)$ converges to α for any initial guess $x_0 \in [a, b]$;

(c) We have the error estimate

$$|\alpha - x_n| \le \frac{\gamma^n}{1 - \gamma}|x_1 - x_0|. \qquad (3.35)$$

3. If g is continuously differentiable on $[a, b]$ with

$$\max_{x \in [a,b]} |g'(x)| = \gamma < 1 \qquad (3.36)$$

then

(a) α *is unique;*

(b) *The iteration* $x_{n+1} = g(x_n)$ *converges to* α *for any initial guess* $x_0 \in [a, b]$;

(c) *We have the error estimate*

$$|\alpha - x_n| \leq \frac{\gamma^n}{1-\gamma}|x_1 - x_0|;$$

(d) *The limit*

$$\lim_{n \to \infty} \frac{\alpha - x_{n+1}}{\alpha - x_n} = g'(\alpha)$$

holds.

Proof: Define $h(x) = g(x) - x$. Then,

$$h(b) = g(b) - b \leq 0$$

and

$$h(a) = g(a) - a \geq 0.$$

Therefore, the Intermediate Value Theorem implies that h has a root α on the interval $[a, b]$; thus, $h(\alpha) = 0$, which implies that $\alpha = g(\alpha)$. This proves (1).

Suppose now that (3.34) holds, and that a second fixed point, β, exists on $[a, b]$. Then we have

$$\alpha = g(\alpha) \tag{3.37}$$

and

$$\beta = g(\beta),$$

so that (from (3.34))

$$|\alpha - \beta| = |g(\alpha) - g(\beta)| \leq \gamma|\alpha - \beta|,$$

which implies that

$$|\alpha - \beta|(1 - \gamma) \leq 0.$$

Since $0 < \gamma < 1$, the only way that this can be true is for $|\alpha - \beta| \leq 0$, and the only way for this to be true is for $\alpha = \beta$, which implies that the fixed point α is unique. (Why?) This proves (2a).

Consider now the iteration (3.31) and the definition of fixed point (3.37). If we subtract and take absolute values, we get

$$|\alpha - x_{n+1}| = |g(\alpha) - g(x_n)| \leq \gamma|\alpha - x_n|.$$

Now write $e_n = |\alpha - x_n|$ so that the above becomes $e_{n+1} \leq \gamma e_n$. The recursion can be solved readily to get $e_n \leq \gamma^n e_0$, from which it follows that $e_n \to 0$ as $n \to \infty$; hence, the iteration converges. This proves (2b).

Finally, we note that

$$|\alpha - x_0| = |\alpha - g(x_0) + x_1 - x_0| \leq |g(\alpha) - g(x_0)| + |x_1 - x_0| \leq \gamma|\alpha - x_0| + |x_1 - x_0|,$$

from which it follows that

$$|\alpha - x_0| \le \frac{1}{1-\gamma}|x_1 - x_0|,$$

so that

$$e_n \le \gamma^n e_0 \le \frac{\gamma^n}{1-\gamma}|x_1 - x_0|,$$

which proves (2c).

We now note that (3.36) implies (3.34), so that (3a)–(3c) follow just as (2a)–(2c) did. It remains only to prove (3d).

We have, from (3.37) and (3.31), that

$$\alpha - x_{n+1} = g(\alpha) - g(x_n) = g'(\xi_n)(\alpha - x_n),$$

so that

$$\frac{\alpha - x_{n+1}}{\alpha - x_n} = g'(\xi_n) \to g'(\alpha)$$

since $\xi_n \to \alpha$ is forced by the convergence of x_n to α. •

Note: It is important to keep in mind here the condition $a \le g(x) \le b$, which is one of our hypotheses. This assumption guarantees that the iterates computed according to $x_{n+1} = g(x_n)$ stay within the interval $[a, b]$, which is the only place where we know that (3.34) or (3.36) holds.

We can illustrate this theorem by using it to analyze the example we looked at previously.

■ **EXAMPLE 3.9**

Consider the iteration

$$x_{n+1} = g(x_n),$$

for $g(x) = \frac{1}{2}e^{-x}$. Since $0 \le g(x) \le \frac{1}{2}$ for all $x > 0$, we have that $g(x) \in [0, \frac{1}{2}]$ for all $x \in [0, \frac{1}{2}]$. Thus, we can take our interval $[a, b]$ as $[0, \frac{1}{2}]$. Since g is continuous on this interval, we know that a fixed point must exist there. Further, since g is continuously differentiable on $[0, \frac{1}{2}]$ and

$$|g'(x)| \le \frac{1}{2} < 1$$

for all $x \in [0, \frac{1}{2}]$, we have that this fixed point is unique. Moreover, the iteration converges linearly and the error estimate (3.35) applies, with $\gamma = \frac{1}{2}$.

The second theorem is a local convergence result; that is, if we only know information about g near the fixed point, we can still deduce that convergence occurs for a sufficiently close initial guess.

Theorem 3.6 (Local Convergence for Fixed-Point Iterations) *Let g be continuously differentiable in an open interval of a fixed point α with $|g'(\alpha)| < 1$; then, for all x_0 sufficiently close to α, the iteration $x_{n+1} = g(x_n)$ converges, and*

$$\lim_{n \to \infty} \frac{\alpha - x_{n+1}}{\alpha - x_n} = g'(\alpha),$$

and

$$|\alpha - x_n| \leq \frac{\gamma^n}{1 - \gamma}|x_1 - x_0|,$$

for some $\gamma < 1$.

Proof: Since g is continuously differentiable in an open interval of the fixed point, with $|g'(\alpha)| < 1$, we can (as we did in the proof of the Newton convergence theorem) find a closed interval J, centered on the fixed point α, such that $|g'(x)| \leq \gamma < 1$ for all $x \in J$. It follows from the definition of the iteration that

$$|\alpha - x_1| = |g(\alpha) - g(x_0)| \leq \gamma|\alpha - x_0|.$$

Therefore, x_1 is closer to the fixed point than x_0 is, and so are all the remaining iterates. Thus, $g(x) \in J$ for all $x \in J$, and we can now apply Theorem 3.5 to complete the proof. •

Finally, the third theorem provides us with a local criterion under which a fixed-point iteration will have a higher than linear order of convergence.

Theorem 3.7 *Consider the fixed-point iteration*

$$x_{n+1} = g(x_n), \tag{3.38}$$

where g is p times continuously differentiable, and $\alpha = g(\alpha)$. If

$$g'(\alpha) = g''(\alpha) = \cdots = g^{(p-1)}(\alpha) = 0$$

but

$$g^{(p)}(\alpha) \neq 0,$$

then the iteration (3.38) converges with order p for x_0 sufficiently close to α.

Proof: The fact that $g'(\alpha) = 0 < 1$ means that the iteration will converge for x_0 sufficiently close to α; this follows from Theorem 3.6. All we have to do is establish the higher convergence rate. We have, by Taylor's Theorem,

$$g(x_n) = g(\alpha) + (x_n - \alpha)g'(\alpha) + \cdots + \frac{(x_n - \alpha)^{p-1}}{(p-1)!}g^{(p-1)}(\alpha) + \frac{(x_n - \alpha)^p}{p!}g^{(p)}(\xi_n),$$

where ξ_n is between x_n and α. Now, all the derivative terms except that in the remainder vanish, so we quickly have that

$$g(x_n) - g(\alpha) = \frac{(x_n - \alpha)^p}{p!}g^{(p)}(\xi_n),$$

so that the iteration then implies that

$$\frac{x_{n+1} - \alpha}{(x_n - \alpha)^p} = \frac{1}{p!}g^{(p)}(\xi_n)$$

from which the convergence with order p follows. •

To illustrate the use of this theorem, we consider Newton's method,

$$x_{n+1} = x_n - \frac{f(x_n)}{f'(x_n)},$$

which can be viewed as a fixed-point iteration with

$$g(x) = x - \frac{f(x)}{f'(x)}.$$

Note that

$$g'(x) = 1 - \left[\frac{[f'(x)]^2 - f(x)f''(x)}{[f'(x)]^2}\right],$$

so

$$g'(\alpha) = 1 - \left[\frac{[f'(\alpha)]^2 - f(\alpha)f''(\alpha)}{[f'(\alpha)]^2}\right] = 1 - 1 = 0,$$

from which we conclude that Newton's method has (local) order of convergence of at least 2. This assumes, of course, that $f'(\alpha) \neq 0$. See §3.11.4 for what happens when this assumption fails to hold.

With the theory that underlies fixed-point iteration behind us, we can now apply this theory to obtain a greater understanding of how certain root-finding methods work, and how to improve them. Much of §3.11 is devoted to this.

Exercises:

1. Do three steps of each of the following fixed-point iterations, using the indicated x_0.

   (a) $x_{n+1} = \cos x_n$, $x_0 = 0$ (be sure to set your calculator in radians);

   (b) $x_{n+1} = e^{-x_n}$, $x_0 = 0$;

   (c) $x_{n+1} = \ln(1 + x_n)$, $x_0 = 1/2$;

   (d) $x_{n+1} = \frac{1}{2}(x_n + 3/x_n)$, $x_0 = 3$.

2. Let $Y = 1/2$ be fixed, and take $h = \frac{1}{8}$. Do three steps of the following fixed-point iteration

   $$y_{n+1} = Y + \frac{1}{2}h\left(-Y \ln Y - y_n \ln y_n\right)$$

   using $y_0 = Y$.

3. Let $Y_0 = 1/2$ and $Y_1 = 0.54332169878500$ be fixed, and take $h = \frac{1}{8}$. Do three steps of the fixed-point iteration

   $$y_{n+1} = \frac{4}{3}Y_1 - \frac{1}{3}Y_0 - 2hy_n \ln y_n$$

   using $y_0 = Y_1$.

4. Consider the fixed-point iteration $x_{n+1} = 1 + e^{-x_n}$. Show that this iteration converges for any $x_0 \in [1, 2]$. How many iterations does the theory predict that it will take to achieve 10^{-5} accuracy?

5. For each function listed below, find an interval $[a, b]$ such that $g([a, b]) \subset [a, b]$. Draw a graph of $y = g(x)$ and $y = x$ over this interval, and confirm that a fixed point exists there. Estimate (by eye) the value of the fixed point, and use this as a starting value for a fixed-point iteration. Does the iteration converge? Explain.

   (a) $g(x) = \frac{1}{2}(x + \frac{2}{x})$;

(b) $g(x) = x + e^{-x} - \frac{1}{4}$;

(c) $g(x) = \cos x$;

(d) $g(x) = 1 + e^{-x}$;

(e) $g(x) = \frac{1}{2}(x^2 + 1)$.

6. Let $h(x) = 1 - x^2/4$. Show that this function has a root at $x = \alpha = 2$. Now, using $x_0 = 1/2$, do the iteration $x_{n+1} = h(x_n)$ to approximate the fixed point of h. Comment on your results.

7. Let $h(x) = 3 - e^{-x}$. Using $x_0 = 2$, perform as many iterations of Newton's method as are needed to accurately approximate the root of h. Then, using the same initial point, do the iteration $x_{n+1} = h(x_n)$ to approximate the fixed point of h. Comment on your results.

8. Use fixed-point iteration to find a value of $x \in [1, 2]$ such that $2\sin \pi x + x = 0$.

9. For $a > 0$, consider the iteration defined by

$$x_{n+1} = \frac{x_n^3 + x_n^2 - x_n a + a}{x_n^2 + 2x_n - a}.$$

(a) For $x_0 = 3/2$ experiment with this iteration for $a = 4$ and $a = 2$. Based on these results, speculate as to what this iteration does. Try to prove this, and use the theorems of this section to establish a convergence rate.

(b) Now experiment with this iteration using $x_0 = 2$ and $a = 5$. Compare your results to Newton's method.

10. Consider the iteration

$$x_{n+1} = \frac{(N-1)x^{N+1} + ax}{Nx^N}.$$

Assume that this converges for integer N and any $a > 0$. What does it converge to? Use the theorems of this section to determine a convergence rate. Experiment with this iteration when $N = 3$ and $a = 8$, using $x_0 = 3/2$.

11. Consider the iteration defined by

$$x_{n+1} = x_n - f(x_n)\left[\frac{f(x_n)}{f(x_n + f(x_n)) - f(x_n)}\right].$$

This is also sometimes known as *Steffenson's method*. Show that it is (locally) quadratically convergent.

12. Apply Steffenson's method to find the root of $f(x) = 2 - e^x$, using $x_0 = 0$. Compare your convergence results to those in the text for the Newton and secant methods.

13. Apply Steffenson's method to $f(x) = x^2 - a$ for the computation of $\sqrt{a}$. For the following values of a, how does the performance compare to Newton's method?

(a) $a = 3$;

(b) $a = 2$;

(c) $a = \pi$.

◁ ● ● ● ▷

3.10   ROOTS OF POLYNOMIALS, PART 1

An important calculation in the analysis or design of certain electronic and/or mechanical devices involves finding *all* the roots of a polynomial:

$$H(s) = a_n s^n + a_{n-1} s^{n-1} + \cdots + a_1 s + a_0.$$

In a typical application we want to know if the real parts of all the roots are positive (or negative) or the size of the imaginary parts (which can represent frequencies of vibration).[12]

One can use any of the methods discussed in this chapter to find the roots of a polynomial, and many of the exercises concern polynomial functions. But there are several aspects to the polynomial root-finding problem that make it rather challenging:

1. Polynomial roots can be *very* sensitive to slight errors in the coefficients;

2. Polynomials with real coefficients can have complex-valued roots;

3. The usual polynomial root-finding problem is to find *all* the roots of the polynomial in question.

In this section we outline a simple, easy-to-program method that can be used to find all the roots of a given polynomial with real coeficients. We can assume, without loss of generality, that the polynomial in question is *monic*, i.e., the coefficient of the highest-order term is 1:

$$p(x) = x^n + a_{n-1} x^{n-1} + \cdots + a_2 x^2 + a_1 x + a_0. \tag{3.39}$$

It will be convenient to gather a number of basic facts about polynomials into a single theorem, which we will not formally prove.

Theorem 3.8 *Consider a polynomial of degree n, $p(x)$, in the form (3.39), with all coefficients $a_k \in \mathbb{R}$. Then the following hold:*

1. *There are exactly n values $\zeta_j \in \mathbb{C}$ such that $p(\zeta_j) = 0$;*

2. *Any complex-valued roots occur in complex-conjugate pairs—i.e., if $\zeta = x + iy$ is a root, so is $\bar{\zeta} = x - iy$;*

3. *All roots are contained within the region of the complex plane defined by*

$$R_{\text{roots}} = \bigcup_{j=1}^{n} R_j,$$

where the sets on the right are defined by

$$\begin{aligned}
R_0 &= \{z \in \mathbb{C} \mid |z| \le |a_0|\} \\
R_j &= \{z \in \mathbb{C} \mid |z| \le (1 + |a_{j-1}|)\}, \quad 2 \le j \le n - 1 \\
R_n &= \{z \in \mathbb{C} \mid |z + a_{n-1}| \le 1\}.
\end{aligned}$$

We will see in Chapter 8 where the localization result comes from.

Some readers may be wondering why we need to develop anything new to solve this problem. Why not simply graph the polynomial to get "eyeball approximations" to the

---

[12]The material in this section requires some understanding of complex arithmetic.

roots, then use Newton's method (or any other method from this chapter) followed by long division (deflation) to get all the roots? This is indeed possible, but it is not always effective. For one thing, it only gets you the real roots. If the "leftover" polynomial is of sufficiently high degree, then finding the complex roots might not be easy. Also, the deflation step (dividing the polynomial by each approximate root found) introduces errors into the coefficients, which can lead to larger errors in the approximate roots. However, for sufficiently low-degree (and well-behaved) examples, this approach is practical, and we will give an example of its use.

■ **EXAMPLE 3.10**

Consider the polynomial

$$p(x) = x^4 - x - 1.$$

This is indeed a monic polynomial, and Fig. 3.11 gives a plot over the interval $[-2, 2]$. A little work with the coefficients shows that the roots can all be found in the region

$$R_{\text{roots}} = \{z \in \mathbb{C} \mid |z| \le 2\}.$$

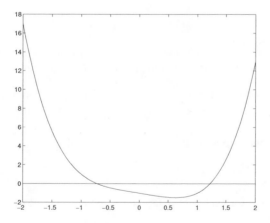

Figure 3.11   Plot of the polynomial $p(x) = x^4 - x - 1$.

It is evident from the graph that we have real roots near $x = 1.25$ and $x = -0.75$, so it is a simple task to apply Newton's method with those starting values. We get the results in Table 3.7.

With these values in hand, we know that the remaining (complex) roots can be found by carrying out the division

$$q(x) = \frac{p(x)}{(x - \zeta_1)(x - \zeta_2)} = x^2 + (0.49625212560524)x + 1.13068544546204;$$

Table 3.7 Simple polynomial example, $p(x) = x^4 - x - 1$; real roots.

k	ζ_1	$p(\zeta_1)$	ζ_2	$p(\zeta_2)$
1	1.22190366972477	0.00729038680243	−0.72529069767442	0.00201571897758
2	1.22074599618017	0.00001199839276	−0.72449275554021	0.00000200816642
3	1.22074408461096	0.00000000003267	−0.72449195900131	0.00000000000200
4	1.22074408460576	0.00000000000000	−0.72449195900052	0.00000000000000

since this is a quadratic, we can get the complex roots directly via the quadratic formula. We thus have the full set of roots:

$$\zeta_1 = 1.22074408460576,$$
$$\zeta_2 = -0.72449195900052,$$
$$\zeta_3 = -0.24812606280262 + 1.03398206097597i,$$
$$\zeta_4 = -0.24812606280262 - 1.03398206097597i.$$

These are very accurate values—they agree completely with those produced by MATLAB's `roots` function—but the process was cumbersome. It required human intervention to deduce the starting values for the Newton iterations, and it would not necessarily have worked for a higher-degree polynomial. For example, if we had considered

$$q(x) = x^6 - x - 1,$$

we would again have found that there were two real roots, but now the remainder polynomial is quartic. How do we get the complex roots[13]?

Perhaps the simplest method that might do the kind of job we want is the Durand–Kerner method. It is based on a very interesting construction from the definition of the polynomial in terms of its roots.

We can use the roots ζ_j to write the original polynomial as

$$p(x) = (x - \zeta_n)(x - \zeta_{n-1}) \cdots (x - \zeta_2)(x - \zeta_1).$$

We can use this to solve for each root in terms of the others as follows:

$$\zeta_1 = x - \frac{p(x)}{(x - \zeta_n)(x - \zeta_{n-1}) \cdots (x - \zeta_3)(x - \zeta_2)},$$

$$\zeta_2 = x - \frac{p(x)}{(x - \zeta_n)(x - \zeta_{n-1}) \cdots (x - \zeta_3)(x - \zeta_1)},$$

$$\vdots$$

$$\zeta_{n-1} = x - \frac{p(x)}{(x - \zeta_n)(x - \zeta_{n-2}) \cdots (x - \zeta_2)(x - \zeta_1)},$$

$$\zeta_n = x - \frac{p(x)}{(x - \zeta_{n-1})(x - \zeta_{n-2}) \cdots (x - \zeta_2)(x - \zeta_1)}.$$

---

[13]There actually is a "quartic formula," but it is clumsy to use and not well known.

This suggests an iteration. Let $\zeta_j^{(k)}$ be the approximation to ζ_j at the k-th iteration; replace x in the right-hand side of the above with $\zeta_j^{(k)}$, and let this define the new approximations $\zeta_j^{(k+1)}$, thus:

$$
\zeta_1^{(k+1)} = \zeta_1^{(k)} - \frac{p(\zeta_1^{(k)})}{(\zeta_1^{(k)} - \zeta_n^{(k)})(\zeta_1^{(k)} - \zeta_{n-1}^{(k)})\cdots(\zeta_1^{(k)} - \zeta_3^{(k)})(\zeta_1^{(k)} - \zeta_2^{(k)})},
$$

$$
\zeta_2^{(k+1)} = \zeta_2^{(k)} - \frac{p(\zeta_2^{(k)})}{(\zeta_2^{(k)} - \zeta_n^{(k)})(\zeta_2^{(k)} - \zeta_{n-1}^{(k)})\cdots(\zeta_2^{(k)} - \zeta_3^{(k)})(\zeta_2^{(k)} - \zeta_1^{(k)})},
$$

$$
\vdots
$$

$$
\zeta_{n-1}^{(k+1)} = \zeta_{n-1}^{(k)} - \frac{p(\zeta_{n-1}^{(k)})}{(\zeta_{n-1}^{(k)} - \zeta_n^{(k)})(\zeta_{n-1}^{(k)} - \zeta_{n-2}^{(k)})\cdots(\zeta_{n-1}^{(k)} - \zeta_2^{(k)})(\zeta_{n-1}^{(k)} - \zeta_1^{(k)})},
$$

$$
\zeta_n^{(k+1)} = \zeta_n^{(k)} - \frac{p(\zeta_n^{(k)})}{(\zeta_n^{(k)} - \zeta_{n-1}^{(k)})(\zeta_n^{(k)} - \zeta_{n-2}^{(k)})\cdots(\zeta_n^{(k)} - \zeta_2^{(k)})(\zeta_n^{(k)} - \zeta_1^{(k)})}.
$$

Note that the iteration appears to be well-defined, except for the possibility of multiple roots. (Readers may want to think about why this is not as great a problem as it appears to be.) The computation looks more complicated than it is, because of the index-heavy notation. There is a variant in which $\zeta_1^{(k+1)}$ is used in the equation for $\zeta_2^{(k+1)}$, and so on. For reasons that should become clear in §7.7, this is sometimes known as a "Gauss–Seidel" variant of the method. To get complex roots, you do have to start with a complex initial value.

Let's look at same examples.

■ **EXAMPLE 3.11**

Again consider

$$
p(x) = x^4 - x - 1.
$$

If we take our initial values as $\zeta_1^{(0)} = 1$, $\zeta_2^{(0)} = -1$, $\zeta_3^{(0)} = i$, and $\zeta_4^{(0)} = -i$, and proceed with the iteration as defined above, we get the results shown in Table 3.8:

Table 3.8   Durand–Kerner polynomial example, $p(x) = x^4 - x - 1$.

k	ζ_1	ζ_2	$\zeta_{3,4}$
1	1.25000000000000	−0.75000000000000	−0.25000000000000 ± 1.00000000000000i
2	1.22055288461538	−0.72343750000000	−0.24855769230769 ± 1.03377403846154i
3	1.22074413864366	−0.72449274739525	−0.24812569562421 ± 1.033981933554441i
4	1.22074408460576	−0.72449195900059	−0.24812606280258 ± 1.03398206097578i
5	1.22074408460576	−0.72449195900052	−0.24812606280262 ± 1.03398206097597i
6	1.22074408460576	−0.72449195900052	−0.24812606280262 ± 1.03398206097597i

These values are as accurate as in Example 3.10, and are obtained much more simply.

■ **EXAMPLE 3.12**

Now let's consider $p(x) = x^6 - x - 1$. According to Theorem 3.8, all the roots lie in the complex disk defined by

$$R_{\text{roots}} = \{z \in \mathbb{C} \mid |z| \le 2\}.$$

We get the set of iterates shown in Table 3.9.

Table 3.9 Durand–Kerner polynomial example, $p(x) = x^6 - x - 1$.

k	ζ_1	$\zeta_{3,4}$
1	1.08155487804878	$-0.30769230769231 \pm 1.70512820512820i$
2	1.13901721798730	$-0.22198006215195 \pm 0.99861639553984i$
3	1.13539113258632	$-1.24651809113936 \pm 0.50899741840019i$
4	1.13500712616363	$-0.85917576678224 \pm 0.84016747596423i$
5	1.13478664030116	$-0.69025787751531 \pm 0.77104060671459i$
6	1.13472702883897	$-0.63470626697919 \pm 0.73992433685472i$
7	1.13472414517347	$-0.62941083605888 \pm 0.73580572524339i$
8	1.13472413840153	$-0.62937242972103 \pm 0.73575595743980i$
9	1.13472413840152	$-0.62937242847031 \pm 0.73575595299978i$
10	1.13472413840152	$-0.62937242847031 \pm 0.73575595299978i$

k	ζ_2	$\zeta_{5,6}$
1	-0.77836538461538	$0.15609756097561 \pm 0.271544715544715i$
2	-0.77822963271006	$0.04158626951333 \pm 1.23875067307858i$
3	-0.77814151833254	$1.06789328401247 \pm 1.25381006341545i$
4	-0.77811064084602	$0.68072752412344 \pm 1.12629581387957i$
5	-0.77809831808509	$0.51191371640728 \pm 1.02410317331166i$
6	-0.77809075367968	$0.45638812939955 \pm 1.00270948401810i$
7	-0.77808961428378	$0.45109357061403 \pm 1.002350003584721i$
8	-0.77808959868072	$0.45105515986063 \pm 1.002364569547111i$
9	-0.77808959867860	$0.45105515860886 \pm 1.002364571587161i$
10	-0.77808959867860	$0.45105515860886 \pm 1.002364571587161i$

Like all root-finding methods, Durand–Kerner will perform poorly for polynomials with multiple roots. To illustrate this, we consider another example.

■ **EXAMPLE 3.13**

Let

$$p(x) = (x^2 + x + 1)(x + 2)^2 = (x^2 + x + 1)(x^2 + 4x + 4) = x^4 + 5x^3 + 9x^2 + 8x + 4$$

which has roots at $\zeta = \frac{1}{2}(-1 \pm \sqrt{3}i), -2, -2$. Our localization isn't much help, because it suggests that we need to look in the very large region defined by $|z| \le 9$. We take $\pm 3 \pm 3i$ as the initial values and get the sequence of iterates given in Table 3.10.

While we have converged to the complex roots in about 12 iterations, after 15 iterations the real roots still have a non-trivial imaginary part. Changing the initial guesses will, of course, change the iterates, but not the slow convergence.

Table 3.10   Durand–Kerner multiple root example, $p(x) = x^4 + 5x^3 + 9x^2 + 8x + 4$.

k	$\zeta_{1,2}$	$\zeta_{3,4}$
1	$0.63425925925926 \pm 2.745370370037037i$	$-3.13425925925926 \pm 2.52314814814815i$
2	$0.41780641520150 \pm 1.87602995645876i$	$-2.91780641520150 \pm 1.72344096231311i$
3	$0.07346662069880 \pm 1.38332434612017i$	$-2.57346662069880 \pm 1.21571014561951i$
4	$-0.20524029076841 \pm 1.08373409475309i$	$-2.29475970923159 \pm 0.84144913620496i$
5	$-0.39076218369095 \pm 0.92341170976034i$	$-2.10923781630905 \pm 0.54298780244318i$
6	$-0.47968786225950 \pm 0.87006386827860i$	$-2.02031213774050 \pm 0.30996703413706i$
7	$-0.49912954857761 \pm 0.86567889432403i$	$-2.00087045142239 \pm 0.15897779863809i$
8	$-0.49999821882186 \pm 0.86601742223755i$	$-2.00000178117814 \pm 0.07954479815244i$
9	$-0.50000001267065 \pm 0.86602539212555i$	$-1.99999998732935 \pm 0.03977228743418i$
10	$-0.50000000000866 \pm 0.86602540378715i$	$-1.99999999999134 \pm 0.01988614333141i$
11	$-0.50000000000000 \pm 0.86602540378444i$	$-2.00000000000000 \pm 0.00994307166561i$
12	$-0.50000000000000 \pm 0.86602540378444i$	$-2.00000000000000 \pm 0.00497153583280i$
13	$-0.50000000000000 \pm 0.86602540378444i$	$-2.00000000000000 \pm 0.00248576791641i$
14	$-0.50000000000000 \pm 0.86602540378444i$	$-2.00000000000000 \pm 0.00124288395823i$
15	$-0.50000000000000 \pm 0.86602540378444i$	$-2.00000000000000 \pm 0.00062144197911i$

■ **EXAMPLE 3.14**

One problem with Durand–Kerner is that the user has to use a complex initial guess to find a complex root. If we make all real initial guesses for the polynomial in Example 3.11, we get the results in Table 3.11. Note that we haven't even converged to the real roots; rather, all the iterates seem to be bouncing around erratically—and taking more iterations doesn't help. This is because Durand–Kerner depends on the accuracy of *all* the iterates for convergence. (This might suggest that *all* your initial guesses should have some non-zero imaginary part.)

Table 3.11   Durand–Kerner convergence failure, $p(x) = x^4 - x - 1$.

k	ζ_1	ζ_2	ζ_3	ζ_4
0	1.00000000000000	-1.00000000000000	2.00000000000000	-2.00000000000000
1	0.83333333333333	-1.16666666666667	0.91666666666667	-0.58333333333333
2	-4.88888888888889	-0.33587301587302	5.56537037037037	-0.34060846560847
3	-2.23215884066467	-5.45557075319540	2.95040908151362	4.73732051234645
4	-2.45596730911110	-2.23199317241196	3.87306538304134	0.81489509848171
5	5.70485505349840	-8.48469654009375	2.01002632527200	0.76981516132335
6	1.63698623298533	-4.718870916709491	2.28686243290975	0.79486050119983
7	2.94334173319389	-2.68401723049215	-1.25565402788743	0.99632952518569
8	1.39771626940761	-0.87276922359398	-1.45862978500119	0.93368273918756
9	0.92623290486333	-1.06128810798049	-0.21334231281918	0.34839751593633
10	1.83566699890128	-0.50150330337031	1.23207846240992	-2.56624215794089

As of the writing of this edition of this text, the "state of the art" method for polynomial root finding is based on a paper by Jenkins and Traub [8], and involves re-writing the polynomial root-finding problem as a matrix eigenvalue problem. We therefore will defer discussion of this until Chapter 8.

Exercises:

1. Use the Durand–Kerner algorithm to find all the roots of the polynomial

$$p(x) = x^4 - 10x^3 + 35x^2 - 50x + 24.$$

   You should get $\zeta_{1,2,3,4} = (1, 2, 3, 4)$.

2. Use the Durand–Kerner algorithm to find all the roots of the following polynomials. (Feel free to use MATLAB'S `roots` command to check your results):

   (a) $p(x) = x^6 + x^5 + x^4 + x^3 + x^2 + x + 1$;

   (b) $p(x) = x^6 - x^5 - 1$;

   (c) $p(x) = x^9 - x^8 - 1$;

   (d) $p(x) = x^5 - 1$.

3. Use Durand–Kerner to find all the roots of the polynomial

$$p(x) = x^7 - x - 1.$$

4. Use Durand–Kerner to find all the roots of the polynomial

$$p(x) = x^8 - x - 1.$$

5. Now consider the polynomial

$$p(x) = x^6 - ax - 1,$$

   where a is a real parameter. We want to investigate how the roots depend on a. For various values of $a \in [-2, 2]$, compute the roots of p and observe how they change as a changes. Can you plot the real roots as a function of a? Try to extend the range of values of a. Does anything interesting happen?

6. Repeat the above for the polynomial

$$p(x) = x^6 - x - b,$$

   where now $b \in [-2, 2]$ is a parameter.

7. Use MATLAB's `rand` function to generate a random polynomial of degree 10. (Remember to make it monic!) Use Durand–Kerner to find the roots of this polynomial, and check your results by using MATLAB's `roots` function.

◁ ● ● ▷

3.11 SPECIAL TOPICS IN ROOT-FINDING METHODS

3.11.1 Extrapolation and Acceleration

One of the more important aspects of numerical analysis is that we can sometimes accelerate or improve the convergence of an algorithm with very little additional effort, simply by using the output of the algorithm to estimate some of the uncomputable quantities in the analysis of the algorithm. The best example of this will come in Chapter 5. Here we use this basic idea to speed up the convergence of a linearly convergent sequence, in a process known as *Aitken extrapolation*.[14]   It can be used to accelerate a linear algorithm into a quadratic one, and thus it makes the fixed-point iteration ideas somewhat competitive with Newton's method. Moreover, we will use these ideas to recover the speed of convergence of Newton's method when the theory of §3.6 does not apply; this is done in §3.11.4.

We begin as we almost always do with fixed-point iteration, by considering the error. If we assume that g is continuously differentiable, then the Mean Value Theorem says that there exists ξ_{n-1} between α and x_{n-1} such that

$$\alpha - x_n = g(\alpha) - g(x_{n-1}) = g'(\xi_{n-1})(\alpha - x_{n-1}).$$

Now consider

$$\alpha - x_n = (\alpha - x_{n-1}) + (x_{n-1} - x_n) = \frac{1}{g'(\xi_{n-1})}(\alpha - x_n) + (x_{n-1} - x_n)$$

so that solving for α implies that

$$\alpha = x_n + \frac{g'(\xi_{n-1})}{1 - g'(\xi_{n-1})}(x_n - x_{n-1}). \tag{3.40}$$

This formula is the starting point for the construction of our extrapolation algorithm. Note that it gives an expression for α in terms of x_n, x_{n-1}, and $g'(\xi_{n-1})$. The problem is that the last quantity is not computable; but it can be *estimated*, and that is the key.

Since we are assuming that $x_n \to \alpha$, we also know that $g'(\xi_n) \to g'(\alpha)$; thus,

$$g'(\xi_{n-1}) \approx g'(\alpha).$$

---

[14] Alexander Craig Aitken (1895–1967) was born at Dunedin, New Zealand, of Scottish parents, and educated at the University of Otago, located in Dunedin. In 1923 he obtained a scholarship for graduate study at the University of Edinburgh in Scotland, where he spent the rest of his life. The idea of accelerating the convergence of a linearly converging sequence, sometimes called the Aitken δ^2 process, is presented in a 1926 paper, "On Bernoulli's numerical solution of algebraic equations," published in the *Proceedings of the Royal Society of Edinburgh*. During World War II he was part of the British effort to decode German ENIGMA messages.

On the other hand, consider the ratio

$$
\begin{aligned}
\gamma_n &= \frac{x_{n-1} - x_n}{x_{n-2} - x_{n-1}}, \\
&= \frac{(\alpha - x_{n-1}) - (\alpha - x_n)}{(\alpha - x_{n-2}) - (\alpha - x_{n-1})}, \\
&= \frac{(\alpha - x_{n-1}) - g'(\xi_{n-1})(\alpha - x_{n-1})}{(\alpha - x_{n-1})/g'(\xi_{n-2}) - (\alpha - x_{n-1})}, \\
&= \frac{1 - g'(\xi_{n-1})}{1/g'(\xi_{n-2}) - 1}, \\
&= g'(\xi_{n-2})\frac{1 - g'(\xi_{n-1})}{1 - g'(\xi_{n-2})}, \\
&\to g'(\alpha).
\end{aligned}
$$

Thus, $\gamma_n \approx g'(\alpha)$ as well. The difference (and it is an important one) is that γ_n is *computable*. We thus can use

$$
\gamma_n \approx g'(\xi_{n-1})
$$

in (3.40) in order to get a completely computable estimate of α:

$$
\alpha \approx x_n + \frac{\gamma_n}{1 - \gamma_n}(x_n - x_{n-1}),
$$

which we use to define a new sequence of approximations:

$$
\bar{x}_n = x_n + \frac{\gamma_n}{1 - \gamma_n}(x_n - x_{n-1}). \tag{3.41}
$$

There are a couple of ways to use Aitken extrapolation in an algorithm. Perhaps the most obvious is to use the ordinary fixed-point iteration to produce the x_n sequence, with the $\bar{x}_n$ sequence being produced passively, as it were.

■ **EXAMPLE 3.15**

Let'e return to our example iteration for which $g(x) = \frac{1}{2}e^{-x}$. Taking $x_0 = 0$, we compute the first several ordinary iterates as follows:

$$
\begin{aligned}
x_1 &= g(x_0) = \frac{1}{2}e^0 = \frac{1}{2}; \\
x_2 &= g(x_1) = \frac{1}{2}e^{-1/2} = 0.3032653299; \\
x_3 &= g(x_2) = \frac{1}{2}e^{-0.3032653299} = 0.369201575.
\end{aligned}
$$

Now, starting with x_2, we can begin to compute the accelerated values $\bar{x}_n$ according to

$$
\bar{x}_n = x_n + \frac{\gamma_n}{1 - \gamma_n}(x_n - x_{n-1}),
$$

where

$$
\gamma_n = \frac{x_{n-1} - x_n}{x_{n-2} - x_{n-1}}.
$$

Proceeding in this fashion we get

$$
\begin{aligned}
\bar{x}_2 &= 0.3032653299 + \left(\frac{-0.33515315365051}{1 + 0.33515315365051} \right)(0.3032653299 - 0.5) \\
&= 0.35265011013472, \\
\bar{x}_3 &= 0.369201575 + \left(\frac{-0.35729286262005}{1 + 0.35729286262005} \right)(0.3692015759 - 0.3032653299) \\
&= 0.35184456205221.
\end{aligned}
$$

Pseudo-code for this might go as in Algorithm 3.4; further computation involving this example is given in Table 3.12.

Algorithm 3.4 *Aitken Extrapolation, Version 1.*

```
x1 = g(x0)
x2 = g(x1)
for k=1 to n do
        if (dabs(x1 - x0) > 1.d-20) then
                gamma = (x2 - x1)/(x1 - x0)
        else
                gamma = 0.0d0
        endif
        xbar = x2 + gamma*(x2 - x1)/(1 - gamma)
        if(abs(xbar - x2) < error) then
                alpha = xbar
                stop
        endif
        x = g(x2)
        x0 = x1
        x1 = x2
        x2 = x
enddo
```

Programming Note: Since division by zero—or a very small number—is possible in the computation of gamma, we put in a conditional test: Only if the denominator is greater than 10^{-20} do we carry through the division.

Although Algorithm 3.4 does result in faster convergence, it does not take as much advantage of the acceleration as is possible. An alternate version actually feeds the extrapolated value back into the computation; this is sometimes known as *Steffenson's iteration*, and is given in Algorithm 3.5, although there is another root-finding method (see the exercises in §3.9) associated with this name, so confusion is possible.

■ **EXAMPLE 3.16**

By using the same function and initial value, we compute as follows:

$$
\begin{aligned}
x_1' &= g(x_0) = \frac{1}{2}e^0 = \frac{1}{2}; \\
x_2' &= g(x_1') = \frac{1}{2}e^{-1/2} = 0.3032653299.
\end{aligned}
$$

Now we do the acceleration using x_0, x_1', and x_2':

$$\gamma = \frac{x_1' - x_2'}{x_0 - x_1'} = -0.3934693402;$$

thus, the next value of the iteration is

$$x_1 = 0.3032653299 + \left(\frac{-0.3934693402}{1 + 0.3934693402} \right)(0.3032653299 - 0.5) = 0.3588166496.$$

We then compute (note that, in this notation, x_2' gets re-defined)

$$x_2' = g(x_1) = \frac{1}{2}e^{-0.3588166496} = 0.3492512051;$$

$$x_3' = g(x_2') = \frac{1}{2}e^{-0.3492512051} = 0.3526079771.$$

Now use Aitken acceleration with x_1, x_2', and x_3':

$$\gamma = \frac{x_2' - x_3'}{x_1 - x_2'} = -0.3509269224.$$

Thus, the next iterate is

$$x_2 = 0.3526079771 + \left(\frac{-0.3509269224}{1 + 0.3509269224} \right)(0.3526079771 - 0.3492512051) = 0.3517359968.$$

Pseudo-code for this algorithm is given in Algorithm 3.5. Note, again, that we checked for division by small numbers before computing gamma.

Algorithm 3.5 *Aitken Extrapolation, Version 2.*

```
for k=1 to n do
        x1 = g(x0)
        x2 = g(x1)
        if (dabs(x1 - x0) > 1.d-20) then
                gamma = (x2 - x1)/(x1 - x0)
        else
                gamma = 0.0d0
        endif
        x0 = x2 + gamma*(x2 - x1)/(1 - gamma)
        if(abs(x0 - x2) < error) then
                alpha = x0
                stop
        endif
enddo
```

Table 3.12 shows more computations with the original fixed-point iteration $x_{n+1} = \frac{1}{2}e^{-x_n}$, together with both Aitken extrapolation algorithms. As a further comparison, we have also included the data for using Newton's method to find the root of the function $f(x) = x - \frac{1}{2}e^{-x}$, whose root is the same as the fixed point of g.

Note that the first Aitken algorithm does converge somewhat more quickly than the unaccelerated fixed-point iteration, but the second Aitken algorithm is *much* faster than both. (The table is somewhat deceptive in this regard. The second Aitken algorithm uses two evaluations of g at each step, so it really is converging in eight iterations instead of four.) Note that Newton and the second Aitken algorithm both converged in four iterations.

Table 3.12   Extrapolation example, $g(x) = \frac{1}{2}e^{-x}$.

n	Fixed-point	Aitken, version 1	Aitken, version 2	Newton
0	0.0	N/A	0.	0.
1	0.50000000000000	N/A	0.35881664959840	0.33333333333333
2	0.30326532985632	0.50000000000000	0.35173599679979	0.35168933155542
3	0.36920157498737	0.35881664959840	0.35173371124943	0.35173371099294
4	0.34564302521408	0.35265011013472	0.35173371124920	0.35173371124920
5	0.35388254815500	0.35184456205221	0.35173371124920	0.35173371124920
6	0.35097870435331	0.35174752132793	0.35173371124920	0.35173371124920
7	0.35199937290228	0.35173541541015	0.35173371124920	0.35173371124920
8	0.35164028150092	0.35173392226949	0.35173371124920	0.35173371124920
9	0.35176657517651	0.35173373734771	0.35173371124920	0.35173371124920
10	0.35172215208802	0.35173371447837	0.35173371124920	0.35173371124920
11	0.35173777701935	0.35173371164868	0.35173371124920	0.35173371124920
12	0.35173228118368	0.35173371129862	0.35173371124920	0.35173371124920
13	0.35173421425181	0.35173371125531	0.35173371124920	0.35173371124920
14	0.35173353432626	0.35173371124995	0.35173371124920	0.35173371124920
15	0.35173377347896	0.35173371124929	0.35173371124920	0.35173371124920

It can be shown, under the correct hypotheses, that the second Aitken algorithm is *quadratically* convergent. In fact, this is one application of Theorem 3.7; see Problem 4.

One problem that we have conveniently avoided discussing so far is that Aitken extrapolation is very susceptible to subtractive cancellation, based on the formula for γ_n as well as the difference $x_{n-1} - x_n$ that appears in (3.41). For this reason the computations in Aitken extrapolation should be done in as high precision as possible. It is sometimes suggested that, after the extrapolated method has converged, a few ordinary iterations be carried out to remove any error caused by the subtractive cancellation.

3.11.2   Variants of Newton's Method

One application of the fixed-point theory developed in §3.9 is to the study of other, Newton-like iterations that can be proposed. In this section, we will look at three such ideas. All three are locally convergent, although the first two are only linear and not quadratic; the third is an example of a cubic convergent method.

The Chord Method   Newton's method works by using the tangent line approximation to find the approximate root of the function, $f(x)$. This, of course, requires the derivative $f'(x)$, and uses two function evaluations in each step of the iteration. The chord method uses the original value of the derivative, $f'(x_0)$, on each iteration, instead of just the first one:

$$x_{n+1} = x_n - \frac{f(x_n)}{f'(x_0)}.$$

By giving up the need to evaluate the derivative at each step of the iteration, we save some cost, but should also expect to lose some convergence and accuracy. The question is: How much do we lose?

If we apply Theorem 3.6 to

$$g(x) = x - \frac{f(x)}{f'(x_0)},$$

we get that

$$g'(x) = 1 - \frac{f'(x)}{f'(x_0)} \Rightarrow g'(\alpha) = 1 - \frac{f'(\alpha)}{f'(x_0)} = \frac{f''(\xi)}{f'(x_0)}(\alpha - x_0),$$

so that $|g'(\alpha)| < 1$—which is what controls the existence of local convergence, according to Theorem 3.6—depends on how close x_0 is to α. Moreover, we won't get quadratic convergence unless $x_0 = \alpha$ (we know this from Theorem 3.7), so we know we have lost the extra speed that Newton's method gives us; the chord method is only linear and only locally convergent. This might prompt the question: Why bother with it? After all, we can use the secant method, which is faster than a linear method, and does not use the derivative either. So what purpose does the chord method serve?

A complete answer would require almost another entire course. The chord method is useful in solving nonlinear *systems* of equations, that is, problems involving 2 or 3 or 20 (or more) equations in 2 or 3 or 20 (or more) unknowns. In this setting, which we briefly discuss in §7.8, the derivative of f is actually a matrix, so every evaluation of f' is actually the evaluation of an $n \times n$ matrix. Having a method, even if it is only linearly convergent, that avoids having to re-evaluate (and invert—since the division by f' in the scalar case becomes matrix inversion in the systems case) that matrix every step of the way is valuable.

One interesting variant of the chord method updates the point at which the derivative is evaluated, but not every iteration. Thus, for the first, say, p iterations, we use

$$x_{n+1} = x_n - \frac{f(x_n)}{f'(x_0)},$$

but for the next p iterations we use

$$x_{n+1} = x_n - \frac{f(x_n)}{f'(x_p)},$$

and so on, periodically updating the derivative evaluation point. Although this method will not converge as quickly as Newton's method, it will converge more quickly than the ordinary chord method, and is less costly than Newton's method.

■ **EXAMPLE 3.17**

Let's use the chord method to find the root of $f(x) = 2 - e^x$, the same example that we have used throughout this chapter. Taking $x_0 = 0$, we have $f'(x_0) = -1$, so

$$
\begin{aligned}
x_1 &= x_0 - \frac{f(x_0)}{f'(x_0)} = 0 - \frac{2-1}{-1} = 1; \\
x_2 &= x_1 - \frac{f(x_1)}{f'(x_0)} = 1 - \frac{2-e}{-1} = 0.2817181715; \\
x_3 &= x_2 - \frac{f(x_2)}{f'(x_0)} = 0.2817181715 - \frac{2 - e^{0.2817181715}}{-1} = 0.9563130411; \\
x_4 &= x_3 - \frac{f(x_3)}{f'(x_0)} = 0.9563130411 - \frac{2 - e^{0.9563130411}}{-1} = 0.3542280558.
\end{aligned}
$$

If we continue the iteration, we find that it does converge, but much more slowly than did Newton's method. For example, $x_{12} = 0.465017583563$, which is still off in the first decimal place, whereas Newton's method had converged in only five iterations.

If we update the computation of f' every, say, three iterations, then we get improved performance. The first three iterates are the same as in the pure chord method:

$$
\begin{aligned}
x_1 &= x_0 - \frac{f(x_0)}{f'(x_0)} = 1; \\
x_2 &= x_1 - \frac{f(x_1)}{f'(x_0)} = 0.2817181715; \\
x_3 &= x_2 - \frac{f(x_2)}{f'(x_0)} = 0.9563130411.
\end{aligned}
$$

But now we compute $f'(0.9563130411) = -2.602084985$, so the next few iterates are

$$
\begin{aligned}
x_4 &= x_3 - \frac{f(x_3)}{f'(x_3)} = 0.7249274450; \\
x_5 &= x_4 - \frac{f(x_4)}{f'(x_3)} = 0.7001083868; \\
x_6 &= x_5 - \frac{f(x_5)}{f'(x_3)} = 0.6947392372,
\end{aligned}
$$

which is a much more accurate approximation (but still not as good as Newton's method).

Other Approximations to the Derivative   In §3.8 we briefly mentioned the method

$$
x_{k+1} = x_k - f(x_k) \frac{h}{f(x_k + h) - f(x_k)}, \tag{3.42}
$$

which is based on using a finite difference approximation to the derivative in Newton's method.

We can now use Theorem 3.7 to analyze the convergence of this method. Here we have

$$
g(x) = x - f(x) \frac{h}{f(x + h) - f(x)},
$$

so that

$$g'(x) = 1 - f'(x)\left(\frac{h}{f(x+h)-f(x)}\right) - f(x)\left(\frac{h}{f(x+h)-f(x)}\right)'.$$

Since $f(\alpha) = 0$, we have

$$g'(\alpha) = 1 - f'(\alpha)\left(\frac{h}{f(\alpha+h)-f(\alpha)}\right).$$

But (2.1) implies that

$$\frac{f(\alpha+h)-f(\alpha)}{h} = f'(\alpha) + \frac{1}{2}hf''(\xi_{\alpha,h}),$$

so that we have

$$g'(\alpha) = 1 - f'(\alpha)\left(\frac{1}{f'(\alpha)+\frac{1}{2}hf''(\xi_{\alpha,h})}\right),$$

which simplifies to

$$g'(\alpha) = \frac{\frac{1}{2}hf''(\xi_{\alpha,h})}{f'(\alpha)+\frac{1}{2}hf''(\xi_{\alpha,h})} = \left(\frac{\frac{1}{2}f''(\xi_{\alpha,h})}{f'(\alpha)+\frac{1}{2}hf''(\xi_{\alpha,h})}\right)h.$$

Thus, for this method, as long as $f'(\alpha) \neq 0$, $g'(\alpha) = \mathcal{O}(h) \neq 0$, and therefore we have only linear convergence, although the rate constant for the linear convergence will typically be small (since h is small). But the secant method will converge faster, in general, and requires only a single function evaluation per step; the method (3.42) requires two each iteration. Thus, this is not a reasonable method to use; it is just as costly as Newton's method, more costly than secant, and slower than both.

Higher-Order Convergence: Halley's Method   Could we perhaps get a method better than Newton's method if we were willing to use the second derivative of f? The answer is yes, and the method is known as Halley's method.[15]

Actually, there are two methods. The most obvious one is to use the quadratic term in a Taylor series expansion of f to define a more accurate x_{n+1} from x_n. This means that we define x_{n+1} as the root of the quadratic

$$0 = f(x_n) + (x - x_n)f'(x_n) + \frac{1}{2}(x - x_n)^2 f''(x_n).$$

This leads to the iteration

$$x_{n+1} = x_n - \frac{2f(x_n)}{f'(x_n) \pm \sqrt{[f'(x_n)]^2 - 2f(x_n)f''(x_n)}}, \qquad (3.43)$$

---

[15]Edmund Halley (1656–1742) was born near London and educated at Oxford. He is most famous for analyzing the orbit of the comet of 1682 and predicting that it would return in 76 years, which it did. To this day the comet is known as Halley's Comet.

Although an astronomer and not a mathematician, Halley had a great influence on the development of mathematics because he encouraged Newton to finally write up and publish his work applying the calculus to problems of planetary and celestial motion, which led to the writing of *Principia Mathematica*, originally published in 1687 with funds from Halley's own pocket. The methods for solving equations that bear his name date from a paper written for the Royal Society of London in 1684. The derivation used here is different from that used by Halley.

where we choose the sign in the denominator to make x_{n+1} closer to x_n, i.e., to maximize the magnitude of the denominator.

The second method is a bit more subtle. Define the rational function

$$R(x) = \frac{a - x}{bx + c}$$

and note that $R(x) = 0 \Rightarrow x = a$. Now, define a, b, and c so that

$$R(x_n) = f(x_n), \ R'(x_n) = f'(x_n), \ R''(x_n) = f''(x_n).$$

Thus, we have defined R as the rational function that matches f and its first two derivatives at x_n (essentially, R is a "rational Taylor approximation" to f); we thus expect that $R \approx f$ near x_n, and we can use the (easily computable) root of R to approximate the root of f. We have a system of three equations in the three unknowns a, b, and c, which we can solve to get

$$a = x - \frac{2f(x)f'(x)}{2[f'(x)]^2 - f(x)f''(x)}.$$

This is the only one that matters; we thus have the iteration

$$x_{n+1} = x_n - \frac{2f(x_n)f'(x_n)}{2[f'(x_n)]^2 - f(x_n)f''(x_n)}. \tag{3.44}$$

Both of these methods converge with order $p = 3$, which can be proved by application of Theorem 3.7. (See Problem 9.) Table 3.13 shows the results of using each of these iterations to find the root of $f(x) = 2 - e^x$. Although this convergence is indeed extremely fast, bear in mind that both versions of Halley's method require a second derivative function, and that each iteration involves *three* function evaluations, whereas Newton requires only two and the secant method only one.

Table 3.13   Halley's method examples, $f(x) = 2 - e^x$.

n	x_n, via (3.44)	Error	x_n, via (3.43)	Error
0	0.00000000	0.693147180560e+00	0.00000000	0.693147180560e+00
1	0.66666667	0.264805138933e−01	0.73205081	0.389036270089e−01
2	0.69314563	0.154727507107e−05	0.69313707	0.101121327991e−04
3	0.69314718	0.000000000000e+00	0.69314718	0.222044604925e−15
4	0.69314718	0.000000000000e+00	0.69314718	0.000000000000e+00

3.11.3   The Secant Method: Theory and Convergence

In this section we will establish the error theory for the secant method. The result here is similar to that for Newton's method, but somewhat more difficult to establish.

Theorem 3.9 *Let f be twice continuously differentiable in a neighborhood of a root α, and assume that $f'(x) \neq 0$ for all x in this neighborhood. Then, for x_0 and x_1 sufficiently close to α, the secant iteration (3.28) converges to α, with*

$$\lim_{n \to \infty} \frac{\alpha - x_{n+1}}{\alpha - x_n} = 0 \tag{3.45}$$

and

$$\lim_{n \to \infty} \frac{\alpha - x_{n+1}}{(\alpha - x_n)^p} = \left(\frac{1}{2} \frac{f''(\alpha)}{f'(\alpha)} \right)^{p-1} \tag{3.46}$$

for $p = (1 + \sqrt{5})/2 \approx 1.618\dots$.

Proof: We begin by establishing an error formula for the secant method similar to the one for Newton's method.

Recall that the secant method is based on finding the exact root of a linear polynomial that interpolates the original function f. Let s be this interpolating polynomial, so that we have

$$s(x) = f(x_n) + (x - x_n) \left(\frac{f(x_n) - f(x_{n-1})}{x_n - x_{n-1}} \right), \tag{3.47}$$

where the next iterate is defined by solving $0 = s(x_{n+1})$. Then the linear interpolation error estimate (2.12) implies that

$$f(x) - s(x) = \frac{1}{2}(x - x_n)(x - x_{n-1})f''(\xi_n)$$

for all x; here ξ_n is in the interval defined by x_n, x_{n-1}, and α. Set $x = \alpha$, so that we have

$$f(\alpha) - s(\alpha) = \frac{1}{2}(\alpha - x_n)(\alpha - x_{n-1})f''(\xi_n).$$

But $f(\alpha) = 0$, so we have

$$-s(\alpha) = \frac{1}{2}(\alpha - x_n)(\alpha - x_{n-1})f''(\xi_n).$$

On the other hand, we know that $s(x_{n+1}) = 0$, so we can substitute this in to get

$$s(x_{n+1}) - s(\alpha) = \frac{1}{2}(\alpha - x_n)(\alpha - x_{n-1})f''(\xi_n).$$

The Mean Value Theorem then implies that there exists a value η_n between α and x_{n+1} such that $s(x_{n+1}) - s(\alpha) = s'(\eta_n)(x_{n+1} - \alpha)$; therefore,

$$s'(\eta_n)(x_{n+1} - \alpha) = \frac{1}{2}(\alpha - x_n)(\alpha - x_{n-1})f''(\xi_n)$$

or

$$x_{n+1} - \alpha = \frac{1}{2}(\alpha - x_n)(\alpha - x_{n-1})\frac{f''(\xi_n)}{s'(\eta_n)}. \tag{3.48}$$

But we can go directly to (3.47) to show that (again appealing to the MVT) there is a θ_n between x_n and x_{n-1} such that

$$s'(\eta_n) = \frac{f(x_n) - f(x_{n-1})}{x_n - x_{n-1}} = f'(\theta_n).$$

Thus, we have that

$$x_{n+1} - \alpha = \frac{1}{2}(\alpha - x_n)(\alpha - x_{n-1})\frac{f''(\xi_n)}{f'(\theta_n)}. \tag{3.49}$$

Note that with this relationship in hand, it quickly follows that if the secant iterates x_n are converging to α, then

$$\lim_{n\to\infty} \frac{\alpha - x_{n+1}}{\alpha - x_n} = -\frac{1}{2} \lim_{n\to\infty} (\alpha - x_{n-1}) \frac{f''(\alpha)}{f'(\alpha)} = 0.$$

Thus, we have (3.45). Finishing the proof then requires two steps: (1) Establishing that convergence occurs, if the initial points are close enough; and, (2) Showing that the limit (3.46) holds.

Convergence: This part of the proof follows the Newton proof very closely. Let I be the neighborhood of α in which f is twice continuously differentiable, and so on. As in the Newton theorem, we can find a closed interval $J \subset I$ such that $J = \{x \mid \alpha - \epsilon \le x \le \alpha + \epsilon\}$ for some $\epsilon > 0$, and the ratio

$$M = \frac{\max_{x\in J} |f''(x)|}{2 \min_{x\in J} |f'(x)|}$$

is bounded. Take x_0 and x_1 in J so that

$$|\alpha - x_i| \le \delta < \min\{\epsilon, 1\}, \ i = 0, 1,$$

and

$$M\delta < 1.$$

Note that this is possible simply by taking both initial points close enough to α. Therefore, from (3.49),

$$|\alpha - x_2| \le M|\alpha - x_1||\alpha - x_0| \le |\alpha - x_0|,$$

so that $x_2 \in J$ is assured. The same argument shows that each secant iterate $x_n \in J$; thus, we have ($e_k = |\alpha - x_k|$):

$$\begin{aligned}
e_2 &\le e_1 e_0 M \le M\delta^2, \\
e_3 &\le e_2 e_1 M \le (e_1 e_0 M) e_1 M = e_1^2 e_0 M^2 < M^2 \delta^3, \\
e_4 &\le e_3 e_2 M \le (e_1^2 e_0 M^2)(e_1 e_0 M)M = e_1^3 e_0^2 M^4 < M^4 \delta^5,
\end{aligned}$$

and an inductive argument easily shows that

$$e_n \le M^{-1} (M\delta)^{q_n}, \tag{3.50}$$

where q_n is the sequence of *Fibonnaci numbers,*[16] defined by

$$q_{n+1} = q_n + q_{n-1}, \quad q_1 = q_0 = 1.$$

---

[16]Leonardo Pisano (1170–ca. 1250) is more widely known by his nickname, Fibonacci. He traveled widely in the Mediterranean region with his father, a diplomat in the service of Pisa. Fibonacci wrote a number of texts, the most famous of which is probably *Liber abbaci*, which appeared in 1202. This contains the famous problem that leads to the Fibonacci sequence:

> A certain man put a pair of rabbits in a place surrounded on all sides by a wall. How many pairs of rabbits can be produced from that pair in a year if it is supposed that every month each pair begets a new pair which from the second month on becomes productive?

Fibonacci also dabbled in geometry and number theory, but his most significant work may have been as simple as spreading the use of the Hindu–Arabic system of numeration.

Since $M\delta < 1$ and $q_n \to \infty$, it follows immediately that $e_n \to 0$ and therefore x_n converges to α.

Order of Convergence: The order of convergence is defined as the value p such that

$$\lim_{n\to\infty} \frac{\alpha - x_{n+1}}{(\alpha - x_n)^p} = C \qquad (3.51)$$

for $C \neq 0$ and finite. Shifting the limit variable yields the equivalent statement

$$\lim_{n\to\infty} \frac{\alpha - x_n}{(\alpha - x_{n-1})^p} = C \qquad (3.52)$$

or, taking powers of both sides,

$$\lim_{n\to\infty} \frac{(\alpha - x_n)^p}{(\alpha - x_{n-1})^{p^2}} = C^p. \qquad (3.53)$$

Multiplying (3.51) and (3.53) then yields that p must satisfy

$$\lim_{n\to\infty} \frac{\alpha - x_{n+1}}{(\alpha - x_{n-1})^{p^2}} = C^{1+p}. \qquad (3.54)$$

At the same time, we note that the convergence of the iterates plus the error estimate (3.49) implies that

$$\lim_{n\to\infty} \frac{\alpha - x_{n+1}}{(\alpha - x_n)(\alpha - x_{n-1})} = -\frac{f''(\alpha)}{2f'(\alpha)}, \qquad (3.55)$$

which, when combined with (3.52), yields

$$\lim_{n\to\infty} \frac{\alpha - x_{n+1}}{(\alpha - x_{n-1})^{1+p}} = -C\frac{f''(\alpha)}{2f'(\alpha)}. \qquad (3.56)$$

Since the order of convergence is unique,[17] it follows from comparing (3.54) and (3.56) that

$$1 + p = p^2$$

and

$$C^{1+p} = -C\frac{f''(\alpha)}{2f'(\alpha)}.$$

Solving these two yields

$$p = \frac{1}{2}(1 + \sqrt{5})$$

and

$$C = \left(-\frac{f''(\alpha)}{2f'(\alpha)}\right)^{1/p},$$

which completes the proof. •

---

[17] We really ought to prove this, but it is not within the intended scope of the book.

3.11.4 Multiple Roots

So far our study of root-finding methods has assumed that the derivative of the function did not vanish at the root: $f'(\alpha) \neq 0$. The rapid convergence of Newton's method and the secant method both depend on this assumption. In this section we investigate what happens if the derivative *does* vanish at the root. We will see that both the Newton and secant methods will continue to converge, but not as rapidly as we expect. Furthermore, there are other problems with trying to find a root when the derivative vanishes.

We start by pointing out some consequences of having $f'(\alpha) = 0$. Geometrically, this means that the graph of the function is tangent to the axis at the root.

Lemma 3.1 *If f is k times continuously differentiable in a neighborhood of α, and*

$$f(\alpha) = f'(\alpha) = \cdots = f^{(k-1)}(\alpha) = 0,$$

that is, if the function and first $k - 1$ derivatives vanish at α, but $f^{(k)}(\alpha) \neq 0$, then we can write

$$f(x) = (x - \alpha)^k F(x), \tag{3.57}$$

where $F(\alpha) \neq 0$. Similarly, if we can write f in the form (3.57), where $F(\alpha) \neq 0$, then it follows that the first $k - 1$ derivatives vanish at α.

Proof: See Problem 17. ●

If $k = 2$, so that both the function and the first derivative are zero at the root, then α is called a *double root*. If more derivatives are zero, then the root is generally called a *multiple root*.

■ **EXAMPLE 3.18**

Let $f(x) = \cos^2 x$, which has a root at $x = \frac{1}{2}\pi$. Since $f'(x) = -2 \sin x \cos x$, it follows that the derivative also vanishes at $x = \frac{1}{2}\pi$, so f has a double root. We can write f in the form called for in the lemma by the simple device of writing

$$f(x) = \cos^2 x = \left(x - \frac{\pi}{2}\right)^2 F(x),$$

where

$$F(x) = \frac{\cos^2 x}{\left(x - \frac{\pi}{2}\right)^2}.$$

As long as $x \neq \frac{\pi}{2}$, $F(x)$ is well-defined. What happens at $x = \frac{\pi}{2}$? We can use L'Hôpital's rule to determine that

$$\lim_{x \to \frac{\pi}{2}} F(x) = \lim_{x \to \frac{\pi}{2}} \frac{-2 \sin x \cos x}{2 \left(x - \frac{\pi}{2}\right)} = \lim_{x \to \frac{\pi}{2}} \frac{-2 \cos^2 x + 2 \sin^2 x}{2} = 1.$$

Thus, we would define

$$F(x) = \begin{cases} \frac{\cos^2 x}{\left(x - \frac{\pi}{2}\right)^2}, & x \neq \frac{\pi}{2}; \\ 1, & x = \frac{\pi}{2}. \end{cases}$$

We might want to continue our study by conducting some experiments. Consider the function $f(x) = 1 - xe^{1-x}$. Note that $\alpha = 1$ is a root, and that $f'(\alpha) = 0$ as well. If we

Table 3.14   Double-root example; $f(x) = 1 - xe^{1-x}$; computation stopped after 20 iterations.

	Newton		Secant	
n	x_n	$\alpha - x_n$	x_n	$\alpha - x_n$
0	0.000000000000	1.000000000000	0.000000000000	1.000000000000
1	0.367879441171	0.632120558829	0.135335283237	0.864664716763
2	0.626665778573	0.373334221427	0.421192747824	0.578807252176
3	0.792118936917	0.207881063083	0.586457122517	0.413542877483
4	0.889216318667	0.110783681333	0.724548464295	0.275451535705
5	0.942618075030	0.057381924970	0.817997254576	0.182002745424
6	0.970768039664	0.029231960336	0.882864790983	0.117135209017
7	0.985242636648	0.014757363352	0.925476813313	0.074523186687
8	0.992585155212	0.007414844788	0.953105630395	0.046894369605
9	0.996283431246	0.003716568754	0.970673923815	0.029326076185
10	0.998139415613	0.001860584387	0.981742402042	0.018257597958
11	0.999069131113	0.000930868887	0.988663870449	0.011336129551
12	0.999534421170	0.000465578830	0.992973792390	0.007026207610
13	0.999767174462	0.000232825538	0.995649804081	0.004350195919
14	0.999883578197	0.000116421803	0.997308460142	0.002691539858
15	0.999941786839	0.000058213161	0.998335397262	0.001664602738
16	0.999970892852	0.000029107148	0.998970783260	0.001029216740
17	0.999985446281	0.000014553719	0.999363742391	0.000636257609
18	0.999992723105	0.000007276895	0.999606707485	0.000393292515
19	0.999996361541	0.000003638459	0.999756907516	0.000243092484
20	0.999998180783	0.000001819217	0.999849751283	0.000150248717

apply Newton's method and the secant method to this example, with $x_0 = 0$ for Newton and secant, and $x_{-1} = -1$ for secant, we get the results shown in Table 3.14.

The table suggests that both iterations are converging, but also that neither one is converging as rapidly as we might have expected, based on previous results. Can we explain this?

Clearly, the fact that $f'(\alpha) = 0$ will have an effect on both Newton and secant. The error formulas (3.12) and (3.49) and limits (3.24) and (3.46) both require that $f'(\alpha) \neq 0$ in order to hold. Can we find anything more in the way of an explanation?

Theorem 3.7 tells us when a fixed-point iteration exhibits quadratic convergence, and we applied it to Newton's method in §3.9. However, that exercise was done under the assumption that $f'(\alpha) \neq 0$. Let us repeat that exercise, this time under the assumption that f has a double root, and therefore

$$f(x) = (x - \alpha)^2 F(x).$$

Then the iteration function for Newton's method is

$$g(x) = x - \frac{f(x)}{f'(x)} = x - \frac{(x-\alpha)^2 F(x)}{2(x-\alpha)F(x) + (x-\alpha)^2 F'(x)} = x - \frac{(x-\alpha)F(x)}{2F(x) + (x-\alpha)F'(x)}$$

and we can compute (many details skipped; see Problem 21)

$$g'(\alpha) = 1 - \frac{1}{2} = \frac{1}{2} \neq 0. \tag{3.58}$$

Note that we no longer have $g'(\alpha) = 0$; therefore (according to Theorem 3.7), we no longer have quadratic convergence. The simple fact that the function is tangent to the axis at the root costs us our rapid convergence. (Note that we still have linear convergence, based on Theorem 3.6.)

Note also that (3.58) gives us a clue as to how to recover the lost quadratic convergence. If we change the Newton iteration to be

$$x_{n+1} = x_n - 2\frac{f(x_n)}{f'(x_n)},$$

then an application of Theorem 3.7 shows that we now have $g'(\alpha) = 0$ and therefore we again have quadratic convergence for Newton's method. More generally, it is not difficult to show that, for a k-fold root, we would use

$$x_{n+1} = x_n - k\frac{f(x_n)}{f'(x_n)} \tag{3.59}$$

as the (modified) Newton iteration.

The problem with this technique is that it requires that we know the degree of multiplicity of the root ahead of time, and this is generally not the case in practice. So an alternative is needed. Given f, with a multiple root of unknown order at $x = \alpha$, define

$$u(x) = \frac{f(x)}{f'(x)}. \tag{3.60}$$

It can be shown (see Problem 19) that u has only a single root at $x = \alpha$. Thus, Newton's method applied to find a root of u will avoid any problems of multiple roots. The drawback of this method is that applying Newton's method to u will require that we have a formula for the *second* derivative of f.

Since the secant method is not a single-point iteration, it does not fit into the theory of §3.9, so we have no way to analyze its behavior rigorously in the presence of multiple roots. Clearly, the results of Table 3.14 suggest that some degradation of performance does occur; this is not surprising, since the secant method is based on approximating Newton's method. However, it is worth noting that we can replace $f(x)$ with $u(x) = f(x)/f'(x)$ in the secant method as well as in Newton's method. This removes the multiple root, and with the secant method we only need the additional function for the first derivative.

Finally, we might note that the linear convergence that Theorem 3.7 predicts for Newton's method in the presence of multiple roots can be dealt with by using Aitken extrapolation.

To illustrate some of these ideas, we return to our example function $f(x) = 1 - xe^{1-x}$. In Table 3.15 we show the results of applying several different algorithms to this example:

1. Newton's method as modified in (3.59); this is x_n in the table.

2. Newton's method based on the change of function (3.60); this is y_n in the table.

3. Newton's method with Aitken extrapolation; this is z_n in the table.

4. The secant method based on the change of function (3.60); this is s_n in the table.

Table 3.15   Multiple-root computations for $f(x) = 1 - xe^{1-x}$.

	Modified Newton's method (3.59)		Newton's method for $u(x) = f(x)/f'(x)$	
n	x_n	$\log_{10}(\alpha - x_n)$	y_n	$\log_{10}(\alpha - y_n)$
0	0.000000000000e+00	0.000	0.000000000000e+00	0.000
1	0.735758882343e+00	−0.578	0.139221119118e+01	−0.406
2	0.978185253678e+00	−1.661	0.104801490853e+01	−1.319
3	0.999842233626e+00	−3.802	0.100076234718e+01	−3.118
4	0.999999991704e+00	−8.081	0.100000019370e+01	−6.713
5	0.999999991704e+00	−8.081	0.999999999938e+00	−10.208
6	0.999999991704e+00	−8.081	0.999999999876e+00	−9.907
7	0.999999991704e+00	−8.081	0.999999999752e+00	−9.606
8	0.999999991704e+00	−8.081	0.999999999752e+00	−9.606
9	0.999999991704e+00	−8.081	0.999999999752e+00	−9.606
10	0.999999991704e+00	−8.081	0.999999999752e+00	−9.606
	Newton's method with Aitken acceleration		Secant Method for $u(x) = f(x)/f'(x)$	
n	z_n	$\log_{10}(\alpha - z_n)$	s_n	$\log_{10}(\alpha - s_n)$
0	0.000000000000e+00	0.000	0.000000000000e+00	0.000
1	0.124054847245e+01	−0.619	0.677393774677e+00	−0.491
2	0.100856591658e+01	−2.067	0.111944928199e+01	−0.923
3	0.100001218112e+01	−4.914	0.986957103091e+00	−1.885
4	0.100000000001e+01	−11.147	0.999485342581e+00	−3.288
5	0.100000000001e+01	−11.147	0.100000224007e+01	−5.650
6	0.100000000001e+01	−11.147	0.999999999674e+00	−9.487
7	0.100000000001e+01	−11.147	0.100000052208e+01	−6.282
8	0.100000000001e+01	−11.147	0.100000029546e+01	−6.530
9	0.100000000001e+01	−11.147	0.999999999965e+00	−10.456
10	0.100000000001e+01	−11.147	0.999999999965e+00	−10.456

Some comments might be in order here. Note that all four modified methods converged to near the root $\alpha = 1$ much faster than did the original Newton and secant iterations in Table 3.14. Note, however, that there appears to be a limit to how accurate any of these iterations can get, and this accuracy is not as good as in the past. What is going on?

There is a problem inherent in the geometry of a multiple root that makes it difficult—if not outright impossible—to approximate the root to high accuracy. To illustrate the problem, let's look at a graph of the polynomial

$$p(x) = x^5 - (5/2)x^4 + (5/2)x^3 - (5/4)x^2 + (5/16)x - (1/32) = \left(x - \frac{1}{2}\right)^5$$

in a very narrow region around the root $\alpha = \frac{1}{2}$. Figure 3.12 shows a plot of 8,000 points from this curve on the interval $[0.45, 0.55]$, computed in floating-point arithmetic. Note that we do not have a smooth curve but, rather, a "cloud" of values, distributed *along* the x-axis. Thus, there are many values of x near the exact root which yield very small values of $p(x)$, and all of the iterations we are using will yield $x_{n+1} - x_n$ very small if $f(x_n)$ is very small. So, once the iteration lands on one of the points that produces a small function value, it essentially stays there, even if it is not all that close to the exact root. Thus, we get "premature convergence" to a value that is not a good approximation to the root. This is not a flaw in the root-finding method, it is a difficulty inherent in the problem of finding a root when the derivative is also zero at the root, using finite-precision arithmetic. We should note that this effect does depend on the way that the polynomial is computed; see Problem 25.

What this tells us is that we need to be very careful how we organize the computations when working near a multiple root, and we need to work in as high a level of precision as possible when doing so, or simply accept that our accuracy is going to be less than we are used to for simple roots.

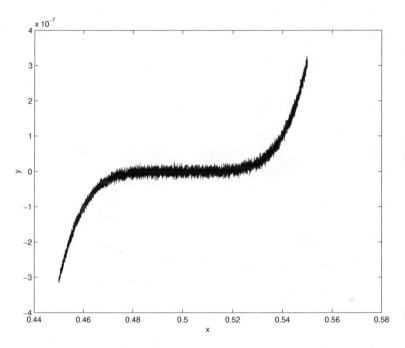

Figure 3.12   Close look near the root of the polynomial $p(x) = x^5 - (5/2)x^4 + (5/2)x^3 - (5/4)x^2 + (5/16)x - (1/32) = \left(x - \frac{1}{2}\right)^5$, computed in floating-point arithmetic.

3.11.5 In Search of Fast Global Convergence: Hybrid Algorithms

We began this chapter with the bisection method, which is perhaps like the tortoise in Aesop's fable: slow but steady and reliable. We then looked at Newton's method and the secant method, which are perhaps both like the rabbit: fast but potentially unreliable. Does there exist a fast *and* reliable method for finding roots of a function?

There is no such "single algorithm;" however, we can create one by combining the features that make the bisection method reliable with the features that make Newton and secant fast. The basic ideas are incorporated, in a somewhat more sophisticated fashion than we present here, in an algorithm known as *Brent's algorithm*.

Given a function f and an interval $[a, b]$, with $f(a)f(b) < 0$, so we know that a root exists in the interval, we begin by doing the secant iteration, using the endpoints of the interval as our initial guesses. As long as the secant iterates are within the bracketing interval, we continue with the secant iteration. At the same time, we use the result of each secant iteration to update the bracketing interval so that it shrinks as much as possible. Finally, if the secant iterate ever leaves the bracketing interval, then we ignore that result, and do bisection in order to get closer to the root before re-starting the secant iteration.

Formally, we have the following sequence of steps:

Algorithm 3.6 *Hybrid Algorithm.*

Given an initial interval $[a, b] = [a_1, b_1]$, *with* $f(a_1)f(b_1) < 0$, *set* $k = 1$, $x_0 = a_1$, $x_1 = b_1$, *and proceed as follows:*

1. *Set* $c = x_k - f(x_k)(x_k - x_{k-1})/(f(x_k) - f(x_{k-1}))$;

2. *If* $c < a_k$ *or* $c > b_k$ *then (do bisection):*

   (a) *Set* $c = a_k + \frac{1}{2}(b_k - a_k)$;

   (b) *If* $f(a_k)f(c) < 0$ *then*

      i. $[a_{k+1}, b_{k+1}] = [a_k, c]$;

      ii. *Set* $x_{k+1} = c$, $x_k = a_k$;

   (c) *If* $f(c)f(b_k) < 0$ *then*

      i. $[a_{k+1}, b_{k+1}] = [c, b_k]$;

      ii. *Set* $x_{k+1} = c$, $x_k = b_k$;

3. *else (update brackets for secant step):*

   (a) *If* $f(a_k)f(c) < 0$ *then* $[a_{k+1}, b_{k+1}] = [a_k, c]$;

   (b) *If* $f(c)f(b_k) < 0$ *then* $[a_{k+1}, b_{k+1}] = [c, b_k]$;

   (c) *Set* $x_{k+1} = c$;

4. *Set* $k = k + 1$ *and go to Step 1.*

Note that although this algorithm does maintain a set of brackets around the root, it is using the secant iterates x_k for the secant step, *not* the bracket values. This is what makes this method different from regula-falsi.

■ **EXAMPLE 3.19**

Algorithm 3.6 is sufficiently complicated that it is worth a close look, via an example.

Let $f(x) = 2xe^{-15} - 2e^{-15x} + 1$; it is a simple matter to compute $f(0) = -1$ and $f(1) = 1$, thus a root exists on the interval $[0, 1]$.

Step 1: The secant prediction is $c = 0.5$, which is indeed in the bracketing interval $[0, 1]$, so we accept the secant step and update as follows:

$$
\begin{aligned}
f(c) &= 0.99889... \Rightarrow [a, b] \leftarrow [0, c] = [0, 0.5], \\
x_2 &= 0.5, \\
x_1 &= 1.0.
\end{aligned}
$$

Step 2: The secant prediction this time is

$$c = x_2 - f(x_2)(x_2 - x_1)/(f(x_2) - f(x_1)) = -0.4511 \times 10^3,$$

which is clearly outside the interval $[0, 0.5]$, so we do a step of bisection, and update as follows

$$
\begin{aligned}
c &= 0.25, \\
f(c) &= 0.95296... \Rightarrow [a, b] \leftarrow [0, c] = [0, 0.25], \\
x_3 &= 0.25, \\
x_2 &= 0.0.
\end{aligned}
$$

Note that in the bisection step, x_2 is redefined to be the other end of the interval from x_3; this ensures that we start the next iteration with the two most recent (hence, most accurate) approximations that "bracket" the root.

In the exercises we ask the student to continue this computation by hand.
Consider another example function

$$f(x) = \frac{20x - 1}{19x}.$$

The root is at $x = 0.05$, of course, and we have to have $x_n \approx 0.07$ before the secant method will begin to converge. Table 3.16 shows the result of applying our algorithm to this function, with $[a, b] = [0.01, 1]$ and $x_0 = a$, $x_1 = b$. The last column shows which algorithm was used to get the current approximate root (S = secant, B = bisection). This is not quite the same as Brent's algorithm, but it is in the same spirit without some of the details.

Table 3.16   Hybrid algorithm applied to $f(x) = \frac{20x-1}{19x}$.

n	x_n	$f(x_n)$	$\alpha - x_n$	a	b	Method
1	0.81000000	0.98765432	−0.7600000E+00	0.010	0.810	S
2	0.41000000	0.92426187	−0.3600000E+00	0.010	0.410	B
3	0.33800000	0.89691685	−0.2880000E+00	0.010	0.338	S
4	0.17400000	0.75015124	−0.1240000E+00	0.010	0.174	B
5	0.14920000	0.69987301	−0.9920000E−01	0.010	0.149	S
6	0.07960000	0.39143084	−0.2960000E−01	0.010	0.080	B
7	0.07368000	0.33830505	−0.2368000E−01	0.010	0.074	S
8	0.03598144	−0.41011085	0.1401856E−01	0.036	0.074	S
9	0.05663919	0.12338844	−0.6639190E−02	0.036	0.057	S
10	0.05186144	0.03778160	−0.1861438E−02	0.036	0.052	S
11	0.04975283	−0.00522940	0.2471688E−03	0.050	0.052	S
12	0.05000920	0.00019369	−0.9201785E−05	0.050	0.050	S
13	0.05000005	0.00000096	−0.4548788E−07	0.050	0.050	S
14	0.05000000	0.00000000	0.8371401E−11	0.050	0.050	S

Exercises:

1. Consider the fixed-point iteration $x_{n+1} = 1 + e^{-x_n}$, with $x_0 = 0$. Do four steps (by hand) and apply both of the Aitken acceleration algorithms to speed up the convergence of the iteration.

2. Repeat the above for the iteration $x_{n+1} = \frac{1}{2}\ln(1 + x_n)$, using $x_0 = 1/2$.

3. Consider the iteration $x_{n+1} = e^{-x}$. Show (by computational experiment) that it coverges for $x_0 = 1/2$; then use both Aitken acceleration algorithms to speed up the convergence. What is the gain in terms of function calls to the iteration function?

4. Write the second Aitken iteration in the form

$$x_{k+1} = G(x_k).$$

   Hint: Take

$$x_1 = g(x_0), \quad x_2 = g(x_1) = g(g(x_0))$$

   and use this to write the updated value of x0 in Algorithm 3.5 entirely in terms of $x_0, g(x_0),$ and $g(g(x_0))$.

5. Consider the iteration

$$x_{k+1} = G(x_k),$$

   where

$$G(x) = g(g(x)) + \frac{H(x)}{1 - H(x)}(g(g(x)) - g(x))$$

   for

$$H(x) = \frac{g(g(x)) - g(x)}{g(x) - x}.$$

   Assume that $\alpha = g(\alpha)$ is a fixed point for g.

   (a) Use L'Hôpital's rule to show that

$$H(\alpha) = g'(\alpha).$$

   (b) Use Theorem 3.7 to show that the fixed-point iteration for G is quadratic.

6. Apply the chord method to each of the functions in Problem 3 of §3.1, using the midpoint of the interval as x_0. Compare your results to what you got for bisection, Newton, and/or the secant method.

7. Repeat the above, but this time apply both of the Aitken acceleration algorithms to improve the convergence of the iteration. Compute the values of the ratio

$$R_n = \frac{\alpha - x_{n+1}}{(\alpha - x_n)^2},$$

   where α is the best value for the root, as found by previous methods. Does the sequence of R_n values appear to be converging to a limit? What does this tell you?

8. Repeat Problem 6, but this time update the value used to compute the derivative in the chord method every other iteration. Comment on your results.

9. Show that both Halley method iterations are of third order. *Hint:* Write them as fixed-point iterations

$$x_{n+1} = g(x_n)$$

and write the iteration function as $g(x) = f(x)G(x)$. The fact that $f(\alpha) = 0$ will make the derivative computations easier.

10. Apply the Halley iteration (3.44) to $f(x) = x^2 - a$, $a > 0$, to derive a cubic method for finding the square root. Test this by using it to find $\sqrt{5}$.

11. Repeat the above for $f(x) = x^N - a$. Test this by finding the cubic, quartic, and quintic roots of 2, 3, 4, and 5.

12. Apply Halley's second method to $f(x) = a - e^x$ to show that

$$x_{n+1} = x_n + 2\left(\frac{ae^{-x_n} - 1}{ae^{-x_n} + 1}\right)$$

is a cubic convergent iteration for $\alpha = \log a$. Analyze the convergence of this iteration, under the assumption that $a \in [\frac{1}{2}, 1]$. Comment on this as a practical means of computing the logarithm of an arbitrary a.

13. Use the secant method with $x_0 = 0$ and $x_1 = 1$ to find the root of $f(x) = 2 - e^x$, which has exact solution $\alpha = \ln 2$. Compute the ratio

$$R = \frac{\alpha - x_{n+1}}{(\alpha - x_n)^p}$$

for $p = (1 + \sqrt{5})/2$. Do we get convergence to the appropriate value?

14. Repeat the experiment in the previous exercise, but use values of p just a bit above and below the correct value. What happens?

15. Consider applying the secant method to find the root of the function $f(x) = 4x - \cos x$. Assume that we want accuracy to within 10^{-8}. Use the secant error estimate (3.49) to show that the iteration will converge for all $x_0 \in [-2, 2]$. How many iterations will be needed for the iteration to converge?

16. Verify by actual computation that your results in the previous exercise were correct; i.e., apply the secant method to $f(x) = 4x - \cos x$. Do your results converge to the specified accuracy in the correct number of iterations, and does this convergence appear to be occurring for all choices of x_0?

17. Prove Lemma 3.1. *Hint:* Expand f in a Taylor series about α.

18. Let $f(x) = 1 - xe^{1-x}$. Write this function in the form (3.57). What is $F(x)$? Use Taylor's Theorem or L'Hôpital's Rule to determine the value of $F(\alpha)$.

19. For $u(x)$ as defined in (3.60), where f has a k-fold root at $x = \alpha$, show that $u(\alpha) = 0$ but that $u'(\alpha) \neq 0$.

20. Use the modified Newton method (3.59) to find the root $\alpha = 1$ for the function $f(x) = 1 - xe^{1-x}$. Is the quadratic convergence recovered?

21. Let f have a double root at $x = \alpha$. Show that the Newton iteration function

$$g(x) = x - f(x)/f'(x)$$

is such that $g'(\alpha) \neq 0$. Provide all details of the calculation.

22. Using a hand calculator, carry out six iterations of the hybrid method for the function

$$f(x) = 2xe^{-15} - 2e^{-15x} + 1.$$

You should be able to match the values generated for x_2 and x_3 in the text. In addition, $x_6 = 0.4541055 \times 10^{-1}$.

23. Write a computer code to implement the hybrid method described in §3.11.5. Apply it to each of the functions given below, using the given interval as the starting interval. On your output, be sure to indicate which iteration method was used at each step.

    (a) $f(x) = x^{1/19} - 19^{1/19}$, $[1, 100]$;
    (b) $f(x) = 2xe^{-5} - 2e^{-5x} + 1$, $[0, 1]$;
    (c) $f(x) = x^2 - (1 - x)^{20}$, $[0, 1]$;
    (d) $f(x) = 2xe^{-20} - 2e^{-20x} + 1$, $[0, 1]$.

24. In Problem 9 of §3.1, we introduced the regula-falsi method. In this exercise, we demonstrate how the hybrid method (Algorithm 3.6) is different from regula-falsi. Let

$$f(x) = e^{-4x} - \frac{1}{10}.$$

    (a) Show that f has a root in the interval $[0, 5]$.

    (b) Using this interval as the starting interval, compute five iterations of regula-falsi; be sure to tabulate the new interval in addition to the new approximate root value predicted by the method at each step. A graph showing the location of each approximate root might be useful.

    (c) Using the same interval as the starting interval, compute five iterations of the hybrid method given in this section. Again, note the new interval as well as the new approximate root value at each step. A graphical illustration might be instructive.

    (d) Based on your results, comment on the difference between regula-falsi and the hybrid method.

25. Write a computer program that evaluates the polynomial $p(x) = (x - 1)^5$ using the following three forms for the computation:

$$
\begin{aligned}
p(x) &= (x - 1)^5, \\
p(x) &= x^5 - 5x^4 + 10x^3 - 10x^2 + 5x - 1, \\
p(x) &= -1 + x(5 + x(-10 + x(10 + x(-5 + x)))).
\end{aligned}
$$

Use this to evaluate the polynomial at 400 equally spaced points on the interval $[0.998, 1.002]$. Comment on the results.

◁ • • • ▷

3.12   VERY HIGH-ORDER METHODS AND THE EFFICIENCY INDEX

A number of very high-order methods have been proposed in the recent literature, and an *efficiency index* exists as a measure of a root-finding method's overall value. In this section we will explore some of these ideas.

In [11] the following simple computation was proposed as a measure of the efficiency of a root-finding method:

$$I = p^{1/d},$$

where I is the efficiency index, p is the theoretical order of the method, and d is the number of function or derivative evaluations needed per step of the method. Thus, for Newton's method, we have $I_{\text{Newton}} = 2^{1/2} = 1.414\ldots$, while for Halley's method (3.44) we have that the order $p = 3$ and the number of evaluations is also 3 (we do not consider repeated use of the same function or derivative evaluation to be different); thus $I_{\text{Halley}} = 3^{1/3} = 1.4422\ldots$, suggesting that Halley's method is (very slightly) more efficient than Newton. The idea is that the order of the method is shrunk somewhat by having a higher number of function/derivative evaluations. Consider now the method [4] defined as follows:

$$(CN) \quad \begin{cases} w_n &= x_n - \dfrac{f(x_n)}{f'(x_n)}, \\[2mm] z_n &= w_n - \dfrac{f(w_n)}{f'(x_n)}\left[1 - \dfrac{f(w_n)}{f(x_n)}\right]^{-2}, \\[2mm] x_{n+1} &= z_n - \dfrac{f(z_n)}{f'(x_n)}\left[1 - \dfrac{f(w_n)}{f(x_n)} - \dfrac{f(z_n)}{f(x_n)}\right]^{-2}. \end{cases}$$

We will henceforth refer to this as the *Chun–Neta method*, after the authors of [4]. This can be proved to be sixth order [4], and it uses only four different function/derivative evaluations, thus its efficiency index is

$$6^{1/4} = 1.5651\ldots,$$

which is higher than for Newton or Halley.

Is this more complicated method really better than Newton? Is Halley really (if only slightly) better than Newton? Let's look at an example. Consider the simple function $f(x) = 2 - e^x$, for which the exact solution is $x_* = \ln 2$. Since we know the exact solution, and all three methods converge, Table 3.17 shows the error at each iteration. Note that Chun–Neta *does* converge more quickly than the others (and Halley is faster than Newton).

Table 3.17   Comparison of Newton and Halley with (CN).

Iteration	Newton	Halley	Chun–Neta
1	0.30685281944005	0.02648051389328	0.02519826516177
2	0.04261170178294	0.00000154727507	0.00000000002376
3	0.00089511935897	0	0
4	0.00000040049983	0	NaN
5	0.00000000000008	0	NaN
6	0	0	NaN

Before we go running off to use procedures like Chun–Neta in all our root-finding computations, we need to note that the total number of function/derivative evaluations used by Newton to get to the sixth iteration, is 12, while Chun–Neta used 12 function/derivative evaluations to reach the third iteration. In other words, it is not obvious from this example that Chun–Neta is really superior. It is worth pointing out that Chun–Neta used only eight evaluations to reach its second iterate, which is accurate to 10 digits; when Newton uses only 8 evaluations, it found an iterate that was accurate to only 6 digits. Thus it appears that Chun–Neta can reach more accurate results faster than Newton.

Many of the high-order methods are rather complicated; one reason for using the Chun–Neta method here is that it is not very complicated. A second high-order method is the so-called *super Halley method* [6], defined by

$$(SH) \qquad x_{n+1} = x_n - \left(1 + \frac{1}{2}\left[\frac{\left(\frac{f(x_n)f''(x_n)}{(f'(x_n))^2}\right)}{1 - \left(\frac{f(x_n)f''(x_n)}{(f'(x_n))^2}\right)}\right]\right)\frac{f(x_n)}{f'(x_n)}.$$

(A list of high-order methods is given in [9].) This is a fourth-order method with three evaluations, so the efficiency index is

$$I_{\text{superHalley}} = 4^{1/3} = 1.5874\ldots,$$

slightly better than the sixth-order Chun–Neta method.

Let's look at some more examples.

■ **EXAMPLE 3.20**

Take $f(x) = x^6 - x - 1$; this has two real roots, at $x_* = 1.13472413840152$, and at $\xi_* = -0.77808959867860$. If we try to find x_* first, using $x_0 = 2$, we get the results shown in Table 3.18.

Table 3.18   Comparison for $f(x) = x^6 - x - 1$, $x_* = 1.13472413840152$.

n	Newton	Halley	Super Halley	Chun–Neta
1	1.68062827225131	1.46655372922485	1.03132927291221	1.33696037969668
2	1.43073898823906	1.18554345302099	1.13907207890072	1.13694126708183
3	1.25497095610944	1.13507468834525	1.13472390794046	1.13472413840156
4	1.16153843277331	1.13472413853078	1.13472413840152	1.13472413840152
5	1.13635327417051	1.13472413840152	1.13472413840152	NaN

Notice that after four iterations (eight evaluations) the Newton iterate is still inexact (the precise error is $|x_* - x_4^{\text{Newton}}| = 0.02681429437179$, which is not especially small), whereas Chun–Neta (16 evaluations) is exact. The super Halley method, for the same cost (three evaluations per iteration), is performing much better than "ordinary" Halley. In fact, super Halley is, in some sense, out-performing Chun–Neta because it is almost as accurate and yet less costly. Both are exact at the fourth iterate, but super Halley got there with only 12 evaluations, whereas Chun–Neta used 16. Given the values of the efficiency index for the two methods, this is not entirely surprising.

If we then try to find the second (real) root, $\xi_* = -0.77808959867860$, we get the results shown in Table 3.19.

Table 3.19   Comparison for $f(x) = x^6 - x - 1$, $\xi_* = -0.77808959867860$.

n	Newton	Halley	Super Halley	Chun–Neta
1	-1.66321243523316	-1.42052750704421	-0.79465796170651	-1.26108380706201
2	-1.38102084158243	-1.00786610328889	-0.77807368302110	-0.83587288749870
3	-1.14600652598945	-0.79590872245371	-0.77808959867862	-0.77809208194547
3	-0.95850422286092	-0.77809390958652	-0.77808959867860	-0.77808959867860

OK, here we have that Chun–Neta is exact after four iterations (16 evaluations) but Newton is not even close. In fact, it took Newton nine iterations (18 evaluations) to find this root exactly, thus Chun–Neta was more efficient in this case. Super Halley was even better, finding the exact root in four iterations (12 evaluations).

It frankly is difficult—almost impossible, in fact—to anticipate which root-finding method will converge most efficiently. The efficiency index is an imprecise measure, as we have seen. It provides some insight into which methods should perform better when considered over a large set of examples, but the variability of performance, especially the dependence on the initial condition, makes it difficult to draw broad conclusions.

Exercises:

1. Write a program to employ both Chun–Neta and super Halley to find the root of a given function. Test it on the following examples, using the given values of x_0.

   (a) $f(x) = \sin x - \frac{1}{2}x$, $x_* = 1.895494267$; $x_0 = 2, 2.5, 3$;

   (b) $f(x) = e^{x^2+7x-30} - 1$, $x_* = 3$, $x_0 = 4, 5, 6$;

   (c) $f(x) = e^x \sin x + \ln(1 + x^2)$, $x_* = 1$, $x_0 = 0, -1, -2$.

2. In §3.7 we employed (and analyzed) Newton's method as an approximator for $\sqrt{a}$ by using it to find the positive root of the function $f(x) = x^2 - a$. In this and some of the following exercises we apply our high-order methods to this task.

   (a) Use Halley's method to find the root of $f(x) = x^2 - a$. Construct an iteration function, as simplified as possible, so that the iteration is $x_{n+1} = G_H(x_n)$. Verify that your construction works by testing it on $a = 0.5$, for which the exact value is $\sqrt{0.5} = 0.70710678118655$. Note the number of arithmetic operations required by your function during each iteration.

   (b) Repeat part (a) for super Halley, obtaining the iteration $x_{n+1} = G_{SH}(x_n)$ for as simple a G_{SH} as possible.

3. Recall that in §3.7 we were able to restrict our attention to values in the interval $[\frac{1}{4}, 1]$, and therefore construct an initial guess by linear interpolation. Modify your codes from the previous exercise to reflect this, and use them to approximate $\sqrt{a}$ for $a = 0.3, 0.6$, and 0.9. In addition, write a code that uses Newton's method (or simply use your code from §3.7), and test it on the same values of a. Which method achieves full accuracy (as measured by MATLAB's sqrt function) in the fewest operations for each a?

4. Modify your codes to step from $a = 0.25$ to $a = 1$ in increments of 0.001, finding $\sqrt{a}$ for each a using each of the three methods. Thus, you will first find $\sqrt{0.25}$, then $\sqrt{0.251}$, and so on, all the way to $a = 1.00$. What is the *average* number of operations used by each method to obtain full accuracy, as measured by the flops command?

5. Repeat the above using some of the alternate initial value generators from Problems 6, 7, and 8 of §3.7.

6. This is more of an essay-type question: As part of your job, you are going to be given a series of difficult root-finding problems that require highly accurate solutions. Of all the methods we have studied in this chapter, including the new methods in this section, which one would you choose? Justify your answer.

7. Think about combining one of these high-order methods with a global method as we did in §3.11.5. Can you design an algorithm that is better (faster, more efficient) than the one in that section?

◁ ● ● ● ▷

3.13   LITERATURE AND SOFTWARE DISCUSSION

The problem of solving a single nonlinear equation numerically is an old one in numerical analysis, with a fairly rich and complete literature. In addition to the treatments in almost any decent numerical analysis text, the monographs by Ostrowski [10], Traub [11], Householder [7], and Wait [12] are worth mentioning for more in-depth discussions of many of the issues.

The standard "state of the art" algorithm is known as Brent's algorithm. The original reference is [5] (which leads some to call it Dekker's method or the Brent–Dekker method) and there is a good discussion in Atkinson's text [2]. Brent also provided an algorithm in [3]. It has been suggested in at least one place [1] that Brent's method might be improved. An implementation of Brent's method may be found in the MATLAB function fzero. Brent's method algorithm is a hybrid method, combining bisection, the secant method, and inverse quadratic interpolation (see Chapter 4) to obtain a very fast and nearly global method.

The development of very high-order methods has become more common in recent years. A sampling of the literature is included in the references.

REFERENCES

1. Alefeld, G. E., Potra, F. A., and Shi, Y., "Enclosing Zeroes of Continuous Functions," *ACM Trans. Math. Software*, vol. 21, no. 3, pp. 327–344, 1995.

2. Atkinson, Kendall, *An Introduction to Numerical Analysis*, John Wiley & Sons, Inc., New York, 1989 (2nd edition).

3. Brent, Richard, *Algorithms for Minimization Without Derivatives*, Prentice-Hall, Englewood Cliffs, NJ, 1973.

4. Chun, Changbum, and Neta, Beny, "A new sixth-order scheme for nonlinear equations," *Appl. Math. Lett.*, vol. 25, pp. 185–189, 2012.

5. Dekker, T., "Finding a zero by means of successive linear interpolation," in *Constructive Aspects of the Fundamental Theorem of Algebra*, Dejon, and Henrici, eds., John Wiley & Sons, Inc. New York, 1969.

6. Gutiérrez, J. M., and Hernández, M. A., "An acceleration of Newton's method: super-Halley method," *Appl. Math. Comput.*, vol. 117, pp. 223-239, 2001.

7. Householder, Alston, *The Numerical Treatment of a Single Nonlinear Equation*, McGraw-Hill, New York, 1970.

8. Jenkins, M., and Traub, J., "A three-stage variable shift algorithm for polynomial zeros and its relation to generalized Rayleigh iteration," *Numer. Math.*, vol. 14, pp. 252-263, 1970.

9. Neta, Beny, Scott, Melvin, and Chun, Changbum, "Basins of attraction for several methods to find simple roots of nonlinear equations," *Appl. Math. Comput.*, vol. 218, pp. 10548–10566, 2012.

10. Ostrowski, Alexander, *Solution of Equations and Systems of Equations*, Academic Press, New York, 1966.

11. Traub, Joseph, *Iterative Methods for the Solution of Equations*, Prentice-Hall, Englewood Cliffs, NJ, 1964.

12. Wait, R. A., *The Numerical Solution of Algebraic Equations*, John Wiley & Sons, Inc., Chichester, UK, 1979.

CHAPTER 4

INTERPOLATION AND APPROXIMATION

One of the oldest problems in mathematics—and, at the same time, one of the most applied—is the problem of constructing an approximation to a given function f from among simpler functions, typically (but not always) polynomials. A slight variation of this problem is that of constructing a smooth function from a discrete set of data points.

In this chapter we will study both of these problems and develop several methods for solving them. We start with a more general treatment of an idea we first saw in Chapter 2.

4.1 LAGRANGE INTERPOLATION

The basic interpolation problem can be posed in one of two ways:

1. Given a set of *nodes* $\{x_i, \quad 0 \le i \le n\}$ and corresponding data values $\{y_i, \quad 0 \le i \le n\}$, find the polynomial $p_n(x)$ of degree less than or equal to n, such that

$$p_n(x_i) = y_i, \quad 0 \le i \le n.$$

2. Given a set of *nodes* $\{x_i, \quad 0 \le i \le n\}$ and a continuous function $f(x)$, find the polynomial $p_n(x)$ of degree less than or equal to n, such that

$$p_n(x_i) = f(x_i), \quad 0 \le i \le n.$$

Note that in the first case we are trying to fit a polynomial to the data, and in the second we are trying to approximate a given function with the interpolating polynomial.

An Introduction to Numerical Methods and Analysis, Second Edition. By James F. Epperson
Copyright © 2013 John Wiley & Sons, Inc.

While the two cases are in fact different, we can always consider the first one to be a special case of the second (by taking $y_i = f(x_i)$, for each i), so we will present most of the material here in terms of the second version of the problem.

It is actually very easy to prove that both versions of the problem have a unique solution. Moreover, the proof shows us how to construct the polynomial.

Theorem 4.1 (Polynomial Interpolation Existence and Uniqueness) *Let the nodes* $x_i \in I$, $0 \le i \le n$ *be given. So long as each of the* x_i *is distinct, i.e.,* $x_i = x_j$ *if and only if* $i = j$, *then there exists a unique polynomial* p_n, *of degree less than or equal to* n, *which satisfies either of*

$$p_n(x_i) = y_i, \quad 0 \le i \le n \tag{4.1}$$

for a given set of data values $\{y_i\}$, *or*

$$p_n(x_i) = f(x_i), \quad 0 \le i \le n \tag{4.2}$$

for a given function $f \in C(I)$.

Proof: We begin by defining the family of functions[1]

$$L_i^{(n)}(x) = \prod_{\substack{k=0 \\ k \ne i}}^{n} \frac{x - x_k}{x_i - x_k}$$

and note that they are polynomials of degree n and have the interesting property that

$$L_i^{(n)}(x_j) = \delta_{ij} = \begin{cases} 1, & i = j; \\ 0, & i \ne j. \end{cases} \tag{4.3}$$

(The symbol δ_{ij} implicitly defined in the above is called the *Kronecker*[2] *delta*.) Figure 4.1 shows examples of these polynomials for five equally spaced nodes on $[-1, 1]$. Based on (4.3), then, if we define the polynomial by

$$p_n(x) = \sum_{k=0}^{n} y_k L_k^{(n)}(x),$$

it follows that

$$p_n(x_i) = \sum_{k=0}^{n} y_k L_k^{(n)}(x_i) = y_i.$$

---

[1]The reader is reminded that the symbol $\prod$ is used for a product in the same way that $\sum$ is used for a sum. Thus

$$\prod_{k=1}^{n} a_k = a_1 \times a_2 \times a_3 \times \cdots \times a_n.$$

[2]Leopold Kronecker (1823–1891) was born in Liegnitz, Prussia, and educated in local schools and the University of Berlin, graduating with a doctorate in 1845. Most of Kronecker's mathematical research was in what would now be called modern algebra and the theory of equations. In 1845, he left academia briefly to enter the business world, but he remained mathematically active even during this interlude. The fortune that he made while working as a banker enabled him to continue as a mathematician despite not having a regular professorship until 1883.

In his later years, Kronecker advocated a somewhat extreme philosophy of mathematics in which only finite numbers and finite operations were valid. The Kronecker delta notation came out of his work on determinants and matrices, done in 1869, under the title "Bemerkungen zur Determinanten-Theorie," published in the *Journal für die reine und angewandte Mathematik*.

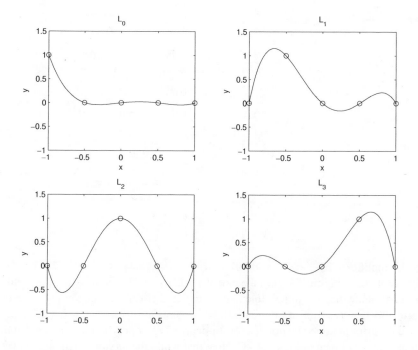

Figure 4.1   Lagrange polynomials for five equally spaced nodes on $[-1, 1]$.

Thus, the interpolatory conditions are satisfied, and it remains only to prove the uniqueness of the polynomial. To this end, assume that there exists a second interpolatory polynomial; call it q, and define

$$r(x) = p_n(x) - q(x).$$

Since both p_n and q are polynomials of degree less than or equal to n, so is their difference. However, we must note that

$$r(x_i) = p_n(x_i) - q(x_i) = y_i - y_i = 0$$

for each of the $n + 1$ nodes. Thus, we have a polynomial of degree less than or equal to n that has $n + 1$ roots. The only such polynomial is the zero polynomial; that is,

$$r(x) \equiv 0 \Rightarrow p_n(x) \equiv q(x),$$

and thus p_n is unique. ●

■ **EXAMPLE 4.1**

To illustrate this construction let $f(x) = e^x$ and consider the problem of constructing an approximation to this function on the interval $[-1, 1]$, using the nodes

$\{-1, -\frac{1}{2}, 0, \frac{1}{2}, 1\}$. We have

$$
\begin{aligned}
p_4(x) &= \frac{(x + \frac{1}{2})(x - 0)(x - \frac{1}{2})(x - 1)}{(-1 + \frac{1}{2})(-1 - 0)(-1 - \frac{1}{2})(-1 - 1)} e^{-1} \\
&+ \frac{(x + 1)(x - 0)(x - \frac{1}{2})(x - 1)}{(-\frac{1}{2} + 1)(-\frac{1}{2} - 0)(-\frac{1}{2} - \frac{1}{2})(-\frac{1}{2} - 1)} e^{-\frac{1}{2}} \\
&+ \frac{(x + 1)(x + \frac{1}{2})(x - \frac{1}{2})(x - 1)}{(0 + 1)(0 + \frac{1}{2})(0 - \frac{1}{2})(0 - 1)} e^{0} \\
&+ \frac{(x + 1)(x + \frac{1}{2})(x - 0)(x - 1)}{(\frac{1}{2} + 1)(\frac{1}{2} + \frac{1}{2})(\frac{1}{2} - 0)(\frac{1}{2} - 1)} e^{\frac{1}{2}} \\
&+ \frac{(x + 1)(x + \frac{1}{2})(x - 0)(x - \frac{1}{2})}{(1 + 1)(1 + \frac{1}{2})(1 - 0)(1 - \frac{1}{2})} e^{1},
\end{aligned}
$$

which simplifies to $p_4(x) = 1.0 + 0.997853749x + 0.499644939x^2 + 0.177347443x^3 + 0.043435696x^4$. Figure 4.2A is a the plot of f and $p_4(x)$. The approximation is sufficiently accurate that it is impossible to discern that there are two curves present, so it is probably better to look at Figure 4.2B, which plots the error $e^x - p_4(x)$. Compare this to the error obtained by doing a fourth-degree Taylor polynomial, centered at the origin, which is given in Figure 4.2C. Note that while the Taylor polynomial is *very* accurate near the middle of the interval, and much less accurate near the ends, the interpolating polynomial has less variation in its error, and a much smaller worst-case error over the same interval.

The construction presented in this section is called *Lagrange*[3] *interpolation*. Other approaches to the interpolation problem are studied in other sections; this one is the most basic and fundamental.

How good is interpolation at approximating a function? We will study this a little more carefully in §§4.3 and 4.12, but we can illustrate some of the issues by a few examples right now. We can already note that the fourth-degree approximation to the exponential function was very accurate, even when compared to the fourth-degree Taylor polynomial. What about some other functions?

Consider now the function $f(x) = (1 + 25x^2)^{-1}$. If we use a fourth-degree interpolating polynomial to approximate this function, the results are as shown in Figure 4.3A. These are not nearly as good as for the exponential function. If we increase the number of points (and therefore the degree of the interpolating polynomial), things improve on part of the interval, but not all of it. Figure 4.3B shows the plots for $n = 8$, and Figure 4.3C shows the $n = 16$ results. In contrast, Figure 4.3D plots the *error* for the 8-degree polynomial interpolation to the exponential function.[4])

---

[3]Joseph-Louis Lagrange (1736–1813) was born in Turin, Italy, of French parents. At a very young age he was made a professor at the Royal Artillery School in Turin. In 1766, he succeeded his friend and mentor Euler as the Director of the Mathematics Section of the Berlin Academy. In 1787 he accepted a similar position at the Paris Academy of Sciences, where he was active in establishing two very influential French schools, the École Polytechnique, and the École Normale. In his later years, he was granted a number of honors by the French emperor, Napoleon Bonaparte.

His results on interpolation were a major part of his work while in Berlin, but the interpolation formula ascribed to him here was not published until 1795, after he had moved to France. Ironically, Lagrange claims to have gotten the idea from some of Newton's work.

[4]The reader should note that all of these examples used equally spaced points to define the nodes. Other choices are possible, and there is a particular choice that is indeed better. (See §4.12.3.

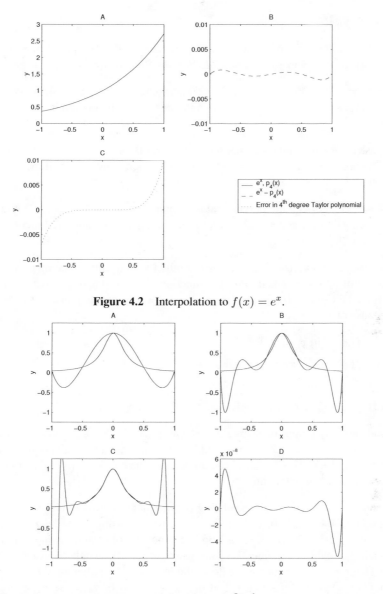

Figure 4.2   Interpolation to $f(x) = e^x$.

Figure 4.3   Polynomial interpolation to $f(x) = (1 + 25x^2)^{-1}$ with $n = 4$ (A), $n = 8$ (B), $n = 16$ (C), and the error in interpolation for $n = 8$ to $f(x) = e^x$(D).

Clearly, there are circumstances ($f(x) = e^x$) in which polynomial interpolation as approximation will work very well, and circumstances ($f(x) = (1 + 25x^2)^{-1}$) in which it will not. More study is therefore required.

The attentive reader will have noted that we have not discussed an algorithm for computing Lagrange interpolating polynomials. The reason for this is quite simple, actually: The Lagrange form of the interpolating polynomial is not well-suited for actual computations, and there is an alternate construction that is far superior to it. So we will defer all discussion of algorithms until Section 4.2.

Exercises:

1. Find the polynomial of degree 2 that interpolates at the data points $x_0 = 0$, $y_0 = 1$, $x_1 = 1$, $y_1 = 2$, and $x_2 = 4$, $y_2 = 2$. You should get $p_2(t) = -\frac{1}{4}t^2 + \frac{5}{4}t + 1$.

2. Find the polynomial of degree 2 that interpolates to $y = x^3$ at the nodes $x_0 = 0$, $x_1 = 1$, and $x_2 = 2$. Plot $y = x^3$ and the interpolating polynomial over the interval $[0, 2]$.

3. Construct the quadratic polynomial that interpolates to $y = 1/x$ at the nodes $x_0 = 1/2$, $x_1 = 3/4$, and $x_2 = 1$. Plot both the interpolate and the function over the interval $[\frac{1}{4}, \frac{5}{4}]$.

4. Construct the quadratic polynomial that interpolates to $y = \sqrt{x}$ at the nodes $x_0 = 1/4$, $x_1 = 9/16$, and $x_2 = 1$. Plot both the interpolate and the function over the interval $[0, 2]$.

5. For each function listed below, construct the Lagrange interpolating polynomial for the set of nodes specified. Plot the function, the interpolating polynomial, and the error, on the interval defined by the nodes. (You may want to plot the error separately, because of scale issues.)

    (a) $f(x) = \ln x$, $x_i = 1, \frac{3}{2}, 2$;

    (b) $f(x) = \sqrt{x}$, $x_i = 0, 1, 4$;

    (c) $f(x) = \log_2 x$, $x_i = 1, 2, 4$;

    (d) $f(x) = \sin \pi x$, $x_i = -1, 0, 1$.

6. Find the polynomial of degree 3 that interpolates $y = x^3$ at the nodes $x_0 = 0$, $x_1 = 1$, $x_2 = 2$, and $x_3 = 3$. (Simplify your interpolating polynomial as much as possible.) *Hint:* This is easy if you think about the implications of the uniqueness of the interpolating polynomial.

7. Construct the Lagrange interpolating polynomial to the function $f(x) = x^2 + 2x$, using the nodes $x_0 = 0$, $x_1 = 1$, $x_2 = -2$. Repeat, using the nodes $x_0 = 2$, $x_1 = 1$, and $x_2 = -1$. (For both sets of nodes, simplify your interpolating polynomial as much as possible.) Comment on your results, especially in light of the uniqueness part of Theorem 4.1, and then write down the interpolating polynomial for interpolating to $f(x) = x^3 + 2x^2 + 3x + 1$ at the nodes $x_0 = 0$, $x_1 = 1$, $x_2 = 2$, $x_3 = 3$, $x_4 = 4$, and $x_5 = 5$. *Hint:* You should be able to do this last part without doing any computations.

8. Let f be a polynomial of degree $\leq n$, and let p_n be a polynomial interpolant to f, at the $n + 1$ distinct nodes $x_0, x_1, \ldots, x_n$. Prove that $p_n(x) = f(x)$ for all x, i.e., that interpolating to a polynomial will reproduce the polynomial, if you use enough nodes. *Hint:* Consider the uniqueness part of Theorem 4.1.

9. Let p be a polynomial of degree $\leq n$. Use the uniqueness of the interpolating polynomial to show that we can write

$$p(x) = \sum_{i=0}^{n} L_i^{(n)}(x)p(x_i)$$

for any distinct set of nodes $x_0, x_1, \ldots, x_n$. *Hint:* See the previous exercise.

10. Show that
$$\sum_{i=0}^{n} L_i^{(n)}(x) = 1$$

for any set of distinct nodes x_k, $0 \le k \le n$. *Hint:* This does not require *any* computation with the sum but, rather, a perceptive choice of polynomial p in the previous exercise.

◁ ● ● ▷

4.2 NEWTON INTERPOLATION AND DIVIDED DIFFERENCES

The Lagrange form of the interpolating polynomial gives us a very tidy construction, but it does not lend itself well to actual computation. The reason is partly because whenever we decide to add a point to the set of nodes, we have to completely recompute all of the $L_i^{(n)}$ functions; we cannot (easily) write p_{n+1} in terms of p_n using the Lagrange construction.

An alternate form of the polynomial, known as the Newton form, avoids this problem, and allows us to easily write p_{n+1} in terms of p_n. The basic result is contained in the following theorem.

Theorem 4.2 (Newton Interpolation Construction) *Let p_n be the polynomial that interpolates f at the nodes x_i, $i = 0, 1, 2, 3, \ldots, n$. Let p_{n+1} be the polynomial that interpolates f at the nodes x_i, $i = 0, 1, 2, 3, \ldots, n, n+1$. Then p_{n+1} is given by*

$$p_{n+1}(x) = p_n(x) + a_{n+1} w_n(x), \qquad (4.4)$$

where

$$w_n(x) = \prod_{i=0}^{n}(x - x_i),$$
$$a_{n+1} = \frac{f(x_{n+1}) - p_n(x_{n+1})}{w_n(x_{n+1})}, \qquad (4.5)$$

and

$$p_0(x) \equiv a_0 = f(x_0).$$

Proof: Since we know that the interpolation polynomial is unique, all we have to do is show that p_{n+1}, as given in (4.4), satisfies the interpolation conditions. For $k \le n$, we have $w_n(x_k) = 0$, so

$$p_{n+1}(x_k) = p_n(x_k) + a_{n+1} w_n(x_k) = p_n(x_k) = f(x_k);$$

hence, p_{n+1} interpolates all but the last point. To check x_{n+1}, we directly compute:

$$\begin{aligned} p_{n+1}(x_{n+1}) &= p_n(x_{n+1}) + a_{n+1} w_n(x_{n+1}) \\ &= p_n(x_{n+1}) + f(x_{n+1}) - p_n(x_{n+1}) \\ &= f(x_{n+1}). \end{aligned}$$

Thus, p_{n+1} interpolates at all the nodes; moreover, it clearly is a polynomial of degree less than or equal to $n + 1$, so we are done. ●

Corollary 4.1 *For $\{a_k\}$ and w_n as defined in Theorem 4.2, we have*

$$
\begin{aligned}
p_n(x) &= a_0 + a_1(x - x_0) + a_2(x - x_0)(x - x_1) + \cdots + a_n(x - x_0)\cdots(x - x_{n-1}) \\
&= \sum_{k=0}^{n} a_k w_{k-1}(x),
\end{aligned}
$$

where we have used $w_{-1}(x) \equiv 1$.

Proof: This is a straight-forward inductive argument; see Problem 5. ●

We can use this Newton form of the interpolating polynomial to construct and evaluate interpolating polynomials in an efficient fashion that is reminiscent of Horner's rule for nested multiplication (§2.1). We have

$$
p_n(x) = a_0 + a_1(x - x_0) + a_2(x - x_0)(x - x_1) + \cdots + a_n(x - x_0)\cdots(x - x_{n-1}),
$$

so that we can write

$$
p_n(x) = a_0 + (x - x_0)(a_1 + (x - x_1)(a_2 + \cdots + (x - x_{n-1})a_n)\cdots)).
$$

■ **EXAMPLE 4.2**

To consider an example, let us construct the quadratic polynomial that interpolates the sine function on the interval $[0, \pi]$, using equally spaced nodes. We have $f(x) = \sin x$, $x_i = 0, \frac{1}{2}\pi, \pi$ and $y_i = 0, 1, 0$. Then

$$
\begin{aligned}
a_0 &= 0 &\Rightarrow&\quad p_0(x) = 0 \\
a_1 &= \frac{\sin\frac{1}{2}\pi - p_0(\frac{1}{2}\pi)}{(\frac{1}{2}\pi - 0)} = \frac{2}{\pi} &\Rightarrow&\quad p_1(x) = \frac{2}{\pi}x; \\
a_2 &= \frac{\sin\pi - p_1(\pi)}{(\pi - 0)(\pi - \frac{1}{2}\pi)} = -\frac{4}{\pi^2} &\Rightarrow&\quad p_2(x) = \frac{2}{\pi}x - \frac{4}{\pi^2}x\left(x - \frac{1}{2}\pi\right).
\end{aligned}
$$

The reader should check that each interpolating polynomial does interpolate at the appropriate points.

The coefficients a_k are called *divided differences*, and we can construct an algorithm for computing them, based on the formula (4.5). It is also possible to slightly modify this algorithm to allow the user to *update* an existing set of divided difference coefficients by adding new points to the set of nodes. See Problem 6.

Algorithm 4.1 *Newton Interpolation (Construction) via Divided Difference Coefficients*

```
input n,x,y
a(0) = y(0)
for k=1 to n do
      w = 1
      p = 0
      for j=0 to k-1 do
            p = p + a(j)*w
            w = w*(x(k) - x(j))
      endfor
      a(k) = (y(k) - p)/w
endfor
```

Algorithm 4.2 *Newton Interpolation (Evaluation) via Divided Difference Coefficients*

```
      input n,xx,x,a
!
!     n = degree of polynomial
!     xx = point at which the polynomial is
!          to be evaluated
!     x = array of nodes
!     a = array of divided difference coefficients
!
      px = a(n)
!
!     px = polynomial value at xx
!
      for k = n-1,0,-1
            xd = xx - x(k)
            px =  a(k) + px*xd
      endfor
```

Programming Hint: We know, from the exercises in §4.1, that polynomial interpolation of sufficiently high degree to a polynomial will reproduce the original polynomial. Thus, one way to check that an interpolation code is working is to see if it can reproduce a polynomial of degree n when asked to interpolate it at $n + 1$ nodes.

An alternate way of arriving at the divided difference coefficients—and, historically, the original way it was done—is by means of a *divided difference table*. Since this is an easy means by which hand calculations can be performed, it is worth looking at here as well.

We begin by defining some terminology and notation. Given a set of nodes and corresponding function values, the function values will be referred to as the *zero-th divided differences* and, in this context, we will write them as $f_0(x_k)$. The *first divided differences* are then formed from the zero-th ones according to

$$f_1(x_k) = \frac{f_0(x_{k+1}) - f_0(x_k)}{x_{k+1} - x_k},$$

and then the *second divided differences* are given by

$$f_2(x_k) = \frac{f_1(x_{k+1}) - f_1(x_k)}{x_{k+2} - x_k},$$

and so on. The general formula is

$$f_j(x_k) = \frac{f_{j-1}(x_{k+1}) - f_{j-1}(x_k)}{x_{k+j} - x_k}.$$

Now using this terminology and notation,[5] we can build a set of tables that can readily be used to construct the interpolating polynomials.

■ **EXAMPLE 4.3**

Take $f(x) = \log_2 x$ as the function to be interpolated, using the nodes $x_0 = 1$, $x_1 = 2$, $x_2 = 4$. We begin by arranging the nodes and corresponding functional data in two columns, as in Table 4.1. Note that we write the functional data using the divided difference notation.

Table 4.1 Initial table for divided differences.

k	x_k	$f_0(x_k)$
0	1	0
1	2	1
2	4	2

The third column of the table is formed by the first divided differences; thus,

$$1 = \frac{f(x_1) - f(x_0)}{x_1 - x_0} = \frac{1 - 0}{2 - 1} = f_1(x_0)$$

and

$$\frac{1}{2} = \frac{f(x_2) - f(x_1)}{x_2 - x_1} = \frac{2 - 1}{4 - 2} = f_1(x_1),$$

so we get Table 4.2.

The fourth (and, in this case, final) column is formed by the second divided difference:

$$-\frac{1}{6} = \frac{1/2 - 1}{4 - 1} = f_2(x_0);$$

see Table 4.3.

(If we had more data, subsequent columns would be formed in a similar fashion.) How do we use this result to construct the interpolating polynomial? The top diagonal

---

[5]We have broken with convention and tradition in our divided difference notation here. Most texts use $f[x_k]$ for the zero-th divided difference, and $f[x_k, x_{k+1}]$ for the first divided difference formed by the nodes x_k and x_{k+1}, $f[x_k, x_{k+1}, x_{k+2}]$ for the second divided difference formed by x_k, x_{k+1}, and x_{k+2}, and so on. This becomes unwieldy when talking about higher and higher differences, and while it is slightly more general, most texts (including this one) do very little to justify that extra generality.

Table 4.2 Table for divided differences after computation of first differences.

k	x_k	$f_0(x_k)$	$f_1(x_k)$
0	1	0	
			1
1	2	1	
			1/2
2	4	2	

Table 4.3 Table for divided differences after computation of second differences.

k	x_k	$f_0(x_k)$	$f_1(x_k)$	$f_2(x_k)$
0	1	0		
			1	
1	2	1		$-\frac{1}{6}$
			$\frac{1}{2}$	
2	4	2		

row of table values gives us the divided difference coefficients based on the nodes as numbered in ascending order. Thus

$$p_2(x) = 0 + (1)(x-1) - \frac{1}{6}(x-1)(x-2) = -\frac{1}{6}(x-1)(x-8).$$

(These are the same values that the pseudocode produces.) The bottom diagonal row gives the divided difference coefficients based on numbering the nodes in a reversed order. Thus we also have

$$p_2(x) = 2 + \frac{1}{2}(x-4) - \frac{1}{6}(x-4)(x-2) = -\frac{1}{6}(x-1)(x-8).$$

Finally, to add a node to the data set and construct the new polynomial is easy, based on these tables. Let's add the new node $x_3 = \frac{1}{2}$ to our current example. We have, initially, the arrangement in Table 4.4. (Note that the nodes do not have to be arranged in ascending order—or, in fact, in any particular order.)

Table 4.4 Table for divided differences with new node added.

k	x_k	$f_0(x_k)$	$f_1(x_k)$	$f_2(x_k)$
0	1	0		
			1	
1	2	1		$-\frac{1}{6}$
			$\frac{1}{2}$	
2	4	2		
3	$\frac{1}{2}$	-1		

Computing forward we get the (new) final result in Table 4.5, from which we conclude that the interpolating polynomial is

$$p_3(x) = 0 + (1)(x-1) - \frac{1}{6}(x-1)(x-2) + \frac{1}{7}(x-1)(x-2)(x-4).$$

The reader should confirm that this does indeed interpolate correctly at the given points, and therefore is the correct polynomial.

Table 4.5   Table for divided differences after computation of third differences.

k	x_k	$f_0(x_k)$	$f_1(x_k)$	$f_2(x_k)$	$f_3(x_k)$
0	1	0			
			1		
1	2	1		$-\frac{1}{6}$	
			$\frac{1}{2}$		$\frac{1}{7}$
2	4	2		$-\frac{5}{21}$	
			$\frac{6}{7}$		
3	$\frac{1}{2}$	-1			

Programming Hint: An interpolating polynomial, if it is correctly constructed, has to reproduce the values at the nodes exactly, so this is an easy test to use in evaluating and debugging your code. If the values at the nodes are computed correctly, then the polynomial has to be correct; if the values at the nodes are not computed correctly, then there is something wrong, either with the construction of the polynomial or with the code that is evaluating the polynomial.

Exercises:

1. Construct the polynomial of degree 3 that interpolates to the data $x_0 = 1$, $y_0 = 1$, $x_1 = 2$, $y_1 = 1/2$, $x_2 = 4$, $y_2 = 1/4$, and $x_3 = 3$, $y_3 = 1/3$. You should get $p(t) = (50 - 35t + 10t^2 - t^3)/24$.

2. For each function listed below, use divided difference tables to construct the Newton interpolating polynomial for the set of nodes specified. Plot the function, the interpolating polynomial, and also the error, on the interval defined by the nodes.

   (a) $f(x) = \sqrt{x}$, $x_i = 0, 1, 4$;

   (b) $f(x) = \ln x$, $x_i = 1, \frac{3}{2}, 2$;

   (c) $f(x) = \sin \pi x$, $x_i = 0, \frac{1}{4}, \frac{1}{2}, \frac{3}{4}, 1$;

   (d) $f(x) = \log_2 x$, $x_i = 1, 2, 4$;

   (e) $f(x) = \sin \pi x$, $x_i = -1, 0, 1$.

3. Let $f(x) = e^x$. Define $p_n(x)$ to be the Newton interpolating polynomial for $f(x)$, using $n + 1$ equally spaced nodes on the interval $[-1, 1]$. Thus we are taking higher and higher degree polynomial approximations to the exponential function. Write a program that computes $p_n(x)$ for $n = 2, 4, 8, 16, 32$, and which samples the error

$f(x) - p_n(x)$ at 501 equally spaced points on $[-1, 1]$. Record the maximum error as found by the sampling, as a function of n, that is, define E_n as

$$E_n = \max_{0 \leq k \leq 500} |f(t_k) - p_n(t_k)|,$$

where $t_k = -1 + 2k/500$, and plot E_n versus n.

4. In §3.7 we used linear interpolation to construct an initial guess for Newton's method as a means of approximating $\sqrt{a}$. Construct the quadratic polynomial that interpolates the square root function at the nodes $x_0 = \frac{1}{4}, x_1 = \frac{4}{9}, x_2 = 1$. Plot the error between p_2 and $\sqrt{x}$ over the interval $[\frac{1}{4}, 1]$ and try to estimate the worst error. What impact will this have on the use of Newton's method for finding $\sqrt{a}$?

5. Prove Corollary 4.1.

6. Write and test a computer code that takes a given set of nodes, function values, and corresponding divided difference coefficients, and computes new divided difference coefficients for new nodes and function values. Test it by recursively computing interpolating polynomials that approximate $f(x) = e^x$ on the interval $[0, 1]$.

7. In 1973 the horse Secretariat became the first (and, so far, only) winner of the Kentucky Derby to finish the race in less than 2 minutes, running the $1\frac{1}{4}$-mile distance in 1 minute, 59.4 seconds, a record that still stands as this edition goes to press in 2013. Remarkably, he ran each quarter mile *faster* than the previous one, as Table 4.6 shows. Here t is the elapsed time (in seconds) since the race began and x

Table 4.6   Data for Problem 7.

x	0.0	0.25	0.50	0.75	1.00	1.25
t	0.0	25.0	49.4	73.0	96.4	119.4

is the distance (in miles) that Secretariat has traveled.

(a) Find the cubic polynomial that interpolates this data at $x = 0, 1/2, 3/4, 5/4$.

(b) Use this polynomial to estimate Secretariat's speed at the finish of the race, by finding $p_3'(5/4)$.

(c) Find the quintic polynomial that interpolates the entire data set.

(d) Use the quintic polynomial to estimate Secretariat's speed at the finish line.

8. The data in Table 4.7 gives the actual thermal conductivity data for the element mercury. Use Newton interpolation and the data for 300 K, 500 K, and 700 K to construct a quadratic interpolate for this data. How well does it predict the values at 400 K and 600 K?

Table 4.7   Data for Problem 8.

Temperature (° K), u	300	400	500	600	700
Conductivity (W/cm ° K), k	0.084	0.098	0.109	0.12	0.127

9. The gamma function, denoted by $\Gamma(x)$, is an important special function in probability, combinatorics, and other areas of applied mathematics. Because it can be shown that $\Gamma(n + 1) = n!$, the gamma function is considered a generalization of the factorial function to non-integer arguments. Table 4.8 gives the values of $\Gamma(x)$ on the interval $[1, 2]$. Use these to construct the fifth-degree polynomial based on the nodes $x = 1, 1.2, 1.4, 1.6, 1.8, 2.0$, and then use this polynomial to estimate the values at $x = 1.1, 1.3, 1.5, 1.7, 1.9$. Plot your polynomial and compare it to the intrinsic gamma function on your computing system or calculator.

Table 4.8   Table of $\Gamma(x)$ values.

x	$\Gamma(x)$
1.00	1.0000000000
1.10	0.9513507699
1.20	0.9181687424
1.30	0.8974706963
1.40	0.8872638175
1.50	0.8862269255
1.60	0.8935153493
1.70	0.9086387329
1.80	0.9313837710
1.90	0.9617658319
2.00	1.0000000000

10. The error function, which we saw briefly in Chapters 1 and 2, is another important special function in applied mathematics, with applications to probability theory and the solution of heat conduction problems. The formal definition of the error function is

$$\mathrm{erf}(x) = \frac{2}{\sqrt{\pi}} \int_0^x e^{-t^2} dt.$$

Table 4.9 gives values of $\mathrm{erf}(x)$ in increments of 0.1, over the interval $[0, 1]$.

Table 4.9   Table of $\mathrm{erf}(x)$ values for Problem 10.

x	$\mathrm{erf}(x)$
0.0	0.00000000000000
0.1	0.11246291601828
0.2	0.22270258921048
0.3	0.32862675945913
0.4	0.42839235504667
0.5	0.52049987781305
0.6	0.60385609084793
0.7	0.67780119383742
0.8	0.74210096470766
0.9	0.79690821242283
1.0	0.84270079294971

(a) Construct the quadratic interpolating polynomial to the error function using the data at the nodes $x_0 = 0$, $x_1 = 0.5$, and $x_2 = 1.0$. Plot the polynomial and the data from Table 4.9 and comment on the observed accuracy.

(b) Repeat part (a), but this time construct the cubic interpolating polynomial using the nodes $x_0 = 0.0$, $x_2 = 0.3$, $x_2 = 0.7$, and $x_3 = 1.0$.

11. As steam is heated up, the pressure it generates is increased. Over the temperature range $[220, 300]$ (degrees Fahrenheit) the pressure, in pounds per square inch, is as given in Table 4.10.[6]

Table 4.10   Temperature–pressure values for steam; Problem 11

T	220	230	240	250	260	270	280	290	300
P	17.188	20.78	24.97	29.82	35.42	41.85	49.18	57.53	66.98

(a) Construct the quadratic interpolating polynomial to this data at the nodes $T_0 = 220$, $T_1 = 260$, and $T_2 = 300$. Plot the polynomial and the data in the table and comment on the accuracy you obtain.

(b) Repeat Part (a), but this time construct the quartic interpolating polynomial using the nodes $T_0 = 220$, $T_1 = 240$, $T_2 = 260$, $T_3 = 280$, and $T_4 = 300$.

(c) Which of the two polynomials would you think it is best to use to get values for $P(T)$ that are not in Table 4.10?

12. Similar data for gaseous ammonia is given in Table 4.11.

Table 4.11   Temperature–pressure values for gaseous ammonia; Problem 12

T	0	5	10	15	20	25	30	35	40
P	30.42	34.27	38.51	43.14	48.21	53.73	59.74	66.26	73.32

(a) Construct the quadratic interpolating polynomial to this data at the nodes $T_0 = 0$, $T_1 = 20$, and $T_2 = 40$. Plot the polynomial and the data in the table and comment on the accuracy you observe.

(b) Repeat part (a), but this time construct the quartic interpolating polynomial using the nodes $T_0 = 0$, $T_1 = 10$, $T_2 = 20$, $T_3 = 30$, and $T_4 = 40$.

(c) Which of the two polynomials would you think is best to use to get values for $P(T)$ that are not in Table 4.11?

13. In Problems 8 of §3.1 and 8 of §3.8, we looked at the motion of a liquid–solid interface under a simplified model of the physics involved, in which the interface moved according to

$$x = 2\beta\sqrt{t}$$

---

[6]Taken from tables in *Introduction to Thermodynamics: Classical and Statistical*, by R. E. Sonntag and G. J. Van Wylen, John Wiley & Sons, Inc., New York, 1971.

for $\beta = \alpha/\sqrt{k}$. Here k is a material property and α is the root of $f(z) = \theta e^{-z^2} - z \operatorname{erf}(z)$, where θ also depends on material properties. Figure 4.4 shows a plot of α versus $\log_{10}\theta$, based on finding the root of $f(z)$. Some of the data used to create this curve is given in Table 4.12.

(a) Use the data at the nodes $\{-6, -4, -2, 0, 2, 4, 6\}$ to construct an interpolating polynomial for this table. Plot the polynomial and compare to the actual graph in Figure 4.4. Use the polynomial to compute values at $\log_{10}\theta = -5, -3, \ldots, 3, 5$. How do these compare to the actual values from Table 4.12?

(b) Compute the higher-degree Newton polynomial based on the entire table of data. Plot this polynomial and compare it to the one generated using only part of the data.

Table 4.12   Data for Problem 13.

$\log_{10}\theta$	α
−6.0000	0.944138E–03
−5.0000	0.298500E–02
−4.0000	0.941277E–02
−3.0000	0.297451E–01
−2.0000	0.938511E–01
−1.0000	0.289450E+00
0.0000	0.767736E+00
1.0000	0.141492E+01
2.0000	0.198151E+01
3.0000	0.245183E+01
4.0000	0.285669E+01
5.0000	0.321632E+01
6.0000	0.354269E+01

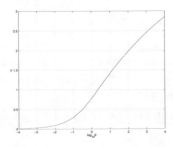

Figure 4.4   Figure for Problem 13.

14. Write a computer program to construct the Newton interpolating polynomial to $f(x) = \sqrt{x}$ using equally spaced nodes on the interval $[0, 1]$. Plot the error $f(x) - p_n(x)$ for $n = 4, 8, 16$ and comment on what you get.

◁ • • • ▷

4.3 INTERPOLATION ERROR

How good an approximation is the interpolating polynomial? We saw in §2.4 that an error estimate was fairly easily derived for linear interpolation, and here we will generalize that result to arbitrary polynomial interpolation. We expect, based on the examples $f(x) = e^x$ and $f(x) = (1 + 25x^2)^{-1}$, that using more points could make the approximation more accurate and yet could also lead to problems.

Theorem 4.3 (Polynomial Interpolation Error Theorem) *Let $f \in C^{n+1}([a,b])$ and let the nodes $x_k \in [a,b]$ for $0 \leq k \leq n$. Then, for each $x \in [a,b]$, there is a $\xi_x \in [a,b]$ such that*

$$f(x) - p_n(x) = \frac{w_n(x)}{(n+1)!} f^{(n+1)}(\xi_x), \tag{4.6}$$

where

$$w_n(x) = \prod_{k=0}^{n} (x - x_k).$$

Proof: The proof is an involved argument using repeated applications of Rolle's Theorem, and is deferred to Appendix A.1. •

This result is too general to be easily interpreted at this point in our study, so we will deal directly with some specific cases. Before doing so, we introduce some notation.

In measuring the error in certain approximations, it will be useful from now on to have some convenient means of measuring the *size* of a function. This is done via the concept of a *norm*. Briefly, a norm is any mapping from functions into the non-negative real numbers (usually denoted by a double-bar notation: $\|f\|$), which satisfies certain basic axioms.

Definition 4.1 (Function norm) *A function norm is any computation, denoted by the symbol $\|f\|$, that satisfies the following conditions:*

1. *$\|f\| > 0$ for any function f that is not identically zero;*

2. *$\|af\| = |a|\|f\|$ for any constant a.*

3. *$\|f + g\| \leq \|f\| + \|g\|$ for any two functions f and g.*

Note that the use of the double vertical bar notation ($\|f\|$) is similar to the use of the single vertical bar notation for the absolute value; this is deliberate, since the notion of a norm plays much the same role for functions as the absolute value does for ordinary numbers: it helps us to measure size, or magnitude.

Some examples of common norms include:

- The *infinity norm* or pointwise norm; if f is continuous on the closed interval $[a,b]$, then we define

$$\|f\|_{\infty,[a,b]} = \max_{x \in [a,b]} |f(x)|.$$

- The *2-norm*; if f is such that its square can be integrated over the interval $[a,b]$, then we define

$$\|f\|_{2,[a,b]} = \left(\int_a^b [f(x)]^2 dx \right)^{1/2}.$$

Note the use of subscripts on the norm notation to distinguish between different norms; the subscript defining the interval will often be dropped or omitted when there is no danger of confusion. Note also that different norms will give different numerical values for the same function, and will convey different kinds of information.

■ **EXAMPLE 4.4**

If $f(x) = e^{-x}$ on the interval $[-1, 1]$, then

$$\|f\|_\infty = e = 2.71828...$$

and

$$\|f\|_2 = \left(\frac{e^2 - e^{-2}}{2}\right)^{1/2} = 1.90443...$$

See Problems 14 and 15 for some further exploration of what different norms mean.

We already know the interpolation error result for $n = 1$: in other words, linear interpolation. In §2.4, we saw that we could prove

$$|f(x) - p_1(x)| \le \frac{1}{8}(x_1 - x_0)^2 \max_{x_0 \le x \le x_1} |f''(x)|, \tag{4.7}$$

which holds for all $x \in [x_0, x_1]$. We can write this in norm notation as

$$\|f - p\|_{\infty,I} \le \frac{1}{8}(x_1 - x_0)^2 \|f''\|_{\infty,I}, \tag{4.8}$$

where $I = [x_0, x_1]$ is the interval defined by the nodes used in the interpolation. What happens for larger values of n?

If $n = 2$ (quadratic interpolation), we have, from (4.6),

$$f(x) - p_2(x) = \frac{1}{6}(x - x_0)(x - x_1)(x - x_2)f'''(\xi_x). \tag{4.9}$$

To get an upper bound like we did for linear interpolation, we assume that the nodes satisfy

$$x_1 - x_0 = x_2 - x_1 = h$$

for some $h > 0$. We thus can write

$$(x - x_0)(x - x_1)(x - x_2) = ((x - x_1) + h)(x - x_1)((x - x_1) - h) = t(t^2 - h^2)$$

for $t = x - x_1$. We now have

$$|f(x) - p_2(x)| \le \frac{1}{6}\left(\max_{-h \le t \le h} |t(t^2 - h^2)|\right) \max_{x_0 \le t \le x_2} |f'''(t)|, \tag{4.10}$$

which holds for all $x \in [x_0, x_2]$. Following the same kind of argument used in §2.4, we can show that

$$\max_{-h \le t \le h} |t(t^2 - h^2)| = \frac{2}{3\sqrt{3}}h^3$$

so that the upper bound on the error becomes

$$|f(x) - p_2(x)| \le \frac{1}{9\sqrt{3}}h^3 \max_{x_0 \le t \le x_2} |f'''(t)|. \tag{4.11}$$

This can be rendered into norm notation as

$$\|f - p_2\|_{\infty,I} \le \frac{1}{9\sqrt{3}}h^3\|f'''\|_{\infty,I} \tag{4.12}$$

where I is again the interval defined by the smallest and largest nodes, and h is the spacing between nodes.

■ **EXAMPLE 4.5**

Let $f(x)$ be the error function, i.e.,

$$f(x) = \operatorname{erf}(x) = \frac{2}{\sqrt{\pi}}\int_0^x e^{-t^2}\,dt.$$

Suppose that we want to construct interpolating polynomial approximations to this function on the interval $[0, 1]$ that are accurate to within 10^{-6}. We have

$$f''(x) = \frac{2}{\sqrt{\pi}}\left(e^{-x^2}\right)' = \frac{2}{\sqrt{\pi}}\left(-2xe^{-x^2}\right),$$

so that a crude upper bound on the second derivative is

$$\|f''\|_{\infty,[0,1]} \le 2 \times 1 \times 2e^{-x^2}/\sqrt{\pi} \le 2 \times 2e^0/\sqrt{\pi} \le 7.1.$$

Therefore the error in linear interpolation is bounded by $(h^2/8)(7.1) = h^2(0.8875)$; thus we need to have $h \le 9.5 \times 10^{-4}$ to get the error less than 10^{-6}. This tells us that the table of error function values must have $N = 1/h > 1,062$ points for linear interpolation between points to achieve the desired accuracy.

On the other hand, we have

$$f'''(x) = \frac{2}{\sqrt{\pi}}\left(-2e^{-x^2} + 4x^2e^{-x^2}\right) = \frac{4}{\sqrt{\pi}}\left(2x^2 - 1\right)e^{-x^2},$$

for which we have (again, crudely)

$$\|f'''\|_{\infty,[0,1]} \le 4 \times (1) \times e^{-x^2}/\sqrt{\pi} \le 4 \times e^0/\sqrt{\pi} \le 5.$$

Hence, the error in quadratic interpolation is bounded by $(h^3/9\sqrt{3})(5) \le 0.4h^3$. To get the error less than the desired 10^{-6}, we then want to have

$$0.4h^3 \le 10^{-6} \Rightarrow h \le (0.4)^{-1/3} \times 10^{-2} \le 1.4 \times 10^{-2}.$$

So we now have to have $N = 1/h > 72$ points in the table to get the desired accuracy.

A similar analysis to that used to get (4.12) shows that the error in cubic interpolation satisfies

$$\|f - p_3\| \le \frac{1}{24}h^4\|f^{(4)}\|_{\infty,I}. \tag{4.13}$$

See Problem 4.

We could of course keep going on with specific cases, but since we will eventually learn, in §4.12.1, that interpolation with high-degree polynomials is not always a good idea, we will stop at this point.

Exercises:

1. What is the error in quadratic interpolation to $f(x) = \sqrt{x}$, using equally spaced nodes on the interval $[\frac{1}{4}, 1]$?

2. Repeat the above for $f(x) = x^{-1}$ on $[\frac{1}{2}, 1]$.

3. Repeat the previous two problems, using cubic interpolation.

4. Show that the error in third-degree polynomial interpolation satisfies

$$\|f - p_3\|_\infty \le \frac{1}{24} h^4 \|f^{(4)}\|_\infty,$$

   if the nodes x_0, x_1, x_2, x_3 are equally spaced, with $x_i - x_{i-1} = h$. *Hint:* Use the change of variable $t = x - x_1 - \frac{1}{2}h$.

5. Show that the error in polynomial interpolation using six equally spaced points (quintic interpolation) satisfies

$$\|f - p_5\|_\infty \le Ch^6 \|f^{(6)}\|_\infty,$$

   where $C \approx 0.0235$. *Hint:* See previous problem.

6. Generalize the derivation of the error bound (4.12) for quadratic interpolation to the case where the nodes are *not* equally spaced. Take $x_1 - x_0 = h$ and $x_2 - x_1 = \theta h$ for some $\theta > 0$.

7. Apply your result for the previous problem to the error in quadratic interpolation to $f(x) = \sqrt{x}$ using the nodes $x_0 = \frac{1}{4}$, $x_1 = \frac{9}{16}$, and $x_2 = 1$.

8. If we want to use a table of exponential values to interpolate the exponential function on the interval $[-1, 1]$, how many nodes are needed to guarantee 10^{-6} accuracy with linear interpolation? Quadratic interpolation?

9. If we want to use a table of values to interpolate the error function on the interval $[0, 5]$, how many points are needed to get 10^{-6} accuracy using linear interpolation? Quadratic interpolation? Would it make sense to use one grid spacing on, say, $[0, 1]$, and another one on $[1, 5]$? Explain.

10. If we want to use a table of values to interpolate the sine function on the interval $[0, \pi]$, how many points are needed for 10^{-6} accuracy with linear interpolation? Quadratic interpolation? Cubic interpolation?

11. Let's return to the computation of the natural logarithm. Consider a computer that stores numbers in the form $z = f \cdot 2^\beta$, where $\frac{1}{2} \le f \le 1$. We want to consider using this, in conjunction with interpolation ideas, to compute the natural logarithm function.

    (a) Using piecewise linear interpolation over a grid of equally spaced points, how many table entries would be required to accurately approximate $\ln z$ to within 10^{-14}?

    (b) Repeat part (a), using piecewise quadratic interpolation.

    (c) Repeat part (a) again, using piecewise cubic interpolation.

Explain, in a brief essay, the importance of restricting the domain of z to the interval $[\frac{1}{2}, 1]$.

12. Assume that for any real c,

$$\lim_{n \to \infty} \frac{c^n}{n!} = 0.$$

Use this to prove that, if p_n is the polynomial interpolate to $f(x) = \sin x$ on the interval $[a, b]$, using *any* distribution of distinct nodes x_k, $0 \le k \le n$, then $\|f - p_n\|_\infty \to 0$ as $n \to \infty$. *Hint:* Can we justify $|w_n(x)| \le c^{n+1}$ for some c?

13. Can you repeat the above for $f(x) = e^x$? Why/why not?

14. Define the norms

$$\|f\|_\infty = \max_{x \in [0,1]} |f(x)|$$

and

$$\|f\|_2 = \left(\int_0^1 [f(x)]^2 dx \right)^{1/2}.$$

Compute $\|f\|_\infty$ and $\|f\|_2$ for the following list of functions:

(a) e^{-ax}, $a > 0$;

(b) $\sin n\pi x$, n integer;

(c) $\sqrt{x}e^{-ax^2}$, $a > 0$;

(d) $1/\sqrt{1+x}$.

In parts (b) and (c), a is a constant parameter.

15. Define

$$f_n(x) = \begin{cases} 1 - nx, & 0 \le x \le \frac{1}{n}; \\ 0, & \frac{1}{n} \le x \le 1. \end{cases}$$

Show that

$$\lim_{n \to \infty} \|f_n(x)\|_\infty = 1,$$

but

$$\lim_{n \to \infty} \|f_n(x)\|_2 = 0.$$

Hint: Draw a graph of $f_n(x)$.

16. Show that

$$\|f\|_{1,[a,b]} = \int_a^b |f(x)| dx$$

defines a function norm.

◁ ● ● ▷

4.4 APPLICATION: MULLER'S METHOD AND INVERSE QUADRATIC INTERPOLATION

We can use the idea of interpolation to develop more sophisticated root-finding methods.

The secant method can be viewed as finding the root of a linear interpolating polynomial. Our error results tell us that a quadratic (second-degree) polynomial should be a better approximation to the function; hence, its root ought to be a better approximation to the root of the original function. Can we take this outline of an idea and use it to construct a practical method for finding roots?

The answer is yes, and the method is known as *Muller's method*. Given three points, x_0, x_1, and x_2, we find the quadratic polynomial $p(x)$ such that $p(x_i) = f(x_i)$, $i = 0, 1, 2$; and then define x_3 as the root of p that is closest to x_2. The actual implementation of Muller's method does present a few problems that need to be resolved, however.

The best way to write the interpolating polynomial is in the Newton form, as[7]

$$p(x) = f(x_2) + A(x - x_2) + B(x - x_2)(x - x_1), \qquad (4.14)$$

where A and B are the divided differences (see Problem 7)

$$A = \frac{f(x_2) - f(x_1)}{x_2 - x_1} \qquad (4.15)$$

and

$$B = \frac{1}{x_2 - x_0} \left(\frac{f(x_2) - f(x_1)}{x_2 - x_1} - \frac{f(x_1) - f(x_0)}{x_1 - x_0} \right). \qquad (4.16)$$

Note that these coefficients will have to be re-computed for each iteration.

Finding the root of (4.14) via the quadratic formula requires that some care be taken. Since we want to find the root nearest x_2, we rewrite p as a quadratic in $x - x_2$:

$$
\begin{aligned}
p(x) &= f(x_2) + A(x - x_2) + B(x - x_2)(x - x_1) \\
&= f(x_2) + A(x - x_2) + B(x - x_2)(x - x_2) + B(x - x_2)(x_2 - x_1) \\
&= f(x_2) + (A + B(x_2 - x_1))(x - x_2) + B(x - x_2)(x - x_2) \\
&= f(x_2) + C(x - x_2) + B(x - x_2)^2,
\end{aligned}
$$

where $C = A + B(x_2 - x_1)$. The quadratic formula then says that the root of p is defined by

$$x - x_2 = \frac{1}{2B} \left(-C \pm \sqrt{C^2 - 4f(x_2)B} \right),$$

where we choose the sign to minimize the right side of the equation. To avoid the pitfalls of subtractive cancellation, we rationalize the numerator to get

$$x - x_2 = -\frac{2f(x_2)}{C \pm \sqrt{C^2 - 4f(x_2)B}},$$

where the sign is chosen to maximize the denominator. Thus, the next point in the iteration is given by

$$x_3 = x_2 - \frac{2f(x_2)}{C + \operatorname{sgn}(C)\sqrt{C^2 - 4f(x_2)B}},$$

---

[7]Note that here we are treating x_2 as the "first" node, instead of x_0. This is possible because the order of the nodes is irrelevant in the computation, although this does affect the formulas somewhat.

where sgn is the "sign" function, defined by

$$\text{sgn}(x) = \begin{cases} -1, & x < 0; \\ 1, & x > 0; \\ 0, & x = 0. \end{cases}$$

Thus, the general case is given by

$$x_{n+1} = x_n - \frac{2f(x_n)}{C_n + \text{sgn}(C_n)\sqrt{C_n^2 - 4f(x_n)B_n}},$$

where

$$B_n = \frac{1}{x_n - x_{n-2}} \left(\frac{f(x_n) - f(x_{n-1})}{x_n - x_{n-1}} - \frac{f(x_{n-1}) - f(x_{n-2})}{x_{n-1} - x_{n-2}} \right)$$

and

$$C_n = \frac{f(x_n) - f(x_{n-1})}{x_n - x_{n-1}} + B_n(x_n - x_{n-1}).$$

Table 4.13 shows the result of applying Muller's method to the example we used thoughout Chapter 3, $f(x) = 2 - e^x$, so that the exact root is $\alpha = \ln 2$. Compare the number of iterations used here to that of the secant method (Table 3.5); note that we can code Muller's method to use only a single new function evaluation in each step.

Table 4.13   Muller's method, $f(x) = 2 - e^x$.

n	x_n	$f(x_n)$	$\alpha - x_n$	$\log_{10}(\alpha - x_n)$
0	0.000000000000	0.100000e+01	0.693147e+00	−0.1592
1	0.500000000000	0.351279e+00	0.193147e+00	−0.7141
2	1.000000000000	−0.718282e+00	0.306853e+00	−0.5131
3	0.687259367753	0.117410e−01	0.588781e−02	−2.2300
4	0.693087847691	0.118662e−03	0.593329e−04	−4.2267
5	0.693147161269	0.385812e−07	0.192906e−07	−7.7147
6	0.693147180560	−0.222045e−14	0.122125e−14	−14.9132
7	0.693147180560	0.000000e+00	0.111022e−15	−15.9546

An alternative to Muller's method, which avoids the difficulties associated with the quadratic formula, is *inverse quadratic interpolation*. In this case, we find the quadratic polynomial $q(y)$ such that

$$q(f(x_i)) = x_i, \quad i = 0, 1, 2.$$

Direct computation with the Newton form shows that

$$q(y) = x_2 + a(y - y_2) + b(y - y_2)(y - y_1),$$

where $y_k = f(x_k)$,

$$a = \frac{x_2 - x_1}{y_2 - y_1}, \tag{4.17}$$

and

$$b = \frac{1}{y_2 - y_0} \left(\frac{x_2 - x_1}{y_2 - y_1} - \frac{x_1 - x_0}{y_1 - y_0} \right). \tag{4.18}$$

Thus, q is quadratic in y and still passes through the points $(x_i, f(x_i))$; we have essentially reversed the notion of ordinate and abscissa in our coordinate system. The approximate root is the value of $q(y) = x$ corresponding to $y = 0$; thus, it is obtained by the simple expedient of evaluating q at 0:

$$x_3 = q(0) = x_2 - ay_2 + by_2y_1.$$

In the general case, we have

$$x_{n+1} = x_n - a_n f(x_n) + b_n f(x_n) f(x_{n-1}),$$

where a_n and b_n are computed from appropriately generalized versions of (4.17) and (4.18). Table 4.14 shows the result of applying inverse quadratic interpolation to the example $f(x) = 2 - e^x$.

Table 4.14   Inverse quadratic interpolation, $f(x) = 2 - e^x$.

n	x_n	$f(x_n)$	$\alpha - x_n$	$\log_{10}(\alpha - x_n)$
0	0.000000000000	0.100000e+01	0.693147e+00	–0.1592
1	0.500000000000	0.351279e+00	0.193147e+00	–0.7141
2	1.000000000000	–0.718282e+00	0.306853e+00	–0.5131
3	0.708748678748	–0.314477e–01	0.156015e–01	–1.8068
4	0.692849949759	0.594373e–03	0.297231e–03	–3.5269
5	0.693146743432	0.874255e–06	0.437127e–06	–6.3594
6	0.693147180561	–0.134603e–11	0.673128e–12	–12.1719
7	0.693147180560	0.000000e+00	0.111022e–15	–15.9546

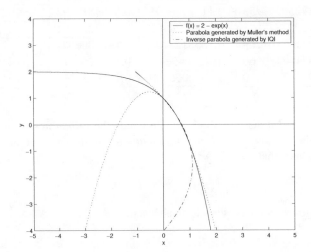

Figure 4.5   Muller's method and inverse quadratic interpolation.

Figure 4.5 shows a graph of $f(x) = 2 - e^x$, along with the graphs of the parabola generated by Muller's method and the inverse parabola generated by inverse quadratic interpolation, using $x_0 = 0$, $x_1 = 1$, and $x_2 = 1/2$ as the initial values in each case.

The convergence theory for both Muller's method and inverse quadratic interpolation is beyond the intended scope of this text, so we will content ourselves with reporting that both methods are locally convergent and superlinear, with $p \approx 1.84$. Note that this order of convergence is not much greater than that for the secant method.

One great utility of Muller's method is that it is able to find complex roots of real-valued functions, because of the square root in the computation. (The same is true of Halley's method (3.43).)

Inverse quadratic interpolation is used as part of the Brent–Dekker root-finding algorithm, which is a commonly implemented automatic root-finding program.

Exercises:

1. Do three steps of Muller's method for $f(x) = 2 - e^x$, using the initial points $0, 1/2$, and 1. Make sure that you reproduce the values in Table 4.13.

2. Repeat the above for inverse quadratic interpolation.

3. Let $f(x) = x^6 - x - 1$; show that this has a root on the interval $[0, 2]$ and do three steps of Muller's method using the ends of the interval plus the midpoint as the initial values.

4. Repeat the above using inverse quadratic interpolation.

5. Let $f(x) = e^x$ and consider the nodes $x_0 = -1$, $x_1 = 0$, and $x_2 = 1$. Let p_2 be the quadratic polynomial that interpolates f at these nodes, and let q_2 be the quadratic polynomial (in y) that *inverse*-interpolates f at these nodes. Construct $p_2(x)$ and $q_2(y)$ and plot both, together with $f(x)$.

6. Repeat the above using $f(x) = \sin \pi x$ and the nodes $x_0 = 0$, $x_1 = \frac{1}{4}$, and $x_2 = \frac{1}{2}$.

7. Show that the formulas (4.15) and (4.16) for the divided-difference coefficients in Muller's method (4.14) are correct.

8. Show that the formulas (4.17) and (4.18) for the coefficients in inverse quadratic interpolation are correct.

9. Apply Muller's method and inverse quadratic interpolation to the functions in Problem 3 of §3.1, and compare your results to those obtained with Newton's method and/or the secant method.

10. Refer back to the discussion of the error estimate for the secant method in §3.11.3. Adapt this argument to derive an error estimate for Muller's method.

◁ ● ● ● ▷

4.5 APPLICATION: MORE APPROXIMATIONS TO THE DERIVATIVE

We can use polynomial interpolation to derive more formulas for approximating the derivative. Essentially, since interpolation error theory shows that polynomials can approximate more complicated functions reasonably well, we infer that the derivative of a polynomial interpolate might approximate the derivative of the complicated function reasonably well.

The interpolation error theorem gives us that

$$f(x) - p_n(x) = \frac{1}{(n+1)!} w_n(x) f^{(n+1)}(\xi_x),$$

and the fact that this is an equality is important, for we can now differentiate both sides to get

$$f'(x) - p_n'(x) = \frac{1}{(n+1)!} \frac{d}{dx} \left[w_n(x) f^{(n+1)}(\xi_x) \right].$$

We have to be careful evaluating the derivative of the bracketed term; recall that ξ_x depends on x, so that

$$\frac{d}{dx} \left[w_n(x) f^{(n+1)}(\xi_x) \right] = w_n'(x) f^{(n+1)}(\xi_x) + w_n(x) \frac{d}{dx} \left[f^{(n+1)}(\xi_x) \right]. \qquad (4.19)$$

The problem in using this formula is that we do not know how ξ_x depends on x; thus, we cannot estimate or predict the size of the derivative in the second term. However, if we evaluate the approximation at one of the nodes, x_i, then we will have $w_n(x_i) = 0$ and the second term on the right side of (4.19) will vanish, leaving us with

$$f'(x_i) - p_n'(x_i) = \frac{1}{(n+1)!} w_n'(x_i) f^{(n+1)}(\xi_i), \qquad (4.20)$$

where we use $\xi_i = \xi_{x_i}$ for simplicity of notation.[8]

We can illustrate this by using a quadratic polynomial to approximate f, and thus derive three formulas for approximating the first derivative. Let the nodes be denoted by x_0, x_1, and x_2, and assume that they are equally spaced, so we have $x_1 - x_0 = h$ and $x_2 - x_1 = h$, and so on. Then the interpolating polynomial in Lagrange form is

$$p_2(x) = f(x_0) L_0^{(2)}(x) + f(x_1) L_1^{(2)}(x) + f(x_2) L_2^{(2)}(x).$$

The approximate derivative is then

$$f'(x_i) \approx p_2'(x_i) = f(x_0)(L_0^{(2)})'(x_i) + f(x_1)(L_1^{(2)})'(x_i) + f(x_2)(L_2^{(2)})'(x_i),$$

where $i = 0, 1, 2$. The error is given as in (4.20); thus,

$$f'(x_i) - p_2'(x_i) = \frac{1}{6} w_2'(x_i) f'''(\xi_i)$$

and we only need to evaluate $(L_j^{(2)})'(x_i)$ and $w_2'(x_i)$ for $i, j = 0, 1, 2$, to complete the construction of the approximations. We get

$$(L_0^{(2)})'(x_0) = -3/2h, \quad (L_1^{(2)})'(x_0) = 2/h, \quad (L_2^{(2)})'(x_0) = -1/2h,$$
$$(L_0^{(2)})'(x_1) = -1/2h, \quad (L_1^{(2)})'(x_1) = 0, \quad (L_2^{(2)})'(x_1) = 1/2h,$$
$$(L_0^{(2)})'(x_2) = 1/2h, \quad (L_1^{(2)})'(x_2) = -2/h, \quad (L_2^{(2)})'(x_2) = 3/2h,$$

and

$$w_2'(x_0) = 2h^2, \quad w_2'(x_1) = -h^2, \quad w_2'(x_2) = 2h^2,$$

---

[8]Note that we have assumed here that ξ_x is a differentiable function of x. This is indeed the case, but actually proving it is a nontrivial exercise that we omit for the sake of a brief and concise presentation.

so that the approximations are

$$f'(x_0) \approx \frac{1}{2h}\left(-f(x_2) + 4f(x_1) - 3f(x_0)\right), \tag{4.21}$$

$$f'(x_1) \approx \frac{1}{2h}\left(f(x_2) - f(x_0)\right), \tag{4.22}$$

$$f'(x_2) \approx \frac{1}{2h}\left(3f(x_2) - 4f(x_1) + f(x_0)\right), \tag{4.23}$$

with errors

$$f'(x_0) - \frac{1}{2h}\left(-f(x_2) + 4f(x_1) - 3f(x_0)\right) = \frac{1}{3}h^2 f'''(\xi_0), \tag{4.24}$$

$$f'(x_1) - \frac{1}{2h}\left(f(x_2) - f(x_0)\right) = -\frac{1}{6}h^2 f'''(\xi_1), \tag{4.25}$$

$$f'(x_2) - \frac{1}{2h}\left(3f(x_2) - 4f(x_1) + f(x_0)\right) = \frac{1}{3}h^2 f'''(\xi_2). \tag{4.26}$$

We recognize the approximation (4.22) as the same central difference formula derived in §2.2; the other two formulas are new. All three are second-order accurate, in that the error is $\mathcal{O}(h^2)$, but (4.21) and (4.23) require three function evaluations, whereas (4.22) requires only two. In some sense, this makes (4.22) a "better" approximation to use, but we will see situations where (4.21) and (4.23) are more useful in §§5.2 and 6.6.2.

Consider, for example, a function defined only on a grid of equally spaced points. Thus we have $y_k = f(x_k)$ for only the nodes x_k. Suppose further that we need an accurate approximation to $f'(x_0)$ (the derivative at the left endpoint) or $f'(x_n)$ (the derivative at the right endpoint). We can't use the centered difference approximation (2.5) at the endpoints, but we can use (4.24) and (4.26) to get

$$f'(x) = \left(\frac{-f(x + 2h) + 4f(x + h) - 3f(x)}{2h}\right) + \mathcal{O}(h^2)$$

and

$$f'(x) = \left(\frac{3f(x) - 4f(x - h) + f(x - 2h)}{2h}\right) + \mathcal{O}(h^2).$$

This will be useful when we need derivative approximations at the ends of intervals for constructing spline approximations (§4.8) or for improving the trapezoid rule (§5.2).

Exercises:

1. Apply the derivative approximations (4.21) and (4.23) to the approximation of $f'(x)$ for $f(x) = x^3 + x^2 + 1$ for $x = 1$ and $h = \frac{1}{8}$.

2. Apply the derivative approximations (4.21) and (4.23) to the same set of functions as in Problem 3 of §2.2, using a decreasing sequence of mesh values, $h^{-1} = 2, 4, 8, \ldots$. Do we achieve the expected rate of accuracy as h decreases?

3. Derive a version of (4.24) under the assumption that $x_1 - x_0 = h$, but $x_2 - x_1 = \eta = \theta h$ for some real, positive, θ. Be sure to include the error estimate as part of your work, and confirm that when $\theta = 1$ you get the same results as in the text.

4. Repeat the above for (4.25).

5. Repeat the above for (4.26).

6. Use the derivative approximations from this section to construct a table of values for the derivative of the gamma function, based on the data in Table 4.8.

7. Try to extend the ideas of this section to construct an approximation to $f''(x_k)$. Is it possible? What happens?

◁ ● ● ▷

4.6 HERMITE INTERPOLATION

So far, we have studied interpolation that involved only knowledge of function values. What happens if we include information about the derivative as well?

To be specific, we want to find a polynomial $p(x)$ that interpolates $f(x)$ in the sense that[9]

$$p(x_i) = f(x_i), \quad p'(x_i) = f'(x_i), \quad 1 \le i \le n.$$

This is known as the *Hermite*[10] *interpolation problem*. Can we do this? The answer is yes, as seen in the following theorem.

Theorem 4.4 (Hermite Interpolation Theorem) *Given the n nodes x_i, $1 \le i \le n$, and a differentiable function $f(x)$, if the nodes are distinct, then there exists a unique polynomial H_n of degree less than or equal to $2n - 1$, such that*

$$H_n(x_i) = f(x_i), \quad H_n'(x_i) = f'(x_i), \quad 1 \le i \le n. \tag{4.27}$$

Proof: This is basically the same as the proof of Theorem 4.1. We define the two families of polynomials

$$h_k(x) = [1 - 2[L_k^{(n)}]'(x_k)(x - x_k)][L_k^{(n)}(x)]^2$$

and

$$\tilde{h}_k(x) = (x - x_k)[L_k^{(n)}(x)]^2;$$

now observe that

$$h_k(x_j) = \delta_{kj}, \quad h_k'(x_j) = 0,$$

and

$$\tilde{h}_k(x_j) = 0, \quad \tilde{h}_k'(x_j) = \delta_{kj}.$$

Therefore, the polynomial

$$H_n(x) = \sum_{k=1}^{n} \left(f(x_k)h_k(x) + f'(x_k)\tilde{h}_k(x) \right)$$

---

[9]Note the change in indexing convention: We are now indexing from 1 to n, instead of from 0 to n.

[10]Charles Hermite (1822–1901) was born in Dieuze, France, on Christmas Eve. A poor test-taker, Hermite was not able to pass the exams for his bachelor's degree until the age of nearly 26. He held professional positions at a number of French schools, most notably the École Polytechnique, the École Normale, and the Sorbonne.

Perhaps his best-known mathematical result is the first proof that e is a transcendental number (published in 1873). His name is attached to a number of mathematical ideas and concepts, including the Hermite differential equation, Hermite polynomials, and Hermitian matrices. The idea of interpolating to the derivative as well as to the function values was part of a paper he published in 1878, *Sur la formule d'interpolation de Lagrange*, which actually considers interpolation not only of the first derivative, but of higher derivatives as well. Hermite interpolation is sometimes called *osculatory interpolation*.

satisfies the interpolating conditions (4.27) and is clearly of degree less than or equal to $2n - 1$. Uniqueness follows by the same argument as used before. ●

Like the Lagrange form in ordinary interpolation, this form of the Hermite polynomial is not conducive to efficient computation. A variation of the Newton approach, including the use of divided difference tables, can be constructed; however, since the most common use of Hermite interpolation is the cubic case (which involves only two nodes, $x = a$ and $x = b$), we will forgo the more general development and show only the cubic construction.

We begin by setting up a divided difference table (Table 4.15), much as we did in §4.2. Note, however, that we enter the functional data twice and use the derivative data for part of the first differences column.

Table 4.15   Divided difference table for Hermite interpolation: initial setup.

k	x_k	$f_0(x_k)$	$f_1(x_k)$
1	a	$f(a)$	
			$f'(a)$
1	a	$f(a)$	
2	b	$f(b)$	
			$f'(b)$
2	b	$f(b)$	

We then complete the table just as we would an ordinary divided difference table, getting the results shown in Table 4.16.

Table 4.16   Divided difference table for Hermite interpolation: final form.

k	x_k	$f_0(x_k)$	$f_1(x_k)$	$f_2(x_k)$	$f_3(x_k)$
1	a	$f(a)$			
			$f'(a)$		
1	a	$f(a)$		B	
			A		D
2	b	$f(b)$		C	
			$f'(b)$		
2	b	$f(b)$			

The elements denoted by letters are defined as follows:

$$A = \frac{f(b) - f(a)}{b - a}, \qquad B = \frac{A - f'(a)}{b - a},$$

$$C = \frac{f'(b) - A}{b - a}, \qquad D = \frac{C - B}{b - a}.$$

The polynomial is then given by

$$H_2(x) = f(a) + f'(a)(x - a) + B(x - a)^2 + D(x - a)^2(x - b). \qquad (4.28)$$

It can easily be checked that this does interpolate to f and f' at both a and b (see Problem 1).

■ **EXAMPLE 4.6**

As an illustration of Hermite interpolation, consider the problem of interpolating to the exponential function on the interval $[-1, 1]$ using data only at the endpoints. We have $x_1 = -1$, $x_2 = 1$, and the divided difference table is shown as Table 4.17, so that the polynomial is

$$H_2(x) = 2.718281828 + 2.718281828(x - 1) \\ + 0.771540317(x - 1)^2 + 0.1839397203(x - 1)^2(x + 1),$$

or (after some simplification),

$$H_2(x) = 0.1839397206x^3 + 0.5876005967x^2 + 0.9912614728x + 0.9554800379.$$

Figure 4.6 shows a graph of both the exponential and H_2 (they are indistinguishable); Figure 4.7 shows the error $e^x - H_2(x)$.

Table 4.17   Divided difference table for cubic Hermite interpolation to $f(x) = e^x$.

k	x_k	$f_0(x_k)$	$f_1(x_k)$	$f_2(x_k)$	$f_3(x_k)$
1	1	2.718281828			
			2.718281828		
1	1	2.718281828		0.771540317	
			1.175201194		0.1839397203
2	-1	0.3678794412		0.4036608764	
			0.3678794412		
2	-1	0.3678794412			

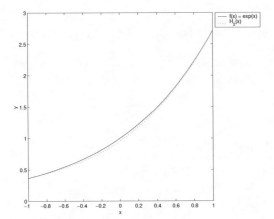

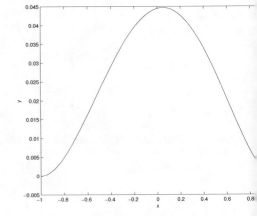

Figure 4.6   Hermite interpolation to $f(x) = e^x$.

Figure 4.7   Error in Hermite interpolation to $f(x) = e^x$.

The accuracy of Hermite interpolation is based on a result very similar to Theorem 4.3.

Theorem 4.5 (Hermite Interpolation Error Theorem) *Let* $f \in C^{2n}([a,b])$ *and let the nodes* $x_k \in [a,b]$ *for all* $k,\quad 1 \le k \le n$. *Then, for each* $x \in [a,b]$, *there is a* $\xi_x \in [a,b]$ *such that*

$$f(x) - H_n(x) = \frac{\psi_n(x)}{(2n)!} f^{(2n)}(\xi_x), \tag{4.29}$$

where

$$\psi_n(x) = \prod_{k=1}^{n} (x - x_k)^2.$$

Proof: Follows essentially the same argument as Theorem 4.3; see Appendix A.1 for details. •

Exercises:

1. Show that H_2, as defined in (4.28), is the cubic Hermite interpolate to f at the nodes $x = a$ and $x = b$.

2. Construct the cubic Hermite interpolate to $f(x) = \sin x$ using the nodes $a = 0$ and $b = \pi$. Plot the error between the interpolate and the function.

3. Construct the cubic Hermite interpolate to $f(x) = \sqrt{1+x}$ using the nodes $a = 0$ and $b = 1$. Plot the error between the interpolate and the function.

4. Show that the error in cubic Hermite interpolation at the nodes $x = a$ and $x = b$ is given by

$$\|f - H_2\|_\infty \le \frac{(b-a)^4}{384} \|f^{(4)}\|_\infty.$$

5. Construct the cubic Hermite interpolate to $f(x) = \sqrt{x}$ on the interval $[\frac{1}{4}, 1]$. What is the maximum error on this interval, as predicted by theory? What is the maximum error that actually occurs (as determined by observation; no need to do a complete calculus max/min problem)?

6. Construct the cubic Hermite interpolate to $f(x) = 1/x$ on the interval $[\frac{1}{2}, 1]$. What is the maximum error as predicted by theory? What is the actual (observed) maximum error?

7. Construct the cubic Hermite interpolate to $f(x) = x^{1/3}$ on the interval $[\frac{1}{8}, 1]$. What is the maximum error as predicted by theory? What is the actual (observed) maximum error?

8. Construct the cubic Hermite interpolate to $f(x) = \ln x$ on the interval $[\frac{1}{2}, 1]$. What is the maximum error as predicted by theory? What is the actual (observed) maximum error?

9. Extend the divided difference table for cubic Hermite interpolation to *quintic* Hermite interpolation, using the three nodes $x = a$, $x = b$, and $x = c$.

10. Construct the quintic Hermite interpolate to $f(x) = \ln x$ on the interval $[\frac{1}{2}, 1]$; use $x = 3/4$ as the third node.

11. What is the error in quintic Hermite interpolation?

12. Extend the ideas of §4.5 to allow us to compute second derivative approximations using Hermite interpolation.

◁ ● ● ● ▷

4.7 PIECEWISE POLYNOMIAL INTERPOLATION

The results of Figure 4.3 suggest that interpolation might be unsatisfactory as an approximation tool. This is true if we insist on letting the *order* of the polynomial get larger and larger (see §4.12). However, if we keep the order of the polynomial fixed, and use different polynomials over different intervals, with the length of the intervals getting smaller and smaller, then interpolation can be a very accurate and powerful approximation tool. These ideas reach their full potential in §4.8 on splines; here we confine ourselves to the mere basics.

Consider the problem of constructing an approximation to a function $f(x)$, defined only by 101 equally spaced points on the interval $[-1, 1]$. If we use ordinary interpolation, we will have to construct a 100-degree polynomial, and we suspect that this might produce unsatisfactory results. However, we could use linear interpolation over each subinterval $[x_{k-1}, x_k]$ or quadratic interpolation over each subinterval $[x_{2k-2}, x_{2k}]$. That is, we define the approximating function as

$$q_d(x) = p_{d,k}(x), \ x_{d(k-1)} \le x \le x_{dk},$$

where each $p_{d,k}$ is a polynomial of degree d. The accuracy of the approximation comes from the error estimates (4.7), (4.11), or (4.13), since the distance between nodes is small.

We illustrate this by approximating $f(x) = (1 + 25x^2)^{-1}$ over $[-1, 1]$ using the seven equally spaced points $x_k = -1 + k/3, k = 0, 1, \ldots, 6$. The results are shown in Figure 4.8 for piecewise linear, quadratic, and cubic interpolation. For the piecewise quadratic case, we have (see Problem 3)

$$q_2(x) = \begin{cases} 0.0385 + 0.1323(x - x_0) + 0.6211(x - x_0)(x - x_1), & x_0 \le x \le x_2; \\ 0.2647 + 2.2059(x - x_2) - 6.6176(x - x_2)(x - x_3), & x_2 \le x \le x_4; \\ 0.2647 - 0.5464(x - x_4) + 0.6211(x - x_4)(x - x_5), & x_4 \le x \le x_6. \end{cases}$$
$$(4.30)$$

Note that all of these piecewise approximations are much more accurate than the 16-degree interpolating polynomial that was constructed in Figure 4.3C. This contrast is heightened even more when we take lots of points. Figures 4.9 and 4.10 show the results when we use 33 equally spaced points and piecewise quadratic approximation; compare this to Figure 4.3C.

The use of different polynomials over different intervals complicates the entire process somewhat. We now have one set of points (the nodes, $x_k, 0 \le k \le n$) that define the interpolation conditions, and a second set of points, called *knots*, that define the subintervals on which the separate polynomial pieces are defined. (In our presentation here we will assume that the knots are a subset of the nodes, but this doesn't have to be the case.) For a piecewise polynomial of degree d, the knots will be $x_{dj}, 0 \le j \le m$, where m is the number of polynomial pieces in the approximation. Thus, for the quadratic example defined in (4.30), the knots are x_0, x_2, x_4, and x_6.

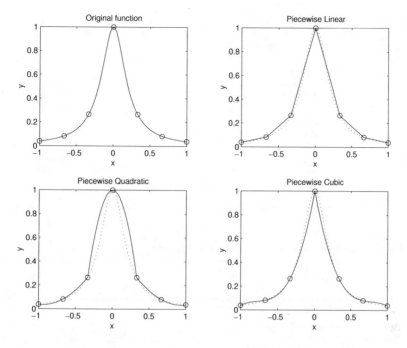

Figure 4.8 Piecewise interpolation to $f(x) = (1 + 25x^2)^{-1}$. The nodes are denoted by circles.

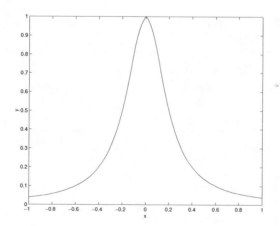

Figure 4.9 Piecewise quadratic interpolation to $f(x) = (1 + 25x^2)^{-1}$ using 33 nodes.

■ **EXAMPLE 4.7**

Consider the problem of constructing a piecewise quadratic approximation to $f(x) = \sin \pi x$, using the nodes $x_k = k/6$, $k = 0, 1, 2, \ldots, 6$. Thus there will be three separate polynomial approximations, and the knots will be $x_0, x_2, x_4,$ and x_6.

We set up three divided difference tables (Table 4.18), one for each polynomial piece:

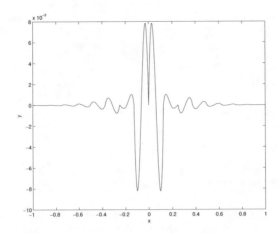

Figure 4.10   Error in 33-node piecewise quadratic interpolation to $f(x) = (1 + 25x^2)^{-1}$.

Table 4.18   Divided difference tables for piecewise quadratic approximation to $f(x) = \sin \pi x$.

x_k	$f_0(x_k)$	$f_1(x_k)$	$f_2(x_k)$
0	0.0000000		
		3.0000000	
$\frac{1}{6}$	0.5000000		–2.4115427
		2.1961524	
$\frac{1}{3}$	0.8660254		

x_k	$f_0(x_k)$	$f_1(x_k)$	$f_2(x_k)$
$\frac{1}{3}$	0.8660254		
		0.80384758	
$\frac{1}{2}$	1.0000000		–4.8230855
		–0.080384758	
$\frac{2}{3}$	0.8660254		

x_k	$f_0(x_k)$	$f_1(x_k)$	$f_2(x_k)$
$\frac{2}{3}$	0.8660254		
		–2.1961524	
$\frac{5}{6}$	0.5000000		–2.4115427
		–3.0000000	
1	0.0000000		

Thus we know that the three polynomials are given by

$$p_{2,1}(x) = 3.0000000x - 2.4115427x\left(x - \frac{1}{6}\right),$$

$$p_{2,2}(x) = 0.8660254 + 0.8038476\left(x - \frac{1}{3}\right) - 4.8230855\left(x - \frac{1}{3}\right)\left(x - \frac{1}{2}\right),$$

$$p_{2,3}(x) = 0.8660254 - 2.1961524\left(x - \frac{2}{3}\right) - 2.4115427\left(x - \frac{2}{3}\right)\left(x - \frac{5}{6}\right),$$

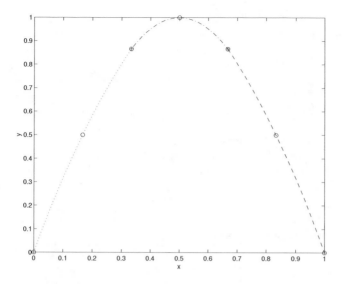

Figure 4.11   Piecewise polynomial approximation to $y = \sin \pi x$ over $[0, 1]$; a circle ("o") denotes a node, a plus sign ("+") denotes a knot.

so that the piecewise polynomial is given by

$$q_2(x) = \begin{cases} 3.0000000x - 2.4115427x(x - \frac{1}{6}), & 0 \le x \le \frac{1}{3}; \\ 0.8660254 + 0.803847576(x - \frac{1}{3}) - 4.8230855(x - \frac{1}{3})(x - \frac{1}{2}), & \frac{1}{3} \le x \le \frac{2}{3}; \\ 0.8660254 - 2.1961524(x - \frac{2}{3}) - 2.4115427(x - \frac{2}{3})(x - \frac{5}{6}), & \frac{2}{3} \le x \le 1. \end{cases}$$

Figure 4.11 shows a plot of this function, using a different line style for each polynomial piece. The nodes and knots are also marked on the graph.

It might be argued that approximating the sine function over $[0, \pi]$ is not much of a challenge, that even simple polynomial interpolation ought to do a decent job of approximating such a function; this is fair criticism, and the student should do the necessary computations to show that Newton interpolation will match the sine function over $[0, \pi]$ very well. But recall that we were able to match the troublesome function $f(x) = 1/(1 + 25x^2)$ very well with piecewise polynomial approximation. This bodes well for its ultimate utility.

Piecewise polynomial approximation is, in fact, extraordinarily useful as an approximation technique, because the use of different polynomials on different parts of the domain allows us to more closely mimic the range of function behavior that is possible. To accurately model more complicated functions, we need to make sure that the various pieces are joined in a way that maintains as much smoothness as possible. This leads us, very naturally, to s discussion of *splines*.

An algorithm for piecewise polynomial approximation will generally consist of two parts: One that constructs the approximation, usually in the form of an array of polynomial coefficients or divided difference coefficients (which is what we did in Example 4.7 above), and a second part that uses the output of the first part to evaluate the approximation. Pseudo-code for each of these is given in Algorithm 4.3.

Algorithm 4.3 *Pseudo-code for Constructing Piecewise Polynomial Approximation*

```
input n, (x(i),y(i),i=0,n), d
!
    kount = 0
    for k=0 to n by d do
        kount = kount + 1
        for i=0 to d do
            xc(i) = x(k+i)
            yc(i) = y(k+i)
        endfor
!
!   compute divided difference coefficients
!   for the xc and yc arrays
!
        ac(0) = yc(0)
        for i=1 to d do
            w = 1
            p = 0
            for j=0 to i-1 do
                p = p + ac(j)*w
                w = w*(xc(i) - xc(j))
            endfor
            ac(i) = (yc(i) - p)/w
        endfor
        for i=0 to d do
            a(kount,i) = ac(i)
        endfor
!
!   This code evaluates the pcw poly at the point xx
!   First, find the subinterval containing the evaluation
!   point xx
!
    input n, d, kount, (x(i),i=0,n), (a(k,i),k=1,kount,i=0,d), xx
    j = search(x,xx)
!
!   Next, extract the correct nodes and DD coefficients
!
    for k=0 to d
        ac(k) = a(j,k)
        xc(k) = x(j+k)
    endfor
    yy = ac(j,d)
    for k = d-1 to 0 by -1
        xd = xx - xc(k)
        yy =  ac(k) + yy*xd
    endfor
```

Programming Hints: Note the line j = search(x,xx) in the evaluation part of Algorithm 4.3. Evaluating a piecewise polynomial requires, first of all, that we find out which "piece" of the polynomial to use for each given x value for which we want to compute $q_d(x)$, and this pseudo-code assumes that this is done in the separate routine called search. If the mesh spacing is uniform, the computation

$$j = \left\lfloor \frac{x - x_0}{h} \right\rfloor$$

guarantees that $x_j \leq x \leq x_{j+1}$. Here $\lfloor \cdot \rfloor$ is the *floor function*, defined as the greatest integer less than or equal to the argument. From this it is an easy task to figure out which polynomial piece to use to compute q_d. If the mesh spacing is not uniform, then a more general search procedure has to be used to find the indices such that $x_{(d-1)k} \leq x \leq x_{dk}$.

MATLAB has a number of routines for constructing and evaluating piecewise polynomial approximations within a specialized data structure:

- pchip—piecewise cubic Hermite polynomial;

- spline—spline construction (see §4.8);

- ppval—evaluates a piecewise polynomial;

- mkpp—Construct a piecewise polynomial from the knots and the polynomial coefficients.

For example, defining the polynomial coefficient vector pc = [1 -2 1; -1 0 0] and then executing pp = mkpp([0 1 2],pc) creates the piecewise quadratic shown in Figure 4.12—the circles mark the knots. (The reader should be aware that MATLAB assumes some shifting in the definitions of the polynomial pieces—the right half of Fig. 4.12 is $y = -(x - 1)^2$, which ordinarily would be rendered in MATLAB as [-1 2 -1], but had to be rendered as [-1 0 0] to achieve the desired effect.)

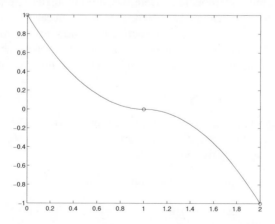

Figure 4.12   Very simple piecewise polynomial created in MATLAB.

The underlying error theorem is easy to state and prove.

Theorem 4.6 (Piecewise Polynomial Interpolation Error Estimate) *Let f be sufficiently differentiable on the closed interval* $[a, b]$, *and let* q_d *be the piecewise polynomial interpolate of degree* d *to* f *on* $[a, b]$, *using* $n + 1$ *equally spaced nodes* x_k, $0 \leq k \leq n$, $n = md$. *Then*

$$\|f - q_d\|_\infty \leq C_d h^{d+1} \|f^{(d+1)}\|_\infty, \ d = 1, 2, 3,$$

where

$$C_1 = 1/8; \ C_2 = 1/9\sqrt{3}; \ C_3 = 1/24.$$

Proof: Follows directly from the separate error estimates (4.8), (4.12), and (4.13). We have

$$
\begin{aligned}
\|f - q_d\|_\infty &= \max_{a \leq x \leq b} |f(x) - q_d(x)|, \\
&= \max_{1 \leq k \leq n} \left(\max_{x_{k-1} \leq x \leq x_k} |f(x) - q_d(x)| \right), \\
&= \max_{1 \leq k \leq n} \left(\max_{x_{k-1} \leq x \leq x_k} |f(x) - p_{d,k}(x)| \right), \\
&= \max_{1 \leq k \leq n} \|f - p_{d,k}\|_{\infty, [x_{k-1}, x_k]}, \\
&\leq \max_{1 \leq k \leq n} C_d h^{d+1} \|f^{(d+1)}\|_{\infty, [x_{k-1}, x_k]}, \\
&= C_d h^{d+1} \|f^{(d+1)}\|_{\infty, [a,b]}, \\
&= C_d h^{d+1} \|f^{(d+1)}\|_\infty,
\end{aligned}
$$

and we are done. ●

While piecewise polynomial approximation is heavily used in a variety of applications (we used it, essentially, to construct the composite trapezoid rule in Chapter 2), it is most important, perhaps, as a precursor to the development of spline approximations, our next main topic.

One interesting and useful form of piecewise polynomial approximation is the use of piecewise cubic Hermite interpolation. Here, we use cubic Hermite polynomials in an obvious way to define the polynomial approximation between each pair of nodes x_{k-1} and x_k. Because we are also interpolating to the derivative values, the resulting approximation is much smoother than ordinary piecewise interpolation and is less prone to having "kinks" in the graph. Some of the exercises ask the student to look at this type of approximation.

Exercises:

1. Use divided difference tables to construct the separate parts of the piecewise quadratic polynomial $q_2(x)$ that interpolates to $f(x) = \cos \frac{1}{2}\pi x$ at $x = 0, \frac{1}{4}, \frac{1}{2}, \frac{3}{4}, 1$. Plot the approximation and the error $\cos \frac{1}{2}\pi x - q_2(x)$.

2. Repeat the above using $f(x) = \sqrt{x}$ with the nodes $x = \frac{1}{5}, \frac{2}{5}, \frac{3}{5}, \frac{4}{5}, 1$.

3. Confirm that (4.30) is the correct piecewise quadratic approximation to $f(x) = 1/(1 + 25x^2)$ using the nodes $x_0 = -1$, $x_1 = -2/3$, $x_2 = -1/3$, $x_3 = 0$, $x_4 = 1/3$, $x_5 = 2/3$, and $x_6 = 1$.

4. Using the data in Table 4.8, construct a piecewise cubic interpolating polynomial to the gamma function, using the nodes $x = 1.0, 1.2, 1.3, 1.5$ for one piece, and the

nodes $x = 1.5, 1.7, 1.8, 2.0$ for the other piece. Use this approximation to estimate $\Gamma(x)$ for $x = 1.1, 1.4, 1.6$ and 1.9. How accurate is the approximation?

5. Using the results of Problem 6 from §4.6, together with the data of function values from Problem 9 of §4.2, construct a piecewise cubic Hermite interpolating polynomial for the gamma function, using the nodes $x = 1.0, 1.3, 1.7, 2.0$. Test the accuracy of the interpolation by using it to approximate $\Gamma(x)$ for $x = 1.1, 1.2, 1.4, 1.5, 1.6, 1.8, 1.9$.

6. Construct a piecewise cubic interpolating polynomial to $f(x) = \ln x$ on the interval $[\frac{1}{2}, 1]$, using the nodes

$$x_k = \frac{1}{2} + \frac{k}{18}, \quad 0 \leq k \leq 9.$$

Compute the value of the error $\ln x - p(x)$ at 500 equally spaced points on the interval $[\frac{1}{2}, 1]$, and plot the error. What is the maximum sampled error?

7. Repeat the above, using piecewise cubic Hermite interpolation over the same grid.

8. Construct piecewise polynomial approximations of degree $d = 1, 2, 3$ to the data in Table 4.12, using only the nodes $\log_{10} \theta_k = -6, -4, -2, 0, 2, 4, 6$. Plot the resulting curve and compare it to the ordinary interpolating polynomial found in Problem 13 of §4.2. Test how well this approximation matches the tabulated values at $\log_{10} \theta = -5, -3, -1, 1, 3, 5$.

9. Show that the error in piecewise cubic Hermite interpolation satisfies

$$\|f - H_2\|_\infty \leq \frac{1}{384} h^4 \|f^{(4)}\|_\infty,$$

where we have assumed uniformly spaced nodes with $x_k - x_{k-1} = h$.

10. Given a grid of points

$$a = x_0 < x_1 < x_2 < \cdots < x_n = b,$$

define the piecewise linear functions ϕ_k^h, $1 \leq k \leq n - 1$, according to

$$\phi_k^h(x) = \begin{cases} \frac{x - x_{k-1}}{x_k - x_{k-1}}, & x_{k-1} \leq x \leq x_k; \\ \frac{x_{k+1} - x}{x_{k+1} - x_k}, & x_k \leq x \leq x_{k+1}; \\ 0, & \text{otherwise.} \end{cases}$$

Define the function space

$$S_0^h = \{f \in C([a, b]), f(a) = f(b) = 0, f \text{ is piecewise linear on the given grid}\}.$$

Show that the ϕ_k^h are a basis for S_0^h, i.e., that every element of the space S_0^h can be written as a linear combination of the ϕ_k^h functions.

11. Implement a routine for approximating the natural logarithm using piecewise polynomial interpolation, i.e., a table look-up scheme. Assume that the table of (known) logarithm values is uniformly distributed on the interval $[\frac{1}{2}, 1]$. Choose enough points in the table to guarantee 10^{-10} accuracy for any x. Use:

(a) Piecewise linear interpolation;

(b) Piecewise cubic Hermite interpolation.

Test your routine against the intrinsic logarithm function on your computer by evaluating the error in your approximation at $5,000$ equally spaced points on the interval $\left[\frac{1}{10}, 10\right]$. Use the way that the computer stores floating-point numbers to reduce the logarithm computation to the interval $\left[\frac{1}{2}, 1\right]$, as long as $\ln 2$ is known to high accuracy.

◁ ● ● ● ▷

4.8   AN INTRODUCTION TO SPLINES

Piecewise polynomial approximation (§4.7) allows us to construct highly accurate approximations, but it does not always produce approximations that are pleasing to the eye, and this is actually a problem in some applications. This occurs because the approximating function is not smooth at the junctions between separate pieces of the piecewise polynomial approximation. Although we have that q_d is continuous, it is *not* continuously differentiable on the entire interval of approximation. The graph of the interpolate can have "kinks" in it, and some of these are apparent in Figure 4.8.

Splines are an attempt to address this problem. The basic idea is to construct a piecewise polynomial approximation that not only interpolates given data or function values, but which also is "smooth," meaning continuously differentiable to some degree.

A complete theory of splines is beyond the scope of this book. Interested readers are referred to the book by Carl de Boor, *A Practical Guide to Splines* [8]. Another good reference is Paddy Prenter's *Splines and Variational Methods* [14]. We will outline two different spline constructions here and state some of the more basic theoretical results.[11]

4.8.1   Definition of the Problem

We first have to properly define the problem before we can solve it. Suppose that we are given a set of *nodes* $\{x_i, \ 0 \le i \le n\}$ at which we wish to interpolate a given function $f(x)$ with a spline of degree d. The problem is then to find a piecewise polynomial function, q_d, which satisfies the following conditions:

---

[11]The history of the development of splines—and even the origin of the word itself—might be of interest to readers.

Many years ago, one of the author's students suggested that the word "splines" was constructed from "spliced lines;" while this is indeed an interesting observation, the truth of the matter is that the mathematical use of the word "spline" comes from the days when being an engineer often meant being an accomplished draftsman. One draftsman's tool was known as a "spline," a thin, very flexible piece of metal that was used to help draw a smooth curved line based on a few discrete points marked on a drawing.

One of the historical uses of splines was in the automobile industry, where they were used to define the shape of sheet-metal pieces for car bodies. Much of the mathematical development work was done at General Motors in the early 1960s, and also at Renault and Citroen in France, at about the same time. Curiously, one of the original papers on splines [16] was written with applications to actuarial data in mind. Some of the impetus for the development of the theory also came from the British aircraft industry during World War II, as well as the ship-building industry.

The work at GM is detailed nicely by Birkhoff [2] and in the retrospective by Young [19]. Paul Davis summarized some of this material in *SIAM News* in 1996; see [5]. Carl DeBoor maintains a spline bibliography on the Internet at http://www.cs.wisc.edu/~deboor/bib/bib.html.

1. *Interpolation:*

$$q_d(x_k) = f(x_k), \quad 0 \le k \le n.$$

Here d is called the *degree of approximation* of the spline.

2. *Smoothness:*

$$\lim_{x \to x_k^-} q_d^{(i)}(x) = \lim_{x \to x_k^+} q_d^{(i)}(x)$$

for $0 \le i \le N$. We call N the *degree of smoothness* of the spline.

3. *Interval of definition:* q_d is a polynomial of degree $\le d$ on each subinterval $[x_{k-1}, x_k]$.

We need to investigate the relationship (if any) between the degree of approximation, d, and the degree of smoothness, N. The degree of the polynomials is related to the number of unknown coefficients, i.e., the degrees of freedom, in the problem; whereas N is related to the number of constraints. We expect that the degrees of freedom and the number of constraints have to balance in order for the spline to be well-defined.

Since there are n subintervals, each being the domain of definition for a separate polynomial of degree d, we have a total of $K_f = n(d+1)$ degrees of freedom. On the other hand, there are $n + 1$ interpolation conditions and $n - 1$ junction points (again called *knots*) between subintervals (note that in the spline construction, neither x_0 nor x_n are considered knots) with $N + 1$ continuity conditions being imposed at each of them. Thus, there are $K_c = n+1+(n-1)(N+1)$ constraints. If we consider the difference $K_f - K_c$, we get

$$K_f - K_c = n(d+1) - [n+1+(n-1)(N+1)] = nd - n - nN + N = n(d-1-N) + N.$$

We can make the first term vanish by setting $d - 1 - N = 0$. This establishes a relationship between the polynomial degree of the spline and the smoothness degree. For example, if we consider the common case of cubic splines, then $d = 3$ and $N = 2$. However, we will not have the number of constraints equal the number of degrees of freedom: $K_f - K_d = N$. Thus, we need to add N additional constraints. Partly for this reason, odd polynomial order splines are preferred, because if d is odd then $N = d - 1$ is even and the additional constraints can be imposed equally at the two endpoints (which is the typical choice) of the interval.

It is worth noting, briefly, that we can consider a continuous piecewise linear function to be a spline with degree of smoothness $N = 0$. See Problem 18.

4.8.2 Cubic B-Splines

The construction we use here is based on the *B-spline*. This idea uses a single exemplar function from which a basis of splines is formed, and the approximation is then defined in terms of this basis. The notion of B-splines is much more general than is presented here, where we restrict ourselves to the (widely used) case of cubic B-splines, with coincident nodes and knots. For simplicity's sake, we assume at first a uniform grid

$$a = x_0 < x_1 < x_2 < \cdots < x_{n-1} < x_n = b \tag{4.31}$$

with $x_k - x_{k-1} = h$ for all $k \ge 1$. We will also need to define the extra grid points

$$x_{-3} = a - 3h, \quad x_{-2} = a - 2h, \quad x_{-1} = a - h, \tag{4.32}$$

and

$$x_{n+1} = b + h, \quad x_{n+2} = b + 2h, \quad x_{n+3} = b + 3h. \tag{4.33}$$

Then define the function

$$B(x) = \begin{cases} 0, & x \leq -2; \\ (x+2)^3, & -2 \leq x \leq -1; \\ 1 + 3(x+1) + 3(x+1)^2 - 3(x+1)^3, & -1 \leq x \leq 0; \\ 1 + 3(1-x) + 3(1-x)^2 - 3(1-x)^3, & 0 \leq x \leq 1; \\ (2-x)^3, & 1 \leq x \leq 2; \\ 0, & 2 \leq x; \end{cases} \tag{4.34}$$

and note that B is a cubic spline with nodes/knots at the points $x = -2, -1, 0, 1, 2$.

Wait a minute. How do we know that $B(x)$ is a cubic spline function? Well, a spline function is a piecewise polynomial with a certain amount of derivative continuity between the pieces, the amount of derivative continuity being related to the polynomial degree by $N = d - 1$, as we showed above. Thus, to check that $B(x)$ is a cubic spline, we simply note that it clearly is a piecewise cubic polynomial, and then we compute the one-sided derivatives at the knots:

$$B'_-(x_k) = \lim_{x \to x_k^-} B'(x), \quad B'_+(x_k) = \lim_{x \to x_k^+} B'(x),$$

and similarly for the second derivative (and the function value). If the one-sided values are equal to each other, then the first and second derivatives are continuous, hence B is a cubic spline. If only the first derivative was continuous, it would fail to be a spline because *cubic splines require second derivative continuity.*

Figure 4.13 shows a graph of B; note the bell shape of the curve. Note also that

$$B(0) = 4, \quad B(\pm 1) = 1, \quad B(\pm 2) = 0, \tag{4.35}$$

$$B'(0) = 0, \quad B'(\pm 1) = \mp 3, \quad B'(\pm 2) = 0, \tag{4.36}$$

$$B''(0) = -12, \quad B''(\pm 1) = 6, \quad B''(\pm 2) = 0. \tag{4.37}$$

Finally, note that B is only "locally defined," meaning that it is non-zero on only a small interval.[12] This local definition property is important in the utility of B-splines as an approximation tool.

We can use B to construct a spline approximation to an arbitrary function f using the grid defined in (4.31), (4.32), and (4.33). Define the sequence of functions

$$B_i(x) = B((x - x_i)/h), \quad -1 \leq i \leq n + 1. \tag{4.38}$$

Each B_i is similar in shape to B, but is centered at x_i instead of at 0. Figure 4.14 shows plots of some of the B_i for a specific grid. Note that

$$B_i(x_i) = B(0), \quad B_i(x_{i\pm 1}) = B(\pm 1), \quad B_i(x_{i\pm 2}) = B(\pm 2), \tag{4.39}$$

---

[12]The correct terminology is to say that B is a function of "compact support," but using the phrase "locally defined" conveys the meaning more clearly at this level.

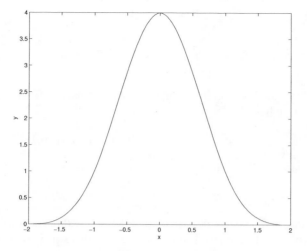

Figure 4.13   Original B-spline.

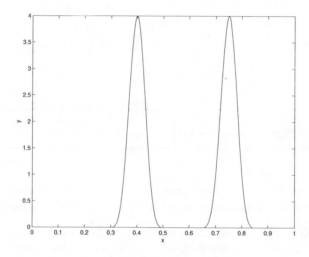

Figure 4.14   B-splines on a grid.

and similarly for the derivatives:

$$B_i'(x_i) = 0, \quad B_i'(x_{i\pm1}) = \mp3/h, \quad B_i'(x_{i\pm2}) = 0, \tag{4.40}$$

$$B_i''(x_i) = -12/h^2, \quad B_i''(x_{i\pm1}) = 6/h^2, \quad B_i''(x_{i\pm2}) = 0. \tag{4.41}$$

To construct a spline interpolant to a given function f, we define the spline as a linear combination of the B_i functions, $-1 \leq i \leq n+1$:

$$q_3(x) = \sum_{i=-1}^{n+1} c_i B_i(x) \tag{4.42}$$

and seek the coefficients c_i, $-1 \leq i \leq n+1$, such that

$$f(x_k) = \sum_{i=-1}^{n+1} c_i B_i(x_k), \quad 0 \leq k \leq n. \tag{4.43}$$

It is at this point that the local definition of B becomes so useful. If we look at (4.43) for $k = 0$, we see that we have

$$f(x_0) = c_{-1} B_{-1}(x_0) + c_0 B_0(x_0) + c_1 B_1(x_0)$$

since $B_i(x_0) = 0$ for all $i \geq 2$. In general, then, we have that

$$f(x_k) = c_{k-1} B_{k-1}(x_k) + c_k B_k(x_k) + c_{k+1} B_{k+1}(x_k).$$

Thus, each equation in the system defined by (4.43) has only three non-zero terms—it is a *tridiagonal* system, just the kind we learned how to solve back in §2.6. Moreover, we can use (4.35) to show that (4.43) simplifies to

$$f(x_k) = c_{k-1} + 4c_k + c_{k+1}.$$

Written in matrix–vector form, the system is

$$\begin{bmatrix} 1 & 4 & 1 & & & \\ & 1 & 4 & 1 & & \\ & & \ddots & \ddots & \ddots & \\ & & & 1 & 4 & 1 \end{bmatrix} \begin{bmatrix} c_{-1} \\ c_0 \\ \vdots \\ c_{n+1} \end{bmatrix} = \begin{bmatrix} f(x_0) \\ f(x_1) \\ \vdots \\ f(x_n) \end{bmatrix}. \tag{4.44}$$

Of course, this is still a system of $n+1$ equations in $n+3$ unknowns; we need to come up with two additional constraints in order to eliminate two of the unknowns. Two common choices are:

1. *The Natural Spline.* Here we impose $q_3''(x_0) = q_3''(x_n) = 0$. This leads to a very simple construction, but also leads to higher error near the endpoints.

2. *The Complete Spline.* Here we impose $q_3'(x_0) = f'(x_0)$ and $q_3'(x_n) = f'(x_n)$. This leads to better approximation properties, and does not actually require that we know the derivative at the endpoints, since we can use the function values to approximate it via one of the formulas from §4.5.

The Natural Spline   We can directly compute, using (4.42) and (4.36), that we have

$$q_3''(x_0) = h^{-2}(6(c_{-1} + c_1) - 12c_0), \quad q_3''(x_n) = h^{-2}(6(c_{n+1} + c_{n-1}) - 12c_n).$$

We therefore achieve $q_3''(x_0) = q_3''(x_n) = 0$ simply by imposing

$$c_{-1} + c_1 = 2c_0 \Rightarrow c_{-1} = 2c_0 - c_1 \tag{4.45}$$

and

$$c_{n+1} + c_{n-1} = 2c_n \Rightarrow c_{n+1} = 2c_n - c_{n-1}. \tag{4.46}$$

If we substitute these into the system (4.44), we find that the first and last equations simplify to

$$6c_0 = f(x_0), \quad 6c_n = f(x_n), \tag{4.47}$$

so that we have the new system

$$
\begin{array}{cccccc}
4c_1 & + & c_2 & & & = & f(x_1) - \frac{1}{6}f(x_0), \\
c_1 & + & 4c_2 & + & c_3 & = & f(x_2), \\
& \ddots & & \ddots & & \vdots & \\
& & c_{n-2} & + & 4c_{n-1} & = & f(x_{n-1}) - \frac{1}{6}f(x_n).
\end{array}
\tag{4.48}
$$

Not only is this now square, in addition to being tridiagonal, but it is also diagonally dominant. Therefore, as discussed in §2.6, the matrix is nonsingular and the solution algorithm of §2.6 can be applied to compute the coefficients.

The Complete Spline   Here we impose derivative data at the endpoints. Direct calculation shows that

$$
q_3'(x_0) = -3h^{-1}(c_{-1} - c_1), \quad q_3'(x_n) = -3h^{-1}(c_{n-1} - c_{n+1}),
$$

so that the conditions $q_3'(x_0) = f'(x_0)$ and $q_3'(x_n) = f'(x_n)$ imply that

$$
c_{-1} = c_1 - \frac{1}{3}hf'(x_0), \quad c_{n+1} = c_{n-1} + \frac{1}{3}hf'(x_n).
\tag{4.49}
$$

When substituted into the system (4.44) we again lose the dependence on c_{-1} and c_{n+1} and get the square system

$$
\begin{array}{cccccc}
4c_0 & + & 2c_1 & \cdots & & = & f(x_0) + \frac{1}{3}hf'(x_0), \\
& & & & & \vdots & \\
c_{k-1} & + & 4c_k & + & c_{k+1} & = & f(x_k), \\
& & & & & \vdots & \\
& & 2c_{n-1} & + & 4c_n & = & f(x_n) - \frac{1}{3}hf'(x_n),
\end{array}
\tag{4.50}
$$

which is (again) diagonally dominant, so the tridiagonal solution algorithm can be applied. Note that in the natural spline case, the system had $n - 1$ equations in the $n - 1$ unknowns $c_i, 1 \le i \le n-1$, whereas here we have a system of $n+1$ equations in the $n+1$ unknowns $c_i, 0 \le i \le n$.

Evaluation   The natural or complete spline can be easily constructed, in the sense that the coefficients c_k are defined, by solving (4.48) or (4.50). How do we use these values to evaluate the spline q_3, defined by (4.42)?

For any $x \in [x_{k-1}, x_k]$, the fact that each B_i is only locally non-zero means that we have

$$
q_3(x) = c_{k-2}B_{k-2}(x) + c_{k-1}B_{k-1}(x) + c_k B_k(x) + c_{k+1}B_{k+1}(x),
\tag{4.51}
$$

so evaluation of $q_3(x)$ requires only that we find the index k such that $x_{k-1} \le x \le x_k$. For a uniform grid this is easily accomplished:

$$
k = \left\lfloor \frac{x - x_0}{h} \right\rfloor + 1.
$$

For a nonuniform grid a more general search routine would have to be employed. Fig. 4.15 illustrates this by graphing B-splines centered at $x = 1, 2, 3, 4$ and marking the image of $x = 2.6$ on each of the four curves.

Programming Hint: Note that (4.51) implies that

$$
q_3(x_k) = f(x_k) = c_{k-1}B_{k-1}(x_k) + c_k B_k(x_k) + c_{k+1}B_{k+1}(x_k) = c_{k-1} + 4c_k + c_{k+1}.
$$

This can be a useful check on the correctness of the spline coefficients.

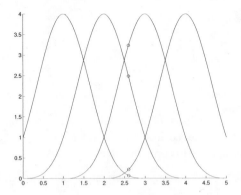

Figure 4.15 Demonstration of local definition of B-splines.

■ EXAMPLE 4.8

We start with a simple example that can be carried out by hand. Suppose that we want to compute a cubic spline approximation to $f(x) = \sin \pi x$ on the interval $[0, \frac{1}{2}]$, using the nodes $x_0 = 0$, $x_1 = \frac{1}{4}$, and $x_2 = \frac{1}{2}$. Thus, $n = 2$ and the spline will be defined by

$$q_3(x) = \sum_{k=-1}^{3} c_k B_k(x).$$

For the natural spline, (4.47) implies that

$$c_0 = \frac{1}{6} \sin 0 = 0$$

and

$$c_2 = \frac{1}{6} \sin \frac{\pi}{2} = \frac{1}{6},$$

and instead of a system of equations to solve, we have the single equation

$$c_0 + 4c_1 + c_2 = f(x_1) \Rightarrow c_1 = \frac{1}{4} \left(f(x_1) - \frac{1}{6} f(x_0) - \frac{1}{6} f(x_2) \right),$$

from which we get

$$c_1 = \frac{1}{4} \left(\frac{\sqrt{2}}{2} - \frac{1}{6} \right) = 0.1351100286.$$

Finally, then, (4.45) and (4.46) imply that

$$c_{-1} = \frac{1}{4} \left(\frac{1}{6} - \frac{\sqrt{2}}{2} \right) = -0.1351100286, \quad c_3 = \frac{1}{3} - \frac{\sqrt{2}}{8} + \frac{1}{24} = 0.1982233048,$$

so the spline approximation is now given by

$$q_3(x) = c_{-1} B_{-1}(x) + c_1 B_1(x) + c_2 B_2(x) + c_3 B_3(x).$$

To compute values of the spline is fairly simple. Suppose that we want to find $q_3(1/6) \approx \sin \pi/6 = 1/2$. We first determine that $x = 1/6$ is between $x_0 = 0$ and $x_1 = \frac{1}{4}$; this means that we want to compute

$$q_3(1/6) = c_{-1}B_{-1}(1/6) + c_1 B_1(1/6) + c_2 B_2(1/6);$$

the B_3 term vanishes because B_3 is identically zero on the subinterval $[x_0, x_1]$. We then have that

$$
\begin{aligned}
q_3(1/6) &= c_{-1}B_{-1}(1/6) + c_1 B_1(1/6) + c_2 B_2(1/6) \\
&= = c_{-1}B(5/3) + c_1 B(-1/3) + c_2 B(-4/3) \\
&= (-0.1351100286)(0.0370370370) + (0.1351100286)(3.4444444444) \\
&\quad + (0.1666666667)(0.2962962967) \\
&= 0.5097576284,
\end{aligned}
$$

which is a decent approximation to the exact value. The entire spline is graphed in Figure 4.16. The error is given in Figure 4.17.

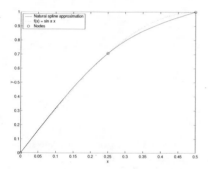

Figure 4.16 Natural spline approximation to part of the sine function.

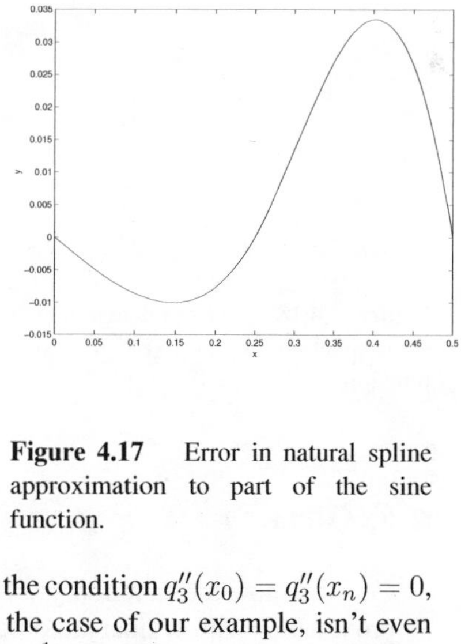

Figure 4.17 Error in natural spline approximation to part of the sine function.

Recall, however, that using a natural spline imposes the condition $q_3''(x_0) = q_3''(x_n) = 0$, which might not be the correct value to use (and, in the case of our example, isn't even a good approximation to the correct value at $x_2 = b = \frac{1}{2}$). If we construct the complete spline approximation, then we are led, via (4.49) and (4.50), to the 3×3 linear system

$$
\begin{bmatrix} 4 & 2 & 0 \\ 1 & 4 & 1 \\ 0 & 2 & 4 \end{bmatrix}
\begin{bmatrix} c_0 \\ c_1 \\ c_2 \end{bmatrix} =
\begin{bmatrix} \frac{1}{12}\pi \\ \frac{1}{2}\sqrt{2} \\ 1 \end{bmatrix},
$$

which we can easily solve to get

$$c_0 = 0.00017369124366, \quad c_1 = 0.13055231141225, \quad c_2 = 0.18472384429387.$$

Thus, the complete set of coefficients is

$$c_{-1} = -0.13124707638690, \quad c_0 = 0.00017369124366, \quad c_1 = 0.13055231141225,$$

$$c_2 = 0.18472384429387, \quad c_3 = 0.13055231141225,$$

and the spline is given by

$$q_3(x) = c_{-1}B_{-1}(x) + c_0 B_0(x) + c_1 B_1(x) + c_2 B_2(x) + c_3 B_3(x).$$

The spline itself is graphed in Figure 4.18 and the error is in Figure 4.19. Note that the approximation is much more accurate than was the case for the natural spline. This is because the natural spline imposed the inaccurate condition $q_3''(x) = 0$ at $x = \frac{1}{2}$, whereas the complete spline used the correct values of the first derivative at both endpoints. Just as a single point of comparison, this time we get

$$q_3(1/6) = 0.4999381524,$$

which is a much better approximation to the correct value of $\frac{1}{2}$ than we got for the natural spline.

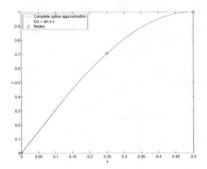

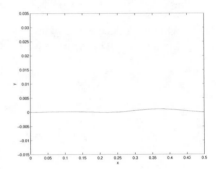

Figure 4.18 Complete spline approximation to part of the sine function.

Figure 4.19 Error in complete spline approximation to part of the sine function (same vertical scale as in Figure 4.17).

■ **EXAMPLE 4.9**

Consider now the function $f(x) = (1 + 25x^2)^{-1}$, which we have had trouble approximating with ordinary interpolation. We can construct both natural and complete spline approximations to this function, using equally spaced points on the interval $[-1, 1]$, with $h^{-1} = 2, 4, 8, 16$. Figures 4.20 and 4.21 show the plots of the natural spline and associated error, while Figures 4.22 and 4.23 do the same for the complete spline and its error. Compare these plots to the simple piecewise polynomial fits shown in Figures 4.8 to 4.10.

Theoretical Results   The error theory for spline approximations is somewhat more involved than that for ordinary piecewise polynomial interpolation, so we will here refer to the main result without proof and then comment on it. See Hall's article [10] for the details, or the article by de Boor [7].

Theorem 4.7 (Spline Approximation) *If $f \in C^4([a, b])$ and q is a cubic spline interpolant to f on a grid*

$$a = x_0 < x_1 < x_2 < \cdots < x_{n-1} < x_n = b$$

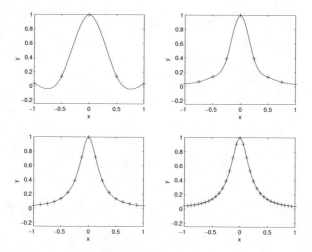

Figure 4.20   Natural Spline Interpolation to $f(x) = (1 + 25x^2)^{-1}$. The nodes are marked by plus signs.

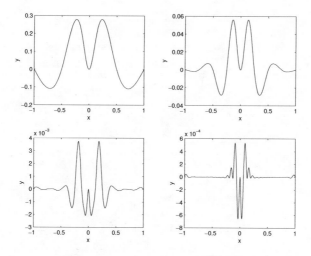

Figure 4.21   Error in Natural Spline Interpolation to $f(x) = (1 + 25x^2)^{-1}$.

with $\max(x_i - x_{i-1}) \le h$ *for all* i, *then*

$$\|f - q\|_\infty \le \frac{5}{384}h^4\|f^{(4)}\|_\infty.$$

Moreover, there exist constants C_k, $1 \le k \le 3$, *such that*

$$\|f^{(k)} - q^{(k)}\|_\infty \le C_k h^{4-k}\|f^{(4)}\|_\infty.$$

Note that the spline interpolant is no more accurate, in terms of the exponent on the mesh size, than ordinary piecewise polynomial approximation (note that the constant in the estimate is smaller). The advantage of spline interpolation lies in the smoothness of the approximation.

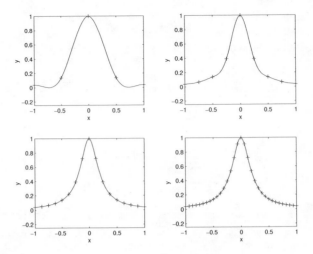

Figure 4.22   Complete Spline Interpolation to $f(x) = (1 + 25x^2)^{-1}$. The nodes are marked by plus signs.

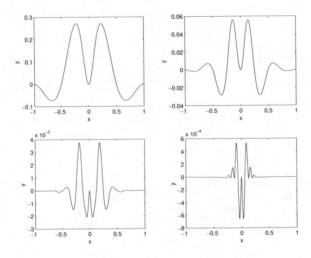

Figure 4.23   Error in Complete Spline Interpolation to $f(x) = (1 + 25x^2)^{-1}$.

Exercises:

1. Given the set of nodes $x_0 = 0$, $x_1 = 1/2$, $x_2 = 1$, $x_3 = 3/2$, and $x_4 = 2$, we construct the cubic spline function

$$q_3(x) = 0.15B_{-1}(x) + 0.17B_0(x) + 0.18B_1(x) + 0.22B_2(x) + 0.30B_3(x) \\ + 0.31B_4(x) + 0.32B_5(x),$$

where each B_k is computed from the exemplar B spline according to (4.38). Compute q_3 and its first derivative at each node.

2. Is the function

$$p(x) = \begin{cases} 0, & x \leq 0; \\ x^2, & 0 \leq x \leq 1; \\ -2x^2 + 6x + 3, & 1 \leq x \leq 2; \\ (x-3)^2, & 2 \leq x \leq 3; \\ 0, & x \geq 3 \end{cases}$$

a spline function? Why or why not?

3. For what value of k is the following a spline function?

$$q(x) = \begin{cases} kx^2 + (3/2), & 0 \leq x \leq 1; \\ x^2 + x + (1/2), & 1 \leq x \leq 2. \end{cases}$$

4. For what value of k is the following a spline function?

$$q(x) = \begin{cases} x^3 - x^2 + kx + 1, & 0 \leq x \leq 1; \\ -x^3 + (k+2)x^2 - kx + 3 & 1, \leq x \leq 2. \end{cases}$$

5. For what value of k is the following a spline function?

$$q(x) = \begin{cases} x^3 + 3x^2 + 1, & -1 \leq x \leq 0; \\ -x^3 + kx^2 + 1, & 0 \leq x \leq 1. \end{cases}$$

6. Construct the natural cubic spline that interpolates to $f(x) = 1/x$ at the nodes $1/2, 5/8, 3/4, 7/8, 1$. Do this as a hand calculation. Plot your spline function and f on the same set of axes, and also plot the error.

7. Repeat the above using the complete spline.

8. Construct a natural spline interpolate to the mercury thermal conductivity data (Table 4.7), using the 300 K, 500 K, and 700 K values. How well does this predict the values at 400 K and 600 K?

9. Confirm that the function $B(x)$, defined in (4.34), is a cubic spline.

10. Construct a natural cubic spline to the gamma function, using the data in Table 4.8, and the nodes $x = 1.0, 1.2, 1.4, 1.6, 1.8,$ and 2.0. Use this approximation to estimate $\Gamma(x)$ at $x = 1.1, 1.3, 1.5, 1.7,$ and 1.9.

11. Repeat the above using the complete spline approximation, and use the derivative approximations from §4.5 for the required derivative endpoint values.

12. Repeat Problem 6 of §4.7, but this time construct the complete cubic spline interpolate to $f(x) = \ln x$, using the same set of nodes. Plot the approximation and the error. What is the maximum sampled error in this case?

13. Recall Problem 7 from §4.2, in which we constructed polynomial interpolates to timing data from the 1973 Kentucky Derby, won by the horse Secretariat. For simplicity, we repeat the data in Table 4.19. Here t is the elapsed time (in seconds) since the race began and x is the distance (in miles) that Secretariat has traveled.

    (a) Construct a natural cubic spline that interpolates this data.

Table 4.19   Data for Problem 13.

x	0.0	0.25	0.50	0.75	1.00	1.25
t	0.0	25.0	49.4	73.0	96.4	119.4

(b) What is Secretariat's speed at each quarter-mile increment of the race? (Use miles per hour as your units.)

(c) What is Secretariat's "initial speed," according to this model? Does this make sense?

Note: It is possible to do this exercise using a uniform grid. Construct the spline that interpolates t as a function of x, then use your knowledge of calculus to find $x'(t)$ from $t'(x)$.

14. Show that the complete cubic spline problem can be solved (approximately) without having an explicit expression for f' at the endpoints. *Hint:* Consider the material from §4.5.

15. Construct the natural cubic spline that interpolates the data in Table 4.12 at the nodes defined by $\log_{10} \theta = -6, -4, \ldots, 4, 6$. Test the accuracy of the approximation by computing $q_3(x)$ for $x = \log_{10} \theta = -5, -3, \ldots, 3, 5$ and comparing the results to the actual data in the table.

16. Construct the exemplar quadratic B-spline; that is, construct a piecewise quadratic function that is C^1 over the nodes/knots $x = -1, 0, 1, 2$, and which vanishes for x outside the interval $[-1, 2]$.

17. Construct the exemplar *quintic* B-spline.

18. For a linear spline function we have $d = 1$, which forces $N = 0$. Thus, a linear spline has no derivative continuity, only function continuity, and no additional conditions are required at the endpoints. Show that the exemplar B-spline of first degree is given by

$$B(x) = \begin{cases} 0, & x \leq -1; \\ x + 1, & -1 \leq x \leq 0; \\ 1 - x, & 0 \leq x \leq 1; \\ 0, & 1 \leq x. \end{cases}$$

Describe, in your own words, how to construct and evaluate a linear spline interpolant using this function as the basic B-spline.

19. Discuss, in your own words, the advantages or disadvantages of spline approximation compared to ordinary piecewise polynomial approximation.

20. Write an essay that compares and contrasts piecewise cubic Hermite interpolation with cubic spline interpolation.

21. The data in Table 4.20 gives the actual thermal conductivity data for the element nickel. Construct a natural spline interpolate to this data, using only the data at 200 K, 400 K, ..., 1400 K. How well does this spline predict the values at 300 K, 500 K, and so on?

Table 4.20   Data for Problem 21.

Temperature (K), u	200	300	400	500	600
Conductivity (watts/cm K), k	1.06	0.94	0.905	0.801	0.721
Temperature (K), u		700	800	900	1000
Conductivity (watts/cm K), k		0.655	0.653	0.674	0.696

22. Construct a natural spline interpolate to the thermal conductivity data in Table 4.21, below. Plot the spline and the nodal data.

Table 4.21   Data for Problem 22

Temperature K), u	200	300	400	500	600	700
Conductivity (watts/cm K), k	0.94	0.803	0.694	0.613	0.547	0.487

◁ ● ● ▷

4.9   APPLICATION: SOLUTION OF BOUNDARY VALUE PROBLEMS

The range of application of spline approximations is incredibly wide, including such diverse things as scalable computer printer fonts, which are typically stored as spline approximations; and modern computer special effects in movies, which are often done by using spline approximations to complicated surfaces which can then be easily manipulated and moved about on the screen by the digital special effects artist. But some of the oldest and most important applications of spline functions are to the accurate solution of differential equations. In §2.7 we briefly touched on the approximate solution of boundary value problems using finite difference approximations. In this section, we will build on that experience by using splines to construct our approximation.

Consider the two-point boundary value problem

$$\begin{aligned} -u'' + a^2 u &= f(x), \quad 0 \le x \le 1, \\ u(0) &= g_0, \\ u(1) &= g_1, \end{aligned}$$

which is slightly more general than the one we looked at in §2.7. We construct the grid of points

$$0 = x_0 < x_1 < x_2 < \cdots x_{n-1} < x_n = 1,$$

where we assume (for simplicity, only) that the grid spacing is uniform; i.e., $x_k - x_{k-1} = h$ for all k, $1 \le k \le n$. We now look for our approximation in the form of a cubic spline defined on this grid. That is, we consider the function

$$u_h(x) = \sum_{k=-1}^{n+1} c_k B_k(x),$$

where the coefficients c_k are yet to be determined. The advantage of this approach over what we did in §2.7 is that here we get a continuous *smooth* function as our approximation, whereas in §2.7 all we got was a set of discrete points. We find the coefficients by requiring that u_h satisfy the differential equation at each of the nodes x_k, $0 \le k \le n$. (Note that the smoothness of the spline function is essential for this to even make sense. We have to have $u_h \in C^2$ to even talk about it satisfying the ODE.) Because we know the values of B_k and its derivatives at each of the nodes, we can easily reduce this to the system of equations

$$T'c' = b', \tag{4.52}$$

where

$$T' = \begin{bmatrix} \alpha & \beta & \alpha & 0 & \cdots & 0 \\ 0 & \alpha & \beta & \alpha & \vdots & 0 \\ \vdots & \ddots & \ddots & 0 & & \ddots \\ 0 & \cdots & 0 & \alpha & \beta & \alpha \end{bmatrix}$$

for

$$\alpha = -6h^{-2} + a^2, \quad \beta = 12h^{-2} + 4a^2,$$

and

$$c' = \begin{bmatrix} c_{-1} \\ c_0 \\ \vdots \\ c_{n+1} \end{bmatrix}, \quad b' = \begin{bmatrix} f(x_0) \\ f(x_1) \\ \vdots \\ f(x_n) \end{bmatrix}.$$

Since this is a system of only $n + 1$ equations in $n + 3$ unknowns, we can't construct a solution yet. We can eliminate the two extra unknowns by imposing the boundary conditions on the approximation:

$$u_h(0) = g_0 \Rightarrow c_{-1} + 4c_0 + c_1 = g_0 \Rightarrow c_{-1} = g_0 - 4c_0 - c_1 \tag{4.53}$$

and

$$u_h(1) = g_1 \Rightarrow c_{n-1} + 4c_n + c_{n+1} = g_1 \Rightarrow c_{n+1} = g_1 - 4c_n - c_{n-1}. \tag{4.54}$$

If we substitute these into the first and last equations of the rectangular system, then we get

$$(-6h^{-2} + a^2)(g_0 - 4c_0 - c_1) + (12h^{-2} + 4a^2)c_0 + (-6h^{-2} + a^2)c_1 = f(x_0)$$

and

$$(-6h^{-2} + a^2)c_{n-1} + (12h^{-2} + 4a^2)c_n + (-6h^{-2} + a^2)(g_1 - 4c_n - c_{n-1}) = f(x_n).$$

The c_1 and c_{n-1} terms drop out here, so we can conclude that

$$c_0 = \frac{h^2}{36}\left(f(x_0) + \left(\frac{6}{h^2} - a^2\right)g_0\right), \quad c_n = \frac{h^2}{36}\left(f(x_n) + \left(\frac{6}{h^2} - a^2\right)g_1\right). \tag{4.55}$$

With these two values known, we can look at the $(n - 1) \times (n - 1)$ system created by using (4.55) to define c_0 and c_n, and dropping the first and last equations in the rectangular system. We are then left with the square system

$$Tc = b$$

where

$$T = \text{tridiag}(-6h^{-2} + a^2, 12h^{-2} + 4a^2, -6h^{-2} + a^2),$$

(which is diagonally dominant),

$$c = (c_1, c_2, \ldots, c_{n-1})^T,$$

and

$$b = \begin{bmatrix} f(x_1) + \frac{1}{6}(1 - \frac{1}{6}a^2h^2)f(x_0) + h^{-2}(1 - \frac{1}{6}h^2a^2)^2 g_0 \\ f(x_2) \\ \vdots \\ f(x_{n-2}) \\ f(x_{n-1}) + \frac{1}{6}(1 - \frac{1}{6}a^2h^2)f(x_n) + h^{-2}(1 - \frac{1}{6}h^2a^2)^2 g_1 \end{bmatrix}.$$

Programming Note: We have used a grid of $n + 1$ points, with $h = 1/n$, but the system we solve is $n - 1 \times n - 1$.

The diagonal dominance of T means that we can use the solution algorithm from §2.6. Once the system is solved, we can get the rest of the coefficients from (4.53), (4.54), and (4.55).

■ **EXAMPLE 4.10**

Consider the example problem

$$\begin{aligned} -u'' + \pi^2 u &= 2\pi^2 \cos(\pi x), \quad 0 \le x \le 1, \\ u(0) &= 1, \\ u(1) &= -1, \end{aligned}$$

which has exact solution $u(x) = \cos \pi x$. The linear system for $h = \frac{1}{8}$ is

$$Tc = b$$

for

$$T = \text{tridiag}(-384 + \pi^2, 768 + 4\pi^2, -384 + \pi^2)$$

and

$$\begin{bmatrix} b_1 \\ b_2 \\ b_3 \\ b_4 \\ b_5 \\ b_6 \\ b_7 \end{bmatrix} = \begin{bmatrix} 2\pi^2 \cos\left(\frac{\pi}{8}\right) + 64\left(1 - \frac{\pi^2}{384}\right)\left(1 + \frac{\pi^2}{192}\right) \\ 2\pi^2 \cos\left(\frac{2\pi}{8}\right) \\ 2\pi^2 \cos\left(\frac{3\pi}{8}\right) \\ 2\pi^2 \cos\left(\frac{4\pi}{8}\right) \\ 2\pi^2 \cos\left(\frac{5\pi}{8}\right) \\ 2\pi^2 \cos\left(\frac{6\pi}{8}\right) \\ 2\pi^2 \cos\left(\frac{7\pi}{8}\right) + 64\left(1 - \frac{\pi^2}{384}\right)\left(1 + \frac{\pi^2}{192}\right) \end{bmatrix}.$$

This has the solution

$$c = \begin{bmatrix} 0.15763927907189 \\ 0.12053536253339 \\ 0.06520274709146 \\ 0.00000000000000 \\ -0.06520274709146 \\ -0.12053536253339 \\ -0.15763927907189 \end{bmatrix}.$$

From this we get that

$$c_0 = 0.17095034913242, \quad c_{-1} = 0.15855932439844,$$

$$c_8 = -0.17095034913242, \quad c_9 = -0.15855932439844,$$

and the spline approximation is therefore given by

$$
\begin{aligned}
u_h(x) \ = \ & 0.15855932439844 B_{-1}(x) + 0.17095034913242 B_0(x) + 0.15763927907189 B_1(x) \\
+ \ & 0.12053536253339 B_2(x) + 0.06520274709146 B_3(x) + 0 B_4(x) \\
- \ & 0.06520274709146 B_5(x) - 0.12053536253339 B_6(x) - 0.15763927907189 B_7(x) \\
- \ & 0.17095034913242 B_8(x - 0.15855932439844 B_9(x)
\end{aligned}
$$

and the error $\cos(\pi x) - u_h(x)$ is plotted in Figure 4.24. If we take a sequence of grids with $h^{-1} = 8, 16, 32, \ldots, 128$ and compute the approximate solutions, we find that the norm of the error, as estimated by sampling at 200 discrete points on the interval, is as indicated in Table 4.22; note that the error goes down like a factor of 4 as h is cut in half, indicating (but not proving) that this scheme is $\mathcal{O}(h^2)$ accurate. Note that this means that the spline solution of the boundary value problem is *less* accurate than a direct spline approximation of the exact solution (which, according to the estimate given in §4.8, would be $\mathcal{O}(h^4)$ accurate).

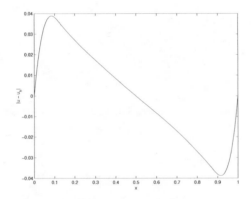

Figure 4.24   Error in spline approximation to boundary value problem for $h = 1/8$.

Table 4.22   Estimated error for spline approximation to BVP solution.

h^{-1}	Estimated error
8	0.03881321103858
16	0.01100251675792
32	0.00295824984969
64	0.00076798902568
128	0.00019487261990
256	0.00004945300189

It can be shown that this approximation is indeed $\mathcal{O}(h^2)$ accurate, but the analysis is beyond our intended scope.

What we have done in this section is an example of what is called *collocation*, a broadly used technique for solving differential equations (approximately) using expansions in terms of basis functions, and then solving for the coefficients in the expansion. Another method that uses a basis function expansion (and which uses a more complicated means of constructing the system for the expansion coefficients) is the finite element method. The book by Prenter [14] has some discussion of collocation by splines. We will briefly discuss spectral collocation in Chapter 10, and the finite element method in §6.10.3 and Chapter 9.

Exercises:

1. Set up the linear system for solving the boundary value problem

$$-u'' + u = 1, \quad u(0) = 1, \quad u(1) = 0,$$

using $h = \frac{1}{4}$. You should get

$$\begin{bmatrix} 196 & -95 & 0 \\ -95 & 196 & -95 \\ 0 & -95 & 196 \end{bmatrix} \begin{bmatrix} c_1 \\ c_2 \\ c_3 \end{bmatrix} = \begin{bmatrix} 101/6 \\ 1 \\ 671/576 \end{bmatrix}.$$

2. Solve the system in the previous problem and find the coefficients for the spline expansion of the approximate solution. The exact solution is

$$u(x) = 1 - \frac{e}{e^2 - 1} e^x + \frac{e}{e^2 - 1} e^{-x};$$

plot your approximation and the error.

3. Use the method from this section to approximate the solution to each of the following boundary value problems using $h^{-1} = 8, 16, 32$. Estimate the maximum error in each case by sampling the difference between the exact and approximate solutions at 200 equally spaced points on the interval.

    (a) $-u'' + u = (\pi^2 + 1) \sin \pi x, \quad u(0) = u(1) = 0; u(x) = \sin \pi x;$

    (b) $-u'' + u = \pi(\pi \sin \pi x + 2 \cos \pi x)e^{-x}, \quad u(0) = u(1) = 0; u(x) = e^{-x} \sin \pi x;$

    (c) $-u'' + u = 3 - \frac{1}{x} - (x^2 - x - 2) \log x, \quad u(0) = u(1) = 0; u(x) = x(1 - x) \log x.$

    (d) $-u'' + u = 4e^{-x} - 4xe^{-x}, \quad u(0) = u(1) = 0; u(x) = x(1 - x)e^{-x};$

    (e) $-u'' + \pi^2 u = 2\pi^2 \sin(\pi x), \quad u(0) = 1, \quad u(1) = 0, u(x) = \sin(\pi x)$

    (f) $-u'' + u = \frac{x^2 + 2x - 1}{(1+x)^3}, \quad u(0) = 1, \quad u(1) = 1/2, u(x) = (1 + x)^{-1}.$

4. Try to extend the method from this section to the more general two-point boundary value problem defined by

$$\begin{aligned} -u'' + bu' + u &= f(x), \\ u(0) = u(1) &= 0. \end{aligned}$$

Is the resulting linear system diagonally dominant for all values of b?

5. Try to extend the method from this section to the more general two-point boundary value problem defined by

$$
\begin{aligned}
-a(x)u'' + u &= f(x), \\
u(0) = u(1) &= 0.
\end{aligned}
$$

Is the resulting linear system diagonally dominant for all choices of $a(x)$?

◁ ● ● ● ▷

4.10 TENSION SPLINES

Splines are a wonderful tool for approximation, but they can still exhibit some poor behavior. Consider the data set plotted in Fig. 4.25. Obviously, this represents a function with a severe jump near $x = 0.5$, but there is no sign of oscillatory behavior. However, a B-spline representation of this data (Fig. 4.26) shows small wiggles on either side of a sharp front. This is fundamentally an artifact of the steep gradient in the data, but in other contexts a spline fit can display behavior that does not match the "sense" of the data. One way to avoid the problem is the notion of a *taut spline* or *tension spline*, an idea that appears to have been first published by Schweikert [17], but which also owes a lot to the work of A. K. Cline [4]; we relied heavily on a short paper of Marušić and Rogina [12] in our presentation here.

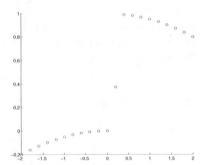

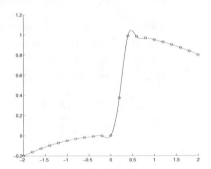

Figure 4.25   Data set for tensioned spline illustration.

Figure 4.26   B-spline fit to data in Fig. 4.25.

Imagine that the curve in Fig 4.26 is a piece of string that is constrained to pass through small loops at the data points. If we were to pull the string taut, we would smooth out the spurious oscillations in the curve. This amounts to studying the mechanical properties of a cable hanging between two supports. More prosaically, we construct our spline from the new basis set $\{1, x, \cosh px, \sinh px\}$, where $p > 0$ is the tension parameter: $p = 0$ means no tension, and it can be shown that this corresponds to the pure spline approximation; $p \to \infty$ gives us a piecewise linear approximation. (We will not attempt to justify either of these statements other than by examples and exercises.)

The reader may well be wondering how practical this scheme might be. After all, we have traded a set of polynomial basis functions for a set of transcendental basis functions. Not only is this going to make execution of any program more expensive, but it leaves open the entire question of how to construct the approximation. The basic idea is the same as in §4.8: First, we construct a primary basis function, as in (4.34):

$$\tau(x) = \frac{2}{\alpha} \begin{cases} \sinh px \cosh 2p + \cosh px \sinh 2p - px - 2p, & x \in [-2, -1] \\ -(\sinh px - px)\beta - 2\cosh px \sinh p + \gamma & x \in [-1, 0] \\ (\sinh px - px)\beta - 2\cosh px \sinh p + \gamma & x \in [0, 1] \\ -\sinh px \cosh 2p + \cosh px \sinh 2p + px - 2p, & x \in [1, 2] \end{cases} \qquad (4.56)$$

where $\alpha = p \cosh p - \sinh p$, $\beta = (1 + 2\cosh p)$, and $\gamma = 2p \cosh p$. Confirmation of this formula is deferred to the exercises. A plot of τ is given in Fig. 4.27, for $p = 4$; note that it does not look very different from Fig. 4.13.

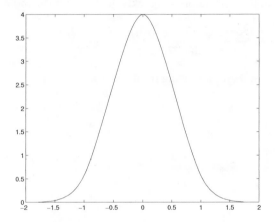

Figure 4.27   Original taut B-spline, $p = 4$.

Construction of a tension spline follows precisely the same recipe as in §4.8: Given data $(x_i, y_i)_{i=1}^{i=n}$, we look for an approximation in the form

$$s(x) = \sum_{i=-1}^{n+1} c_i \tau_i(x),$$

where the c_i values are coefficients to be determined, and

$$\tau_i(x) = \tau\left(\frac{x - x_i}{h}\right).$$

Note that, as in §4.8, we have added additional grid points, which will again require the imposition of additional conditions. We will explicitly cover the *natural* spline case, leaving the complete spline case to the exercises.

For each original grid point we have the equation

$$\sum_{i=-1}^{n+1} c_i \tau_i(x) = y_j.$$

Because of the local nature of the τ_i functions, this becomes

$$c_{j-1}\tau_{j-1}(x_j) + c_j\tau_j(x_j) + c_{j+1}\tau_{j+1}(x_j) = y_j.$$

We quickly have that

$$\tau_j(x_j) = \tau\left(\frac{x_j - x_j}{h}\right) = \tau(0) = 4,$$

and

$$\tau_{j-1}(x_j) = \tau_{j+1}(x_j) = \tau(\pm 1) = \theta(p),$$

for

$$\theta(p) = \frac{2[(\sinh p - p)(1 + 2\cosh p) - 2\cosh p \sinh p + 2p\cosh p]}{p\cosh p - \sinh p}.$$

This certainly looks imposing, but it also looks like some simplification ought to be possible, and if we multiply out the numerator, we quickly get

$$\theta(p) = \frac{2[(\sinh p - p)(1 + 2\cosh p) - 2\cosh p \sinh p + 2p\cosh p]}{p\cosh p - \sinh p} = 2\left(\frac{\sinh p - p}{p\cosh p - \sinh p}\right).$$

This is plotted in Fig. 4.28, for p running from 0 to 10.

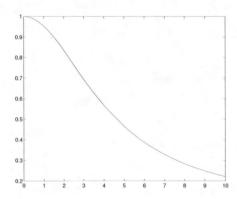

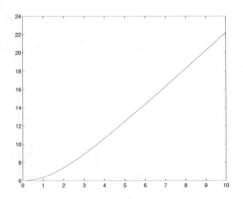

Figure 4.28 $\theta(p) = \tau(\pm 1)$ as a function of p.

Figure 4.29 $\delta_{2,1}(p) = \tau''(\pm 1)$ as a function of p.

The important thing for our purposes is that $0 < \theta(p) \le 1$. The system of linear equations for the taut spline is, initially, as follows:

$$\begin{bmatrix} \theta(p) & 4 & \theta(p) & & & \\ & \theta(p) & 4 & \theta(p) & & \\ & & \ddots & \ddots & \ddots & \\ & & & \theta(p) & 4 & \theta(p) \end{bmatrix} \begin{bmatrix} c_{-1} \\ c_0 \\ \vdots \\ c_{n+1} \end{bmatrix} = \begin{bmatrix} f(x_0) \\ f(x_1) \\ \vdots \\ f(x_n) \end{bmatrix}. \qquad (4.57)$$

This, of course, is a system of $n+1$ equations in $n+3$ unknowns. Following the derivation for the natural spline in §4.8, we have the boundary conditions

$$\tau''(x_0) = 0 \Rightarrow c_{-1}\tau''_{-1}(x_{-1}) + c_0\tau''_0(x_0) + c_1\tau''_1(x_1) = 0,$$

and

$$\tau''(x_n) = 0 \Rightarrow c_{n-1}\tau''_{n-1}(x_n) + c_n\tau''_n(x_n) + c_{n+1}\tau''_{n+1}(x_n) = 0.$$

Continuing with $x = x_0$, we get that

$$c_{-1}\tau''(1) + c_0\tau''(0) + c_1\tau''(-1) = 0. \tag{4.58}$$

We still need to compute τ'' at various points. It is an involved, but not arduous computation, to show that

$$\tau''(0) = \delta_{2,0}(p) = \frac{-4p^2\tanh p}{p - \tanh p},$$

and

$$\tau''(\pm 1) = \delta_{2,1}(p) = \frac{2p^2\tanh p}{p - \tanh p} = -\frac{1}{2}\delta_{2,0}(p)$$

We have plotted $\delta_{2,1}(p)$ in Fig. 4.29. We therefore have, from (4.58),

$$c_{-1} = 2c_0 - c_1.$$

Similarly,

$$c_{n+1} = 2c_n - c_{n-1}.$$

Note that this looks a lot like what we got in §4.8 for the ordinary natural B-spline; we can eliminate c_{-1}, c_0, c_n, and c_{n+1}, so the system (4.57) becomes (after a bit of work)

$$Kc = F,$$

where

$$K = \text{tridiag}(\theta(p), 4, \theta(p)), \tag{4.59}$$

$$c = (c_1, c_2, \ldots, c_{n-1})^T,$$

and

$$F = (f(x_1) - \theta(p)c_0, f(x_2), \ldots, f(x_{n-2}), f(x_{n-1}) - \theta(p)c_n)^T,$$

with

$$c_{-1} = 2c_0 - c_1,$$

$$c_0 = \frac{f(x_0)}{4 - \frac{\theta(p)\delta_{2,0}(p)}{\delta_{2,1}(p)}} = \frac{f(x_0)}{4 + 2\theta(p)},$$

$$c_n = \frac{f(x_n)}{4 - \frac{\theta(p)\delta_{2,0}(p)}{\delta_{2,1}(p)}} = \frac{f(x_n)}{4 + 2\theta(p)},$$

and

$$c_{n+1} = 2c_n - c_{n-1}.$$

Note that the matrix is tridiagonal and diagonally dominant, so we know how to solve it.

■ **EXAMPLE 4.11**

As an obvious illustration of this, let's fit a series of tension splines to the data set in Fig. 4.25. Once the coefficients $c_{-1}, c_0, \ldots, c_n, c_{n+1}$ are computed, the spline is evaluated in the same way as for an ordinary B-spline. Fig. 4.30 shows the data and spline curve for $p = 0.01$; obviously, with such a small tension value we do not expect much difference, and Fig. 4.31, which plots the difference between the pure polynomial spline and the tension spline, confirms this (although the lower "overshoot" does appear to be significantly affected). If we take $p = 1$, we get Figs. 4.32-4.33 which shows that the lower "overshoot" is indeed beginning to damp out, but the upper one is much the same. Finally, for $p = 6$, we get Figs. 4.34-4.35; this is perhaps the smallest value of p for which both overshoots are gone.

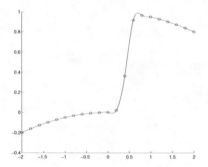

Figure 4.30 Tension spline fit, $p = 0.01$.

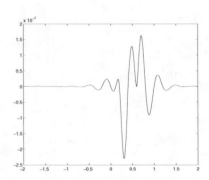

Figure 4.31 Difference between pure polynomial spline and tension spline, $p = 0.01$

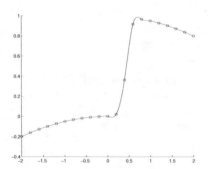

Figure 4.32 Tension spline fit, $p = 1$.

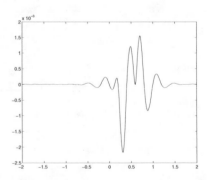

Figure 4.33 Difference between pure polynomial spline and tension spline, $p = 1$

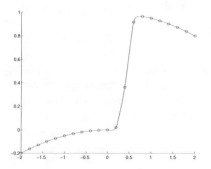

Figure 4.34   Tension spline fit, $p = 6$.

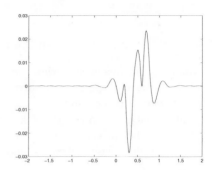

Figure 4.35   Difference between pure polynomial spline and tension spline, $p = 6$.

Exercises:

1. Show that the piecewise function defined in (4.56) is, indeed, continuous and has continuous first and second derivatives.

2. Fill in the details of the natural spline construction. In particular, confirm the expressions for $\delta_{2,0}(p)$ and $\delta_{2,1}(p)$ (and that $\tau''(1) = \tau''(-1)$) as well as the form of the final linear system (4.59).

3. Derive the linear system for the construction of a complete taut spline, by following what was done in §4.8.

4. Consider the dataset in Table 4.23:

Table 4.23   Data for Problem 4

x	600	650	700	750	800	850	900	950	1000	1050	1100
y	0.64	0.65	0.66	0.69	0.91	2.2	1.2	0.62	0.6	0.61	0.61

Plot the data, and construct a (natural) polynomial spline fit to it. Note the "wiggles" to the left of the peak, which appear to be contrary to the sense of the data, which is increasing monotonically towards the peak near $x = 850$. Find the smallest value of p in a taut natural spline fit to this data which yields a monotone curve to the left of the peak.

5. Repeat the previous problem using complete splines. Use a simple finite difference approximation based on the data to get the necessary derivative values.

◁ ● ● ▷

4.11   LEAST SQUARES CONCEPTS IN APPROXIMATION

4.11.1   An Introduction to Data Fitting

An important area in approximation is the problem of fitting a curve to experimental data. Since the data is experimental, we must assume that it is polluted with some degree of error, most commonly measurement error ("noise"), so we do not necessarily want to construct a curve that goes through every data point. (In fact, the material in §4.12.1 suggests that this would be a disastrous way to proceed.) Rather, we want to construct a function that represents the "sense of the data" and which is, in some sense, a close approximation to the data.

The most common approach is known as *least squares* data fitting. Consider Figure 4.36; this shows an example set of data for which the general trend is clearly a straight line. But *which* straight line do we use to represent the data?

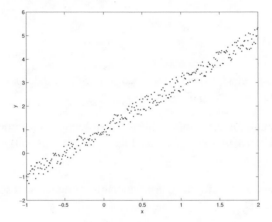

Figure 4.36   Example plot of data.

The least squares approach defines the "correct" straight line as the line that minimizes the sum of the squares of the distances between the data points and the line. Let the experimental data be defined as pairs (x_k, y_k), $1 \leq k \leq n$ for some n. Thus, we want to find the coefficients m and b in the equation $y = mx + b$ such that

$$F(m, b) = \sum_{k=1}^{n} \left(y_k - (mx_k + b) \right)^2$$

is minimized. This is a straight forward problem from multivariable calculus: We compute the partial derivatives F_m and F_b and find where they both vanish, and this will define a critical point. It can be shown that this critical point defines a global minimum for F. Thus, m and b are defined by the two equations

$$\frac{\partial F}{\partial m} = -2 \sum_{k=1}^{n} \left(y_k - (mx_k + b) \right) x_k = 0,$$

$$\frac{\partial F}{\partial b} = -2 \sum_{k=1}^{n} \left(y_k - (mx_k + b) \right) = 0,$$

which can be simplified to a system of two equations in two unknowns

$$\sum_{k=1}^{n} \left(y_k - (mx_k + b) \right) x_k = 0,$$

$$\sum_{k=1}^{n} \left(y_k - (mx_k + b) \right) = 0,$$

or,

$$m \left(\sum_{k=1}^{n} x_k^2 \right) + b \left(\sum_{k=1}^{n} x_k \right) = \sum_{k=1}^{n} x_k y_k,$$

$$m \left(\sum_{k=1}^{n} x_k \right) + b \left(\sum_{k=1}^{n} 1 \right) = \sum_{k=1}^{n} y_k.$$

The solution here is then

$$m = \frac{n \sum_{i=1}^{n} x_k y_k - \left(\sum_{i=1}^{n} x_k \right) \left(\sum_{i=1}^{n} y_k \right)}{n \sum_{k=1}^{n} x_k^2 - \left(\sum_{i=1}^{n} x_k \right)^2},$$

$$b = \frac{\left(\sum_{i=1}^{n} x_k^2 \right) \left(\sum_{i=1}^{n} y_k \right) - \left(\sum_{i=1}^{n} x_k \right) \left(\sum_{i=1}^{n} x_k y_k \right)}{n \left(\sum_{k=1}^{n} x_k^2 \right) - \left(\sum_{i=1}^{n} x_k \right)^2},$$

which, for the data in Figure 4.36, produces the straight line graph shown in Figure 4.37.

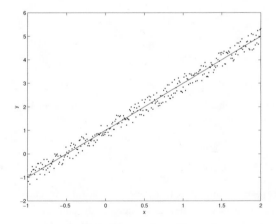

Figure 4.37   Straight line fit to data.

The notion of a least squares data fit can be generalized beyond simply fitting a straight line to data. We can look at higher degree polynomials and we can also look at higher dimensional data sets. The exercises include some examples of more involved least squares data fit problems.

■ **EXAMPLE 4.12**

Consider the data in Table 4.24. If we plot this, we get what appears to be a straight line as the general trend of the data, so we look for the equation of the line $y = mx + b$

which best fits this data in the least squares sense. Forming the separate sums gives us

$$\sum_{i=1}^{6} x_i = 15, \quad \sum_{i=1}^{6} y_i = 367, \quad \sum_{i=1}^{6} x_i^2 = 55, \quad \sum_{i=1}^{6} x_i y_i = 1303,$$

from which it follows that

$$m = 22.0286, \quad b = 6.0952$$

and the line is plotted, along with the data, in Figure 4.38.

Table 4.24   Data for Example 4.12.

x	0.0	1.0	2.00	3.00	4.00	5.00
y	10.0	25.0	51.0	66.0	97.0	118

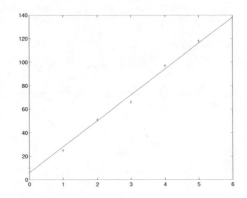

Figure 4.38   Least squares fit to the data in Table 4.24.

■ **EXAMPLE 4.13**

We don't have to restrict ourselves to linear or quadratic models to make good use of the idea of least squares data fits. Consider the y_k data in Table 4.25. When we plot this data, we get the curve shown in Figure 4.39. Generally, this looks like an exponential growth curve. Ordinarily this would require us to do a fit to a curve of the form $y_k = Ae^{rx_k}$, which will lead to a nonlinear system for the parameters A and r. However, if the raw data is exponential, then the logarithm of the data is linear, since we have

$$z_k = \ln y_k = rx_k + (\ln A),$$

and we can fall back on our existing algorithm to do a fit to the log data.

   To verify this, we look at the logarithm of our example data; this data is plotted in Figure 4.40, and the general trend is indeed linear. So we do a least squares fit to the log data, getting the straight line

$$z = 0.0209x - 0.4842.$$

This is plotted, along with the raw (logarithm) data, in Figure 4.41. It follows, then, that our curve fit to the original data is

$$y = e^{(0.0209x - 0.4842)},$$

which is plotted in Figure 4.42, along with the original data. Note that, except for the last two points, this is a pretty good fit to the data set.

Table 4.25 Data for Example 4.13.

k	x_k	y_k	$\log y_k$	k	x_k	y_k	$\log y_k$
1	0	0.716	−0.3341	8	70	2.500	0.9163
2	10	0.893	−0.1132	9	80	3.151	1.1477
3	20	1.055	0.0535	10	90	4.300	1.4586
4	30	1.134	0.1258	11	100	5.308	1.6692
5	40	1.167	0.1544	12	110	4.966	1.6026
6	50	1.281	0.2476	13	120	10.919	2.3905
7	60	1.994	0.6901				

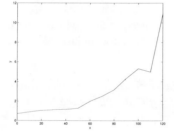

Figure 4.39 Plot of raw data for Example 4.13.

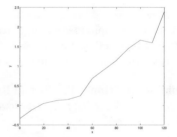

Figure 4.40 Plot of log data for Example 4.13.

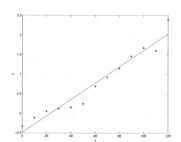

Figure 4.41 Log data plus fitted curve for Example 4.13.

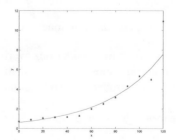

Figure 4.42 Raw data plus fitted curve for Example 4.13.

4.11.2 Least Squares Approximation and Orthogonal Polynomials

The notion of least squares approximation can be extended beyond the data-fitting problem. Consider the problem of finding an approximation to a given function f in terms of a set of basis functions $\{\phi_k, 1 \le k \le n\}$. How do we find the coefficients in the expansion

$$f(x) \approx q_n(x) = \sum_{k=1}^{n} c_k \phi_k(x)?$$

One way to do this is to require that the coefficients c_k produce an approximation that minimizes the error

$$r_n = f - q_n$$

in the least squares sense, i.e., in the sense of the integral 2-norm. Thus we seek c_k such that

$$R_n = \|f - q_n\|_2^2$$

is minimized. Before doing this it will be convenient to introduce the notion of an *inner product*. It is best to do this thoroughly, so we pause in our development of approximations, but only briefly.

Inner Products of Functions   The reader should be familiar with the notion of the dot product of two vectors in $\mathbb{R}^n$:

$$x \cdot y = \sum_{i=1}^{n} x_i y_i.$$

This is an example of a more general operation called an *inner product*, which can be defined on general vector spaces, including spaces of functions, rather than just Euclidean n-space. The formal definition is as follows.

Definition 4.2 (Inner Product on Real Vector Spaces) *Let f and g be elements of a real vector space V. Let (f, g) denote any operation on f and g that satisfies the following three properties:*

1. *$(f, f) > 0$ for all nonzero $f \in V$;*

2. *$(f, \alpha g_1 + \beta g_2) = \alpha(f, g_1) + \beta(f, g_2)$ for all $f, g \in V$ and all scalars α and β;*

3. *$(f, g) = (g, f)$ for all $f, g \in V$.*

Then, (f, g) is called an inner product.

If we want to consider our overlying vector space to be $C([a, b])$, that is, continuous functions on a closed interval, then we can easily establish that the positively weighted integral of a product of two functions will be an inner product.

Theorem 4.8 *Let w be integrable on $[a, b]$ and non-negative, i.e., $\int_a^b w(x)dx$ is defined and $w(x) \ge 0$ for all $x \in [a, b]$. For given f and g in $C([a, b])$, define $(f, g)_w$ as*

$$(f, g)_w = \int_a^b w(x)f(x)g(x)dx. \tag{4.60}$$

Then, $(\cdot, \cdot)_w$ defines an inner product on $C([a, b])$.

Proof: See Problem 6. •

Note that it therefore follows that if $(f, g)_w$ is an inner product, then $\|f\|_w = (f, f)_w^{1/2}$ defines a norm. In the common case when $w(x) \equiv 1$, we will simply write $(f, g)_w = (f, g)$;

i.e., we will drop the subscript, and the norm is the ordinary 2-norm for functions that we defined earlier in this chapter. Problem 7 offers some practice with a norm defined by a weighted inner product.

This definition of inner product will allow us to apply a number of ideas from linear algebra to the construction of approximations, as we will soon see. Of more immediate interest is the fact that we can use the inner product notation to write the residual, R_n, in a very convenient form:

$$R_n = \|f\|_w^2 - 2\sum_{k=1}^n c_k(f, \phi_k) + \sum_{i=1}^n \sum_{j=1}^n c_i c_j(\phi_i, \phi_j).$$

Note that we can regard R_n as a function of the n variables c_k, and thus apply ordinary calculus to the problem of minimizing R_n. After some manipulations (see Problem 8) we find that the c_k are defined by simultaneously solving the set of equations

$$
\begin{array}{ccccccc}
(\phi_1, \phi_1)c_1 & + & (\phi_1, \phi_2)c_2 & + & \cdots & (\phi_1, \phi_n)c_n & = & (f, \phi_1), \\
(\phi_2, \phi_1)c_1 & + & (\phi_2, \phi_2)c_2 & + & \cdots & (\phi_2, \phi_n)c_n & = & (f, \phi_2), \\
\vdots & & \vdots & & \cdots & \vdots & & \vdots \\
(\phi_n, \phi_1)c_1 & + & (\phi_n, \phi_2)c_2 & + & \cdots & (\phi_n, \phi_n)c_n & = & (f, \phi_n).
\end{array}
$$

This system can be organized along matrix–vector lines as

$$
\begin{bmatrix}
(\phi_1, \phi_1) & (\phi_1, \phi_2) & \cdots & (\phi_1, \phi_n) \\
(\phi_2, \phi_1) & (\phi_2, \phi_2) & \cdots & (\phi_2, \phi_n) \\
\vdots & \vdots & \cdots & \vdots \\
(\phi_n, \phi_1) & (\phi_n, \phi_2) & \cdots & (\phi_n, \phi_n)
\end{bmatrix}
\begin{bmatrix} c_1 \\ c_2 \\ \vdots \\ c_n \end{bmatrix}
=
\begin{bmatrix} (f, \phi_1) \\ (f, \phi_2) \\ \vdots \\ (f, \phi_n) \end{bmatrix}.
\tag{4.61}
$$

Solving a system of linear equations is a problem that we do not encounter in the general case until Chapter 7. We can avoid it altogether at this point *if* our basis functions satisfy the *orthogonality* condition

$$(\phi_i, \phi_j) = 0, \quad \text{for all } i \neq j. \tag{4.62}$$

In this case, the matrix in (4.61) is a diagonal matrix and we very easily have

$$c_k = \frac{(f, \phi_k)}{(\phi_k, \phi_k)}.$$

So, to summarize what we have done so far, we can construct an approximation to a given function f from a given basis set $\{\phi_1, \phi_2, \ldots, \phi_n\}$, and the construction is very easy, *if* the basis satisfies the condition (4.62). So, the question becomes: When can we find a basis that satisfies (4.62), and how good is the resulting approximation?

The answer is that we can always find such a basis if we consider polynomial functions for our basis elements, and the resulting approximations are usually quite good. The special basis functions that satisfy (4.62) are called *orthogonal polynomials*. To be more specific with this, we have to introduce some new concepts and notation, and recall a major theorem from linear algebra. But first, one more definition.

Definition 4.3 (Vector Space of Polynomials of Degree $\leq N$) *For any $N \geq 0$ define $\mathcal{P}_N$ as the* vector space *of polynomials of degree $\leq N$. Note that this space has a standard basis consisting of $\{1, x, x^2, x^3, \ldots, x^N\}$, and thus is an $(N+1)$-dimensional space.*

And, now, the theorem:

Theorem 4.9 *Let w be a given non-negative weight function on an interval $[a, b]$, and $(\cdot, \cdot)_w$ the associated inner product, defined as in (4.60). Then there exists a family of orthogonal polynomials $\{\phi_k\}, \phi_k \in \mathcal{P}_k, 0 \le k \le N$, such that*

$$(\phi_i, \phi_j)_w = 0, \quad i \ne j,$$

and

$$(\phi_i, \phi_i)_w > 0, \quad i \ge 0.$$

In addition, the ϕ_k satisfy the following:

1. *The set $\{\phi_0, \phi_1, \ldots, \phi_N\}$ is a basis for $\mathcal{P}_N$;*

2. *If q_k is an arbitrary element of $\mathcal{P}_k$, for $k < N$, then $(q_k, \phi_N)_w = 0$ for all $N > k$ (thus, orthogonal polynomials are orthogonal to all polynomials of strictly lower degree);*

3. *For $j \ge 1$, the roots of each ϕ_j are all in $[a, b]$ and are all distinct.*

Proof: The proof is somewhat lengthy, in part because of the length of the theorem, but it is not difficult.

To establish that the family $\{\phi_k\}$ exists, we will construct it directly, using the Gram–Schmidt process from linear algebra. Take $\phi_0(x) = 1$, and define $\xi_k(x) = x^k$, for $0 \le k \le N$. Then the subsequent ϕ_k can be found according to

$$\phi_k(x) = \xi_k(x) - \sum_{j=0}^{k-1} \frac{(\phi_j, \xi_k)_w}{(\phi_j, \phi_j)_w} \phi_j(x) \tag{4.63}$$

and an inductive argument shows very quickly that the orthogonality holds. In Problem 9 we ask the student to fill in the details of this part of the proof.

Having now proved that the family of orthogonal polynomials exists, we turn our attention to proving each of (1)–(3).

(1) The space $\mathcal{P}_N$ is finite-dimensional with dimension $N + 1$. It therefore follows that any set of $N + 1$ independent elements of $\mathcal{P}_N$ will be a basis. The orthogonality condition (4.62) forces the members of the family $\{\phi_k\}$ to be independent (see Problem 10); therefore, the set $\{\phi_0, \phi_1, \ldots, \phi_N\}$ is a basis for $\mathcal{P}_N$.

(2) Let q_k be an arbitrary polynomial of degree $k < N$. Then we can write

$$q_k = \sum_{i=0}^{k} a_i \phi_i$$

because $\{\phi_0, \phi_1, \ldots, \phi_k\}$ is a basis for $\mathcal{P}_k$. Therefore,

$$(q_k, \phi_N) = \sum_{i=0}^{k} a_i (\phi_i, \phi_N) = 0$$

since ϕ_N is orthogonal to each element of $\{\phi_0, \phi_1, \ldots, \phi_k\}$. In fact, we can write (Problem 11)

$$q_k = \sum_{i=0}^{k} \frac{(q_k, \phi_i)}{(\phi_i, \phi_i)} \phi_i. \tag{4.64}$$

(3) First, suppose that ϕ_j, $j \geq 1$, has no roots in $[a, b]$. This means that ϕ_j does not change sign on the interval, thus

$$(\phi_0, \phi_j) = \int_a^b w(x)\phi_0(x)\phi_j(x)dx \neq 0,$$

since the integrand does not change sign on $[a, b]$. But the orthogonality requires that $(\phi_0, \phi_j) = 0$; hence, we have a contradiction, so there must be at least one root in $[a, b]$.

Now, let x_1 be any root of ϕ_j that lies in $[a, b]$, and suppose that it is a multiple root. Then it follows that

$$\phi_j(x) = (x - x_1)^2 q(x)$$

for some polynomial q; thus,

$$q(x) = (x - x_1)^{-2}\phi_j(x)$$

is a polynomial of degree $j - 2$, therefore, $(q, \phi_j) = 0$, by another part of this theorem. But

$$(q, \phi_j) = \int_a^b w(x)(x - x_1)^{-2}(\phi_j(x))^2 dx > 0,$$

so we have another contradiction. Thus, any roots that lie in $[a, b]$ must be simple roots.

Suppose now that only some of the roots lie in $[a, b]$. Call these roots x_i, $1 \leq i < j$, and note that we can write ϕ_j as

$$\phi_j(x) = \Psi_j(x)(x - x_1) \cdots (x - x_i),$$

where $\Psi_j(x)$ does not change sign in $[a, b]$ and is a polynomial of degree $j - i$. Therefore,

$$\phi_j(x)(x - x_1) \cdots (x - x_i) = \Psi_j(x)(x - x_1)^2 \cdots (x - x_i)^2$$

is also a polynomial that does not change sign in $[a, b]$. Hence, the integral

$$I = \int_a^b \Psi_j(x)(x - x_1)^2 \cdots (x - x_i)^2 dx$$

cannot be zero. However,

$$I = \int_a^b \Psi_j(x)(x - x_1)^2 \cdots (x - x_i)^2 dx = (\phi_j, q)$$

where $q(x) = (x - x_1) \cdots (x - x_i)$, and q is a polynomial of degree $i < j$. Therefore, $(\phi_j, q) = 0$, and we have a contradiction. Thus, $i \geq j$, and since a polynomial of degree j cannot have more than j roots, we must have $i = j$. ●

Families of Orthogonal Polynomials

At this point it might be useful to look at some examples of orthogonal polynomial families. Four of the most common ones are discussed below.

Note that the orthogonality condition (4.62) means that an orthogonal polynomial can be multiplied by an arbitrary nonzero constant and still satisfy (4.62). To avoid the problems of nonuniqueness that this can lead to, it is common to impose a specific scaling on the elements of each family.

1. *Legendre polynomials:* The Legendre[13] polynomials are the orthogonal polynomials on $[-1, 1]$ with no weight function (more correctly, the unit weight function); thus, we have

$$\int_{-1}^{1} P_i(x)P_j(x)dx = 0, \quad i \neq j.$$

The usual scaling is to take $P_n(1) = 1$. The first five Legendre polynomials are:

$$
\begin{aligned}
P_0(x) &= 1, \\
P_1(x) &= x, \\
P_2(x) &= \frac{1}{2}(3x^2 - 1), \\
P_3(x) &= \frac{1}{2}(5x^3 - 3x), \\
P_4(x) &= \frac{1}{8}(35x^4 - 30x^2 + 3).
\end{aligned}
$$

2. *Chebyshev polynomials:* The common notation for the Chebyshev[14] polynomials is $T_n(x)$, and the interval and weight function are defined in the orthogonality relation

$$\int_{-1}^{1} \frac{T_i(x)T_j(x)}{\sqrt{1-x^2}}dx = 0, \quad i \neq j.$$

The common scaling is to set the leading coefficient equal to 2^{n-1}. The first five Chebyshev polynomials are:

$$
\begin{aligned}
T_0(x) &= 1, \\
T_1(x) &= x, \\
T_2(x) &= 2x^2 - 1, \\
T_3(x) &= 4x^3 - 3x, \\
T_4(x) &= 8x^4 - 8x^2 + 1.
\end{aligned}
$$

It can be shown that the Chebyshev polynomials are related in a very simple way to cosines; see Theorem 4.10.

---

[13] Adrien-Marie Legendre (1752–1833) was born and educated in Paris. He contributed greatly to what we now call number theory as well as elliptic function theory. He was the first to publish, in 1805, a description of the method of least squares, although it appears that Gauss had previously worked out much of the same material.

Legendre introduced the polynomials that bear his name in a 1785 paper on the gravitational attraction of spherical bodies. They arise in this context because they are the solutions to an ordinary differential equation that occurs as part of the solution process for the equations of motion in a spherical coordinate system.

[14] Pafnuty Lvovich Chebyshev (1821–1894) was born near Borovsk, Russia, southwest of Moscow, and educated at the University of Moscow, from which he was graduated in 1841. From 1847 until his death he lived in St. Petersburg. Many areas of mathematics, from number theory to the theory of equations, were touched by Chebyshev, but he is perhaps best known, at least in applied mathematics, for his work in approximation theory.

Because of the many different ways that the Russian alphabet can be transliterated into the Roman alphabet, there are several different ways to spell Chebyshev's name. The most common alternative is "Tschebyscheff." A marvelously engaging discussion of the perils of transliterating between the Roman and the Russian (Cyrillic) alphabets, as well as a substantial treatment of Chebyshev's life, is contained in *The Thread*, by Philip J. Davis, a leading mathematician in the area of interpolation and approximation.

3. *Hermite polynomials:* The Hermite polynomials are orthogonal on the entire real line,[15] using the weight function $w(x) = e^{-x^2}$; that is,

$$\int_{-\infty}^{\infty} e^{-x^2} H_i(x) H_j(x) dx = 0, \quad i \neq j.$$

The common scaling is to set the leading coefficient equal to 2^n. The first five Hermite polynomials are:

$$
\begin{aligned}
H_0(x) &= 1, \\
H_1(x) &= 2x, \\
H_2(x) &= 4x^2 - 2, \\
H_3(x) &= 8x^3 - 12x, \\
H_4(x) &= 16x^4 - 48x^2 + 12.
\end{aligned}
$$

4. *Laguerre Polynomials*: The Laguerre[16] polynomials are orthogonal on the positive real line, using the weight function $w(x) = e^{-x}$, that is,

$$\int_{0}^{\infty} e^{-x} L_i(x) L_j(x) dx = 0, \quad i \neq j.$$

The common scaling is to set the leading coefficient equal to $\frac{(-1)^n}{n!}$. The first five Laguerre polynomials are then:

$$
\begin{aligned}
L_0(x) &= 1, \\
L_1(x) &= -x + 1, \\
L_2(x) &= \frac{1}{2}(x^2 - 4x + 2), \\
L_3(x) &= \frac{1}{6}(-x^3 + 9x^2 - 18x + 6), \\
L_4(x) &= \frac{1}{24}(x^4 - 16x^3 + 72x^2 - 96x + 24).
\end{aligned}
$$

We can use any of these orthogonal polynomial families to construct approximations to functions defined on the appropriate interval. These approximations are "best possible" in the sense that they minimize the error in the appropriate weighted 2-norm; i.e.,

$$\| f - q_n \|_w < \| f - p_n \|_w$$

for all $p_n \in \mathcal{P}_n, p_n \neq q_n$.

Consider, as illustrations, the following set of examples.

---

[15]Charles Hermite (1822–1901) was born in Dieuze, France, on Christmas Eve. A poor test-taker, Hermite was not able to pass the exams for his bachelor's degree until the age of nearly 26. He held professional positions at a number of French schools, most notably the École Polytechnique, the École Normale, and the Sorbonne.

Perhaps his best-known mathematical result is the first proof that e is a transcendental number (published in 1873). His name is attached to a number of mathematical ideas and concepts, including the Hermite differential equation, Hermite polynomials, and Hermitian matrices. The idea of interpolating to the derivative as well as to the function values was part of a paper he published in 1878, *Sur la formule d'interpolation de Lagrange*, which actually considers interpolation not only of the first derivative, but of higher derivatives as well. Hermite interpolation is sometimes called *osculatory interpolation*.

[16]Edmond Nicolas Laguerre (1834 – 1886) was born and died in Bar-le-Duc, France. He was educated at the École Polytechnique but did not graduate with a high ranking. He served in the military as an artillery officer for 10 years before returning to the École as a faculty member, where he remained for the rest of his life. Although best known for the orthogonal polynomials that bear his name, and their associated differential equation, Laguerre also published work in analysis, geometry, and abstract linear spaces.

■ **EXAMPLE 4.14**

Let's construct the fourth-degree least squares approximation to the exponential function, $f(x) = e^x$, over the interval $[-1, 1]$, using Legendre polynomials. The approximation is defined by

$$p_4(x) = \sum_{k=0}^{4} \frac{(f, P_k)}{(P_k, P_k)} P_k,$$

where the P_k are the Legendre polynomials. We thus need to compute the integrals

$$I_0 = \int_{-1}^{1} e^x dx, \quad I_1 = \int_{-1}^{1} x e^x dx, \quad I_2 = \int_{-1}^{1} \frac{1}{2}(3x^2 - 1) e^x dx,$$

$$I_3 = \int_{-1}^{1} \frac{1}{2}(5x^3 - 3x) e^x dx, \quad I_4 = \int_{-1}^{1} \frac{1}{8}(35x^4 - 30x^2 + 3) e^x dx,$$

and

$$J_0 = \int_{-1}^{1} dx, \quad J_1 = \int_{-1}^{1} x^2 dx, \quad J_2 = \int_{-1}^{1} \frac{1}{4}(3x^2 - 1)^2 dx,$$

$$J_3 = \int_{-1}^{1} \frac{1}{4}(5x^3 - 3x)^2 dx, \quad J_4 = \int_{-1}^{1} \frac{1}{64}(35x^4 - 30x^2 + 3)^2 dx.$$

In practice, these integrals would be computed using some type of numerical integration routine, such as the trapezoid rule or the more accurate methods we discuss in Chapter 5. For this simple example, though, it is possible to use direct calculus methods or (much more attractive!) a computer algebra package such as Maple or Mathematica. However it is done, to eight digits the integrals are

$$I_0 = 2.3504024, \quad I_1 = 0.73575888, \quad I_2 = 0.14312574,$$

$$I_3 = 0.020130181, \quad I_4 = 0.0022144731,$$

and

$$J_0 = 2.00000000, \quad J_1 = 0.66666667, \quad J_2 = 0.40000000,$$

$$J_3 = 0.28571429, \quad J_4 = 0.22222222,$$

so the polynomial approximation is

$$p_4(x) = \frac{2.3504024}{2.00000000} P_0(x) + \frac{0.73575888}{0.66666667} P_1(x) + \frac{0.14312574}{0.40000000} P_2(x)$$
$$+ \frac{0.020130181}{0.28571429} P_3(x) + \frac{0.0022144731}{0.22222222} P_4(x),$$

which simplifies to

$$p_4(x) = 1.0000309 + 0.99795487x + 0.49935229x^2 + 0.17613908x^3 + 0.043597439x^4.$$

Figure 4.43 shows a plot of the error $e^x - p_4(x)$. Compare this to the error plots for fourth-degree Taylor approximation and fourth degree Lagrange or Newton interpolation from earlier in this chapter. Note, in particular, that the least squares error oscillates back and forth between its maximum and minimum values (or nearly so), several times. It can be shown that this is a necessary and sufficient condition for the approximating polynomial to be the "best" approximation to the function, and is one reason why least squares approximations are considered valuable: they are close to being the best possible approximations.

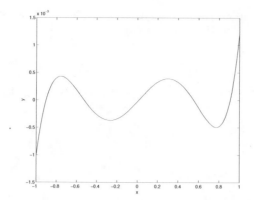

Figure 4.43   Error in fourth-degree least squares approximation to the exponential function.

■ **EXAMPLE 4.15**

Here we construct a Legendre polynomial approximation to the function $f(x) = (1 + 25x^2)^{-1}$ using $n = 8, 16$, and 32 degree polynomials. Figures 4.44A–C show plots of both f and the least squares approximation; Figure 4.44D shows the error in the 32-degree approximation.

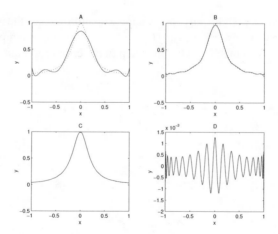

Figure 4.44   Legendre least squares approximation to $f(x) = 1/(1 + 25x^2)^{-1}$. A: $n = 8$; B: $n = 16$; C: $n = 32$; D: error in the $n = 32$ case. In A and B, $f(x)$ is denoted by the dotted curve.

■ **EXAMPLE 4.16**

This time, we use a Chebyshev polynomial approximation to the same f as in the previous example, again using $n = 8, 16$, and 32 degree polynomials. Figures 4.45A– C show plots of both f and the least squares approximation; Figure 4.45D shows the error in the 32-degree approximation. The performance of the Chebyshev and

Legendre approximations are very similar, with the Chebyshev being very slightly better.

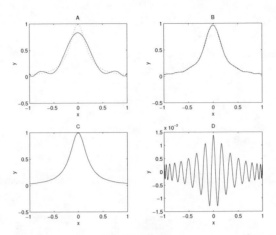

Figure 4.45   Chebyshev least squares approximation to $f(x) = 1/(1 + 25x^2)^{-1}$. A: $n = 8$; B: $n = 16$; C: $n = 32$; D: error in the $n = 32$ case. In A and B, $f(x)$ is denoted by the dotted curve.

■ **EXAMPLE 4.17**

Here we construct a Legendre polynomial approximation to e^x on the interval $[-1, 1]$, in much the same way as was done for $f(x) = (1 + 25x^2)^{-1}$ in Example 2. However, the accuracy here is so much greater that we plot the errors for the $n = 2, 4, 8$ cases in Figure 4.46.

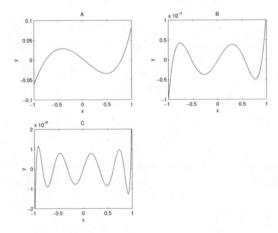

Figure 4.46   Error in Legendre least squares approximation to $f(x) = e^x$. A: error for $n = 2$; B: error for $n = 4$; C: error for $n = 8$.

■ **EXAMPLE 4.18**

Chebyshev approximation to e^x; This is the same as Example 4.17, except we use a Chebyshev expansion instead of a Legendre expansion.

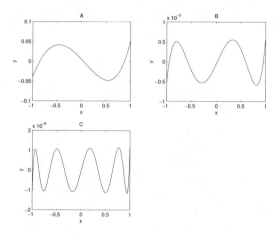

Figure 4.47   Error in Chebyshev least squares approximation to $f(x) = e^x$. A: error for $n = 2$; B: error for $n = 4$; C: error for $n = 8$.

Several comments might be in order here. We first note how much easier it was to obtain a high degree of accuracy for the exponential than it was for $f(x) = (1 + 25x^2)^{-1}$. Second, although it might be difficult to discern in the plots, the results for the Chebyshev approximations were, in each case, very slightly better than for the Legendre approximations. A full discussion of this requires more mathematical machinery than we want to deal with right now, but it is generally true that Chebyshev least squares approximations are superior to those done with any other choice of basis.

Finally, note that to do a least squares approximation, we have to be able to compute the inner products, which are integrals. Thus, we need a tool like the trapezoid rule or, perhaps better, some of the more sophisticated methods to be developed in Chapter 5.

Exercises:

1. Modify the methods of §4.11.1 to compute the linear function of two variables that gives the best least squares fit to the data for this exercise in Table 4.27.

2. The data in Table 4.26 gives the actual thermal conductivity data for the element iron. Construct a quadratic least squares fit to this data and plot both the curve and the raw data. How well does your curve represent the data? Is the fit improved any by using a cubic polynomial?

3. Repeat Problem 2, this time using the data for nickel from Problem 21 in §4.8.

4. Modify the methods of §4.11.1 to compute the quadratic polynomial that gives the best least squares fit to the data in Table 4.27.

5. An astronomical tracking station records data on the position of a newly discovered asteroid orbiting the Sun. The data is reduced to measurements of the radial distance

Table 4.26 Data for Problem 2.

Temperature (K), u	100	200	300	400	500
Conductivity (W/cm K), k	1.32	0.94	0.835	0.803	0.694
Temperature (K), u	600	700	800	900	1000
Conductivity (W/cm K), k	0.613	0.547	0.487	0.433	0.38

Table 4.27 Data for Problems 1 and 4.

Problem 4		Problem 1		
x_n	y_n	x_n	y_n	z_n
-1	0.9747	0	0	0.9573
0	0.0483	0	1	2.0132
1	1.0223	1	0	2.0385
2	4.0253	1	1	1.9773
3	9.0152	0.5	0.5	1.9936

from the Sun (measured in millions of kilometers) and angular position around the orbit (measured in radians), based on knowledge of Earth's position relative to the Sun. In theory, these values should fit into the polar coordinate equation of an ellipse, given by

$$r = \frac{L}{2(1 + \epsilon \cos \theta)},$$

where ϵ is the eccentricity of the elliptical orbit and L is the width of the ellipse (sometimes known as the *latus rectum* of the ellipse) at the focus. (See Figure 4.48.) However, errors in the tracking process and approximations in the transformation to (r, θ) values perturb the data. For the data in Table 4.28, find the eccentricity of the orbit by doing a least squares fit to the data. *Hint:* Write the polar equation of the ellipse as

$$2r(1 + \epsilon \cos \theta) = L,$$

which can then be written as

$$2r = \epsilon(-2r \cos \theta) + L,$$

so $y_k = 2r_k$ and $x_k = -2r_k \cos \theta_k$.

6. Prove Theorem 4.8.

7. Let $w(x) = x$ on the interval $[0, 1]$; compute $\|f\|_w$ for each of the following functions:

   (a) e^{-x};

   (b) $1/\sqrt{x^2 + 1}$;

   (c) $1/\sqrt{x}$.

Compare to the values obtained using the unweighted 2-norm.

8. Derive the linear system (4.61) as the solution to the least squares approximation problem.

Table 4.28   Data for Problem 5.

θ_n	r_n
−0.1289	42895
−0.1352	42911
−0.1088	42851
−0.0632	42779
−0.0587	42774
−0.0484	42764
−0.0280	42750
−0.0085	42744
0.0259	42749
0.0264	42749
0.1282	42894

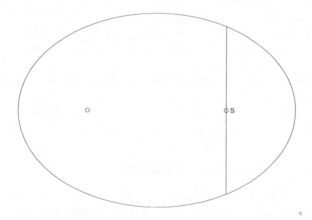

Figure 4.48   Figure for Problem 5. The closed curve is the elliptical orbit, and the vertical line has length L.

9. Provide the missing details to show that the family of polynomials defined in (4.63) is, indeed, orthogonal.

10. Let $\{\phi_k\}$ be a family of orthogonal polynomials associated with a general weight function w and an interval $[a, b]$. Show that the $\{\phi_k\}$ are independent in the sense that
$$0 = c_1\phi_1(x) + c_2\phi_2(x) + \cdots + c_n\phi_n(x)$$
holds for all $x \in [a, b]$ if and only if $c_k = 0$ for all k.

11. Prove the expansion formula (4.64) for a polynomial.

12. Construct the second-degree Legendre least squares approximation to $f(x) = \cos \pi x$ over the interval $[-1, 1]$.

13. Construct the second-degree Laguerre least squares approximation to $f(x) = e^x$ on the interval $[0, \infty)$.

14. Construct the third-degree Legendre least squares approximation to $f(x) = \sin \frac{1}{2}\pi x$ over the interval $[-1, 1]$.

◁ ● ● ● ▷

4.12   ADVANCED TOPICS IN INTERPOLATION ERROR

Based on the examples we have seen so far, the value of interpolation as an approximation tool seems unclear. When we applied it to the exponential function, we got excellent results, but when we applied it to the example $f(x) = (1 + 25x^2)^{-1}$ (known as the *Runge*[17] *example*), we got substantially less accuracy for the same number of nodes. This is not simply a case of using insufficiently many points in a particular example. It can be shown that if we took more and more nodes and used higher-degree interpolating polynomials, the error in interpolating to the Runge example would continue to get larger and larger, when measured in the norm $\|f - p_n\|_\infty$. In fact, the trend that is suggested in Figures 4.3B and C continues: For values of x in the middle of the interval $[-1, 1]$, $|f(x) - p_n(x)|$ actually goes to zero, but near the ends of the interval the polynomial interpolates oscillate wildly and do not converge. What is going on here?

A complete answer, although mathematically very interesting, is beyond our scope here, and so we will skip the details and present only the broader ideas. We also look at a several ways in which the potential problems with interpolation at equally spaced nodes can affect a calculation, and derive a better set of interpolating nodes.

4.12.1   Stability of Polynomial Interpolation

One problem with polynomial interpolation using high-degree polynomials is that it really is a potentially unstable process, in the sense that small changes to the data can lead to large changes in the interpolating polynomial. To see this, consider the effects of rounding error on the interpolation problem. Let $f(x)$ be the exact function that we wish to interpolate, and let $\hat{f}(x)$ be the function polluted by rounding error, and assume that $f(x) - \hat{f}(x) = \epsilon(x)$, where we assume that

$$\|f - \hat{f}\|_\infty = \|\epsilon\|_\infty = \epsilon_0$$

for some small ϵ_0.

The polynomial interpolate that we compute is based on the function that is polluted by rounding error, so we have the Lagrange form

$$\hat{p}_n(x) = \sum_{i=0}^{n} L_i^{(n)}(x)\hat{f}(x_i),$$

---

[17]Carl Runge (1856–1927) was born in Bremen and educated at the University of Munich. He originally intended to study literature but after less than two months switched to mathematics and physics. In 1877 he began attending the University of Berlin, where he received a doctorate in 1880, on differential geometry. He held academic positions at Hanover and Göttingen and retired in 1925. Much of his professional work was more in physics than in mathematics, but he did make contributions in the numerical solution of differential equations (Runge–Kutta methods) and polynomial interpolation theory. The paper in which he outlined the theory behind the so-called "Runge example" appeared in 1901, under the title *Über empirische Funktionen und die Interpolation zwischen äquidistanten Ordinaten*.

and we want to estimate the error $f(x) - \hat{p}(x)$. For convenience we define the "ideal" interpolate, based on the exact function, f, as

$$p_n(x) = \sum_{i=0}^{n} L_i^{(n)}(x) f(x_i).$$

Theorem 4.3 gives us the error between f and p_n:

$$f(x) - p_n(x) = \frac{1}{(n+1)!} \left(\prod_{i=0}^{n} (x - x_i) \right) f^{(n+1)}(\xi_x).$$

We can relate this to the error we want to bound as follows:

$$f(x) - \hat{p}_n(x) = \underbrace{(f(x) - p_n(x))}_{\text{Error due to interpolation}} + \underbrace{(p_n(x) - \hat{p}_n(x))}_{\text{Error due to rounding}}.$$

We now turn our attention to analyzing the error due to rounding. We have

$$|p_n(x) - \hat{p}_n(x)| = \left| \sum_{i=0}^{n} L_i^{(n)}(x)(f(x_i) - \hat{f}(x_i)) \right|,$$

$$\leq \|\epsilon\|_\infty \sum_{i=0}^{n} \|L_i^{(n)}\|_\infty,$$

$$= \epsilon_0 \sum_{i=0}^{n} \Lambda_i,$$

where the norms are both taken over the interval defined by the nodes, and $\Lambda_i = \|L_i^{(n)}\|_\infty$. Let's assume that the rounding error is bounded and small: a very reasonable assumption based on our work in Chapter 1. Thus, we take $\|\epsilon\|_\infty = \epsilon_0 = \mathcal{O}(\mathbf{u})$.

We appear to be in good shape, having bounded the error due to rounding by something small times a sum depending on the norm of the Lagrange functions. The problem is that this sum can grow quite large as n increases. A general theorem (see Theorem 4.6 of [15]) is not necessary, since we can get the essentials by being experimental. Assume that the nodes are equidistant on the interval $[a, b]$, with $x_0 = a$ and $x_n = b$, and with $x_{j+1} - x_j = h$ being the uniform mesh spacing. Then we can write

$$x_j = a + jh, \quad 0 \leq j \leq n,$$

for each node, and

$$x = a + \eta_x h, \quad 0 \leq \eta_x \leq n,$$

where $\eta_x = (x - a)/h$ takes on *real* values between 0 and n. Therefore,

$$L_i^{(n)}(x) = \prod_{\substack{k=0 \\ k \neq i}}^{n} \frac{x - x_k}{x_i - x_k},$$

$$= \prod_{\substack{k=0 \\ k \neq i}}^{n} \frac{\eta_x - k}{i - k}.$$

Hence, the Lagrange functions (and, therefore, the Λ_i) are not dependent on the choice of a, b, or h. They depend entirely on n, η_x (which depends on x), and the distribution of the nodes. This means that we ought to be able to plot them rather easily and get some idea of how large that sum term can be. Figure 4.49 shows plots of

$$L(x) = \sum_{i=0}^{n} |L_i^{(n)}(x)|$$

for various values of n (assuming equally spaced points on the interval $[0, 1]$), and Figure 4.50 shows a plot of

$$M_n = \sum_{i=0}^{n} \Lambda_i$$

as a function of n. Note the *rapid* growth of M_n. If we take larger values of n, the trend continues and the growth becomes clearly exponential, as Figure 4.51 shows.

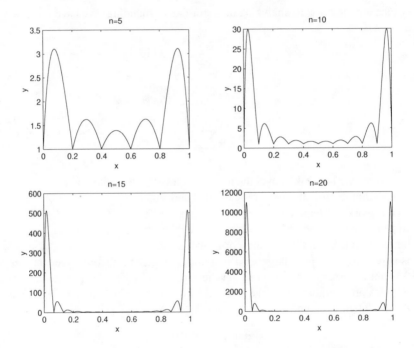

Figure 4.49   Plots of $L(x)$ versus x for $n = 5, 10, 15, 20$.

What does this mean for our original interpolation problem? Recall how the overall error was related to the error in interpolation and the error in rounding:

$$f(x) - \hat{p}_n(x) \quad = \quad (f(x) - p_n(x)) + (p_n(x) - \hat{p}_n(x)) ;$$

from which it follows that

$$\|f - \hat{p}_n\|_\infty \le \|f - p_n\|_\infty + \|p_n - \hat{p}_n\|_\infty \le \|f - p_n\|_\infty + \epsilon_0 M_n. \qquad (4.65)$$

Since we now know that M_n can become quite large, we know that the presence of the small ϵ_0 multiplying the rounding-error term is *not* enough to guarantee that the overall

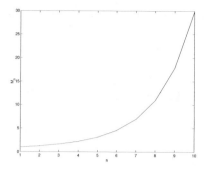

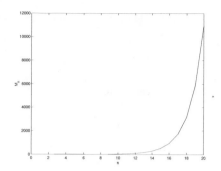

Figure 4.50   Plot of M_n versus n, for $n \leq 10$.

Figure 4.51   Plot of M_n versus n, for $n \leq 20$.

computation will not be badly corrupted by rounding error. On the other hand, if we keep the degree of the polynomial interpolation less than, say, seven, then we have that $M_n \leq 10$ (see Figure 4.50) and we know the overall error is not seriously corrupted by rounding error.

Note that we are not claiming that polynomial interpolation of high degree *has* to be affected by large amounts of rounding error; the inequality in (4.65) goes the wrong way to support that conclusion. However, the estimate (4.65) does *allow* for large amounts of rounding error whenever n is large enough that M_n is very large, and we can avoid this potential problem by taking n so that M_n is small. Thus, we avoid using polynomial interpolates of degree much higher than 7 or so.

4.12.2   The Runge Example

(*Note:* This section of the text requires a bit more background than the others in this chapter, especially in complex arithmetic.)

One is tempted to look at the function $f(x) = (1 + 25x^2)^{-1}$ and say that it is a smooth, well-behaved functon for all x, and this is indeed the case, so long as x is real. But if we allow x to be imaginary, then problems can occur; consider the value of $f(i/5)$, where $i = \sqrt{-1}$. Why does this matter, if we are considering the problem of interpolating to f as a function of the real variable x?

It turns out that it matters a great deal. It can be shown (see §3.4 of [11]) that for functions of the form $f(x) = (1 + a^2x^2)^{-1}$, the error in interpolation can be expressed as

$$f(x) - \overset{\circ}{p}_n(x) = \left(\frac{r_n}{1 + a^2x^2} \right) \left(\frac{w_n(x)}{w_n(ia^{-1})} \right),$$

where $r_n = x$ if n is even, $r_n = ia^{-1}$ if n is odd, and

$$w_n(x) = \prod_{k=0}^{n} (x - x_k).$$

Since

$$\left| \frac{r_n}{1 + a^2x^2} \right| \leq C$$

for all x and n, the question of convergence then comes down to the size of the ratio

$$R_n(x) = \left| \frac{w_n(x)}{w_n(ia^{-1})} \right|.$$

Now we can write

$$R_n(x) = \left| \frac{\sigma_n(x)}{\sigma_n(ia^{-1})} \right|^{n+1}$$

for

$$\sigma_n(z) = \left(\prod_{k=0}^{n} |z - x_k| \right)^{\frac{1}{n+1}}.$$

But we can take logarithms of both sides to get that

$$\log \sigma_n(z) = \frac{1}{n+1} \sum_{k=0}^{n} \log |z - x_i|.$$

Now, if the nodes are equally spaced on the interval $[-1, 1]$, the sum can be interpreted as a Riemann sum for the integral

$$I(f) = \frac{1}{2} \int_{-1}^{1} \log |z - t| dt;$$

therefore, we can write

$$\log \sigma_n(z) \approx \frac{1}{2} \int_{-1}^{1} \log |z - t| dt,$$

and we define the right side of this as $\log \sigma(z)$, thus implicitly defining $\sigma(z)$:

$$\frac{1}{2} \int_{-1}^{1} \log |z - t| dt = \log \sigma(z).$$

The integral can be evaluated, but we have to be careful about doing it, since $z = x + iy$ is complex. Thus, we have

$$\frac{1}{2} \int_{-1}^{1} \log |z - t| dt = \frac{1}{2} \int_{-1}^{1} \log \sqrt{(x-t)^2 + y^2} dt = \frac{1}{4} \int_{-1}^{1} \log[(x-t)^2 + y^2] dt.$$

Careful application of integration by parts yields

$$\begin{aligned} \frac{1}{2} \int_{-1}^{1} \log |z - t| dt &= \frac{1}{4}(1-x) \log((x-1)^2 + y^2) + \frac{1}{4}(1+x) \log((x+1)^2 + y^2) \\ &\quad - \frac{1}{2} y \arctan \frac{x-1}{y} + \frac{1}{2} y \arctan \frac{x+1}{y}. \end{aligned}$$

Thus,

$$\begin{aligned} \sigma_n(z) \approx \sigma(z) &= e^{-1} \left((x-1)^2 + y^2\right)^{(1-x)/4} \left((x+1)^2 + y^2\right)^{(1+x)/4} \\ &\quad \times \exp \left\{ \frac{1}{2} y \left(\arctan \frac{x+1}{y} - \arctan \frac{x-1}{y} \right) \right\}. \end{aligned}$$

Now the error satisfies

$$|f(x) - p_n(x)| = |R_n(x)| \approx C_n \left| \frac{\sigma(x)}{\sigma(ia^{-1})} \right|^{n+1},$$

so convergence depends, in the limit as $n \to \infty$, on the ratio $|\sigma(x)/\sigma(ia^{-1})|$. For real x we get that

$$\sigma(x) = e^{-1}(x-1)^{(1-x)/2}(x+1)^{(1+x)/2},$$

whereas for imaginary arguments we get

$$\sigma(iy) = e^{-1}\sqrt{1+y^2}\exp\left\{y\arctan\frac{1}{y}\right\}.$$

To get convergence on the entire interval $[-1,1]$, we need to have

$$\max_{x\in[-1,1]} |\sigma(x)| = 2e^{-1} = 0.7357588824 < |\sigma(ia^{-1})|.$$

An elementary computation shows that $|\sigma(i/5)| = 0.4937581336...$; therefore, we will not get convergence on the entire interval $[-1,1]$ with $a = 5$, because

$$\frac{\max_{x\in[-1,1]}|\sigma(x)|}{\sigma(i/5)} = \frac{0.7357588824}{0.4937581336} > 1.$$

By decreasing a, we can finally get $0.7357588824 < |\sigma(ia^{-1})|$. For example, for $a = 3/4$ we have $\sigma(i/a) = 1.446...$.

What this shows is that the problem really is that the *singularities* of the function (i.e., the points $\pm i/5$) are "too close" to the interval of interpolation. If we looked at $g(x) = (1 + a^2x^2)^{-1}$ for a smaller value of a, then we would get convergence on the entire interval from $[-1,1]$. An example of this is given in Figure 4.52, using g with $a = 3/4$, and $n = 4, 8, 16$ nodes; the last plot shows the error for the $n = 16$ case. See also Problem 3. Note that the error still shows the development of "spikes" near the endpoints; the difference is that in this case, as $n \to \infty$, the amplitude of the spikes will eventually decay to zero.

It is also possible to work on a more arbitrary interval $[-b, b]$, in which case we learn that the length of this interval also plays a role in the convergence of the interpolation. (See [9] for a discussion of this.) Essentially, convergence will occur if $1 > Cab$, where $C = 0.5255...$.

4.12.3 The Chebyshev Nodes

In §4.12.1 and §4.12.2 we saw that high-degree polynomial interpolation *at equidistant points* can be a bad idea. Is there an alternate choice for the distribution of nodes that will produce better results? The answer is "yes," and the nodes in question are called the *Chebyshev nodes*.

Define the family of functions $T_n(x)$, $n \geq 0$, by the formula

$$T_n(x) = \cos(n\arccos x), \quad x \in [-1,1]. \tag{4.66}$$

Remarkably, these functions are polynomials (in fact, they are the Chebyshev polynomials introduced back in §4.11.2).

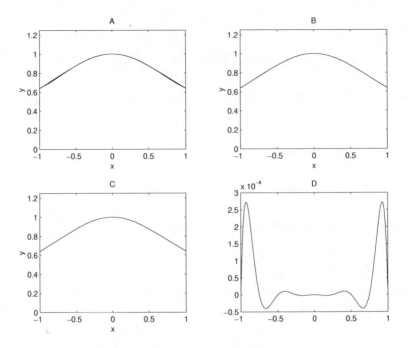

Figure 4.52   Successful interpolation to the Runge-like function, $g(x) = 1/(1 + (9/16)x^2)^{-1}$. A: 4 nodes; B: 8 nodes; C: Error in 16-node interpolation.

Theorem 4.10 *The functions $T_n(x)$ satisfy the following:*

  1. Each T_n is a polynomial of degree n;

  2. For $n \geq 1$, $T_{n+1}(x) = 2xT_n(x) - T_{n-1}(x)$;

  3. $T_n(x) = 2^{n-1}x^n +$ lower-order terms.

Proof: We observe first that both (1) and (3) will follow quickly from (2), so we start by proving (2). Write $x = \cos\theta$ so that (4.66) becomes

$$T_n(x) = \cos n\theta.$$

Elementary trigonometry tells us that

$$\cos n\theta \cos\theta = \frac{1}{2}\left(\cos(n+1)\theta + \cos(n-1)\theta\right),$$

which can be solved to get

$$\cos(n+1)\theta = 2\cos\theta\cos n\theta - \cos(n-1)\theta,$$

from which (2) follows, using $T_n(x) = \cos n\theta$, and so on.

Now, to prove (1), we note that we trivially have

$$\cos(n\arccos x) = 1$$

for $n = 0$, and

$$\cos(n \arccos x) = x$$

for $n = 1$. Now assume that $\cos(n \arccos x)$ is a polynomial for all $n \le k$, and an inductive argument based on the recursion in (2) quickly finishes the proof.

The final part follows more or less directly from (2). We have that $T_0(x) = 1$, and $T_1(x) = x$, so that the recursion yields

$$T_2(x) = 2xT_1(x) - T_0(x) = 2x(x) - 1 = 2x^2 - 1,$$

and we can use induction to formally prove it for all n (see Problem 5). •

It should be noted that all orthogonal polynomial families satisfy a three-term recurrance relation similar to the one in Theorem 4.10, part 2.

Assume now (for convenience) that we are only interested in constructing approxima-tions on the interval $[-1, 1]$. As a polynomial of degree n, each T_n will have exactly n roots; we can show (it actually follows from Theorem 4.9) that these roots will be distinct and all lie in the interval $[-1, 1]$. Denote, then, the roots of the n-degree polynomial T_n as $z_k^{(n)}$, and note that (4.66) implies that

$$z_k^{(n)} = \cos\left(\frac{(2k - 1)\pi}{2n}\right), \quad 1 \le k \le n. \tag{4.67}$$

The Chebyshev nodes are simply the roots of the polynomials T_n. To construct an n-degree polynomial interpolate, use the $n + 1$ roots of T_{n+1}; i.e., take the interpolation nodes to be

$$x_k = z_{k+1}^{(n+1)}, \quad 0 \le k \le n,$$

so that we have the following.

Definition 4.4 (Chebyshev Nodes)

$$x_k = \cos\left(\frac{(2k + 1)\pi}{2(n + 1)}\right), \quad 0 \le k \le n. \tag{4.68}$$

Below, we will show why this works, but first let's look at some examples.

■ **EXAMPLE 4.19**

Let $f(x) = e^x$ on the interval $[-1, 1]$. The Chebyshev nodes for the case $n = 1$ (linear interpolation) are given by

$$x_0 = \cos\left(\frac{(2 \times 0 + 1)\pi}{2(1 + 1)}\right) = \cos\frac{\pi}{4} = \frac{1}{2}\sqrt{2}$$

and

$$x_1 = \cos\left(\frac{(2 \times 1 + 1)\pi}{2(1 + 1)}\right) = \cos\frac{3\pi}{4} = -\frac{1}{2}\sqrt{2},$$

so the linear interpolate to the exponential is given by

$$p_1(x) = 1.260591837 + 1.085441641x.$$

■ **EXAMPLE 4.20**

Let's now look at higher-degree interpolation to the exponential; specifically, we will consider fourth-degree interpolation. The Chebyshev nodes now are

$$x_0 = \cos\left(\frac{(2 \times 0 + 1)\pi}{2(4+1)}\right) = \cos\frac{\pi}{10}$$

$$x_1 = \cos\left(\frac{(2 \times 1 + 1)\pi}{2(4+1)}\right) = \cos\frac{3\pi}{10}$$

$$x_2 = \cos\left(\frac{(2 \times 2 + 1)\pi}{2(4+1)}\right) = \cos\frac{5\pi}{10}$$

$$x_3 = \cos\left(\frac{(2 \times 3 + 1)\pi}{2(4+1)}\right) = \cos\frac{7\pi}{10}$$

$$x_4 = \cos\left(\frac{(2 \times 4 + 1)\pi}{2(4+1)}\right) = \cos\frac{9\pi}{10}.$$

Note that these values are somewhat less simple to work with than the integer or simple fraction nodes we have generally used. But the computer doesn't really care about that, and the Newton algorithm yields the polynomial

$$p_4(x) = 1 + 0.997317240x + 0.4995561859x^2 + 0.177334621x^3 + 0.043434107x^4.$$

Figure 4.53 shows the plot of the error in this interpolation, over the interval $[-1, 1]$. Compare it to Figure 4.2B, which shows the plot of the same error for fourth degree interpolation using equally spaced points. It is difficult to tell because of the scales used, but the Chebyshev interpolation is slightly better in terms of the maximum error.

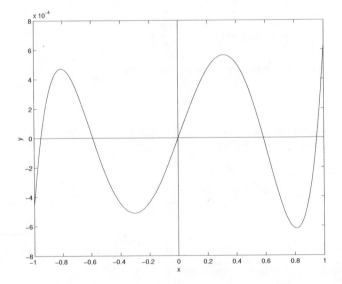

Figure 4.53   Error in fourth degree Chebyshev interpolation to the exponential.

■ EXAMPLE 4.21

Consider again $f(x) = (1 + 25x^2)^{-1}$, the Runge example. If we use the Chebyshev nodes to compute interpolating approximations of degree 8, 16, and 32, we get the plots shown in Figure 4.54. Note that the accuracy of the approximation is much better than was the case using equally spaced nodes, as in Figure 4.3. Figure 4.54C shows that, with only 32 nodes, we cannot distinguish by eye between the interpolate and the original function. Figure 4.54D shows the error in the 32-degree approximation.

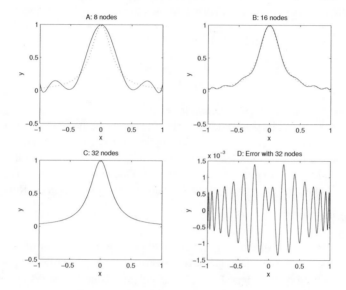

Figure 4.54   Chebyshev interpolation to the Runge example.

Why do these nodes work better? A complete answer would have to be phrased in the vague terms that we used in §4.12.2 to discuss the Runge example, or we would have to bring in a lot of additional mathematical machinery.

A fairly decent answer, however, can be based on the basic interpolation error theorem.

Theorem 4.11 *Let $f \in C^{n+1}([-1, 1])$ be given, and let p_n be the n-degree polynomial interpolate to f, using the Chebyshev nodes (4.68). Then,*

$$\|f - p_n\|_\infty \leq \frac{1}{2^n(n + 1)!}\|f^{(n+1)}\|_\infty.$$

Proof: Recall that we have

$$f(x) - p_n(x) = \frac{w_n(x)}{(n + 1)!}f^{(n+1)}(\xi_x),$$

where

$$w_n(x) = \prod_{i=0}^{n}(x - x_i).$$

Since polynomials are uniquely defined by their roots (up to a constant multiple), and since w_n and T_{n+1} have the same roots, it follows that

$$w_n = c_n T_{n+1}$$

for some constant c_n. Now, w_n is what is known as a *monic polynomial*, i.e., the coefficient of the highest-order term is 1:

$$w_n(x) = x^{n+1} + \text{lower-order terms.}$$

Therefore, c_n is the *reciprocal* of the leading coefficient for T_{n+1}. Therefore, from Theorem 4.10, we have that

$$c_n = 2^{-n}.$$

But each Chebyshev polynomial is a cosine on the interval $[-1, 1]$; this means that

$$|T_{n+1}(x)| \le 1, \quad -1 \le x \le 1.$$

Therefore,

$$|w_n(x)| = |c_n T_{n+1}(x)| \le 2^{-n}$$

for all $x \in [-1, 1]$, and the proof is complete. •

What the theorem (and its proof) shows is that the choice of the Chebyshev nodes allows us to bound the factor in the error theorem that depends on w_n. It can be shown, in fact, that any other choice of nodes would lead to a larger $\|w_n\|_\infty$, so in this sense the Chebyshev nodes are optimal. They are not "perfect," however. One can find functions for which the Chebyshev nodes fail, but these examples are much more involved than the Runge example.

Exercises:

1. Use your computer's random number function to generate a set of random values r_k, $0 \le k \le 8$, with $|r_k| \le 0.01$. Construct the interpolating polynomial to $f(x) = \sin x$ on the interval $[-1, 1]$, using nine equally spaced nodes; call this $p_8(x)$. Then construct the polynomial that interpolates $f(x_k) + r_k$; call this $\hat{p}_8(x)$. How much difference is there between the divided difference coefficients for the two polynomials? Plot both p_8 and $\hat{p}_8$ and comment on your results. (*Note:* It is important here that you look at x values *between* the nodes. Do not produce your plots based simply on the values at the nodes.)

2. Repeat the above, using $f(x) = e^x$.

3. Use the Newton interpolating algorithm to construct interpolating polynomials of degree 4, 8, 16, and 32 using equally spaced nodes on the interval $[-1, 1]$, to the function $g(x) = (1 + 4x^2)^{-1}$. Is the sequence of interpolates converging to g throughout the interval?

4. Use the Newton interpolating algorithm to construct interpolating polynomials of degree 4, 8, 16, and 32 using equally spaced nodes on the interval $[-1, 1]$, to the function $g(x) = (1 + 100x^2)^{-1}$. Is the sequence of interpolates converging to g throughout the interval?

5. Write up a complete proof of Theorem 4.10, providing all the details omitted in the text.

6. Construct interpolating polynomials of degree 4, 8, 16, and 32 on the interval $[-1, 1]$ to $f(x) = e^x$ using equidistant and Chebyshev nodes. Sample the respective errors at 500 equally spaced points and compare your results.

7. The Chebyshev nodes are defined on the interval $[-1, 1]$. Show that the change of variable

$$t_k = a + \frac{1}{2}(b - a)(x_k + 1)$$

will map the Chebyshev nodes to the interval $[a, b]$.

8. Use the formula in Problem 7 to find the Chebyshev nodes for linear interpolation on the interval $[\frac{1}{4}, 1]$. Use them to construct a linear interpolate to $f(x) = \sqrt{x}$. Plot the error in this interpolation. How does it compare to the error estimate used in §3.7?

9. Repeat the above for $f(x) = x^{-1}$ on the interval $[\frac{1}{2}, 1]$.

10. What is the error estimate for Chebyshev interpolation on a general interval $[a, b]$? In other words, how does the change of variable affect the error estimate?

4.13   LITERATURE AND SOFTWARE DISCUSSION

The subject of interpolation and approximation is a very broad one, with an extensive literature. In fact, approximation theory is a significant area of mathematical research, generally distinct from (although obviously related to) numerical analysis. The classic text on interpolation and approximation is Davis's book [6]; more recent and equally good works are those by Cheney [3], Powell [13], and Rivlin [15]. The recent book by Trefethen [18] should also be considered. Generally, approximation theory is distinct from interpolation, which is one reason that Davis's book is so useful, since it covers both. Modern work in what might be called computational approximation tends to center on splines, which is why de Boor's book [8], although old, is a good reference. A rather large collection of FORTRAN codes is given there, which can be found at NETLIB under the library pppack; MATLAB has incorporated many of them into their Splines Toolbox. Another standard spline reference is [1].

Several interesting ideas were not discussed here (the subject of interpolation and approximation is rich enough that an entire course could be based on this one subject). In Padé approximation, we look for *rational function* (i.e., ratios of polynomials) approximations to given functions. This is only marginally more involved than polynomial approximation, but of course allows for more general functional behavior to be approximated accurately, since rational functions can exhibit asymptotic behavior, but polynomials cannot. A good introduction to Padé approximation is given by Cheney in Chapter 5 of [3].

Another missing topic is Fourier (i.e., trigonometric) approximation. Again, an entire course could be based on this subject, and this is done in many engineering departments. Wavelets, related to trigonometric approximation, is also probably not appropriate for a text at this level. Future editions may have to re-think these two omissions.

REFERENCES

1. Ahlberg, J. H., Nilson, E. N., and Walsh, J. L., *The Theory of Splines and Their Applications*, Academic Press, New York, 1967.

2. Birkhoff, Garrett; "Fluid dynamics, reactor computations, and surface representation," in *A History of Scientific Computation* (Steve Nash, editor), 1990.

3. Cheney, E.W., *Approximation Theory*, Chelsea, New York, 1966.

4. Cline, A. K. "Scalar- and planar-valued curve fitting using splines under tension," Commun. ACM vol. 17, pp. 218-220 (1974).

5. Davis, Paul; "B-splines and geometric design," *SIAM News*, vol. 29, no. 5, 1996.

6. Davis, Philip, *Interpolation and Approximation*, Dover, New York, 1975; originally published by Blaisdell Publishing Company (1963).

7. de Boor, Carl, "On uniform approximation by splines," *J. Approx. Theory*, vol. 1, pp. 219–235, 1968.

8. de Boor, Carl, *A Practical Guide to Splines*, Springer-Verlag, New York, 1978.

9. Epperson, James, "On the Runge example," *Am. Math. Monthly*, pp. 329–341, 1987.

10. Hall, C.A., "On error bounds for spline interpolation," *J. Approx. Theory*, vol. 1, pp. 209–218, 1968.

11. Isaacson, E., and Keller, H. B., *Analysis of Numerical Methods*, Dover, New York, 1994; originally published by John Wiley & Sons, Inc., (1966).

12. Marušić, Miljenko, and Rogina, Mladen, "B-spline in tension," *VII Conference on Applied Mathematics*, Univ. Osijek, Osijek, Yugoslavia, 1990, pp. 129-134.

13. Powell, M. J. D., *Approximation Theory and Methods*, Cambridge University Press, Cambridge, UK, 1981.

14. Prenter, Paddy, *Splines and Variational Methods*, Wiley-Interscience, New York, 1975.

15. Rivlin, Theodore J., *An Introduction to the Approximation of Functions*, Dover, New York, 1981; originally published by Blaisdell Publishing Company (1969).

16. Schoenberg, I. J., "Contributions to the problem of approximation of equidistant data by analytic functions," *Quart. Appl. Math.*, vol. 4, pp. 45–99 and 112–141, 1946.

17. Schweikert, D. G., "An interpolating curve using a spline in tension," *J. Math. Phys.*, vol. 45, pp. 312–317, 1966.

18. Trefethen, L. N., *Approximation Theory and Approximation Practice*, SIAM, Philadelphia, 2012.

19. Young, David, "Garrett Birkhoff and applied mathematics," *Notices AMS*, vol. 44, no. 11, pp. 1446–1449, 1997.

CHAPTER 5

NUMERICAL INTEGRATION

In this chapter we are concerned with a common problem from calculus: Find the value of the definite integral

$$I(f) = \int_a^b f(x)dx.$$

We are interested in finding numerical methods that yield an accurate approximation to the exact value of $I(f)$. Typically, the approximation will be of the form

$$I_n(f) = \sum_{i=i_0}^n w_i f(x_i),$$

where $i_0 = 0$ or $i_0 = 1$ is almost always the case. The *weights* w_i and *nodes* or *abscissas* x_i define the method. Schemes for approximating integrals are often called *quadrature rules*, and different schemes use different rules for defining the weights and abscissas. In Chapter 2 we saw perhaps the simplest example of a reasonable quadrature scheme, the trapezoid rule. Here we will look at other methods, some of which are substantially more accurate than the trapezoid rule.

   In addition to constructing various quadrature schemes and analyzing their accuracy, we will also look at ways to estimate and improve the accuracy of existing quadrature methods, as well as the idea of *adaptive quadrature*, by which we try to estimate the value of the integral to within a user-specified tolerance.

An Introduction to Numerical Methods and Analysis, Second Edition. By James F. Epperson
Copyright © 2013 John Wiley & Sons, Inc.

Note that we seem to be suggesting that the integral can be viewed (approximately, at least) as a sum of function values. This is not surprising, considering the theoretical foundations of the integral, which we now review briefly.

5.1 A REVIEW OF THE DEFINITE INTEGRAL

To construct the definite integral of a function, we first define the *mesh points* (or *grid points*) $x_i^{(n)}$ according to

$$a = x_0^{(n)} < x_1^{(n)} < x_2^{(n)} \cdots < x_{n-1}^{(n)} < x_n^{(n)} = b$$

and the *evaluation points* η_i are then each taken from within the appropriate subinterval, i.e., $\eta_i \in [x_{i-1}, x_i]$, $1 \leq i \leq n$, arbitrarily. Now, if we assume that

$$\lim_{n \to \infty} \left[\max_{1 \leq i \leq n} \left(x_i^{(n)} - x_{i-1}^{(n)} \right) \right] = 0$$

(this simply means that the largest distance between adjacent points goes to zero as we take more and more points), then, under very mild conditions on f, it can be shown that the limit

$$L = \lim_{n \to \infty} \sum_{i=1}^{n} f(\eta_i)(x_i^{(n)} - x_{i-1}^{(n)})$$

exists and that its value is independent of the choices made for the mesh points $\{x_i^{(n)}\}$ and evaluation points $\{\eta_i\}$. When this happens, we call this limit value the *definite integral* of f, and we write

$$L = I(f) = \int_a^b f(x)dx.$$

Note that this construction of the integral requires very little. We only need the fact that f is continuous, that the sequence of meshes is such that the distance between adjacent points becomes arbitrarily small, and that the evaluation points are taken from anywhere in each subinterval. A summation of the form

$$R_n(f) = \sum_{i=1}^{n} f(\eta_i)(x_i^{(n)} - x_{i-1}^{(n)})$$

is called a *Riemann sum*, after the German mathematician Georg Bernhard Riemann,[1] whose name is associated with the first development of much of the rigorous theory of the definite integral. Unfortunately, Riemann sums typically converge very slowly—meaning

---

[1]Georg Friedrich Bernhard Riemann (1826–1866) was a child of poor health, the son of a Lutheran minister. He started university studies at Göttingen in 1846, but soon transferred to the University of Berlin. In 1849 he returned to Göttingen to finish his studies under the direction of Gauss, completing his doctoral dissertation in 1851. Albert Einstein would later use some of Riemann's ideas from this dissertation in the development of the theory of relativity. In 1862, Riemann's health began to decline, and he eventually came down with tuberculosis, which led to his death just four years later.

Riemann's brief life was nonetheless one of great mathematical significance. He made substantial contributions to the foundations of geometry and analysis, and virtually invented what we now call analytic number theory. His development of the Riemann sum came in his 1854 "Habilitationschrift" (sort of a postdoctoral dissertation, required to get an academic position in nineteenth century Europe), titled *Über die Darstellbarkeit einer Function durch eine trigonometrische Reihe.*

that it requires a large value of n for the summation value to be a good approximation of the limit value—so we cannot use this as a practical means of approximating the value of the integral. Figure 5.1 shows a graph of the function $y = f(x) = \frac{1}{2} + \sin \pi x$ and the rectangles generated by a Riemann sum approximation to the integral of this f over the interval $[\frac{1}{2}, \frac{3}{2}]$.

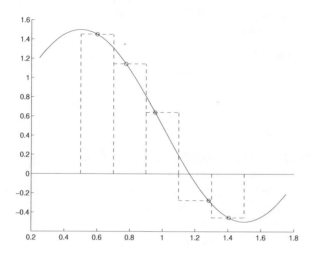

Figure 5.1   Illustration of a Riemann sum, with the evaluation points (marked by circles) chosen randomly.

We usually will dispense with the superscript notation for the abscissas from now on; i.e., instead of writing $x_i^{(n)}$, we will simply write x_i. On a couple of occasions we will revert to it, most notably in §5.6, on Gaussian quadrature.

Since the integral is essentially a limit of a sum of function values, it makes sense to think of approximating the integral by taking *finite* sums of function values. All of our quadrature rules will be based on this idea, as was the trapezoid rule, which we have already seen in Chapter 2.

Exercises:

1. Basic properties of the definite integral show that it is a *linear operator*; that is, it distributes across sums and multiplication by constants:

$$I(\alpha f + \beta g) = \alpha I(f) + \beta I(g).$$

Prove that if

$$I_n(f) = \sum_{i=0}^{n} w_i f(x_i),$$

then I_n is also linear:

$$I_n(\alpha f + \beta g) = \alpha I_n(f) + \beta I_n(g).$$

2. Assume that the quadrature rule I_n integrates all polynomials of degree less than or equal to N exactly:

$$I_n(p) = I(p)$$

for all $p \in \mathcal{P}_N$. Use this to prove that, for any integrand f, the error $I - I_n$ is equal to the error in integrating the Taylor remainder:

$$I(f) - I_n(f) = I(R_N) - I_n(R_N),$$

where $f(x) = p_N(x) + R_N(x)$. Does it really matter that we are using the Taylor polynomial and remainder? In other words, will this result hold for any polynomial approximation and its associated error?

3. Use Problem 1 to prove the following: If a quadrature rule I_n is exact for all powers x^k for $k \leq d$, then it is exact for all polynomials of degree less than or equal to d.

5.2 IMPROVING THE TRAPEZOID RULE

In §2.5 we introduced the trapezoid rule for computing integrals. Recall that, for equally spaced points, the approximation takes the form

$$T_n(f) = \frac{h}{2} \left(f(x_0) + 2f(x_1) + \cdots + 2f(x_{n-1}) + f(x_n) \right)$$

and the error is given by

$$I(f) - T_n(f) = -\frac{b-a}{12} h^2 f''(\xi_h), \tag{5.1}$$

where ξ_h is some value in the interval $[a, b]$.

We can improve the trapezoid rule by some deft analysis of the second derivative term in the error. Let us return to the error estimate (5.1), which we can write in the form

$$I(f) - T_n(f) = -\frac{1}{12} h^3 \sum_{i=1}^{n} f''(\xi_{i,h}).$$

Now let's look at the summation term, carefully. We have

$$h^3 \sum_{i=1}^{n} f''(\xi_{i,h}) = h^2 \sum_{i=1}^{n} h f''(\xi_{i,h})$$

and the sum can be interpreted as a Riemann sum for the integral

$$I(f'') = \int_a^b f''(x) dx = f'(b) - f'(a).$$

Thus, we know that

$$\lim_{n \to \infty} \sum_{i=1}^{n} h f''(\xi_{i,h}) = f'(b) - f'(a),$$

so that

$$\sum_{i=1}^{n} h f''(\xi_{i,h}) \approx f'(b) - f'(a).$$

Therefore,

$$I(f) - T_n(f) \approx -\frac{1}{12}h^2(f'(b) - f'(a)).$$

This relationship can be interpreted and used in two equally important and useful ways:

Error estimation: We can view the quantity $\hat{E}_n(f) = -\frac{1}{12}h^2(f'(b) - f'(a))$ as a "computable estimate of the error" and use it to predict how many points are required to obtain a given accuracy. Bear in mind that this approach does not *guarantee* that we will achieve the desired accuracy.

Improvement of the approximation: Define the new quadrature rule,

$$T_n^C(f) = T_n(f) - \frac{1}{12}h^2(f'(b) - f'(a)),$$

which we will call the *corrected trapezoid rule*. This value will tend to be much more accurate than that obtained using the ordinary trapezoid rule and only marginally more expensive to compute.

■ **EXAMPLE 5.1**

To illustrate the first usage, consider the problem of trying to compute values of the integral

$$I(f) = \int_0^1 e^{-x^2}\, dx.$$

How do we know when we have computed this to within a specified accuracy? Suppose we want

$$|I(f) - I_n(f)| \le 10^{-6};$$

then we could compute trapezoid rule values along with the values

$$E_n(f) = -\frac{h^2}{12}(f'(b) - f'(a)) = \frac{h^2}{6e}.$$

When $|E_n(f)| \le 10^{-6}$ we would accept $T_n(f)$ as a sufficiently accurate approximation to $I(f)$; this occurs in this case for $n = 256$, since

$$|E_n(f)| = \frac{h^2}{6e} \le 10^{-6} \Leftarrow h \le \sqrt{6e} \times 10^{-3} = 4.04 \times 10^{-3}$$

and $1/256 < 4.04 \times 10^{-3}$ but $1/128 > 4.04 \times 10^{-3}$. In practice, though, it would be better to use $E_n(f)$ to estimate the error in $T_n(f)$, but then to accept the more accurate value defined by $T_n^C(f)$.

■ **EXAMPLE 5.2**

As an example of the second usage, consider the Table 5.1 which shows the values generated by approximating the integral $I(f) = \int_0^1 e^x\, dx$ using T_n^C. Note that the error is dropping by about a factor of 16 as the number of points doubles, suggesting that the error in the corrected rule goes like h^4. This is, in fact, the case, as we will see in §5.8.1. Note that we lose the expected factor of 16 decrease in the last two rows; this is not because there is a problem with the method or with the underlying theory, but because of the effects of finite-precision arithmetic. There is so much subtractive cancellation in the computation of $I(f) - T_n^C(f)$ that the value of the error ratio is being affected. We are starting to reach the limits of the accuracy that can be measured by the machine arithmetic.

Table 5.1 Corrected trapezoid rule applied to $f(x) = e^x$, $[a, b] = [0, 1]$.

n	$T_n^C(f)$	$I(f) - T_n^C(f)$	Error ratio
2	1.718133554	0.148274E–03	N/A
4	1.718272520	0.930842E–05	15.9290
8	1.718281246	0.582426E–06	15.9822
16	1.718281792	0.364118E–07	15.9955
32	1.718281826	0.227589E–08	15.9989
64	1.718281828	0.142245E–09	15.9998
128	1.718281828	0.889111E–11	15.9986
256	1.718281828	0.556444E–12	15.9785
512	1.718281828	0.339728E–13	16.3791
1024	1.718281828	0.288658E–14	11.7692
2048	1.718281828	0.666134E–15	4.3333

Finally, we note that, to use this correction, it is not even necessary to have a formula for the derivative of f. Recall two of the derivative approximations from Chapter 4:

$$f'(x) = \frac{-3f(x) + 4f(x+h) - f(x+2h)}{2h} + \frac{h^2}{3} f'''(\xi),$$

and

$$f'(x) = \frac{3f(x) - 4f(x-h) + f(x-2h)}{2h} + \frac{h^2}{3} f'''(\xi).$$

Thus, we can write

$$f'(b) = f'(x_n) = \frac{3f(x_n) - 4f(x_{n-1}) + f(x_{n-2})}{2h} + \frac{h^2}{3} f'''(\xi_b), \qquad (5.2)$$

and

$$f'(a) = f'(x_0) = \frac{-3f(x_0) + 4f(x_1) - f(x_2)}{2h} + \frac{h^2}{3} f'''(\xi_a), \qquad (5.3)$$

so that we can write the corrected trapezoid rule as

$$\begin{aligned} T_n^C(f) &= T_n(f) - \frac{h}{24}(3f(x_n) - 4f(x_{n-1}) + f(x_{n-2}) + 3f(x_0) - 4f(x_1) + f(x_2)) \\ &+ \frac{h^4}{36}(f'''(\xi_a) - f'''(\xi_b)). \end{aligned}$$

Now define

$$\tilde{T}_n^C(f) = T_n(f) - \frac{h}{24}(3f(x_n) - 4f(x_{n-1}) + f(x_{n-2}) + 3f(x_0) - 4f(x_1) + f(x_2))$$

and note that, if the corrected trapezoid rule is $\mathcal{O}(h^4)$ accurate, so is the new "approximate" trapezoid rule, since the additional error term due to the derivative approximations is $\mathcal{O}(h^4)$.

Table 5.2 "Approximate" corrected trapezoid rule applied to $f(x) = e^x$, $[a,b] = [0,1]$.

n	$\tilde{T}_n^C(f)$	$I(f) - \tilde{T}_n^C(f)$	Error ratio
2	1.718861152	−0.579323E−03	N/A
4	1.718386631	−0.104802E−03	5.5278
8	1.718290593	−0.876407E−05	11.9582
16	1.718282447	−0.618963E−06	14.1593
32	1.718281869	−0.409496E−07	15.1152
64	1.718281831	−0.263078E−08	15.5656
128	1.718281829	−0.166666E−09	15.7847
256	1.718281828	−0.104863E−10	15.8938
512	1.718281828	−0.658362E−12	15.9278
1024	1.718281828	−0.406342E−13	16.2022
2048	1.718281828	−0.199840E−14	20.3333

Table 5.2 shows the results of applying this approximation to the same example as before. Note that we do maintain the expected factor of 16 decrease in the error for h sufficiently small.

Exercises:

1. Apply the trapezoid rule and corrected trapezoid rule, with $h = \frac{1}{4}$, to approximate the integral
$$I = \int_0^1 x(1 - x^2)dx = \frac{1}{4}.$$

2. Apply the trapezoid rule and corrected trapezoid rule, with $h = \frac{1}{4}$, to approximate the integral
$$I = \int_0^1 \frac{1}{\sqrt{1 + x^4}}dx = 0.92703733865069.$$

3. Apply the trapezoid rule and corrected trapezoid rule, with $h = \frac{1}{4}$, to approximate the integral
$$I = \int_0^1 \ln(1 + x)dx = 2\ln 2 - 1.$$

4. Apply the trapezoid rule and corrected trapezoid rule, with $h = \frac{1}{4}$, to approximate the integral
$$I = \int_0^1 \frac{1}{1 + x^3}dx = \frac{1}{3}\ln 2 + \frac{1}{9}\sqrt{3}\pi.$$

5. Apply the trapezoid rule and corrected trapezoid rule, with $h = \frac{1}{4}$, to approximate the integral
$$I = \int_1^2 e^{-x^2}dx = 0.1352572580.$$

6. For each integral below, write a program to do the corrected trapezoid rule using the sequence of mesh sizes $h = \frac{1}{2}(b - a), \frac{1}{4}(b - a), \frac{1}{8}(b - a), \ldots, \frac{1}{2048}(b - a)$, where

$b - a$ is the length of the given interval. Verify that the expected rate of decrease of the error is observed.

(a) $f(x) = x^2 e^{-x}$, $[0, 2]$, $I(f) = 2 - 10e^{-2} = 0.646647168$;

(b) $f(x) = 1/(1 + x^2)$, $[-5, 5]$, $I(f) = 2\arctan(5)$;

(c) $f(x) = \ln x$, $[1, 3]$, $I(f) = 3\ln 3 - 2 = 1.295836867$;

(d) $f(x) = e^{-x}\sin(4x)$, $[0, \pi]$, $I(f) = \frac{4}{17}(1 - e^{-\pi}) = 0.2251261368$;

(e) $f(x) = \sqrt{1 - x^2}$, $[-1, 1]$, $I(f) = \pi/2$.

7. Apply the trapezoid rule and corrected trapezoid rule to the approximation of

$$I = \int_0^1 x^2 e^{-2x} dx = 0.0808308960... ,$$

and compare your results in light of the expected error theory for both methods, and comment on what occurs. How does the error behave in each case, as a function of h? How should it have behaved?

8. Repeat the above for

$$I = \int_0^\pi \sin^2 x \, dx = \frac{1}{2}\pi.$$

9. The length of a curve $y = g(x)$, for x between a and b, is given by the integral

$$L(g) = \int_a^b \sqrt{1 + [g'(x)]^2} dx.$$

Use the corrected trapezoid rule to find the length of one "arch" of the sine curve.

10. Use the corrected trapezoid rule to find the length of the exponential function from $x = -1$ to $x = 1$. How small does h have to be for the computation to converge to within 10^{-6}?

11. Repeat the above for the tangent function, from $x = -\pi/4$ to $x = \pi/4$.

12. Define the function

$$F(t) = \int_a^t f(x) dx$$

and note that

$$F(b) = I(f) = \int_a^b f(x) dx.$$

Use Taylor expansions of F and f about $x = a$ to show that

$$I(f) - \frac{1}{2}(b - a)(f(b) + f(a)) - \frac{1}{12}(b - a)^2(f'(b) - f'(a)) = \mathcal{O}((b - a)^5).$$

Use this to show that the corrected trapezoid rule is $\mathcal{O}(h^4)$ when applied over a uniform grid of length h.

13. Construct a version of the "quasi-corrected" trapezoid rule that uses the derivative approximations

$$f'(a) \approx \frac{f(a + h) - f(a)}{h}, \quad f'(b) \approx \frac{f(b) - f(b - h)}{h}.$$

Explain why we should expect this to be less accurate than the rule using the approximations (5.2) and (5.3), and demonstrate that this is the case on the integral

$$I = \int_0^1 x(1 - x^2)dx = \frac{1}{4}.$$

◁ • • ▷

5.3 SIMPSON'S RULE AND DEGREE OF PRECISION

Common sense, plus the results of Chapter 4 on interpolation error, lead us to think that we might be able to do better than the trapezoid rule by using a higher degree of polynomial. Simpson's rule is nothing more than this next logical step, relying on quadratic interpolation to generate the quadrature rule.[2]

We follow the basic outline from our presentation for the trapezoid rule. Thus we start by approximating $I(f)$ by a single quadratic approximation. Let p_2 be the quadratic polynomial that interpolates f at the three points $x_0 = a$, $x_2 = b$, and $x_1 = c = (a+b)/2$, the midpoint. Define the basic Simpson's Rule, then, as

$$S_2(f) = I(p_2) = \int_a^b \left(L_0(x)f(a) + L_1(x)f(c) + L_2(x)f(b) \right) dx,$$

where the Lagrange functions are here defined as

$$L_0(x) = \frac{(x-c)(x-b)}{(a-c)(a-b)}, \quad L_1(x) = \frac{(x-a)(x-b)}{(c-a)(c-b)}, \quad L_2(x) = \frac{(x-a)(x-c)}{(b-a)(b-c)}.$$

Then the quadrature rule is given by

$$S_2(f) = Af(a) + Cf(c) + Bf(b),$$

where

$$A = \int_a^b L_0(x)dx, \quad C = \int_a^b L_1(x)dx, \quad B = \int_a^b L_2(x)dx.$$

To compute these we define $h = b - c = c - a = (b-a)/2$, so that we have

$$A = \int_a^{a+2h} L_0(x)dx, C = \int_a^{a+2h} L_1(x)dx, B = \int_a^{a+2h} L_2(x)dx.$$

---

[2]Thomas Simpson (1710–1761) was born and died in Leicestershire, England. Little is known of his youth and education, and the legend persists to this day that he kept "low company" with whom he would "guzzle porter and gin." He published an early text on Newton's calculus (*The Doctrine and Application of Fluxions*) and also worked in probability theory. Most of his mathematics teaching was done privately.

The quadrature rule that bears his name was originally derived by the Italian Bonaventura Cavalieri in 1639, and was known to James Gregory in the late seventeenth century and to Roger Cotes in the early eighteenth century. It was rediscovered by Simpson and published in 1743 in his paper *Mathematical Dissertations on a Variety of physical and analytical subjects* . . . .

Then

$$A = \int_a^{a+2h} L_0(x)dx$$

$$= \frac{1}{2h^2} \int_a^{a+2h} (x - (a+h))(x - (a+2h))dx$$

$$= \frac{1}{2h^2} \int_{-h}^h u(u-h)du,$$

where we have used the change of variable $u = x - a - h$ in the last step. The computation is now fairly straighforward:

$$A = \frac{1}{2h^2} \int_{-h}^h u(u-h)du,$$

$$= \frac{1}{2h^2} \left(\frac{1}{3}u^3 - \frac{1}{2}hu^2 \right) \Big|_{-h}^h$$

$$= \frac{1}{2h^2} \left(\left(\frac{1}{3}h^3 - \frac{1}{2}h^3 \right) - \left(-\frac{1}{3}h^3 - \frac{1}{2}h^3 \right) \right)$$

$$= \frac{1}{2h^2} \left(-\frac{1}{6}h^3 + \frac{5}{6}h^3 \right)$$

$$= \frac{h}{3}.$$

Similar computations (the student should verify these) yield $B = A$ and $C = 4h/3$. Thus, Simpson's rule becomes

$$S_2(f) = \frac{h}{3}(f(a) + 4f(c) + f(b)),$$

where $h = (b-a)/2$; thus, h is the distance between points in the discretization of the interval $[a, b]$.

■ **EXAMPLE 5.3**

To apply this crude version of Simpson's rule to the function $f(x) = \frac{1}{2} + \sin \pi x$, over the interval $[\frac{1}{4}, \frac{5}{4}]$, we have

$$S_2(f) = \frac{1/2}{3} \left(\frac{1}{2} + \sin \frac{1}{4}\pi + 4 \left(\frac{1}{2} + \sin \frac{3}{4}\pi \right) + \frac{1}{2} + \sin \frac{5}{4}\pi \right)$$

$$= \frac{1}{6} \left(3 + \frac{1}{2}\sqrt{2} + 2\sqrt{2} - \frac{1}{2}\sqrt{2} \right) = 0.9714045208.$$

The exact value is $I(f) = 0.9501581580$, showing that our approximation is decently accurate even though we used only a single approximating parabola. Figure 5.2 shows the application of Simpson's rule to this example.

The composite rule is easily constructed by adding up individual instances of Simpson's rule applied to *pairs* of subintervals. (Note that this means that Simpson's rule requires an even number of subintervals.) We get

$$S_n(f) = \sum_{i=1}^{n/2} \frac{h_i}{3} \left[f(x_{2i-2}) + 4f(x_{2i-1}) + f(x_{2i}) \right], \tag{5.4}$$

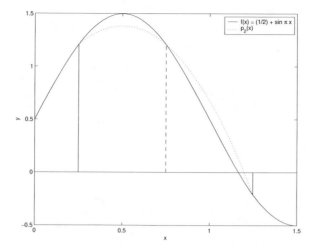

Figure 5.2   Illustration of Basic Simpson's Rule.

where $h_i = (x_{2i} - x_{2i-2})/2$. If we assume a uniform spacing of the mesh points, i.e., $h_i = h$ for all i, then all this simplifies a great deal and we have

$$S_n(f) = \frac{h}{3}\left(f(x_0) + 4f(x_1) + 2f(x_2) + 4f(x_3) + \ldots + 2f(x_{n-2})\right.$$
$$\left. +\quad 4f(x_{n-1}) + f(x_n)\right). \tag{5.5}$$

■ **EXAMPLE 5.4**

If we consider the same example that we used previously, but now use two approximating parabolas, then we get

$$S_4(f) = \frac{1/4}{3}\left(f(1/4) + 4f(1/2) + 2f(3/4) + 4f(1) + f(5/4)\right) = 0.9511844634,$$

which is much more accurate than the S_2 value. Figure 5.3 shows Simpson's rule applied in this case, and the decrease in the error is evident.

Simpson's Rule can be constructed for a nonuniform grid, but the formula corresponding to either of (5.4) or (5.5) is somewhat more involved. See Problem 12.

■ **EXAMPLE 5.5**

If we apply Simpson's rule to the example in which $f(x) = e^x$ over the interval $[0, 1]$, using a sequence of decreasing grids, we get the results shown in Table 5.3.

Note that the error here is decreasing by a factor of roughly 16 as h is halved.[3] Drawing on our experience with the trapezoid rule, we therefore suspect that

$$I(f) - S_n(f) = \mathcal{O}(h^4).$$

---

[3]But again note that the last case is an exception. The reason is because the method is, at this point, too accurate for the finite precision arithmetic to properly resolve.

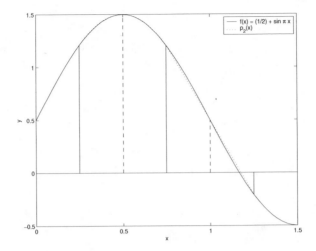

Figure 5.3   Illustration of Simpson's Rule.

However, our work on the interpolation error for quadratic interpolation (4.11), together with the definition of Simpson's rule as the *exact* integral of a piecewise quadratic interpolate, suggest that

$$f - p_2 = \mathcal{O}(h^3) \Rightarrow I(f) - S_n(f) = I(f) - I(p_2) = \int_a^b (f(x) - p_2(x))dx = \mathcal{O}(h^3).$$

Thus Simpson's rule appears to be *more* accurate than the underlying theory (from interpolation) would suggest. What's going on here that causes this?

Table 5.3   Simpson's rule applied to $f(x) = e^x$, $[a, b] = [0, 1]$.

n	$S_n(f)$	$I(f) - S_n(f)$	Error Ratio
2	1.718861151877	−0.579323E−03	N/A
4	1.718318841922	−0.370135E−04	15.6517
8	1.718284154700	−0.232624E−05	15.9113
16	1.718281974052	−0.145593E−06	15.9777
32	1.718281837562	−0.910273E−08	15.9944
64	1.718281829028	−0.568969E−09	15.9986
128	1.718281828495	−0.355611E−10	15.9998
256	1.718281828461	−0.222178E−11	16.0057
512	1.718281828459	−0.137890E−12	16.1127
1024	1.718281828459	−0.910383E−14	15.1463
2048	1.718281828459	0.444089E−15	−20.5000

It is possible to directly prove an error estimate for Simpson's rule, in a manner similar to that for the trapezoid rule. But a more illuminating explanation of *why* Simpson's rule is "more accurate than it ought to be" can be had by looking at the extent to which it integrates polynomials exactly. This leads us to the notion of *degree of precision* for a quadrature rule.

Definition 5.1 (Degree of Precision) *Let I_n be a quadrature rule, and assume that it is exact for all polynomials of degree $\leq p$. Then we say that I_n has degree of precision p.*

We ought to expect that we get better accuracy from integration methods that have a higher degree of precision.

As an example, we note that Problem 21 of §2.5 shows that the trapezoid rule has degree of precision $p = 1$, because it integrates all linear polynomials exactly. Simpson's rule, by construction, clearly ought to have degree of precision $p = 2$. We will show that it actually has degree of precision $p = 3$; i.e., Simpson's rule integrates cubic polynomials exactly, which is something of a bonus beyond what we ought to expect from the construction. We will then show that having degree of precision $p = 3$ leads to an error estimate that is $\mathcal{O}(h^4)$.

Let q_3 be an arbitrary cubic polynomial, which we choose to write as

$$q_3(x) = Ax^3 + q_2(x),$$

where q_2 contains all the lower-order terms from q_3. Then the error in using Simpson's rule to integrate q_3 becomes

$$I(q_3) - S_2(q_3) = A(I(x^3) - S_2(x^3)) + (I(q_2) - S_2(q_2)).$$

But Simpson's rule will clearly integrate the quadratic part of q_3 exactly; thus, Simpson's rule will be exact for all cubic polynomials if and only if it is exact for x^3. Direct computation gives us

$$
\begin{aligned}
S_2(x^3) &= \frac{(b-a)/2}{3}(a^3 + 4((a+b)/2)^3 + b^3) \\
&= \frac{b-a}{6}\left(a^3 + \frac{1}{2}(a+b)^3 + b^3\right) \\
&= \frac{b-a}{12}(2a^3 + (a+b)^3 + 2b^3) \\
&= \frac{b-a}{4}(a^3 + a^2 b + ab^2 + b^3) \\
&= \frac{1}{4}(b^4 - a^4) \\
&= I(x^3);
\end{aligned}
$$

thus, Simpson's rule is exact for all cubic polynomials. More formally, we have proved the following lemma.

Lemma 5.1 *Let $q_3(x)$ be an arbitrary polynomial of degree less than or equal to 3. Then Simpson's rule integrates q_3 exactly; i.e.,*

$$S_2(q_3) = \int_a^b q_3(x)\,dx$$

for any interval $[a, b]$.

Now, how does this give Simpson's rule increased accuracy? Let $q_3 \approx f$, where q_3 now is some cubic polynomial approximation to f. Let the error in this approximation be given by R_3. Then we have

$$I(f) - S_2(f) = I(q_3 + R_3) - S_2(q_3 + R_3).$$

But both the integral and Simpson's rule are *linear*, in that their action distributes across sums (see the exercises in §5.1). Thus, we have $I(q_3 + R_3) = I(q_3) + I(R_3)$ and similarly for S_2. Therefore,

$$I(f) - S_2(f) = I(R_3) - S_2(R_3).$$

Thus, Simpson's rule is as accurate for f as it is for the *error* in *any* cubic polynomial approximation to f. Since we expect the best cubic approximation to be better than the best quadratic approximation, the fact that Simpson's rule integrates cubics exactly suggests a significant increase in accuracy.

   All this is very vague and intuitive. Can we make it more precise? The answer is "yes," as we see in the following theorem.

Theorem 5.1 (Simpson's Rule Error Estimate) *If $f \in C^4([a,b])$, then there exists a point $\xi \in [a,b]$ such that*

$$I(f) - S_2(f) = -\frac{1}{90}\left(\frac{b-a}{2}\right)^5 f^{(4)}(\xi). \tag{5.6}$$

Proof: Let p_3 interpolate f at the nodes $x_0 = a$, $x_1 = c - \epsilon$, $x_2 = c + \epsilon$, $x_3 = b$, where $\epsilon > 0$ is a small parameter. Then for all $x \in [a,b]$, we have

$$f(x) - p_3(x) = R_3(x) = \frac{1}{4!}(x - x_0)(x - x_1)(x - x_2)(x - x_3)f^{(4)}(\xi_x)$$

for some $\xi_x \in [a,b]$. Since Simpson's rule integrates cubic polynomials exactly, we have that

$$I(f) - S_2(f) = I(R_3) - S_2(R_3).$$

Note that the remainder will vanish at $x = a$ and $x = b$, so

$$S_2(R_3) = 4\left(\frac{b-a}{6}\right)\left(\frac{1}{4!}\right)(c - x_0)(c - x_1)(c - x_2)(c - x_3)f^{(4)}(\xi_c).$$

Recall that both x_1 and x_2 depend on the parameter ϵ, and that we have $x_1 \to c$ and $x_2 \to c$ as $\epsilon \to 0$. Thus, in the limit as $\epsilon \to 0$, $R_3(c) \to 0$, and hence $S_2(R_3)$ goes to zero.

   On the other hand, as $\epsilon \to 0$ we have

$$I(R_3) \to \frac{1}{4!}\int_a^b (x - a)(x - c)^2(x - b)f^{(4)}(\xi_x)dx.$$

We therefore can say that

$$I(f) - S_2(f) = \frac{1}{4!}\int_a^b (x - a)(x - c)^2(x - b)f^{(4)}(\xi_x)dx,$$

and, since $(x - a)(x - c)^2(x - b)$ does not change sign for $x \in [a,b]$, the Integral Mean Value Theorem can be used to simplify the integral so that we have

$$I(f) - S_2(f) = \frac{1}{4!}f^{(4)}(\xi)\int_a^b (x - a)(x - c)^2(x - b)dx.$$

Direct evaluation of the integral completes the proof. $\bullet$

   The error for the composite rule now follows just as it did for the trapezoid rule.

Corollary 5.1 *If $f \in C^4([a, b])$ and the grid is uniform, with mesh spacing h, then*

$$I(f) - S_n(f) = -\frac{(b-a)h^4}{180}f^{(4)}(\xi).$$

Proof: We have, for a uniform grid ($x_{i+1} - x_i = (b-a)/n = h$),

$$
\begin{aligned}
I(f) - S_n(f) &= \sum_{i=1}^{n/2}\left(\int_{x_{2i-2}}^{x_{2i}} f(x)dx - \left(\frac{x_{2i} - x_{2i-2}}{6}\right)(f(x_{2i}) + 4f(x_{2i-1}) + f(x_{2i}))\right), \\
&= \sum_{i=1}^{n/2}\left(-\frac{1}{90}\left(\frac{2h}{2}\right)^5 f^{(4)}(\xi_i)\right), \quad \text{[from (5.6)]}, \\
&= -\frac{h^4}{90}\sum_{i=1}^{n/2} hf^{(4)}(\xi_i), \\
&= -\frac{h^4}{90}\sum_{i=1}^{n/2} \frac{b-a}{n}f^{(4)}(\xi_i), \\
&= -\frac{(b-a)h^4}{180}\sum_{i=1}^{n/2} \frac{2}{n}f^{(4)}(\xi_i),
\end{aligned}
$$

and the Discrete Average Value Theorem allows us to replace the sum with a single pointwise evaluation of the fourth derivative:

$$I(f) - S_n(f) = -\frac{(b-a)h^4}{180}f^{(4)}(\xi),$$

which completes the proof. •

■ **EXAMPLE 5.6**

Although we cannot always use the error estimate to predict how small to take h, this is sometimes a useful exercise. Consider now our usual example in which $f(x) = e^x$ and the interval is $[0, 1]$. Suppose that we want the error to be less than 10^{-6}. How small should h be to guarantee that?

The absolute value of the error is

$$|E_n(f)| = |I(f) - S_n(f)| = \frac{1}{180}h^4|f^{(4)}(\xi)|.$$

Therefore, to impose $|E_n(f)| \leq 10^{-6}$, we impose

$$\frac{1}{180}h^4 \max_{x\in[0,1]} |f^{(4)}(x)| \leq 10^{-6}$$

and solve for h. Thus, we have

$$\frac{e}{180}h^4 \leq 10^{-6} \Leftarrow h \leq 0.09020788609.$$

So, taking $h \leq 0.0902...$ will be sufficient for the desired accuracy. Note that for the trapezoid rule we would require that $h \leq 0.002101083838$, which is substantially smaller.

Programming Hint: One should always use the theory underlying a numerical method to help debug a program implementing that method. For example, since Simpson's rule is $\mathcal{O}(h^4)$ accurate, we should see the error go down by a factor of 16 as the mesh size is halved. While this might not hold exactly for larger values of h, it will begin to hold as you take h smaller and smaller. If it doesn't, then it generally means that either the integrand is not smooth enough or your program has an error in it. As was the case with the trapezoid rule, there exists a class of functions for which Simpson's rule is "super-accurate," and for these functions the error will not decrease as rapidly as we expect, because it starts out smaller than it should be. (See §5.8.1 for details.) Always use a few simple examples to predict the error and make sure that the code is performing as expected.

The explanation so far for the extra accuracy of Simpson's rule might seem a bit less than satisfying. Yes, maybe we can show how all the mathematics fits together, but what is really happening here to *cause* the extra accuracy? The reason is hinted at in the proof of Theorem 5.1, and with a little work we can do a fairly complete exposition.

Consider the cubic polynomial

$$
\begin{aligned}
q_3(x) \quad = \quad & f(a) + \left(\frac{f(c) - f(a)}{h} \right) (x - a) \\
& + \left(\frac{f'(c) - \frac{f(c) - f(a)}{h}}{h} \right) (x - a)(x - c) + A(x - a)(x - c)^2,
\end{aligned}
$$

where

$$
A = h^{-2} \left(\frac{f(b) - f(c)}{h} - 2f'(c) + \frac{f(c) - f(a)}{h} \right)
$$

and

$$
h = c - a = b - c.
$$

Now this is an imposing formula, but it ought to be straightforward to show that q_3 is the cubic that interpolates to f at a, b, and c, and that also interpolates to f' at c:

$$
q_3'(c) = f'(c).
$$

Figure 5.4 shows the graph of this function for our example function $\frac{1}{2} + \sin \pi x$. Note that the the requirement that the derivatives match at the midpoint results in a *much* closer approximation to the function than was the case for the quadratic p_2; thus, we expect to get much better approximation to the integral.

Now for the payoff: If we use q_3 as the approximating polynomial to generate a quadrature rule, what do we get? In other words, what is

$$
S_2^*(f) = I(q_3)
$$

in terms of $f(a)$, $f(b)$, $f(c)$, and $f'(c)$? We will omit the details here (see Problem 16, however), but the bottom line is the following: Using q_3 as the approximating polynomial *yields the same quadrature rule as when we use p_2:* $I(q_3) = I(p_2)$. Or, to turn this around a bit, using the quadratic approximation p_2 produces the same quadrature rule as the slightly more accurate cubic q_3. This boost in accuracy is very much tied in to using the midpoint of the interval as the third point. Although we can construct a version of Simpson's rule based on using any point c between a and b, it will be only $\mathcal{O}(h^4)$ accurate when c is the midpoint.

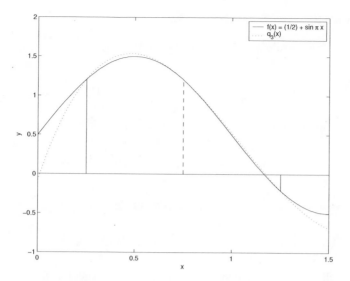

Figure 5.4 Another look at Simpson's rule.

We close this section with a brief code segment that implements Simpson's rule over a uniform grid.

```
Algorithm 5.1  Simpson's Rule
input a, b, n
external f
     simp = f(a) + f(b)
     h = (b - a)/n
     sum4 = 0.0
     for i=1 to n-1 by 2 do
          x = a + i*h
          sum4 = sum4 + f(x)
     endfor
     sum2 = 0.0
     for i=2 to n-2 by 2 do
          x = a + i*h
          sum2 = sum2 + f(x)
     endfor
     simp = (h/3.0)*(simp + 4*sum4 + 2*sum2)
end code
```

The extra accuracy that we get with Simpson's rule is an artifact of using an even-degree polynomial interpolate to define the integration rule. If we construct a method based on cubic interpolation (see Problem 15) we still get a degree of precision equal to 3, and an error estimate that is $\mathcal{O}(h^4)$.

Exercises:

1. Apply Simpson's rule with $h = \frac{1}{4}$, to approximate the integral

$$I = \int_0^1 x(1 - x^2)dx = \frac{1}{4}.$$

2. Apply Simpson's rule with $h = \frac{1}{4}$, to approximate the integral

$$I = \int_0^1 \frac{1}{\sqrt{1 + x^4}}dx = 0.92703733865069.$$

3. Apply Simpson's rule with $h = \frac{1}{4}$, to approximate the integral

$$I = \int_0^1 \ln(1 + x)dx = 2\ln 2 - 1.$$

How small does the error theory say that h has to be to get that the error is less than 10^{-3}? 10^{-6}? How small does h have to be for the trapezoid rule to achieve this accuracy?

4. Apply Simpson's rule with $h = \frac{1}{4}$, to approximate the integral

$$I = \int_0^1 \frac{1}{1 + x^3}dx = \frac{1}{3}\ln 2 + \frac{1}{9}\sqrt{3}\pi.$$

5. Apply Simpson's rule with $h = \frac{1}{4}$, to approximate the integral

$$I = \int_1^2 e^{-x^2}dx = 0.1352572580.$$

6. For each function below, write a program to do Simpson's rule using the sequence of mesh sizes $h = \frac{1}{2}(b - a), \frac{1}{4}(b - a), \frac{1}{8}(b - a), \ldots, \frac{1}{2048}(b - a)$, where $b - a$ is the length of the given interval. Verify that the expected rate of decrease of the error is observed. Comment on any anomolies that are observed.

   (a) $f(x) = \ln x$, $[1, 3]$, $I(f) = 3\ln 3 - 2 = 1.295836867$;
   (b) $f(x) = x^2 e^{-x}$, $[0, 2]$, $I(f) = 2 - 10e^{-2} = 0.646647168$;
   (c) $f(x) = 1/(1 + x^2)$, $[-5, 5]$, $I(f) = 2\arctan(5)$;
   (d) $f(x) = \sqrt{1 - x^2}$, $[-1, 1]$, $I(f) = \pi/2$;
   (e) $f(x) = e^{-x}\sin(4x)$, $[0, \pi]$, $I(f) = \frac{4}{17}(1 - e^{-\pi}) = 0.2251261368$.

7. For each integral Problem 6, how small does h have to be to get an accuracy, according to the error theory, of at least 10^{-3}? 10^{-6}? Compare to the value of h required by the trapezoid rule for this accuracy. (Feel free to use a computer algebra system to help you with the computation of the derivatives.)

8. Since the area of the unit circle is $A = \pi$, it follows that

$$\frac{\pi}{2} = \int_{-1}^1 \sqrt{1 - x^2}dx.$$

Therefore, we can approximate π by approximating this integral. Use Simpson's rule to compute approximate values of π in this way and comment on your results.

9. If we wanted to use Simpson's rule to approximate the natural logarithm function on the interval $[\frac{1}{2}, 1]$ by approximating

$$\ln x = \int_1^x \frac{1}{t} dt,$$

how many points would be needed to obtain an error of less than 10^{-6}? How many points for an error of less than 10^{-16}? What are the corresponding values for the trapezoid rule?

10. Use Simpson's rule to produce a graph of $E(x)$, defined to be the *length* of the exponential curve from 0 to x, for $0 \le x \le 3$. See Problem 14 of §2.5.

11. Let $f(x) = |x|$; use the trapezoid rule (with $n = 1$), corrected trapezoid rule (with $n = 1$), and Simpson's rule (with $n = 2$), to compute

$$I(f) = \int_{-1}^1 f(x) dx$$

and compare your results to the exact value. Explain what happens in light of our error estimates for the trapezoid and Simpson's rules.

12. Write out the expression for Simpson's rule when c is *not* the midpoint of the interval $[a, b]$. To simplify matters, take $c - a = h$, $b - c = \theta h$.

13. What is the degree of precision of the corrected trapezoid rule, T_1^C? What about the n-subinterval version, T_n^C?

14. Prove that if we want to show that the quadrature rule $I_n(f)$ has degree of precision p, it suffices to show that it will exactly integrate x^k, $0 \le k \le p$ over the integral $(0, 1)$.

15. Construct the analogue of Simpson's rule based on exactly integrating a cubic interpolate at equally spaced points on the interval $[a, b]$.

16. Show that $I(q_3) = S_2(f)$.

17. Consider the quadrature rule defined by exactly integrating a cubic Hermite interpolate:
$$I_1(f) = I(H_2).$$

Write down the quadrature formula for both the basic and composite settings, and state and prove an error estimate, using the error results for Hermite interpolation from Chapter 4.

18. Consider a quadrature rule in the form

$$I_n(f) = \sum_{k=1}^n a_k f(x_k),$$

where the coefficients $a_k > 0$ and the grid points x_k are all known. Assume that I_n integrates the trivial function $w(x) = 1$ exactly:

$$I_n(w) = \sum_{k=1}^{n} a_k = I(w) = \int_a^b dx = b - a,$$

and that this holds for all intervals (a, b). Consider now the effects of rounding error on integrating an arbitrary function f. Let $\hat{f}(x) = f(x) + \epsilon(x)$ be f polluted by rounding error, with $|\epsilon(x)| \leq C\mathbf{u}$ for some constant $C > 0$, for all $x \in [a, b]$. Show that

$$|I_n(f) - I_n(\hat{f})| \leq C\mathbf{u}(b - a).$$

Comment on this in comparison to the corresponding result for numerical differentiation, as given in §2.2.

19. The *normal probability distribution* is defined as

$$p(x) = \frac{1}{\sigma\sqrt{2\pi}} e^{-(x-\mu)^2/2\sigma^2},$$

where μ is the mean, or average, and σ is the variance. This is the famous bell-shaped curve that one hears so much about; the mean gives the center of the bell and the variance gives its width. If x is distributed in this fashion, then the probability that $a \leq x \leq b$ is given by the integral

$$P(a \leq x \leq b) = \int_a^b p(x)dx.$$

(a) Use the change of variable $z = (x - \mu)/\sigma$ to show that

$$P(-m\sigma \leq x \leq m\sigma) = \frac{1}{\sqrt{2\pi}} \int_{-m}^{m} e^{-z^2/2}dz.$$

(b) Compute values of $P(-m\sigma \leq x \leq m\sigma)$ for $m = 1, 2, 3$, using Simpson's rule.

20. Use Simpson's rule to solve Problem 9 from §5.2.

21. Use Simpson's rule to solve Problem 10 from §5.2.

5.4   THE MIDPOINT RULE

An interesting and very simple quadrature rule is the midpoint rule, based on exactly integrating a linear Taylor approximation to the integrand. This results in a rule that is actually *more* accurate than the trapezoid rule. We proceed in much the same manner as with the trapezoid and Simpson rules, except that since this is the third time, we skip some of the details.

First consider the integral

$$I(f) = \int_a^b f(x)dx$$

and the Taylor approximation

$$p_1(x) = f(c) + (x - c)f'(c),$$

where $c = (a + b)/2$, the midpoint of the interval. We define the numerical method by exactly integrating p_1:

$$M_1(f) = \int_a^b p_1(x)dx = (b - a)f(c).$$

The composite rule then becomes

$$M_n(f) = h \sum_{i=1}^n f(a + (i - 1/2)h).$$

Note that the term involving the derivative is not present in the quadrature rule. This is because we have

$$\int_a^b (x - c)f'(c)dx = 0$$

when c is the midpoint of the interval. This rule is called the *midpoint rule* because it evaluates the integrand at the midpoint of each subinterval in order to construct the quadrature rule. Figure 5.5 shows a single-interval example, and Figure 5.6 shows a composite example. Both illustrations use $f(x) = \frac{1}{2} + \sin \pi x$, integrated over the interval $[\frac{1}{4}, \frac{5}{4}]$.

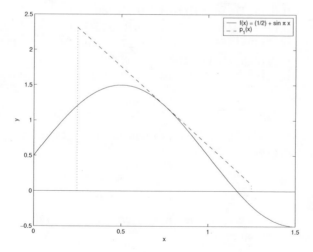

Figure 5.5 Single interval midpoint rule.

The error estimate comes rather easily from the construction of the approximation. We have $M_1(f) = I(p_1)$, so

$$I(f) - M_1(f) = I(f) - I(p_1) = I(f - p_1) = I(R_1),$$

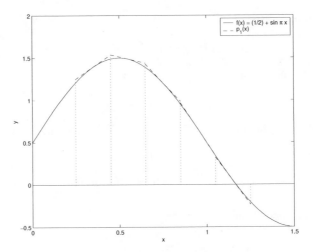

Figure 5.6   Midpoint rule.

Table 5.4   Midpoint rule applied to $f(x) = e^x$, $[a, b] = [0, 1]$.

n	$M_n(f)$	$I(f) - M_n(f)$	Error ratio
2	0.17005127166502D+01	0.1776911D–01	N/A
4	0.17138152797711D+01	0.4466549D–02	3.9783
8	0.17171636649957D+01	0.1118163D–02	3.9945
16	0.17180021920527D+01	0.2796364D–03	3.9986
32	0.17182119133839D+01	0.6991508D–04	3.9997
64	0.17182643493169D+01	0.1747914D–04	3.9999
128	0.17182774586502D+01	0.4369809D–05	4.0000
256	0.17182807360054D+01	0.1092454D–05	4.0000
512	0.17182815553455D+01	0.2731135D–06	4.0000
1024	0.17182817601807D+01	0.6827838D–07	4.0000
2048	0.17182818113894D+01	0.1706960D–07	4.0000

where $R_1(x) = \frac{1}{2}(x - c)^2 f''(\xi)$ is the Taylor remainder. We can then apply the Integral Mean Value Theorem to get

$$I(f) - M_1(f) = \frac{1}{24}(b - a)^3 f''(\xi),$$

which is essentially *half* the error of the single-interval trapezoid rule. For the composite midpoint rule the error becomes

$$I(f) - M_n(f) = \frac{b - a}{24} h^2 f''(\xi).$$

■ **EXAMPLE 5.7**

Let's look at using the midpoint rule to approximate

$$I(f) = \int_0^1 e^x\,dx$$

using $h = \frac{1}{2}$. Then we will have

$$M_2(f) = \frac{1}{2} \times f(0+(1/2)(1/2)) + \frac{1}{2} \times f(1/2+(1/2)(1/2)) = \frac{1}{2} \times e^{1/4} + \frac{1}{2} \times e^{3/4} = 1.700512717.$$

If we continue with smaller and smaller values of h, then we get the results presented in Table 5.4.

Exercises:

1. Apply the midpoint rule with $h = \frac{1}{4}$, to approximate the integral

$$I = \int_0^1 x(1-x^2)\,dx = \frac{1}{4}.$$

How small does h have to be to get that the error is less than 10^{-3}? 10^{-6}?

2. Apply the midpoint rule with $h = \frac{1}{4}$, to approximate the integral

$$I = \int_0^1 \frac{1}{\sqrt{1+x^4}}\,dx = 0.92703733865069.$$

How small does h have to be to get that the error is less than 10^{-3}? 10^{-6}?

3. Apply the midpoint rule with $h = \frac{1}{4}$, to approximate the integral

$$I = \int_0^1 \ln(1+x)\,dx = 2\ln 2 - 1.$$

How small does h have to be to get that the error is less than 10^{-3}? 10^{-6}?

4. Apply the midpoint rule with $h = \frac{1}{4}$, to approximate the integral

$$I = \int_0^1 \frac{1}{1+x^3}\,dx = \frac{1}{3}\ln 2 + \frac{1}{9}\sqrt{3}\pi.$$

How small does h have to be to get that the error is less than 10^{-3}? 10^{-6}?

5. Apply the midpoint rule with $h = \frac{1}{4}$, to approximate the integral

$$I = \int_1^2 e^{-x^2}\,dx = 0.1352572580.$$

How small does h have to be to get that the error is less than 10^{-3}? 10^{-6}?

6. Show that the midpoint rule can be derived by integrating, exactly, a polynomial interpolate of degree *zero*.

7. Apply the midpoint rule to each of the following functions, integrated over the indicated interval. Use a sequence of grids $h = (b-a), (b-a)/2, (b-a)/4, \ldots$ and confirm that the approximations are converging at the correct rate. Comment on any anomolies that you observe.

   (a) $f(x) = \ln x$, $[1,3]$, $I(f) = 3\ln 3 - 2 = 1.295836867$;
   (b) $f(x) = x^2 e^{-x}$, $[0,2]$, $I(f) = 2 - 10e^{-2} = 0.646647168$;
   (c) $f(x) = \sqrt{1 - x^2}$, $[-1,1]$, $I(f) = \pi/2$;
   (d) $f(x) = 1/(1 + x^2)$, $[-5,5]$, $I(f) = 2\arctan(5)$;
   (e) $f(x) = e^{-x}\sin(4x)$, $[0,\pi]$, $I(f) = \frac{4}{17}(1 - e^{-\pi}) = 0.2251261368$.

8. For each integral in the previous problem, how small does h have to be to get accuracy, according to the error theory, of at least 10^{-3}? 10^{-6}?

9. State and prove a formal theorem concerning the error estimate for the midpoint rule over n subintervals. You may want to state and prove a formal theorem for the single-subinterval rule first, and then use this in the more general theorem.

10. Let T_1 be the trapezoid rule using a single subinterval, M_1 be the midpoint rule using a single subinterval, and S_2 be Simpson's rule using a single quadratic interpolant (hence, a single *pair* of subintervals). Show that for any continuous function f, and any interval $[a,b]$,

$$S_2(f) = \frac{1}{3}T_1(f) + \frac{2}{3}M_1(f).$$

◁ ● ● ▷

5.5 APPLICATION: STIRLING'S FORMULA

Stirling's formula[4] is an interesting and useful way to approximate the factorial function, $n!$, for large values of n. Standard derivations require a substantial background in classical analysis, but in fact the rule can be derived in a fairly general form based on nothing more than the trapezoid rule and its error estimate.

Theorem 5.2 (Stirling's Formula) *For all $n \geq 2$, there exists a value C_n, $2.37 < C_n < 2.501$, such that*

$$n! = C_n\sqrt{n}(n/e)^n.$$

Proof: We start with the observation that

$$\ln n! = \sum_{k=1}^{n} \ln k,$$

$$= \sum_{k=1}^{n-1} \frac{1}{2}(\ln k + \ln(k+1)) + \frac{1}{2}\ln n,$$

---

[4]James Stirling (1692–1770), a Scotsman, began his studies at Oxford in 1710, but was expelled for political reasons before finishing his degree. Nonetheless, by 1717 he had already published his first paper, an extension of some of Newton's work on plane curves. For the next few years he was in Italy, trying to obtain a position, and in 1722 he returned to Britain. In 1726, while living in London, he published a monograph on infinite series and other topics called *Methodus Differentialis*, which contains what we call Stirling's formula as one of its examples.

and the summation is exactly the trapezoid rule for

$$I(\ln) = \int_1^n \ln x \, dx = n \ln n - n + 1$$

using $n - 1$ subintervals of length 1. The error estimate for the trapezoid rule then gives

$$\ln n! = \left(n + \frac{1}{2}\right) \ln n - n + 1 - \frac{1}{12} \sum_{k=1}^{n-1} \xi_k^{-2}$$

for some set of $\xi_k \in [k, k+1]$. Now exponentiate both sides and let σ_n be the summation, to get

$$n! = \sqrt{n}(n/e)^n e^{1 - \frac{1}{12}\sigma_n} = \sqrt{n}(n/e)^n C_n$$

and it remains only to establish the bounds on C_n. To this end, note that we can bound the sum as follows:

$$1 \le \sum_{k=1}^{n-1} \xi_k^{-2} \le \sum_{k=1}^{n-1} k^{-2} \le \sum_{k=1}^{\infty} k^{-2} = \frac{\pi^2}{6}.$$

Thus, we have

$$2.37 < \exp\left(1 - \frac{1}{12}\frac{\pi^2}{6}\right) \le e^{1+\sigma_n} \le \exp\left(1 - \frac{1}{12}\right) < 2.501.$$

This estimate on C_n is not as sharp as others, but it is sufficient for most purposes. The sharpest well-known result ([6], p. 467) states that there is a value $c_n \in [0, 1]$ such that

$$n! = \sqrt{2\pi n} \left(\frac{n}{e}\right)^n e^{c_n/12n}.$$

Exercises:

1. Use Stirling's formula to show that

$$\lim_{n \to \infty} \left(\frac{x^n}{n!}\right) = 0$$

   for all x.

2. For $x = 10$ and $\epsilon = 10^{-3}$, how large does n have to be for

$$\left|\frac{x^n}{n!}\right| \le \epsilon$$

   to hold? Repeat for $\epsilon = 10^{-6}$. Use Stirling's formula here—don't just plug numbers into a calculator.

3. Use Stirling's formula to determine the value of

$$\lim_{n \to \infty} \frac{(n!)^p}{(pn)!},$$

where $p \geq 2$ is an integer.

4. Use Stirling's formula to get an upper bound on the ratio

$$R_n = \frac{1 \cdot 3 \cdot 5 \cdots (2n-1)}{2^{n+1}n!}$$

as a function of n.

◁ ● ● ● ▷

5.6 GAUSSIAN QUADRATURE

Gaussian[5] quadrature is a very powerful tool for approximating integrals. It is derived in a very different way than the trapezoid, Simpson, and midpoint rules (although the midpoint rule can be considered a special case of Gaussian quadrature), and a complete discussion of Gaussian quadrature usually requires a substantial background in approximation theory.[6]

Before getting into the details of the derivation of the method, it might be a good idea to give some idea of how accurate Gaussian quadrature can be. The quadrature rules are all based on special values of *weights* and *abscissas* (evaluation points, commonly called "Gauss points") that are pre-computed and stored. They are available in most standard mathematics tables, and can also be computed by standard, available computer codes. Table 5.5 gives some of the values for a few cases, and the text website at www.wiley.com has links to several files containing the weights and abscissas for a range of values of n[7].

The quadrature rule is written in the form

$$G_n(f) = \sum_{i=1}^{n} w_i^{(n)} f(x_i^{(n)}) \approx \int_{-1}^{1} f(x)dx, \tag{5.7}$$

where the weights $w_i^{(n)}$ and Gauss points $x_i^{(n)}$ will be determined later. For now, we will simply assume that they are available, as is, in fact, the case.[8]

---

[5]Carl Friedrich Gauss (1777–1855) is widely regarded as one of the greatest mathematicians of all time, in a class with Archimedes, Newton, and Euler. A child prodigy who taught himself to read and do arithmetic before beginning elementary school, Gauss attended the Collegium Carolinum Brunswick and the University of Göttingen, graduating in 1800. His doctoral dissertation was accepted in 1801. Briefly supported by his patron, the Duke of Brunswick, Gauss accepted an appointment at Göttingen as director of the observatory in 1804 and as professor of astronomy in 1807.

Like Euler, there are few areas of mathematics that are untouched by Gauss's mind and Gauss's name. Unlike Euler, who published a lot, Gauss published sparingly, often not at all, leaving some of his most significant results in his personal diary. Nonetheless, Gauss has to his credit results in physics, astronomy (including the calculation of the orbit of the planetoid Ceres), number theory, and non-Euclidean geometry. His collected works come to 12 volumes. Gaussian quadrature bears his name because of a paper he presented to the Göttingen Society in 1814, titled *Methodus nova itegralium valores per approximationem inveniendi*.

[6]The treatment we use here—which avoids a lot of the need for approximation theory—is based on an idea due to Roy Mathias, published in *Int. J. Math. Educ. Sci. Tech.*, vol. 28, pp. 134–137 (1997).

[7]The website may be found at www.wiley.com/go/epperson2edition.

[8]Integrals on general intervals $[a, b]$ can be treated by a change of variable in the integrand, as we will see later in this section.

Table 5.5 Gaussian quadrature nodes and weights.

n	$x_i^{(n)}$	$w_i^{(n)}$
1	0.0000000000000000E+00	0.2000000000000000E+01
2	$\pm$ 0.5773502691896257E+00	1.0000000000000000E+00
4	$\pm$ 0.3399810435848563E+00	0.6521451548625464E+00
	$\pm$ 0.8611363115940526E+00	0.3478548451374476E+00
8	$\pm$ 0.1834346424956498E+00	0.3626837833783620E+00
	$\pm$ 0.5255324099163290E+00	0.3137066458778874E+00
	$\pm$ 0.7966664774136268E+00	0.2223810344533745E+00
	$\pm$ 0.9602898564975362E+00	0.1012285362903697E+00
16	$\pm$ 0.9501250983763744E-01	0.1894506104550685E+00
	$\pm$ 0.2816035507792589E+00	0.1826034150449236E+00
	$\pm$ 0.4580167776572274E+00	0.1691565193950024E+00
	$\pm$ 0.6178762444026438E+00	0.1495959888165733E+00
	$\pm$ 0.7554044083550030E+00	0.1246289712555339E+00
	$\pm$ 0.8656312023878318E+00	0.9515851168249290E-01
	$\pm$ 0.9445750230732326E+00	0.6225352393864778E-01
	$\pm$ 0.9894009349916499E+00	0.2715245941175185E-01

■ **EXAMPLE 5.8**

Consider the example integral

$$I(f) = \int_{-1}^{1} e^x dx = e^1 - e^{-1} = 2.3504023872876029138...$$

Using $n = 2$, we get the approximation

$$G_2(f) = (1)e^{-0.5773502691896257} + (1)e^{0.5773502691896257} = 2.342696087910$$

which is very easily computed, and very accurate as well, for the amount of effort put into the computation. Table 5.6 shows the approximate values given by using Gaussian quadrature at 4, 8, and 16 points, in addition to this $n = 2$ case. For comparison's sake, the table also gives the approximate values (and associated errors) given by Simpson's rule. Note the extraordinary accuracy of Gaussian quadrature compared to Simpson's rule. How do we get such high accuracy?

Table 5.6 Gaussian quadrature applied to integrating $f(x) = e^x$, over the interval $[a, b] = [-1, 1]$.

n	$G_n(f)$	$I(f) - G_n(f)$	$S_n(f)$	$I(f) - S_n(f)$
2	2.342696087910	0.77062993778738D–02	2.362053756543	–0.11651369255893D–01
4	2.350402092156	0.29513124299996D–06	2.351194831880	–0.79244459265215D–03
8	2.350402387288	0.19539925233403D–13	2.350453017242	–0.50629954676307D–04
16	2.350402387288	0.15543122344752D–13	2.350405569305	–0.31820170360852D–05

290 NUMERICAL INTEGRATION

The trapezoid and Simpson rules—and midpoint, as well—were derived by exactly integrating a polynomial approximation to the integrand. In Gaussian quadrature, we construct an approximate integral that is "best possible" in the sense that it has the highest possible degree of precision. Thus we pose the following question:

What are the coefficients $\{w_i^{(n)}\}$ *and* Gauss points $\{x_i^{(n)}\}$ such that the quadrature rule

$$\int_{-1}^{1} f(x)dx \approx \sum_{i=1}^{n} w_i^{(n)} f(x_i^{(n)})$$

is *exact* for as large a degree of polynomial as possible?

To put it another way (see Problem 3), we want to find $\{w_i^{(n)}\}$ and $\{x_i^{(n)}\}$ such that

$$\int_{-1}^{1} x^k dx = \sum_{i=1}^{n} w_i^{(n)} (x_i^{(n)})^k \tag{5.8}$$

for $k = 0, 1, 2, ..., N$, for N as large as possible.

The high accuracy of Gaussian quadrature then comes from the fact that it integrates very high degree polynomials exactly. In the trapezoid method, Simpson's rule, and the midpoint method, we use a fixed, predetermined set of grid points, and only the weights are defined by the method. A fixed degree of polynomial approximation is used over each subinterval, so a fixed degree of polynomial is integrated exactly, no matter how many points are used. In Gaussian quadrature, both the grid points (i.e., the abscissas or Gauss points) and the weights are chosen to maximize the degree of precision, in other words, the degree of polynomial that is integrated exactly. The degree of polynomial that is exactly integrated goes to infinity as the number of points used in the quadrature rule goes to infinity.

Since (5.8) is a system of (nonlinear) equations, we could just rely on that as a basis for computation, but this approach is unsatisfactory for a number of reasons, the most important one being that it is next to impossible to gain any broad understanding of the numerical method.

Note that we have to find, not only the weights and abscissas, but also the value of N. Intuitively, we suspect that $N = 2n - 1$. This is because a polynomial of degree $2n - 1$ has $2n$ coefficients, and thus the number of unknowns (n weights plus n abscissas) equals the number of equations (one for each power in the polynomial). We will show that taking $N = 2n$ yields a contradiction, and then we will construct a solution for $N = 2n - 1$.

Lemma 5.2 *If $N = 2n$, then there are no weights and Gauss points such that (5.8) is satisfied for all $k = 0, 1, 2, \ldots, N$.*

Proof: Let $\{w_i^{(n)}\}$ and $\{x_i^{(n)}\}$ satisfy (5.8), and assume that (5.8) holds for all $k = 0, 1, 2, \ldots, N$; i.e., the quadrature rule is exact for all polynomials of degree less than or equal to $N = 2n$. Define

$$L(x) = \prod_{j=1}^{n} (x - x_j^{(n)})^2.$$

Thus, $L(x) \geq 0$ and

$$\int_{-1}^{1} L(x)dx > 0.$$

However,

$$\sum_{i=1}^{n} w_i^{(n)} L(x_i^{(n)}) = 0.$$

Therefore, we have a contradiction, since the quadrature rule (5.7) is supposed to be exact for all polynomials of degree $2n$, and it is clearly not exact for the particular polynomial of degree $2n$ given by $L(x)$. Hence, our assumption that a solution exists must be false, and the lemma is proved. •

Now we turn to the $N = 2n - 1$ case. We will derive necessary conditions for the weights and abscissas, and then show that these necessary conditions do imply that the rule is exact for all polynomials of degree $\leq 2n - 1$. Since the weights are the easier of the two, we will start there.

Lemma 5.3 *Let $\{w_i^{(n)}\}$ be a set of weights and $\{x_i^{(n)}\}$ a set of Gauss points, such that (5.8) is satisfied for $k = 0, 1, 2, \ldots, N = 2n - 1$. Then the weights must satisfy*

$$w_i^{(n)} = \int_{-1}^{1} L_i^{(n)}(x)dx, \qquad (5.9)$$

where

$$L_i^{(n)}(x) = \prod_{\substack{k = 1 \\ k \neq i}}^{n} \frac{x - x_k^{(n)}}{x_i^{(n)} - x_k^{(n)}}.$$

Proof: Note that the $L_i^{(n)}$ are of degree $n - 1 \leq 2n - 1$, and recall that $L_i^{(n)}(x_j) = \delta_{ij}$. Thus, the fact that the quadrature rule is supposed to be exact for *all* polynomials of degree $\leq 2n - 1$ forces

$$\int_{-1}^{1} L_i^{(n)}(x)dx = \sum_{j=1}^{n} w_j^{(n)} L_i^{(n)}(x_j^{(n)}) = w_i^{(n)},$$

and this completes the proof. •

Now that we understand how the weights are related to the Gauss points, we can concentrate on finding the Gauss points. To this end, define

$$P_n(x) = \prod_{i=1}^{n}(x - x_i^{(n)}), \qquad (5.10)$$

where the $x_i^{(n)}$ are the Gauss points, which we assume to be distinct, and note that P_n is a polynomial of degree n, since it has n roots. Let $Q(x)$ be *any* polynomial of degree $\leq n - 1$, so that the product $P_n Q$ has degree $\leq 2n - 1$. This means that the quadrature rule must integrate the product exactly, i.e.,

$$\int_{-1}^{1} P_n(x)Q(x)dx = \sum_{i=1}^{n} w_i^{(n)} P_n(x_i^{(n)})Q(x_i^{(n)}).$$

But the construction of P_n implies that $P_n(x_i^{(n)}) = 0$; thus, we have that

$$\int_{-1}^{1} P_n(x)Q(x)dx = 0$$

for *all* polynomials Q of degree $\leq n - 1$. Since the quadrature rule has to be exact when applied to the product $P_n Q$, this means that the polynomial P_n is one member of a family of *orthogonal polynomials* (§4.11.2), and the Gauss points are the roots of P_n. At this point it might be appropriate to recall some results from Chapter 4 on orthogonal polynomials:

Theorem 5.3 *For any $k > 1$ there exists a family of polynomials $\phi_i \in \mathcal{P}_i, 0 \leq i \leq k$, such that*

1. For all $i \neq j$,

$$\int_{-1}^{1} \phi_i(x)\phi_j(x)dx = 0;$$

2. For all $q \in \mathcal{P}_j, j < i$,

$$\int_{-1}^{1} \phi_i(x)q(x)dx = 0;$$

3. All of the roots of each of the ϕ_i are distinct and lie in the interval $[-1, 1]$.

With these results in hand, we can state and prove our main result.

Theorem 5.4 (Construction of Gaussian Quadrature) *For $N = 2\hat{n} - 1$, there exists a set of Gauss points $x_i^{(n)}$ and weights $w_i^{(n)}$, with the weights given by (5.9), such that (5.8) holds for all $k = 0, 1, 2, \ldots, N$.*

Proof: We begin by defining the Gauss points $x_i^{(n)}$ as the roots of the Legendre family of orthogonal polynomials.[9]

Let $P(x)$ be an arbitrary polynomial of degree $\leq 2n - 1$. Note that we can write

$$P(x) = P_n(x)Q(x) + R(x),$$

where both Q and R are polynomials of degree $\leq n - 1$ and P_n is defined in (5.10). (This follows from ordinary polynomial division considerations.) It therefore follows (Problem 9 of §4.1), that

$$R(x) = \sum_{i=1}^{n} R(x_i^{(n)})L_i^{(n)}(x).$$

In addition, note that $P_n(x_i^{(n)}) = 0$ implies that $P(x_i^{(n)}) = R(x_i^{(n)})$. Then we have

$$\int_{-1}^{1} P(x)dx = \int_{-1}^{1} \left(P_n(x)Q(x) + R(x) \right) dx,$$

$$= \int_{-1}^{1} P_n(x)Q(x)dx + \int_{-1}^{1} R(x)dx.$$

---

[9] If §4.11.2 (Least Squares Approximation and Orthogonal Polynomials) was not covered prior to this, it would be wise at this point to review Theorem 4.9 for background material on orthogonal polynomials, and the subsequent discussion for material on the Legendre family.

The orthogonality property of P_n then implies that the first integral is zero. Hence,

$$
\begin{aligned}
\int_{-1}^{1} P(x)dx &= \int_{-1}^{1} R(x)dx, \\
&= \int_{-1}^{1} \sum_{i=1}^{n} R(x_i^{(n)})L_i(x)dx, \\
&= \sum_{i=1}^{n} R(x_i^{(n)}) \int_{-1}^{1} L_i(x)dx, \\
&= \sum_{i=1}^{n} R(x_i^{(n)})w_i^{(n)}, \\
&= \sum_{i=1}^{n} P(x_i^{(n)})w_i^{(n)},
\end{aligned}
$$

and we have proved that the quadrature rule (5.7) is exact for an arbitrary polynomial of degree $\leq 2n - 1$. This completes the proof. •

We indicated before that the accuracy of Gaussian quadrature comes from its ability to integrate, exactly, polynomials of high degree. Let's give some indication of why this is so. What we know is that G_n will integrate any polynomial of degree $\leq 2n - 1$ exactly. So, given the problem of integrating f, let p_{2n-1} be *any* polynomial of degree $\leq 2n - 1$. Then,

$$ I(f) - G_n(f) = I(p_{2n-1}) + I(f - p_{2n-1}) - G_n(p_{2n-1}) - G_n(f - p_{2n-1}), \quad (5.11) $$

where all we have done is added and subtracted p_{2n-1} and used the linearity of the integral and the quadrature rule. But since the quadrature rule is exact when applied to p_{2n-1}, we have that

$$ I(p_{2n-1}) - G_n(p_{2n-1}) = 0, $$

so that (5.11) becomes

$$ I(f) - G_n(f) = I(f - p_{2n-1}) - G_n(f - p_{2n-1}). $$

Thus, the error in Gaussian quadrature is equal to the error in integrating the polynomial approximation *error*. Note that this will hold for *any* polynomial approximation of degree $\leq 2n - 1$.

The question then becomes: Can we find a specific polynomial approximation that yields a concise error formula? The answer is, yes, we can.

Let $p_{2n-1} = H_n$, the Hermite interpolate to f, which uses the Gauss points as the nodes. Then, from Theorem 4.5 we have that

$$ f(x) - p_{2n-1}(x) = f(x) - H_n(x) = \frac{1}{(2n)!}\psi_n(x)f^{(2n)}(\xi_x), $$

where

$$ \psi_n(x) = \prod_{i=1}^{n}(x - x_i^{(n)})^2. \quad (5.12) $$

Therefore,

$$ G_n(f - p_{2n-1}) = G_n(f - H_n) = \sum_{k=1}^{n} \frac{w_k^{(n)}}{(2n)!}\psi_n(x_k^{(n)})f^{(2n)}(\xi_k); $$

but (5.12) implies that $\psi(x_k^{(n)}) = 0$; hence, we have that $G_n(f - H_n) = 0$, so

$$I(f) - G_n(f) = I(f - H_n) = \int_{-1}^{1} \frac{1}{(2n)!} \psi_n(x) f^{(2n)}(\xi_x) dx.$$

Since $\psi_n(x) \geq 0$ for all x, we can apply the Integral Mean Value Theorem to get that there exists a value $\eta_n \in [-1, 1]$ such that

$$I(f) - G_n(f) = \frac{1}{(2n)!} \left(\int_{-1}^{1} \psi_n(x) dx \right) f^{(2n)}(\eta_n).$$

Recall now that the Gauss points or nodes $x_i^{(n)}$ are the roots of the n-degree Legendre polynomial, P_n. Thus, $\psi_n(x) = (r_n P_n(x))^2$, where r_n is the reciprocal of the leading coefficient of P_n. We then have

$$I(f) - G_n(f) = \frac{r_n^2}{(2n)!} \left(\int_{-1}^{1} P_n^2(x) dx \right) f^{(2n)}(\eta_n) = \frac{r_n^2}{(2n)!} \|P_n\|_2^2 f^{(2n)}(\eta_n).$$

The values of r_n and $\|P_n\|_2$ can be determined with a little effort (see Problem 14) and from this we get the final error statement

$$I(f) - G_n(f) = \frac{2^{2n+1}(n!)^4}{(2n+1)[(2n)!]^3} f^{(2n)}(\eta_n). \tag{5.13}$$

It is not immediately obvious how rapidly this decays, because of the competing effects of the factorials in the numerator and denominator. Thus, we appeal to Stirling's formula (§5.5) to gain some insight. We have that

$$n! = C_n \sqrt{n}(n/e)^n$$

and

$$(2n)! = C_{2n} \sqrt{2n}(2n/e)^{2n},$$

so that

$$\frac{2^{2n+1}(n!)^4}{(2n+1)[(2n)!]^3} = \frac{2^{2n+1} C_n^4 n^2 (n/e)^{4n}}{(2n+1) C_{2n}^3 (2n)^{3/2} (2n/e)^{6n}} = K_n \frac{\sqrt{n}}{n + \frac{1}{2}} \left(\frac{e}{16n} \right)^{2n},$$

where $K_n = \frac{C_n^4}{2^{3/2} C_{2n}^3}$ lies between 0.7 and 1.04. Note that this estimate says that the error goes down exponentially with n, as long as the integrand is smooth enough. Just as an example, we note that for $n = 16$ we have

$$\frac{2^{2n+1}(n!)^4}{(2n+1)[(2n)!]^3} = K_n \frac{\sqrt{n}}{n + \frac{1}{2}} \left(\frac{e}{16n} \right)^{2n} = K_n \times 1.653181645 \times 10^{-64},$$

which goes a long way toward explaining the rapid convergence we saw in Table 5.6.

Other intervals, other rules   What if the integral is not posed over the interval $[-1, 1]$? This is not an obstacle to the application of Gaussian quadrature, since we can apply a simple change of variable to rewrite any integral over $[a, b]$ as an integral over $[-1, 1]$. We have

$$\int_a^b g(x) dx = \frac{1}{2}(b - a) \int_{-1}^{1} g(a + \frac{1}{2}(b - a)(z + 1)) dz = \int_{-1}^{1} f(z) dz,$$

where

$$f(z) = \frac{1}{2}(b-a)g(a + \frac{1}{2}(b-a)(z+1)).$$

(See Problem 15.) This affects the error estimate because of the factors of $\frac{1}{2}(b-a)$ in the change of variable. Instead of (5.13), we get

$$I(f) - G_n(f) = \frac{(b-a)^{2n+1}(n!)^4}{(2n+1)[(2n)!]^3} f^{(2n)}(\eta_n).$$

See Problem 16 for the reduction of this via Stirling's formula to a more manageable expression.

■ **EXAMPLE 5.9**

Suppose that we want to compute

$$I(f) = \int_1^2 x \ln x \, dx = 0.6362943610.$$

The change of variable gives us

$$x = 1 + \frac{1}{2}(z+1),$$

so the integral now is

$$I(f) = \frac{1}{2} \int_{-1}^1 \left(1 + \frac{1}{2}(z+1)\right) \ln \left(1 + \frac{1}{2}(z+1)\right) dz.$$

For simplicity's sake we define the new integrand as

$$F(z) = \frac{1}{2} \left(1 + \frac{1}{2}(z+1)\right) \ln \left(1 + \frac{1}{2}(z+1)\right).$$

The $n = 2$ Gaussian quadrature approximation is then computed quite simply:

$$G_2(F) = w_1 F(x_1) + w_2 F(x_2) = 0.6361494997.$$

There are also other Gaussian quadrature rules. In fact, any family of orthogonal polynomials will generate a Gaussian quadrature rule according to the following "template." The rules we have developed here are usually called Gauss–Legendre quadrature, because they are based on the Legendre family of orthogonal polynomials.

Theorem 5.5 (General Gaussian Quadrature Rule) *Let $w(x) \geq 0$ be a weight function on the interval $[a, b]$, and let $\{\phi_k(x)\}$ be the family of orthogonal polynomials with respect to this weight function and this interval. Define the quadrature rule*

$$G_n(f) = \sum_{i=1}^n w_i^{(n)} f(x_i^{(n)})$$

for $x_i^{(n)}$ the roots of ϕ_n, and with $w_i^{(n)}$ given by

$$w_i^{(n)} = \int_a^b w(x) \left(\prod_{\substack{k=1 \\ k \neq i}}^n \frac{x - x_k^{(n)}}{x_i^{(n)} - x_k^{(n)}} \right) dx.$$

Then $G_n(p)$ is exact for all polynomials $p \in \mathcal{P}_{2n-1}$, and there exists $\xi_n \in [a,b]$ such that

$$\int_a^b w(x)f(x)dx - G_n(f) = \frac{1}{(2n)!} \left(\int_a^b \psi_n(x)dx \right) f^{(2n)}(\xi_n),$$

for all $f \in C^{2n}([a,b])$, where

$$\psi_n(x) = \prod_{k=1}^n \left(x - x_k^{(n)} \right)^2.$$

Proof: See Problem 17. •

■ **EXAMPLE 5.10**

We can apply this theorem to construct a Gaussian quadrature rule for integrals of the form

$$I(f) = \int_0^\infty e^{-x} f(x)dx.$$

Here the weight function is $w(x) = e^{-x}$, so the orthogonal polynomial family that we need to use is the Laguerre family. To keep the task tractable, we will construct only the $n = 2$ rule. The second-order Laguerre polynomial is (see §4.11.2)

$$L_2(x) = \frac{1}{2}(x^2 - 4x + 2),$$

so the Gauss points are

$$x_1 = 2 - \sqrt{2} = 0.5857864376, \quad x_2 = 2 + \sqrt{2} = 3.414213562,$$

and the weights are

$$w_1 = \int_0^\infty e^{-x} \left(\frac{x - x_2}{x_1 - x_2} \right) dx = \frac{1}{4}(2 + \sqrt{2}) = 0.8535533903,$$

and

$$w_2 = \int_0^\infty e^{-x} \left(\frac{x - x_1}{x_2 - x_1} \right) dx = \frac{1}{4}(2 - \sqrt{2}) = 0.1464466092.$$

We can check that these are the correct values by confirming that they can be used to integrate, exactly, any polynomial of degree ≤ 3 in $I(f)$. It suffices to use $f(x) = x^3$, so we look at

$$\int_0^\infty e^{-x} x^3 dx = 6.$$

Our Gauss–Laguerre approximation is

$$\begin{aligned} G_2(f) &= w_1 f(x_1) + w_2 f(x_2) \\ &= 0.8535533903 \times (0.5857864376)^3 + 0.1464466092 \times (3.414213562)^3 \\ &= 5.99999999 \end{aligned}$$

Considering the fixed precision of the computations for the Gauss points and weights, we ought to consider this adequate confirmation of our work. Note, by the way, that there is no explicit computation with the weight *function* in the evaluation of G_2; $w(x)$ does not appear in the formula, only $f(x)$.

Exercises:

1. Apply Gaussian quadrature with $n = 4$ to approximate each of the following integrals.

   (a)
   $$I = \int_{-1}^{1} \ln(1 + x^2)dx = 2\ln 2 + \pi - 4;$$

   (b)
   $$I = \int_{-1}^{1} \sin^2 \pi x \, dx = 1;$$

   (c)
   $$I = \int_{-1}^{1} (x^8 + 1)dx = 20/9;$$

   (d)
   $$I = \int_{-1}^{1} e^{-x^2} dx = 1.493648266;$$

   (e)
   $$I = \int_{-1}^{1} \frac{1}{1 + x^4} dx = 1.733945974;$$

2. Apply Gaussian quadrature with $n = 4$ to approximate each of the following integrals. Remember that you have to do a change of variable to $[-1, 1]$ first.

   (a)
   $$I = \int_{0}^{1} \ln(1 + x)dx = 2\ln 2 - 1;$$

   (b)
   $$I = \int_{0}^{1} \frac{1}{\sqrt{1 + x^4}} dx = 0.92703733865069;$$

   (c)
   $$I = \int_{0}^{1} x(1 - x^2)dx = \frac{1}{4};$$

   (d)
   $$I = \int_{0}^{1} \frac{1}{1 + x^3} dx = \frac{1}{3}\ln 2 + \frac{1}{9}\sqrt{3}\pi;$$

(e)

$$I = \int_1^2 e^{-x^2} dx = 0.1352572580.$$

3. Show that (5.8) is both necessary and sufficient to make the quadrature exact for all polynomials of degree less than or equal to $2N - 1$.

4. Write a program that does Gaussian quadrature, using the weights and Gauss points given in the text. Apply this to each of the integrals below, and compare your results to those for the other quadrature methods in the exercises. Remember that you will have to do a change of variable if the interval is not $[-1, 1]$. Be sure to run your program for more than one value of n.

   (a) $f(x) = \sqrt{1 - x^2}$, $[-1, 1]$, $I(f) = \pi/2$;

   (b) $f(x) = x^2 e^{-x}$, $[0, 2]$, $I(f) = 2 - 10e^{-2} = 0.646647168$;

   (c) $f(x) = \ln x$, $[1, 3]$, $I(f) = 3 \ln 3 - 2 = 1.295836867$;

   (d) $f(x) = 1/(1 + x^2)$, $[-5, 5]$, $I(f) = 2 \arctan(5)$;

   (e) $f(x) = e^{-x} \sin 4x$, $[0, \pi]$, $I(f) = \frac{4}{17}(1 - e^{-\pi}) = 0.2251261368$.

5. Let $P(x) = 6x^3 + 5x^2 + x$, and let $P_2(x) = 3x^2 - 1$ (this is the quadratic Legendre polynomial). Find linear polynomials $Q(x)$ and $R(x)$ such that $P(x) = P_2(x)Q(x) + R(x)$. Verify that $I(P) = I(R)$.

6. Let $P(x) = x^3 + x^2 + x - 1$, and repeat Problem 5.

7. Let $P(x) = 3x^3 + x^2 - 6$, and repeat Problem 5.

8. Verify that the weights for the $n = 2$ Gaussian quadrature rule satisfy the formula (5.9).

9. Repeat the above for the $n = 4$ rule.

10. Show, by direct computation, that the $n = 2$ and $n = 4$ Gaussian quadrature rules are exact for the correct degree of polynomials.

11. The quadratic Legendre polynomial is $P_2(x) = (3x^2 - 1)/2$. Show that it is orthogonal (over $[-1, 1]$) to all linear polynomials.

12. The cubic Legendre polynomial is $P_3(x) = (5x^3 - 3x)/2$. Show that it is orthogonal (over $[-1, 1]$) to all polynomials of degree less than or equal to 2.

13. The quartic Legendre polynomial is $P_4(x) = (35x^4 - 30x^2 + 3)/8$. Show that it is orthogonal (over $[-1, 1]$) to all polynomials of degree less than or equal to 3.

14. The first two Legendre polynomials are

$$P_0(x) = 1, \quad P_1(x) = x,$$

and it can be shown that the others satisfy the recurrance relation

$$(n + 1)P_{n+1}(x) = (2n + 1)P_n(x) - nP_{n-1}(x).$$

Use this to show (by induction) that the leading coefficient for the Legendre polynomials satisfies

$$k_n = \frac{(2n)!}{2^n (n!)^2}$$

and the 2-norm of the Legendre polynomials satisfies

$$\|P_n\|_2^2 = \frac{2}{2n+1}.$$

15. Let

$$I(f) = \int_a^b f(x)dx.$$

Show that the change of variable $x = a + \frac{1}{2}(b-a)(z+1)$ gives that

$$I(f) = \int_{-1}^1 F(z)dz$$

for $F(z) = \frac{1}{2}(b-a)f(a + \frac{1}{2}(b-a)(z+1))$.

16. Show that the error for Gaussian quadrature applied to

$$I(f) = \int_a^b f(x)dx$$

is $\mathcal{O}([(b-a)(e/8n)]^{2n})$.

17. Prove Theorem 5.5. *Hint:* Simply generalize what was done in the text for the special case of

$$I(f) = \int_{-1}^1 f(x)dx.$$

18. Once again, we want to consider the approximation of the natural logarithm function, this time using numerical quadrature. Recall that we have

$$\ln x = \int_1^x \frac{1}{t}dt.$$

Recall also that it suffices to consider $x \in [\frac{1}{2}, 1]$.

   (a) How many grid points are required for 10^{-16} accuracy using the trapezoid rule? Simpson's rule?

   (b) How many grid points are required if Gauss–Legendre quadrature is used?

19. Write a computer program that uses Gaussian quadrature for a specified number of points to compute the natural logarithm over the interval $[\frac{1}{2}, 1]$, to within 10^{-16} accuracy. Compare the accuracy of your routine to the intrinsic logarithm function on your system.

20. Write a brief essay, in your own words, of course, which explains the importance of the *linearity* of the integral and quadrature rule in the development of Gaussian quadrature.

◁ ● ● ● ▷

5.7 EXTRAPOLATION METHODS

One of the most important ideas in computational mathematics is that we can take the information from a few approximations and use that to both estimate the error in the approximation and to generate a significantly improved approximation. We have already seen examples of this with Aitken extrapolation for converging sequences (§3.11.1) and the corrected trapezoid rule (§5.2). In this section, we will embark on a more detailed study of some of these ideas.

Given an integral $I(f)$, consider a generic quadrature rule that we will denote by $I_n(f)$. Assume an error relationship of the form

$$I(f) - I_n(f) \approx Cn^{-p}. \tag{5.14}$$

We note in passing that this is true for all three of the trapezoid ($p = 2$), midpoint ($p = 2$), and Simpson's ($p = 4$) rules, but *not* for Gaussian quadrature. We want to use this approximate equality to do three different things:

1. Estimate the value of p.

2. Construct a computable estimate of the error $I(f) - I_n(f)$.

3. Construct an improved approximation of $I(f)$.

We can, in fact, do all three without a lot of work, and it leads us to a very efficient scheme for approximate integration.

Estimating p We begin by applying (5.14) for three cases: n, $2n$, and $4n$ points:

$$
\begin{aligned}
I(f) - I_n(f) &\approx Cn^{-p}, \\
I(f) - I_{2n}(f) &\approx C(2n)^{-p}, \\
I(f) - I_{4n}(f) &\approx C(4n)^{-p}.
\end{aligned}
$$

Now, consider the ratio

$$r_{4n} = \frac{I_n - I_{2n}}{I_{2n} - I_{4n}},$$

where we have dropped the explicit argument f from the integration rule for notational simplicity. We have

$$
\begin{aligned}
r_{4n} &= \frac{(I_n - I) + (I - I_{2n})}{(I_{2n} - I) + (I - I_{4n})} \approx \frac{-Cn^{-p} + C(2n)^{-p}}{-C(2n)^{-p} + C(4n)^{-p}}, \\
&= \frac{(2n)^{-p} - n^{-p}}{(4n)^{-p} - (2n)^{-p}} = \frac{2^{-p} - 1}{4^{-p} - 2^{-p}}, \\
&= 2^p.
\end{aligned}
$$

Thus, the computable ratio r_{4n} is approximately equal to 2^p. Therefore we can estimate the value of p by solving to get

$$p \approx \frac{\log r_{4n}}{\log 2}.$$

The importance of this ability to estimate p is that we can use it to verify that a computer program is working, or to estimate convergence rates when the integrand is not smooth enough to apply our error theory. Let's look at some examples that illustrate each of these ideas.

■ **EXAMPLE 5.11**

Consider the problem of evaluating the integral

$$I = \int_0^1 e^{-x^2}\,dx.$$

This integral cannot be evaluated in closed form (there is no simple antiderivative for e^{-x^2})so we cannot compare our estimates to an exact solution. But we can observe that the integrand is infinitely smooth, so the error theory for Simpson's rule applies, thus, approximations using Simpson's rule ought to yield that p is nearly 4. Table 5.7 shows the result of doing this computation, and we see that

$$r_{256} = \frac{0.746824133300 - 0.746824132843}{0.746824132843 - 0.746824132814} = 15.76...\,,$$

which implies that $p \approx 4$, as expected. If this had not been the case, we would have for the most part had only two possible conclusions to draw: (1) The integrand is not as smooth as we thought it was; or, (2) The program is not working properly. There is a third possibility, that the function is periodic, in which case Simpson's rule will be "super-accurate" and thus the error may not decrease very rapidly at all, because it starts out very small. (See §5.8.1 and Problems 8 and 9 in this section.) Table 5.7 shows the complete set of computations for n ranging from 4 to 2048. (Note that we begin to lose accuracy at the end of the computation: The estimated value of p begins to wander away from 4. This occurs because $S_n(f)$ is so accurate as an approximation to $I(f)$ that the computations in estimating p are badly corrupted by rounding error for large n. The student ought to think about how this happens.)

Table 5.7 Estimation of p.

n	$S_n(f)$	p
4	0.746855379791E+00	N/A
8	0.746826120527E+00	N/A
16	0.746824257436E+00	3.973123
32	0.746824140607E+00	3.995232
64	0.746824133300E+00	3.998911
128	0.746824132843E+00	3.999734
256	0.746824132814E+00	3.999953
512	0.746824132813E+00	3.999232
1024	0.746824132812E+00	4.002964
2048	0.746824132812E+00	4.063215

■ **EXAMPLE 5.12**

Next, consider the integral

$$I(f) = \int_0^1 \sqrt{x}\,dx = \frac{2}{3}.$$

Since the integrand does not have a bounded second derivative on the interval of integration, we cannot expect the trapezoid rule or Simpson's rule to be as accurate as theory predicts, something that is borne out by computational experiment (see Table 5.8). But this only begs the question: How accurate are these methods when applied to this problem? In other words, what sort of mesh exponent *do* we get?

Using the computation outlined above, we can estimate the exponent p for this example, based on the data in Table 5.8. For both the trapezoid rule and Simpson's rule, we get that $p \approx 1.50$ (for this example only; in general, of course, we expect $p = 4$ for Simpson's rule and $p = 2$ for the trapezoid rule). See Table 5.9.

Table 5.8   Trapezoid and Simpson's rules applied to $f(x) = \sqrt{x}$.

n	$T_n(f)$	$S_n(f)$
2	0.6035533906E+00	0.6380711875E+00
4	0.6432830462E+00	0.6565262648E+00
8	0.6581302216E+00	0.6630792801E+00
16	0.6635811969E+00	0.6653981886E+00
32	0.6655589363E+00	0.6662181827E+00
64	0.6662708114E+00	0.6665081031E+00
128	0.6665256573E+00	0.6666106059E+00
256	0.6666165490E+00	0.6666468462E+00
512	0.6666488815E+00	0.6666596591E+00
1024	0.6666603622E+00	0.6666641891E+00
2048	0.6666644336E+00	0.6666657907E+00
4096	0.6666658761E+00	0.6666663570E+00

Table 5.9   Estimation of p from data in Table 5.8.

n	p (Trapezoid rule)	p (Simpson's rule)
8	0.1420027799E+01	0.1420027799E+01
16	0.1445602214E+01	0.1445602214E+01
32	0.1462662030E+01	0.1462662030E+01
64	0.1474156297E+01	0.1474156297E+01
128	0.1481998892E+01	0.1481998892E+01
256	0.1487405107E+01	0.1487405107E+01
512	0.1491159895E+01	0.1491159895E+01
1024	0.1493781623E+01	0.1493781623E+01
2048	0.1495619062E+01	0.1495619062E+01
4096	0.1496910216E+01	0.1496910216E+01

Error estimation and an improved approximation   To directly estimate the error, we note that our assumptions imply that

$$I - I_{2n} \approx C(2n)^{-p} = 2^{-p}\left(Cn^{-p}\right) \approx 2^{-p}(I - I_n)$$

and we can then solve the approximate equality to get

$$I \approx \frac{2^p I_{2n} - I_n}{2^p - 1}.$$

We then define

$$R_{2n} = \frac{2^p I_{2n} - I_n}{2^p - 1}, \tag{5.15}$$

which we call *Richardson's*[10] *extrapolated value*; and

$$E_{2n} = R_{2n} - I_{2n} = \frac{I_{2n} - I_n}{2^p - 1} \tag{5.16}$$

which is our *computable estimate of the error* in I_{2n} as an approximation to $I = I(f)$. The idea is to take the extrapolated value R_{2n} as the exact value. Of course, having computed R_{2n}, it would make sense to use it as the approximate value of $I(f)$.

■ **EXAMPLE 5.13**

Refer back to the data in Table 5.7. From this we have

$$S_8(f) = 0.746826120527, \quad S_{16}(f) = 0.746824257436,$$

so that

$$R_{16}(f) = 0.7468241335$$

and

$$E_{16}(f) = 0.12421 \times 10^{-6}.$$

■ **EXAMPLE 5.14**

We now return once again to our example involving $\int_0^1 e^x dx$, using the trapezoid rule as the basic integration scheme. Table 5.10 shows the extrapolated approximation and the estimated error, based on (5.15) and (5.16), along with the actual error. We note that the estimated error E_n tracks very well with the actual trapezoid error, $I(f) - T_n(f)$. Note that the error $I(f) - R_n(f)$ decreases by a factor of 16 as the mesh size is halved, suggesting that the $\mathcal{O}(h^2)$ trapezoid rule has been improved to $\mathcal{O}(h^4)$ by the extrapolation, something that can in fact be made rigorous; see §5.8.1.

We can also use the extrapolation error E_{2n} as a computable estimate of the error in a computation, and therefore use this to decide when an approximation is sufficiently accurate.

---

[10]Lewis Fry Richardson (1881–1953) was born in Newcastle upon Tyne in Great Britain, and attended several different schools before finishing his education at King's College, Cambridge, in 1903. His professional career spanned a number of different posts in industry, academia, and government science laboratories. He was the first person to suggest using mathematical techniques to predict the weather, by solving the fluid equations that would govern temperature, air pressure, etc. He first did this during World War I, while serving as an ambulance driver in France, long before the development of modern high-speed computers, and it was because of this that he developed the notion of extrapolation methods for the accurate numerical approximation of solutions based on cruder approximations.

Table 5.10 Richardson extrapolation applied to $f(x) = e^x$, $[a, b] = [0, 1]$.

n	$R_n(f)$	$|I(f) - R_n(f)|$	$|I(f) - T_n(f)|$	$|E_n|$
8	1.718272532150	0.929631E–05	0.223676E–02	0.224606E–02
16	1.718281246223	0.582236E–06	0.559300E–03	0.559882E–03
32	1.718281792050	0.364088E–07	0.139832E–03	0.139868E–03
64	1.718281826183	0.227585E–08	0.349584E–04	0.349607E–04
128	1.718281828317	0.142247E–09	0.873962E–05	0.873977E–05
256	1.718281828450	0.889089E–11	0.218491E–05	0.218492E–05
512	1.718281828458	0.553779E–12	0.546227E–06	0.546228E–06
1024	1.718281828459	0.364153E–13	0.136557E–06	0.136557E–06
2048	1.718281828459	0.266454E–14	0.341392E–07	0.341392E–07

■ **EXAMPLE 5.15**

Consider the integral

$$I(f) = \int_0^1 e^{-x^3} dx.$$

There is no closed-form antiderivative for the integrand, so there is no way to compare our approximation with an exact solution. However, if we compute Simpson's rule values for this integral, we get

$$I_{16}(f) = 0.807512351889, \quad I_{32}(f) = 0.807511254956,$$

for which the estimated error is

$$E_{32} = 0.731 \times 10^{-7}.$$

We thus can say, with some confidence, that $|I(f) - I_{32}(f)| \leq 10^{-7}$, because the estimated error is that small. We expect, of course, that the extrapolated value $R_{32} = 0.807511182$ would be even more accurate, so we use E_{2n}, which is a computable estimate of the error $I(f) - I_{2n}(f)$, to tell us when to accept the extrapolated value $R_{2n}(f)$ as the approximate value of the integral.

Exercises:

1. Apply Simpson's rule with $h = \frac{1}{2}$ and $h = \frac{1}{4}$ to approximate the integral

$$I = \int_0^1 \frac{1}{\sqrt{1 + x^4}} dx = 0.92703733865069,$$

and use Richardson extrapolation to obtain the improved value of the approximation. What is the *estimated* value of the error in S_4, compared to the actual error?

2. Repeat Problem 1, for

$$I = \int_0^1 x(1 - x^2) dx = \frac{1}{4}.$$

3. Repeat Problem 1, for

$$I = \int_0^1 \ln(1+x)dx = 2\ln 2 - 1.$$

4. Repeat Problem 1, for

$$I = \int_0^1 \frac{1}{1+x^3}dx = \frac{1}{3}\ln 2 + \frac{1}{9}\sqrt{3}\pi.$$

5. Repeat Problem 1, for

$$I = \int_1^2 e^{-x^2}dx = 0.1352572580.$$

6. Write a trapezoid rule program to compute the value of the integral

$$I = \int_0^1 e^{-x^2}dx.$$

Take h small enough to justify a claim of accuracy to within 10^{-6}, and explain how the claim is justified. (There are several ways of doing this.)

7. Define

$$I(f) = \int_0^1 e^x dx$$

and consider the approximation of this integral using Simpson's rule together with extrapolation. By computing a sequence of approximate values S_2, S_4, S_8, and so on, and analyzing the ratios of the errors, determine experimentally the accuracy of the extrapolated rule

$$R_{2n} = (16S_{2n} - S_n)/15.$$

8. Consider the integral

$$I(f) = \int_0^\pi \sin^2 x \, dx = \frac{1}{2}\pi.$$

Write a trapezoid rule or Simpson's rule program to approximate this integral, using Richardson extrapolation to improve the approximations, and comment on your results. In particular, comment upon the rate at which the error decreases as h decreases, and on the amount of improvement obtained by extrapolation.

9. Repeat the above, this time for the integral

$$I(f) = \int_0^{3\pi/4} \sin^2 x \, dx = \frac{1}{4} + \frac{3}{8}\pi.$$

10. The data in Table 5.11 supposedly comes from applying the midpoint rule to a smooth, non-periodic function. Can we use this data to determine whether or not the program is working properly? Explain.

Table 5.11   Data for Problem 10.

n	$M_n(f)$
4	–0.91595145
8	–0.95732875
16	–0.97850187
32	–0.98921026
64	–0.99459496
128	–0.99729494
256	–0.99864683
512	–0.99932326
1024	–0.99966159

11. The error function, defined as

$$\text{erf}(x) = \frac{2}{\sqrt{\pi}} \int_0^x e^{-t^2} dt,$$

is an important function in probability theory and heat conduction. Use Simpson's rule with extrapolation to produce values of the error function that are accurate to within 10^{-8} for x ranging from 0 to 5 in increments of $1/4$. Check your values against the intrinsic error function on your computer or by looking at a set of mathematical tables such as *Handbook of Mathematical Functions* [1].

12. Bessel functions appear in the solution of heat conduction problems in a circular or cylindrical geometry, and can be defined as a definite integral, thus:

$$J_k(x) = \frac{1}{\pi} \int_0^\pi \cos(x \sin t - kt) dt.$$

Use Simpson's rule plus extrapolation to produce values of J_0 and J_1 that are accurate to within 10^{-8} over the interval $[0, 6]$ in increments of $1/4$. Check your values by looking at a set of tables; for example, *Handbook of Mathematical Functions* [1].

13. Apply the trapezoid rule to approximate the integral

$$I = \int_0^{\pi^2/4} \sin \sqrt{x} \, dx = 2.$$

Use Richardson extrapolation to estimate how rapidly your approximations are converging, and comment on your results in light of the theory for the trapezoid rule.

14. Show that for any function f,

$$S_2(f) = (4T_2(f) - T_1(f))/3.$$

Comment on the significance of this result in light of this section.

◁ ● ● ▷

5.8 SPECIAL TOPICS IN NUMERICAL INTEGRATION

5.8.1 Romberg Integration

If we specialize now to the case of the trapezoid rule, we can make much of §5.7 more precise, and, at the same time, develop a very accurate recursive procedure for approximating integrals. A fundamental result is the Euler–Maclaurin formula,[11] which we now state, without proof.

Theorem 5.6 (Euler–Maclaurin Formula) *If f is sufficiently differentiable, then, for any $N > 0$ there exists a set of constants c_k, $1 \leq k \leq N + 1$, such that, for some $\xi \in [a, b]$, the error in the trapezoid rule satisfies*

$$
\begin{aligned}
I(f) - T_n(f) \quad = \quad & \gamma_1 h^2 + \gamma_2 h^4 + \cdots + \gamma_N h^{2N} \\
& + c_{N+1}(b - a)h^{2N+2} f^{(2N+2)}(\xi),
\end{aligned} \tag{5.17}
$$

where $\gamma_k = c_k(f^{(2k-1)}(b) - f^{(2k-1)}(a))$.

The significance of the Euler–Maclaurin formula is that it allows us to write the error in the trapezoid rule as a series of powers of the mesh spacing h, and then use this series to derive new quadrature rules that are more and more accurate. Note that the series expansion of the error can be carried out as far as f is differentiable, but has to be terminated; in general, it cannot be taken as an infinite series.

The first few constants are

$$
c_1 = -\frac{1}{12}, \quad c_2 = \frac{1}{720}, \quad c_3 = -\frac{1}{30,240}.
$$

There is a general formula relating these to the so-called Bernoulli numbers.

One consequence of the Euler–Maclaurin formula is that the trapezoid rule is shown to be extraordinarily accurate when applied to periodic functions over full periods. In this case, the derivatives at the endpoints will be equal, and thus (5.17) becomes

$$
I(f) - T_n(f) \quad = \quad c_{N+1}(b - a)h^{2N+2} f^{(2N+2)}(\xi), \tag{5.18}
$$

where N is arbitrary, restrained only by the smoothness of f. See Problem 10 for an illustration of this.

Note that if we look at the $N = 1$ case, we have that

$$
I(f) - T_n(f) \quad = \quad -\frac{1}{12}\left(f'(b) - f'(a)\right)h^2 + c_2(b - a)h^4 f^{(4)}(\xi),
$$

which shows that the corrected trapezoid rule is $\mathcal{O}(h^4)$, which we saw experimentally in §5.2.

To construct an even more accurate quadrature rule from the error expansion, simply write down (5.17) for twice as many subintervals (replace h by $h/2$):

$$
I(f) - T_{2n}(f) \quad = \quad \frac{1}{4}\gamma_1 h^2 + \frac{1}{16}\gamma_2 h^4 + \cdots + \frac{1}{4^N}\gamma_N h^{2N} + \mathcal{O}(h^{2N+2}). \tag{5.19}
$$

---

[11]Maclaurin developed the Euler–Maclaurin formula independently of Euler, in about 1737, but did not publish it prior to including it in his book, *Treatise of fluxions*.

If we multiply (5.19) by 4, subtract it from (5.17), and then solve for $I(f)$, we get

$$I(f) - \frac{4T_{2n}(f) - T_n(f)}{3} = b_2 h^4 + b_3 h^6 + \cdots + b_N h^{2N} + \mathcal{O}(h^{2N+2}), \quad (5.20)$$

where $b_k = -\frac{1}{3}\left(1 - 4^{1-k}\right) c_k (f^{(2k-1)}(b) - f^{(2k-1)}(a))$.

Note what we have done here. The value $\frac{1}{3}(4T_{2n}(f) - T_n(f))$ is nothing more than Richardson's extrapolated value (5.15) for $T_n(f)$, and the expression (5.20) shows that it is $\mathcal{O}(h^4)$ accurate. But there is no reason to stop here. We rewrite (5.20) as

$$I(f) - R_{2n}(f) = b_2 h^4 + b_3 h^6 + \cdots + b_N h^{2N} + \mathcal{O}(h^{2N+2}), \quad (5.21)$$

for which we also then have

$$I(f) - R_{4n}(f) = \frac{1}{16} b_2 h^4 + \frac{1}{128} b_3 h^6 + \cdots + \frac{1}{4^N} b_N h^{2N} + \mathcal{O}(h^{2N+2}), (5.22)$$

so that, multiplying (5.22) now by 16, subtracting from (5.21), and solving for $I(f)$ now yields

$$I(f) - \frac{16R_{4n} - R_{2n}}{15} = a_3 h^6 + \cdots + a_k h^{2k} + \cdots + a_N h^{2N} + \mathcal{O}(h^{2N+2})$$

where $a_k = -\frac{1}{15}(1 - 4^{2-k})b_k$. Thus, $\frac{1}{15}(16R_{4n} - R_{2n})$ can be viewed as yet another extrapolated approximation for the integral, and this one is $\mathcal{O}(h^6)$ accurate.

There is no reason not to keep going. Each step of the *extrapolation* process yields a new quadrature method that is more accurate than the preceding one. The process can be systematized to yield an algorithm known as *Romberg integration*.[12] Continuing much farther requires that we define some notation.

Let $T_n^{(0)}(f)$ denote the trapezoid rule values; this is the first column of what will become a triangular array. We denote the second column by $T_n^{(1)}(f)$. These values are computed from the trapezoid values according to the Richardson extrapolation formula:

$$T_{2n}^{(1)}(f) = R_{2n}(f) = \frac{4T_{2n}^{(0)}(f) - T_n^{(0)}(f)}{3}.$$

Note that the indexing means that the second column will be one entry shorter than the first. Generally, then, each column is computed from the preceding one according to the formula

$$T_{2n}^{(j+1)}(f) = \frac{4^{j+1}T_{2n}^{(j)}(f) - T_n^{(j)}(f)}{4^{j+1} - 1}, \quad (5.23)$$

---

[12]Werner Romberg (1909–2003) was born in Berlin, and educated at Ludwig-Maximilian-Universität in Munich, where he got his doctorate in 1933. In 1938, he joined the faculty of the University of Oslo; he spent most of the rest of his working life in Norway. In 1949, he joined the Norwegian Institute of Technology in Trondheim as an associate professor of physics. His paper on what we call Romberg integration was published in 1955. It was not until the late 1960s that the method attracted a lot of attention. See the paper by Jacques Dutka, "Richardson extrapolation and Romberg integration," *Hist. Math.*, vol. 11 (1984), pp. 3–21, for some of the history of this method.

and the array looks like this:

$$T_n^{(0)}(f)$$
$$T_{2n}^{(0)}(f) \quad T_{2n}^{(1)}(f)$$
$$T_{4n}^{(0)}(f) \quad T_{4n}^{(1)}(f) \quad T_{4n}^{(2)}(f)$$
$$\vdots \qquad \vdots \qquad \vdots \qquad \ddots$$
$$T_{2^k n}^{(0)}(f) \quad T_{2^k n}^{(1)}(f) \quad \cdots \quad \cdots \quad T_{2^k n}^{(k)}(f).$$

If the integrand is smooth enough, then each step across a row of the array is eliminating another power of h^2 from the error expansion; i.e., the first column has error that is $\mathcal{O}(h^2)$, the second column error is $\mathcal{O}(h^4)$, the third column error is $\mathcal{O}(h^6)$, and so on. Meanwhile, going *down* the array is decreasing h by a factor of 2 for each row: If we start with $h = \frac{1}{2}$, then in the sixth row we have $h = 1/64$. The upshot of all this is that the diagonal elements of the Romberg array are very accurate approximations to the integral. A formal statement is the following, which we present without proof (see p. 328 of [3]).

Theorem 5.7 (Romberg Integration) *Assume that f is sufficiently differentiable on $[a, b]$, and let $\Theta_k(f)$ be the kth diagonal element of the Romberg array: $\Theta_k(f) = T_{2^k n}^{(k)}(f)$. If the mesh size for the initial trapezoid rule $T_n^{(0)}$ was h, then*

$$I(f) - \Theta_k(f) = \mathcal{O}\left(4^{-k} h^{2k+2}\right).$$

In addition, at the same time that Romberg integration is producing accurate approximations, it can also be used to estimate the error, since for each entry in the Romberg array we can compute an estimate of the error using Richardson extrapolation:

$$I(f) - T_{2n}^{(k)} = E_{2n}^{(k)} = \frac{T_{2n}^{(k)} - T_n^{(k)}}{4^k - 1}.$$

(This assumes, of course, that the integrand is sufficiently smooth.) This can be used to stop the Romberg process when the estimated error is sufficiently small.

Note that the bulk of the computational work in Romberg integration is involved in computing the first column—the trapezoid rule values—since the computation of the subsequent columns, done by (5.23), involves only a few simple operations. The ultimate efficiency therefore depends on our ability to rapidly compute the trapezoid rule values recursively, i.e., compute $T_{2n}^{(0)}(f)$ from $T_n^{(0)}(f)$ without wasted effort. Fortunately, this is very easy, as the next result shows.

Theorem 5.8 *Let $T_n(f)$ denote the trapezoid rule applied to a given function f over a given interval $[a, b]$, using n subintervals, with uniform mesh spacing $h = (b - a)/n$. Then we can compute $T_{2n}(f)$, the trapezoid rule using twice as many subintervals, according to*

$$T_{2n}(f) = \frac{1}{2} T_n(f) + \left(\frac{b - a}{2n}\right) \sum_{j=1}^{n} f(a + (2j - 1)(b - a)/(2n)). \qquad (5.24)$$

Proof: The key step is to recognize that we can write the trapezoid rule for n subintervals as

$$T_n(f) = \left(\frac{b - a}{n}\right) \left(\frac{1}{2} f(a) + \frac{1}{2} f(b) + \sum_{i=1}^{n-1} f(a + i(b - a)/n)\right).$$

Thus, the rule for $2n$ subintervals is

$$T_{2n}(f) = \left(\frac{b-a}{2n}\right)\left(\frac{1}{2}f(a) + \frac{1}{2}f(b) + \sum_{i=1}^{2n-1} f(a + i(b-a)/(2n))\right).$$

Now manipulate:

$$
\begin{aligned}
T_{2n}(f) &= \left(\frac{b-a}{2n}\right)\left(\frac{1}{2}f(a) + \frac{1}{2}f(b) + \sum_{i=1}^{2n-1} f(a + i(b-a)/(2n))\right), \\
&= \frac{1}{2}\left(\frac{b-a}{n}\right)\left(\frac{1}{2}f(a) + \frac{1}{2}f(b) + \sum_{i=2,4,6,\dots}^{2n-1} f(a + i(b-a)/(2n))\right. \\
&\qquad\qquad \left. + \sum_{i=1,3,5,\dots}^{2n-1} f(a + i(b-a)/(2n))\right), \\
&= \frac{1}{2}\left(\frac{b-a}{n}\right)\left(\frac{1}{2}f(a) + \frac{1}{2}f(b) + \sum_{j=1}^{n-1} f(a + 2j(b-a)/(2n))\right. \\
&\qquad\qquad \left. + \sum_{j=1}^{n} f(a + (2j-1)(b-a)/(2n))\right), \\
&= \frac{1}{2}\left(\frac{b-a}{n}\right)\left(\left[\frac{1}{2}f(a) + \frac{1}{2}f(b) + \sum_{j=1}^{n-1} f(a + j(b-a)/n)\right]\right. \\
&\qquad\qquad \left. + \sum_{j=1}^{n} f(a + (2j-1)(b-a)/(2n))\right), \\
&= \frac{1}{2}T_n(f) + \left(\frac{b-a}{2n}\right)\left(\sum_{j=1}^{n} f(a + (2j-1)(b-a)/(2n))\right),
\end{aligned}
$$

and we are done. ●

The point of this result is that we can compute the entire first column of the Romberg array using the minimum number of function evaluations. A naive implementation, in which $T_{2n}^{(0)}(f)$ was computed from scratch, would require that we recompute n values of f that we had already computed in finding $T_n^{(0)}(f)$. As a result of this theorem, we can compute $T_{2^k n}^{(k)}(f)$ using no more function evaluations than for the ordinary trapezoid rule, yet we achieve much more accuracy.

Table 5.12 shows the application of Romberg integration to our standard example of integrating e^x over the interval $[0, 1]$. Note the extremely rapid decay of the error.

We close this section with an algorithm (in outline form) for Romberg integration.

Table 5.12   Romberg integration of $f(x) = e^x$, $[a, b] = [0, 1]$.

$n = 2^k$	$\Theta_k(f)$	$I(f) - \Theta_k(f)$
1	1.85914091422952	0.14085908577048e+00
2	1.71886115187659	0.57932341754729e–03
4	1.71828268792476	0.85946571215523e–06
8	1.71828182879453	0.33548497313518e–09
16	1.71828182845908	0.32640556923980e–13
32	1.71828182845905	0.44408920985006e–15
64	1.71828182845905	–0.44408920985006e–15

Algorithm 5.2 *Romberg Integration*

1. *Compute $T_n^{(0)}(f) = T_n(f)$; this is the initial trapezoid rule computation, using n subintervals. Often $n = 1$, but this is not required.*

2. *For k from 1 to N do*

   (a) *Compute $T_{2^k n}(f) = T_{2^k n}^{(0)}(f)$; this is the first entry on a new row of the Romberg array.*

   (b) *Extrapolate across the row: for j from 0 to $k - 1$ do*

      i. *Compute $T_{2^k n}^{(j+1)}(f) = (4^{j+1} T_{2^k n}^{(j)}(f) - T_{2^{k-1} n}^{(j)}(f))/(4^{j+1} - 1)$.*

Programming Hint: Note that it is easy to encode this using only enough storage for two rows in the Romberg array; we don't need to store the entire array. In fact, if we are careful, we can get away with storing only a single row of the array and overwriting it with the new values as they are computed.

■ **EXAMPLE 5.16**

To illustrate the computation with a minimum of extraneous effort, let's refer to Table 5.13, which gives trapezoid rule values (i.e., $T_n^{(0)}(f)$ values) for approximating

$$I(f) = \int_0^1 \frac{dx}{1 + x^4} = 0.8669729871.$$

Table 5.13   Trapezoid rule integration of $f(x) = (1 + x^4)^{-1}$, $[a, b] = [0, 1]$.

$n = 2^k$	$T_k(f)$
1	0.7500000000
2	0.8455882353
4	0.8617323343

The first extrapolation produces the first Romberg value:

$$\Theta_1(f) = T_2^{(1)}(f) = \frac{4T_2^{(0)}(f) - T_1^{(0)}(f)}{3} = 0.8774509800.$$

It takes two extrapolations to produce the second Romberg value:

$$T_4^{(1)}(f) = \frac{4T_4^{(0)}(f) - T_2^{(0)}(f)}{3} = 0.8616289989,$$

$$\Theta_2(f) = T_4^{(2)}(f) = \frac{16T_4^{(1)}(f) - T_2^{(1)}(f)}{15} = 0.8605742004.$$

The Richardson error estimate for $T_4^{(1)}(f)$ and $T_2^{(1)}(f)$ is given by

$$E_4^{(1)}(f) = \frac{T_2^{(1)}(f) - T_4^{(1)}(f)}{15} = 0.00105479874;$$

thus, we are confident that the Romberg value $\Theta_2(f)$ is accurate to within about 10^{-3} or so, even though it was produced with only *five* function evaluations.

5.8.2 Quadrature with Non-smooth Integrands

(Much of this section is drawn from material in Kendall Atkinson's text, *An Introduction to Numerical Analysis*, 2nd Edition, Wiley, New York, 1989. See also de Doncker and Piessens [4].)

So far we have assumed, in most of our developments, that the function being integrated (the *integrand*) was as smooth as we needed it to be. But this will not always be the case, and we need to understand the implications of this.

We begin with an example. Let $f(x) = \sqrt{x}$ and consider

$$I(f) = \int_0^1 \sqrt{x}\,dx = \frac{2}{3}.$$

Let's apply trapezoid, Simpson, and Gaussian quadrature to this integral; the results are summarized in Table 5.14. Note that none of the three methods is converging to the exact value as rapidly as we would expect, based on our previous examples. Why?

Recall that all of our error estimates include a factor that is based on the evaluation of some derivative of the integrand at an unspecified point in the interval of integration. For the trapezoid rule, we require that f be twice continuously differentiable; for Simpson's rule we require that it be four times continuously differentiable; and for the n-point Gaussian rule we require that f be $2n$ times continuously differentiable. However, on the interval $[0, 1]$, $f(x) = \sqrt{x}$ does not even have a continuous *first* derivative. The practical effect of this is that the contribution to the error estimate from the derivative of f can become arbitrarily large as we take more and more points. This growth will not be large enough to prevent convergence to the correct answer, but it will be large enough to destroy the expected convergence *rate*. Note that this is not a violation of our carefully developed error theory; the integrand in this case does not satisfy the hypotheses of the error theorems, and so they do not apply.

An integrand that exhibits the kind of behavior shown in our example is called *singular*, with $x = 0$ being the specific *point of singularity*. In this particular case, we can actually

Table 5.14   Integration of $f(x) = x^{1/2}$, $[a, b] = [0, 1]$.

n	$T_n(f)$	$S_n(f)$	$G_n(f)$
4	0.643283046243e+00	0.656526264793e+00	0.667827645375e+00
8	0.658130221624e+00	0.663079280085e+00	0.666835580100e+00
16	0.663581196877e+00	0.665398188628e+00	0.666689631499e+00
32	0.665558936279e+00	0.666218182746e+00	0.666669667368e+00
64	0.666270811379e+00	0.666508103078e+00	0.666667050398e+00
128	0.666525657297e+00	0.666610605936e+00	0.666666715190e+00
256	0.666616548977e+00	0.666646846203e+00	0.666666672768e+00
512	0.666648881550e+00	0.666659659074e+00	0.666666667431e+00
1024	0.666660362219e+00	0.666664189109e+00	0.666666666762e+00
2048	0.666664433593e+00	0.666665790718e+00	0.666666666679e+00

remove the problem by making the change of variable $x = u^2$ in the integral. However, this sort of trick will not always work,[13] so we must be prepared to deal with singular integrands in a more sophisticated manner.

To be specific, we consider the integral

$$I(f) = \int_a^b f(x)dx,$$

where f is assumed to be smooth on the open interval (a, b), but to have some kind of singular (non-smooth) behavior at one or both of the endpoints. Introduce the change of variable

$$x = \phi(t), \quad \phi(t) = a + \frac{b-a}{\gamma}\Psi(t), \quad \Psi(t) = \int_{-1}^t \psi(u)du, \tag{5.25}$$

where

$$\psi(t) = \exp\left(-\frac{c}{1-t^2}\right), \quad \gamma = \Psi(1)$$

and $c > 0$ is arbitrary.

This probably looks rather imposing, so let's go over it a bit more carefully. The first point to make is that ψ has been constructed to be extremely smooth near $t = \pm 1$; note that the argument to the exponential goes to $-\infty$ as $t \to \pm 1$; thus, ψ and *all* its derivatives will vanish at the endpoints. This means that Ψ and hence ϕ will be very smooth at the endpoints: smooth enough to compensate for almost any non-smooth behavior in f. Second, the change of variables from x to t will map the original interval $[a, b]$ to $[-1, 1]$. In fact, the new integral becomes (Problem 5)

$$I(f) = \int_a^b f(x)dx = \int_{-1}^1 f(\phi(t))\phi'(t)dt.$$

Since all derivatives of ϕ vanish at the new endpoints, it follows from the Euler–Maclaurin formula that the trapezoid rule will be a very accurate quadrature rule to use here. The

---

[13]Although it is always a good idea to consider a clever change of variable as a way to avoid a singularity.

involved nature of the change of variable makes the application of the trapezoid rule appear difficult, but it really isn't. Let $\{t_k\}$ be a set of $n+1$ equally spaced grid points on $[-1, 1]$,

$$-1 = t_0 < t_1 < t_2 < \cdots < t_n = 1$$

with $t_k = -1 + 2/n$, $0 \le k \le n$. Then the trapezoid rule applied to the transformed integral becomes (remember that the integrand vanishes at the endpoints)

$$I(f) \approx I_n(f) = \frac{b-a}{2} \sum_{k=1}^{n-1} w_k f(x_k),$$

where

$$x_k = \phi(t_k), \quad w_k = \frac{4}{n\gamma} \exp\left(-\frac{c}{1-t_k^2}\right).$$

The evaluation of the weights is a straightforward calculation with the exponential function. The evaluation of the x_k values requires the computation of $\phi(t_k)$, which is itself an integral. However, the values of that integral can be computed once and saved, much as the weights and Gauss points from Gaussian quadrature, and thus this is only a momentary difficulty. Alternatively, we could use the approximation techniques from Chapter 4 to construct a highly accurate polynomial approximation to Ψ, and simply evaluate that as needed. Table 5.15 shows the values of $\Psi(t)/\Psi(1)$ for the $n = 4, 8$, and 16 cases, computed using Romberg integration.

■ **EXAMPLE 5.17**

To illustrate the method, consider the following example:

$$I(f) = \int_0^1 \left(\log \frac{1}{x}\right)^{\frac{1}{2}} dx = \frac{\sqrt{\pi}}{2} = 0.8862269255.$$

This has singularities at both endpoints, since $\log 0$ is undefined and the square root will have undefined derivatives at $x = 1$ since $\log 1 = 0$. The $n = 4$ version of this quadrature is written (using $c = 4$) quite simply as

$$I_4(f) = \frac{1}{2}\left(\frac{1}{\gamma}\exp\left(\frac{-4}{1-t_1^2}\right)\left(\log \frac{1}{x_1}\right)^{1/2} + \frac{1}{\gamma}\exp\left(\frac{-4}{1-t_2^2}\right)\left(\log \frac{1}{x_2}\right)^{1/2}\right.$$
$$\left. + \frac{1}{\gamma}\exp\left(\frac{-4}{1-t_3^2}\right)\left(\log \frac{1}{x_3}\right)^{1/2}\right).$$

Now we have (from (5.25) and Table 5.15)

$$x_1 = \Psi(t_1)/\Psi(1) = 0.3175495761647776 \times 10^{-1},$$
$$x_2 = \Psi(t_2)/\Psi(1) = 0.5000000000000000,$$
$$x_3 = \Psi(t_3)/\Psi(1) = 0.9682450390677892,$$

and

$$\gamma = \int_{-1}^1 \exp\left(\frac{-4}{1-t^2}\right) dt = 1.4059716861223 \times 10^{-2},$$

Table 5.15 Table of $\Psi(t_k)/\Psi(1)$ for $-1 \le t_k \le 1$.

n	k	t_k	$\Psi(t_k)/\Psi(1)$
4	0	−0.1000000000000000E+01	0.0000000000000000E+00
	1	−0.5000000000000000E+00	0.3175495761647776E−01
	2	0.0000000000000000E+00	0.5000000000000001E+00
	3	0.5000000000000000E+00	0.9682450390677892E+00
	4	0.1000000000000000E+01	0.1000000000000000E+01
8	0	−0.1000000000000000E+01	0.0000000000000000E+00
	1	−0.7500000000000000E+00	0.1962922276951938E−03
	2	−0.5000000000000000E+00	0.3175495761647776E−01
	3	−0.2500000000000000E+00	0.2004317448343370E+00
	4	0.0000000000000000E+00	0.5000000000000001E+00
	5	0.2500000000000000E+00	0.7995682518334509E+00
	6	0.5000000000000000E+00	0.9682450390677892E+00
	7	0.7500000000000000E+00	0.9998037051191319E+00
	8	0.1000000000000000E+01	0.1000000000000000E+01
16	0	−0.1000000000000000E+01	0.0000000000000000E+00
	1	−0.8750000000000000E+00	0.1930534427638750E−07
	2	−0.7500000000000000E+00	0.1962922276951938E−03
	3	−0.6250000000000000E+00	0.5462420695964468E−02
	4	−0.5000000000000000E+00	0.3175495761647776E−01
	5	−0.3750000000000000E+00	0.9512852921910396E−01
	6	−0.2500000000000000E+00	0.2004317448343370E+00
	7	−0.1250000000000000E+00	0.3405226277340978E+00
	8	0.0000000000000000E+00	0.5000000000000001E+00
	9	0.1250000000000000E+00	0.6594773707084767E+00
	10	0.2500000000000000E+00	0.7995682518334509E+00
	11	0.3750000000000000E+00	0.9048714673344923E+00
	12	0.5000000000000000E+00	0.9682450390677892E+00
	13	0.6250000000000000E+00	0.9945375763409888E+00
	14	0.7500000000000000E+00	0.9998037051191319E+00
	15	0.8750000000000000E+00	0.9999999792694836E+00
	16	0.1000000000000000E+01	0.1000000000000000E+01

so the approximation becomes

$$
\begin{aligned}
I_4(f) &= \left(\frac{0.5}{1.4059716861223 \times 10^{-2}}\right) \times (0.00482794999383144 \times 1.85733853145623 \\
&\quad + 0.0183156388887342 \times 0.8325546111576978 \\
&\quad + 0.00482794999383144 \times 0.179638760313930) \\
&= \frac{0.5 \times 0.0250831941163514}{1.4059716861223 \times 10^{-2}} \\
&= 0.892023444139597,
\end{aligned}
$$

for which the error is approximately 0.58×10^{-2}. Table 5.16 shows the result of applying the algorithm to this example (with $c = 4$) for larger values of n. Note that the high accuracy of the trapezoid rule for this kind of example is borne out. Note also that the accuracy stagnates a bit for the finer meshes. This is probably due to inaccuracies in the computation of the $\Psi(t_k)/\Psi(1)$ values; to avoid this problem we would need to compute these in very high precision.

Table 5.16   Singular quadrature method applied to $f(x) = \left(\log \frac{1}{x}\right)^{\frac{1}{2}}$, $[a, b] = [0, 1]$.

n	$I_n(f)$	$I(f) - I_n(f)$
4	0.892023444140	0.579652E–02
8	0.885813277564	0.413648E–03
16	0.886218502855	0.842260E–05
32	0.886226915463	0.999072E–08
64	0.886226924278	0.117666E–08
128	0.886226924425	0.102882E–08
256	0.886226924497	0.957054E–09
512	0.886226924526	0.928343E–09
1024	0.886226924527	0.925976E–09

Although this method was designed to work on finite intervals with integrands having endpoint singularities, it can be easily modified to compute integrals on infinite intervals as well. Suppose that we want to compute

$$I(f) = \int_a^\infty f(x)dx.$$

The change of variable

$$x = a + \frac{1+z}{1-z}$$

gives us the new integral

$$I(f) = \int_a^\infty f(x)dx = \int_{-1}^1 F(z)dz,$$

where $F(z) = 2f(a + (1+z)(1-z)^{-1})(1-z)^2$. This new integral now has the type of endpoint singularity (at $z = 1$), that the previous development was designed to attack, so we can apply the same ideas to get the quadrature rule

$$I(f) \approx I_n(f) = \frac{b-a}{2} \sum_{k=1}^{n-1} W_k F(z_k) \qquad (5.26)$$

with

$$z_k = \phi(t_k), \quad W_k = \frac{4}{n\gamma} \exp\left(-\frac{c}{1-t_k^2}\right), \qquad (5.27)$$

where $\phi(t)$ is now given by

$$\phi(t) = -1 + \frac{2}{\gamma}\Psi(t).$$

■ **EXAMPLE 5.18**

Consider now the following example.

$$I(f) = \int_0^\infty e^{-x^2} dx = \frac{1}{2}\sqrt{\pi}.$$

Table 5.17 shows the approximations as computed by (5.26)–(5.27). Note that the error again stagnates.

Table 5.17 Singular quadrature method applied to $f(x) = e^{-x^2}$, $[a, b] = [0, \infty)$.

n	$I_n(f)$	$I(f) - I_n(f)$
4	1.141419627988	0.255193E+00
8	0.939027865252	0.528009E–01
16	0.884525323277	0.170160E–02
32	0.886229670193	0.274474E–05
64	0.886226922961	0.249179E–08
128	0.886226922944	0.250830E–08
256	0.886226922943	0.250951E–08
512	0.886226922942	0.251117E–08
1024	0.886226922941	0.251128E–08

5.8.3 Adaptive Integration

It is possible—rather easy, in fact—to write a general-purpose quadrature routine based on any of the methods discussed so far in this chapter. However, we often want to have firm knowledge that the error is "small enough" for the needs of the application. We could write a code that simply takes more and more points until some computable estimate of the error (such as we get from Richardson extrapolation) is $\leq \tau$, where τ is a user-supplied tolerance. Romberg integration is perhaps the best choice for this, incidentally.

However, such a process is often needlessly inefficient, for the simple reason that we do not want to take *too many* points in those regions of the interval where the integrand is smooth and therefore high accuracy is easily achieved. Consider the graph in Figure 5.7; to accurately integrate this function over the interval (say) $[-5, 5]$, we will need to use many points near the origin, but fewer away from it. Using something like Romberg integration will put a lot of points uniformly throughout the interval, thus wasting effort.

For this reason, most automatic quadrature routines are *adaptive*; that is, they do not work with a uniform mesh, but adjust the mesh points to achieve a given overall accuracy. There are many ways to accomplish this task; what we outline here is one of the simpler techniques.

At the core of adaptive methods (whether applied to numerical integration or other areas of computation) is the use of two (or more, but usually two) approximations to produce a computable estimate of the error in order to gauge the accuracy of the computation. Roughly speaking, if this computable estimate of the error is small enough, then we cease the calculation and move on. In the development presented here, we will use the Richardson extrapolation results from §5.7 to estimate the error.

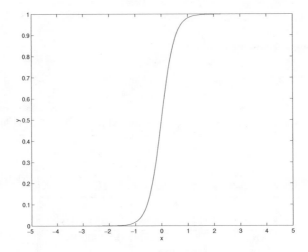

Figure 5.7   Function with rapid change near the origin.

An illustration might be useful at this point. Suppose we wish to compute

$$I = \int_a^b f(x)dx$$

to an accuracy of τ; that is, we want to find an approximate value $Q \approx I$ such that $|I - Q| \le \tau$. We will use Simpson's rule as the quadrature rule at the core of our method—many other choices are possible.

We start by computing two different approximations to the integral, one based on applying Simpson's rule over the entire interval, one based on applying it over two subintervals. Thus, we have

$$S_1 = \frac{(b-a)/2}{3}\left(f(a) + 4f((a+b)/2) + f(b)\right)$$

and

$$\begin{aligned} S_2 &= \frac{(b-a)/4}{3}\left(f(a) + 4f((a+b)/4) + f((a+b)/2)\right) \\ &+ \frac{(b-a)/4}{3}\left(f((a+b)/2) + 4f(3(a+b)/4) + f(b)\right). \end{aligned}$$

We can apply Richardson extrapolation to estimate the absolute error in S_2:

$$E_2 = |S_2 - S_1|/15.$$

If $E_2 \le \tau$ then we use the Richardson extrapolated value $R_2 = (16S_2 - S_1)/15$ as our approximation to I, and we are confident that this approximation is to the specified accuracy.[14]

However, we do not really expect to compute our integral to any realistic accuracy with only two approximations; we are going to have to continue the process. Basically, we next

---

[14]Note that we are only "confident," not "certain." Since we are using a computable *estimate* of the error, we do not *know* that the error is less than τ.

try to compute, accurately, half of the integral,

$$I' = \int_{(a+b)/2}^{b} f(x)dx$$

using the same ideas, then half of this integral, and so on. Describing this in any sort of precise, formal manner, that is ammenable to being put into an algorithm, is going to require some notation. Define

$$J(\alpha, \beta) = \int_{\alpha}^{\beta} f(x)dx;$$

thus, the value we want to compute is $I = J(a, b)$. We apply a similar notation to the Simpson's rule approximations:

$$S(\alpha, \beta) = \frac{(\beta - \alpha)/2}{3} \left[f(\alpha) + 4f((\alpha + \beta)/2) + f(\beta) \right].$$

It will also be useful to define $\theta_k = 2^{-k+1}(b - a)$, so $\theta_1 = (b - a)$, $\theta_2 = (b - a)/2$, and so on. Thus, our first step involved the computation of

$$S_1 = S(a, b) = S(b - \theta_1, b)$$

and

$$\begin{aligned} S_2 &= S(a, (a + b)/2) + S((a + b)/2, b) \\ &= S(b - \theta_1, b - \theta_2) + S(b - \theta_2, b). \end{aligned}$$

The complete algorithm continues this process, recursively, until we have computed *part* of the integral to the specified accuracy. It is simplest to organize the algorithm so that the values of the integral accumulate from right to left; i.e., we next try to approximate

$$J(b - \theta_2, b) = J((a + b)/2, b) = \int_{(a+b)/2}^{b} f(x)dx$$

using the two approximate values

$$S_1 = S(b - \theta_2, b) = S((a + b)/2, b)$$

and

$$S_2 = S(b - \theta_2, b - \theta_3) + S(b - \theta_3, b).$$

Once again we compute the Richardson extrapolation values $E = |S_1 - S_2|/15$ and $R = (16S_2 - S_1)/15$. The step is evaluated in the following way:

If $E \leq \frac{1}{2}\tau$, then we accept R as the approximate value of $J((a + b)/2, b)$.

We use $\frac{1}{2}\tau$ because the integral in question involves only *half* the full interval of integration. We continue this process, at the kth step trying to approximate $J(b - \theta_k, b)$ to an estimated accuracy of $2^{-k+1}\tau$.

Eventually, we will find a subinterval small enough that the desired fractional error tolerance is satisfied. When this happens we accumulate the approximate value of the integral and move on to try to compute the next portion of the integral.

A key element in making this an efficient process is the re-use of function values that were previously computed. As the computation proceeds, the program maintains a list of grid points and the associated function values that have already been computed. Since new points are introduced as the algorithm proceeds, the grid is getting redefined and re-indexed at each step. What makes this a difficult algorithm to encode (and explain!) is the recursion and the need for careful management of the data to ensure that we do not use any unnecessary function evaluations. This algorithm has been implemented in a number of places, including in the old FORTRAN package QUADPACK and in the MATLAB routine quad. In languages that support recursion it is much easier to encode, but much more difficult to follow.

■ **EXAMPLE 5.19**

Let's continue the illustration with a specific example, rather than a generic one. We will try to approximate the value of the integral

$$I = J(-5, 1) = \int_{-5}^{1} \frac{1}{1 + e^{-4x}} dx = 1.0045374814641640301,$$

which is the function graphed in Figure 5.7, using a tolerance of $\tau = 10^{-2}$. Thus, $f(x) = \frac{1}{1+e^{-4x}}$, $a = -5$, $b = 1$, and $\tau = 0.01$. (Since we are using such a crude error for this illustration, we will display our answers using only five digits, although the computations are being done using 10 digits.)

Older versions of MATLAB (circa 1994) included a pair of adaptive quadrature routines, quad and quad8. Modern (2012) versions of MATLAB include several more sophisticated adaptive routines.

The author's code produced the plot in Fig 5.8, and used a total of 29 function calls. The (very old) MATLAB code quad produced the plot in Fig 5.9, and used 109 function calls. This does not suggest that the author's programming skills are superior to those of the MATLAB staff but, rather, that adaptive routines, written to be general, can be led into wasting a lot of effort. The (old) routine quad8 produced the plot in Fig. 5.10, but used only 33 function evaluations.

If we repeat the same computation, but this time specify that $\tau = 10^{-4}$, we find that the (old) quad used 445 evaluations and that quad8 used 97.[15] (See Fig. 5.11.) Table 5.18 summarizes the results (the number of function evaluations needed to obtain a given accuracy) of running a variety of adaptive routines on this example.[16]

Table 5.18   Adaptive Quadrature

τ	quad (old)	quad8 (old)	quad (2012)	quadl (2012)
0.01	109	33	13	18
0.0001	445	97	29	48
1.e−6	1825	145	57	108
1.e−8	2049	289	121	168

---

[15]The author's program was lost when the author changed jobs in 2001, and so is not, alas, available for further testing. There seems little point in rewriting it, given the easy-to-use routines in MATLAB.

[16]The old version of quad reached a recursion limit and generated a couple of screens' worth of warning messages, before reporting an accurate value of the integral and the function evaluation count of 2049.

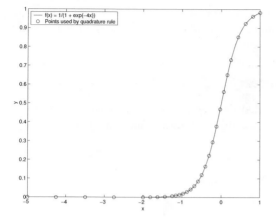

Figure 5.8   Adaptive quadrature (author's routine).

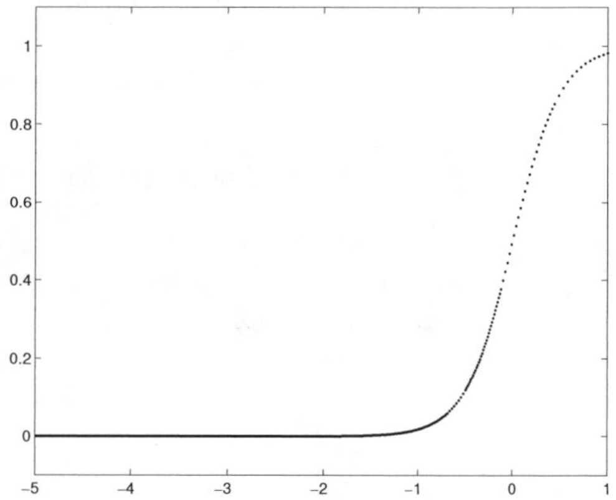

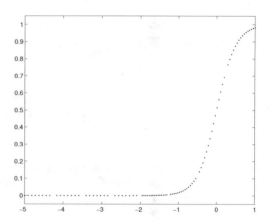

Figure 5.9   Adaptive quadrature using MATLAB's quad routine (very old version).

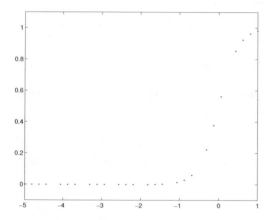

Figure 5.10   Adaptive quadrature using MATLAB's quad8 routine (very old version).

Figure 5.11   Adaptive Quadrature using MATLAB's quad8 routine but with $\tau = 0.0001$.

Now, it is fair to ask: What are these new routines, and what do they do? The new version of quad is probably self-explanatory: It is an updated version of the method outlined at the beginning of this section. The routine quadl uses what is known as Lobatto quadrature, which can best be summarized quickly as an attempt to make Gaussian quadrature more amenable to adaptive ideas. There is also quadgk, which uses Gauss–Kronrod quadrature, another idea for making Gaussian quadrature efficiently adaptive. The implementation of this routine does not have the facility to return the function count, so it was left out of this comparison. The MATLAB documentation has appropriate references, some of which are given at the end of the chapter. The first thing to notice is that the newer version of quad is much more efficient than either of the older routines, as well as the one written by the author. It appears to be consistently superior to quadl, but such a conclusion cannot be substantiated by a single example. Which adaptive routine works best on a given example will depend on the specifics of the example. (It should be noted that all of the routines integrated our example to the expected accuracy.)

5.8.4   Peano Estimates for the Trapezoid Rule

Recall that the trapezoid rule is constructed by exactly integrating a linear polynomial approximation to the integrand. We can use this to derive an interesting error formula, known as a *Peano*[17] *estimate*.

Consider

$$I(f) = \int_a^b f(x)dx$$

and

$$T_1(f) = \frac{1}{2}(b-a)(f(b)+f(a)),$$

the basic trapezoid rule. Since T_1 is derived from exactly integrating a linear approximation, it is *exact* when applied to any and all linear functions:

$$T_1(p) = I(p)$$

for all functions p of the form $p(x) = Ax + B$. (See Problem 14.) In addition, we can view both I and T_1 as *linear operators*, meaning that

$$I(\alpha f + \beta g) = \alpha I(f) + \beta I(g)$$

for constants α and β, and similarly for $T_1(\alpha f + \beta g)$. (See Problem 15.) Now, for any given integrand f, use Taylor's Theorem to write

$$f(x) = p_1(x) + R_1(x),$$

where p_1 is the linear Taylor polynomial and R_1 is the integral form of the remainder. Since T_1 is exact for linear polynomials, we know that

$$I(p_1) = T_1(p_1).$$

Therefore,

$$
\begin{aligned}
I(f) - T_1(f) &= I(p_1 + R_1) - T_1(p_1 + R_1), \\
&= I(R_1) - T_1(R_1), \\
&= \int_a^b \int_a^x (x-t)f''(t)dt\,dx - \frac{1}{2}(b-a)\int_a^b (b-t)f''(t)dt.
\end{aligned}
$$

We can interchange the order of integration in the double integral to get

$$
\int_a^b \int_a^x (x-t)f''(t)dt\,dx = \int_a^b \int_t^b (x-t)f''(t)dx\,dt = \int_a^b \left(\int_t^b (x-t)dx \right) f''(t)dt
$$

$$
= \int_a^b \frac{1}{2}(b-t)^2 f''(t)dt,
$$

---

[17] Giuseppe Peano (1858–1932) entered the University of Turin, in Italy, in 1876 and graduated with a doctorate in mathematics in 1880. Known primarily for his work in mathematical logic and the foundations of mathematics, he published the development of the Peano kernel in two papers that appeared in Italian journals in 1913–1914.

so that we can finally get

$$
\begin{aligned}
I(f) - T_1(f) &= \int_a^b \frac{1}{2}(b-t)^2 f''(t)dt - \frac{1}{2}(b-a)\int_a^b (b-t)f''(t)dt \\
&= \frac{1}{2}\int_a^b (b-t)[(b-t)-(b-a)](f''(t)dt \\
&= \frac{1}{2}\int_a^b (a-t)(b-t)f''(t)dt.
\end{aligned}
$$

This is known as the *Peano estimate* for the error, and the function $K(t) = \frac{1}{2}(a-t)(b-t)$ is known as the *Peano kernel* for the (basic) trapezoid rule. When extended to the composite case (Problem 17), we have

$$
I(f) - T_n(f) = \int_a^b K(t)f''(t)dt,
$$

where

$$
K(t) = \frac{1}{2}(x_{i-1}-t)(x_i-t), \quad x_{i-1} \le t \le x_i. \tag{5.28}
$$

There are a couple of reasons for looking at the Peano form of the error estimate. One is that it allows us to bound the error when the integrand is not smooth in the pointwise sense. Note that we can write

$$
\begin{aligned}
|I(f) - T_n(f)| &= \left| \int_a^b K(t)f''(t)dt \right|, \\
&\le \max_{a \le t \le b} |K(t)| \int_a^b |f''(t)|dt, \\
&\le \frac{1}{8}h^2 \int_a^b |f''(t)|dt,
\end{aligned}
$$

where we have assumed a uniform mesh spacing of h. (See Problem 19.) Thus, even if f is not pointwise twice continuously differentiable, we can still get an error estimate if f'' is integrable in the absolute sense. For example, consider the problem of integrating $f(x) = x^{3/2}$ over the interval $[0,1]$. We know that $f''(x)$ will be proportional to $x^{-1/2}$, and therefore the pointwise estimates of §2.5 will not hold. However, $x^{-1/2}$ is absolutely integrable, so the Peano form of the error tells us that we can expect second-order accuracy when using the trapezoid rule. Table 5.19 confirms this conclusion.

A second reason to use the Peano form is that it allows us to obtain lower-order estimates. If f has only a single derivative (whether in the pointwise or absolutely integrable sense), we can use integration by parts to obtain the estimate

$$
I(f) - T_1(f) = \frac{1}{2}\int_a^b (a-t)(b-t)f''(t)dt = \frac{1}{2}\int_a^b (t-c)f'(t)dt,
$$

where $c = (a+b)/2$. This suggests that the composite kernel in this case is

$$
K(t) = \left(t - \frac{1}{2}(x_i + x_{i-1}) \right), \quad x_{i-1} \le t \le x_i,
$$

Table 5.19 Trapezoid rule applied to $f(x) = x^{3/2}$, $[a, b] = [0, 1]$.

$n = h^{-1}$	$T_n(f)$	$I(f) - T_n(f)$
2	0.4267766953E+00	–0.2677669530E–01
4	0.4070181109E+00	–0.7018110858E–02
8	0.4018124648E+00	–0.1812464800E–02
16	0.4004634013E+00	–0.4634013020E–03
32	0.4001176712E+00	–0.1176712098E–03
64	0.4000297399E+00	–0.2973986253E–04
128	0.4000074919E+00	–0.7491908991E–05
256	0.4000018830E+00	–0.1883044172E–05
512	0.4000004725E+00	–0.4725406820E–06
1024	0.4000001184E+00	–0.1184497712E–06
2048	0.4000000297E+00	–0.2966805679E–07

from which an estimate that is $\mathcal{O}(h)$ follows quickly. (See Problem 18.)

In fact, this explains to us the order of convergence that we observe when using the trapezoid rule to integrate $f(x) = \sqrt{x}$; although the second derivative is not integrable, the first derivative $f'(x) = \frac{1}{2}x^{-1/2}$ is, so we see that the error is $\mathcal{O}(h)$, as can be confirmed from the data in Table 5.14. (See Problem 16.)

Peano estimates can be constructed for any of the other quadrature rules, but they are sometimes a bit more complex. See Problem 20.

Exercises:

1. Use the Euler–Maclaurin formula to state and prove a formal theorem that the corrected trapezoid rule, T_n^C, is $\mathcal{O}(h^4)$ accurate.

2. Using a hand calculator, compute $T_1^{(0)}(f)$, $T_2^{(0)}(f)$, $T_4^{(0)}(f)$, and $T_8^{(0)}(f)$ for each of the following functions, then use Romberg integration to compute $\Theta_3(f)$. *Note:* Be sure to use Theorem 5.8 to minimize the work in computing the first column of the Romberg array.

(a)
$$I = \int_0^1 \frac{1}{\sqrt{1 + x^4}} dx = 0.92703733865069;$$

(b)
$$I = \int_0^1 \ln(1 + x) dx = 2\ln 2 - 1;$$

(c)
$$I = \int_0^1 x(1 - x^2) dx = \frac{1}{4};$$

(d)
$$I = \int_0^1 \frac{1}{1 + x^3} dx = \frac{1}{3}\ln 2 + \frac{1}{9}\sqrt{3}\pi;$$

(e)
$$I = \int_1^2 e^{-x^2} dx = 0.1352572580.$$

3. Write a program to do Romberg integration. Be sure to use Theorem 5.8 to minimize the number of function evaluations. Test your program by applying it to the following example problems.

(a) $f(x) = \ln x$, $[1, 3]$, $I(f) = 3 \ln 3 - 2 = 1.295836867$;

(b) $f(x) = x^2 e^{-x}$, $[0, 2]$, $I(f) = 2 - 10e^{-2} = 0.646647168$;

(c) $f(x) = \sqrt{1 - x^2}$, $[-1, 1]$, $I(f) = \pi/2$;

(d) $f(x) = 1/(1 + x^2)$, $[-5, 5]$, $I(f) = 2 \arctan(5)$;

(e) $f(x) = e^{-x} \sin(4x)$, $[0, \pi]$, $I(f) = \frac{4}{17}(1 - e^{-\pi}) = 0.2251261368$.

For each example, compute

$$N_f = \frac{\text{number of function evaluations}}{- \log_{10} |\text{error}|}.$$

This measures the number of function evaluations needed to produce each correct decimal digit in the approximation.

4. Write a computer program that uses Romberg integration for a specified number of points to compute the natural logarithm over the interval $[\frac{1}{2}, 1]$, to within 10^{-16} accuracy. Compare the accuracy of your routine to the intrinsic logarithm function on your system.

5. Show that the change of variable $x = \phi(t)$, where ϕ is as given in §5.8.2, transforms the integral

$$I(f) = \int_a^b f(x) dx$$

into the integral

$$I(f) = \int_{-1}^1 f(\phi(t))\phi'(t) dt.$$

6. Apply the singular integral technique of §5.8.2, with $n = 4$, to estimate the value of each of the following integrals. Do this with a hand calculator, using the values in Table 5.15.

(a)
$$I(f) = \int_0^1 \frac{\ln x}{1 - x^2} dx = -\frac{\pi^2}{8};$$

(b)
$$I(f) = \int_0^1 \frac{\ln x}{1 - x} dx = -\frac{\pi^2}{6};$$

(c)
$$I(f) = \int_0^1 x \ln(1 - x) dx = -\frac{3}{4};$$

(d)

$$I(f) = \int_0^1 \left(\ln \frac{1}{x} \right)^{-1/2} dx = \sqrt{\pi}.$$

7. Repeat the previous problem, but this time use a computer program together with the values in Table 5.15 to compute the $n = 16$ approximations.

8. We have looked at the gamma function in a number of exercises in previous chapters. The formal definition of $\Gamma(x)$ is the following:

$$\Gamma(x) = \int_0^\infty e^{-t} t^{x-1} dt.$$

Use the infinite interval algorithm from §5.8.2 to construct a table of values for the gamma function over the interval $[1, 2]$. Compare your results to the values you get for $\Gamma(x)$ on your computer or from a standard book of tables, such as [1]

9. Modify your Romberg integration program to compute values of $\Psi(t_k)/\Psi(1)$ for $t_k \in [-1, 1]$, and use this to extend the values in Table 5.15 to the $n = 64$ case.

10. Apply the trapezoid rule to each of the following functions, integrated over the indicated intervals, and interpret the results in terms of the Euler–Maclaurin formula.

    (a) $f(x) = 1 + \sin \pi x$, $[a, b] = [0, 2]$;
    (b) $f(x) = \sin^2 x$, $[a, b] = [0, \pi]$.

11. Using a hand calculator and τ as indicated, perform the adaptive quadrature scheme outlined in §5.8.3 on each of the following integrals. Be sure to present your results in an orderly fashion so that the progress of the calculation can be followed.

    (a) $\tau = 5 \times 10^{-6}$;

    $$I = \int_0^1 \ln(1 + x) dx = 2 \ln 2 - 1;$$

    (b) $\tau = 10^{-5}$;

    $$I = \int_0^1 \frac{1}{1 + x^3} dx = \frac{1}{3} \ln 2 + \frac{1}{9} \sqrt{3} \pi;$$

    (c) $\tau = 10^{-5}$;

    $$I = \int_1^2 e^{-x^2} dx = 0.1352572580;$$

    (d) $\tau = 10^{-4}$;

    $$I = \int_0^1 \frac{1}{\sqrt{1 + x^4}} dx = 0.92703733865069.$$

12. Apply the MATLAB routines quad, quadl, and quadgk to each of the following integrations, with a tolerance $\tau = 1.e-8$ in each case. Then repeat the computations over the left half of the interval, only.

    (a)

    $$I = \int_0^1 \frac{1}{1 + 1023 e^{-16t}} dt = 0.56679020695363;$$

(b)
$$I = \int_0^{2\pi} e^{\sin 4\pi x} dx = 8.11767960946423;$$

(c)
$$I = \int_0^1 \sin(e^{\pi x}) dx = 0.20499307668744;$$

(d)
$$I = \int_0^1 \frac{1}{1+x^3} dx = \frac{1}{3}\ln 2 + \frac{1}{9}\sqrt{3}\pi;$$

(e)
$$I = \int_0^1 \frac{1}{\sqrt{1+x^4}} dx = 0.92703733865069.$$

13. This is an experimental or research problem. Try to find a specific quadrature problem

$$I = \int_a^b f(x) dx$$

such that quad outperforms quadl consistently as the tolerance τ decreases. Then try to find a different one such that quadl outperforms quad.

14. Show that the trapezoid rule $T_1(f)$ is exact for all functions of the form $f(x) = Ax + B$.

15. Show that

$$I(\alpha f + \beta g) = \alpha I(f) + \beta I(g)$$

for constants α and β, and similarly for $T_n(\alpha f + \beta g)$.

16. Show that the data in Table 5.14 confirms that the trapezoid and Simpson's rules applied to $f(x) = \sqrt{x}$ are both $\mathcal{O}(h)$ accurate.

17. Confirm that (5.28) is the correct Peano kernel for the composite trapezoid rule.

18. Show that if f' is integrable over $[a, b]$, but f'' is not, then the trapezoid rule is $\mathcal{O}(h)$ accurate.

19. Show that the Peano Theorem implies an error estimate for the trapezoid rule of the form

$$|I(f) - T_n(f)| \leq \frac{1}{8}h^2 \int_a^b |f''(t)| dt.$$

Be sure to provide all details missing from the development in the text.

20. Derive the Peano kernel for Simpson's rule.

◁ • • ▷

5.9 LITERATURE AND SOFTWARE DISCUSSION

The quadrature problem has a long history and a fairly mature literature. The standard monograph is perhaps Davis and Rabinowitz [3], and decent treatments can be found in any good numerical analysis text.

The trapezoid, Simpson, and midpoint rules are special cases of a broad class of methods known as Newton–Cotes formulas, for which a unified error theory is available. See Atkinson [2] for a standard treatment of this.

The main deficiency of Gaussian quadrature has been the need to recompute f at all grid points whenever the grid is refined. A variation of Gaussian quadrature, known as Gauss–Kronrad quadrature, addresses this issue and has become popular in recent years. Extensive work on a similar idea was done by Patterson and some of these ideas are incorporated in the QUADPACK package. See Piessens [7] or Davis and Rabinowitz [3]. The article by Gander and Gautschi [5] outlines many of the adaptive routines used in MATLAB.

Multiple integrals are usually treated by obvious extensions of the ideas presented here, although the fact that higher-dimensional geometry is potentially much more involved than the simple intervals in $\mathbb{R}$ does complicate matters. Multiple integrals are sometimes treated by probabilistic methods known generally as *Monte Carlo methods*.

REFERENCES

1. Abramowitz, M., and Stegun, I. A., *Handbook of Mathematical Functions*, Government Printing Office, Washington, DC, 1964. This is a classic resource, and an updated online version exists at http://dlmf.nist.gov/.

2. Atkinson, Kendall, *An Introduction to Numerical Analysis*, John Wiley & Sons, Inc., New York, 1989 (2nd edition).

3. Davis, P., and Rabinowitz, P., *Methods of Numerical Integration*, 2nd edition, Academic Press, New York, 1984.

4. de Doncker, E., and Piessens, R., "Algorithm 32: Automatic Computation of integrals with singular integrand, over a finite or infinite range," *Computing*, vol. 17, pp. 265–279, 1976.

5. Gander, W. and Gautschi, W.; "Adaptive Quadrature Revisited," *BIT*, Vol. 40, pp. 84–101, 2000.

6. Jeffreys, Harold, and Jeffreys, Bertha, *Methods of Mathematical Physics*, Cambridge University Press, Cambridge, UK, 1972.

7. Piessens, R., et al, *QUADPACK: A Subroutine Package for Automatic Integration*, Springer-Verlag, New York, 1983.

CHAPTER 6

NUMERICAL METHODS FOR ORDINARY DIFFERENTIAL EQUATIONS

We are concerned here with the problem of solving differential equations, numerically. At first we concentrate on the so-called *initial value problem* (IVP): Find a function $y(t)$ such that

$$\frac{dy}{dt} = f(t, y(t)), \quad y(t_0) = y_0,$$

where f is a known function of two variables, and t_0 and y_0 are known values. This is called the initial value problem because (as the notation suggests) we can view the independent variable t as time, and the equation as modeling a process that moves forward from some initial time t_0 with initial state y_0. (Very often, $t_0 = 0$.) The dependent variable y, the unknown function, may be a scalar function or, possibly, a vector function defined as

$$y(t) = (y_1(t), y_2(t), \dots, y_N(t))^T.$$

In §2.3 we developed Euler's method for approximating solutions to initial value problems; in this chapter we will not only review Euler's method, but we will also look at more sophisticated (and therefore, we hope, more accurate) methods for solving this type of problem. Later we will tackle the boundary value problem (BVP), which can be written as

$$-\frac{d^2u}{dx^2} = F\left(x, u, \frac{du}{dx}\right), \quad a < x < b; \tag{6.1}$$

$$u(a) = g_0, \tag{6.2}$$

$$u(b) = g_1. \tag{6.3}$$

An Introduction to Numerical Methods and Analysis, Second Edition. By James F. Epperson
Copyright © 2013 John Wiley & Sons, Inc.

Here the unknown function is u with independent variable x, F is a known function of three variables, and g_0 and g_1 are known data values. Very often the interval $(a, b) = (0, 1)$.

In both cases we want to find an unknown function. We will do this by approximating individual points on the graph of the function, as we did in §2.7 with Euler's method for initial value problems. Thus, we will seek (in the case of the IVP) a set of values y_k such that $y_k \approx y(t_k)$ for some set of (known) grid points t_k, or (in the case of the BVP) a set of values u_k such that $u_k \approx u(x_k)$ for some (known) grid points x_k. Note that this means that our approximation is only defined on the grid points, although we could use the approximation methods of Chapter 4 to construct continuous approximate solutions to differential equations—this, in fact, is something that is often done, and we showed one example of it in §4.9, where we used splines to solve two-point boundary value problems. In §6.10.3 we will introduce the finite element method for BVPs, which also uses the notion of expanding the approximation as a linear combination of simple functions. In Chapter 9 we revisit this idea (and extend it to some partial differential equations—PDEs), and in Chapter 10 we will introduce spectral methods for BVPs and some PDEs. But in this chapter we will concentrate on the basics.

It should be noted that the numerical solution of ordinary differential equations—whether IVP's or BVP's—is a very active on-going area of research. What we present here is a selection of the basic algorithms and underlying theory. The reader is referred to the list of references at the end of this chapter for more in-depth treatments of this material.

6.1 THE INITIAL VALUE PROBLEM: BACKGROUND

Consider the ordinary differential equation

$$\frac{dy}{dt} = f(t, y(t)), \quad y(t_0) = y_0, \tag{6.4}$$

where f is a function from $\mathbb{R}^{N+1}$ into $\mathbb{R}^N$ for some $N > 0$ (if $N = 1$, then we have a scalar equation; otherwise, a vector equation); t_0 is a given scalar value, often taken to be $t_0 = 0$, and known as the *initial point*; and y_0 is a known vector in $\mathbb{R}^N$, known as the *initial value*. We want to find the unknown function $y(t)$, which solves (6.4) in the sense that

$$y'(t) - f(t, y(t)) = 0$$

for all $t > t_0$, and $y(t_0) = y_0$.

Some examples will be useful at this point.

■ EXAMPLE 6.1

Consider the simple problem

$$y' = -2ty, \quad y(0) = 1.$$

Here $f(t, y) = -2ty, t_0 = 0, y_0 = 1$, and we can apply methods from a standard ordinary differential equations (ODE) course to show that $y(t) = e^{-t^2}$ is the solution.

■ **EXAMPLE 6.2**

Consider now the simple *system of equations*

$$y_1' = -4y_1 + y_2, \quad y_1(0) = 1;$$
$$y_2' = y_1 - 4y_2, \quad y_2(0) = 0.$$

In this case, $y(t) = (y_1(t), y_2(t))^T$ is a *vector* of two components, and $f(t, y)$ is defined by

$$f(t, y) = (f_1(t, y_1, y_2), f_2(t, y_1, y_2))^T,$$

where

$$f_1(t, y_1, y_2) = -4y_1 + y_2,$$
$$f_2(t, y_1, y_2) = y_1 - 4y_2;$$

the initial point is still $t_0 = 0$, but the initial value is now the vector $y_0 = (1, 0)^T$. The solution, again obtained from a standard ODE course, is

$$y_1 = \frac{1}{2}e^{-3t} + \frac{1}{2}e^{-5t},$$
$$y_2 = \frac{1}{2}e^{-3t} - \frac{1}{2}e^{-5t}.$$

■ **EXAMPLE 6.3**

What about the *second*-order equation

$$y'' + 3y' + y = 0, \quad y(0) = 1, y'(0) = 0? \tag{6.5}$$

It does not appear to fit the standard form for our initial value problems, yet we can put it into this form without a lot of work. Define the vector $w(t) = (w_1(t), w_2(t))^T$, where $w_1(t) = y(t), w_2(t) = y'(t)$. Then $w_1' = w_2$ and (6.5) becomes $w_2' = -3w_2 - w_1$. Thus we have the *first-order system*

$$w_1' = w_2, \qquad w_1(0) = 1;$$
$$w_2' = -w_1 - 3w_2, \qquad w_2(0) = 0.$$

■ **EXAMPLE 6.4**

More generally, we can take any scalar equation of order N:

$$y^{(n)} = g(t, y, y', y'', \dots, y^{(n-1)}),$$

and write it as a first-order system of N equations in N unknowns, according to

$$w_1 = y, \; w_2 = y', \; \dots, w_n = y^{(n-1)},$$

where

$$w_1' = w_2, \quad w_1(0) = y_0;$$
$$w_2' = w_3, \quad w_2(0) = y_0';$$
$$w_3' = w_4, \quad w_3(0) = y_0'';$$
$$\vdots = \vdots$$
$$w_n' = g(t, w_1, w_2, \dots, w_{n-1}, w_n), \quad w_n(0) = y_0^{(n-1)}.$$

Thus, we have the single vector equation

$$w' = f(t, w), \quad w(0) = w_0,$$

where

$$f(t, w) = \begin{pmatrix} w_2 \\ w_3 \\ w_4 \\ \vdots \\ g(t, w_1, w_2, w_3, \ldots, w_n) \end{pmatrix}, \quad w_0 = \begin{pmatrix} y_0 \\ y_0' \\ y_0'' \\ \vdots \\ y_0^{(n-1)} \end{pmatrix}.$$

To keep the presentation simple, we will state our results in language and notation appropriate for scalar equations, but the reader should be aware that corresponding results hold for vector equations, and we will spend some time discussing vector equations later in the chapter, in §6.8.

The nature of the solution of the differential equation—indeed, the very question of whether or not a solution exists—as well as our ability to approximate the solution accurately, is very much connected with the nature of the function f. Basically, if f is "smooth enough," then a solution will exist and be unique, and we will therefore be able to approximate it accurately with a wide variety of methods. However, there are different ways of expressing what we mean by "smooth enough." We will use two of them: (a) Lipschitz continuity; and, (b) smooth and uniformly monotone decreasing. The former is the condition generally used to construct solutions of initial value problems; it is, generally, the weakest condition that we can get away with. The monotone decreasing condition describes a wide class of physically significant problems; the "monotone decreasing" part of the condition corresponds to dissipation or decay in the physical system being modeled.

Definition 6.1 (Lipschitz Continuity) *Let g be a given function from $\mathbb{R}$ to $\mathbb{R}$. We say that g is* Lipschitz[1] continuous *on an interval I if there exists a constant K such that*

$$|g(x_1) - g(x_2)| \le K|x_1 - x_2|$$

for all $x_1, x_2 \in I$.

Definition 6.2 (Smooth and Uniformly Monotone Decreasing) *Let g be a given function from $\mathbb{R}$ to $\mathbb{R}$. We say that g is smooth and uniformly monotone decreasing if g is differentiable and the derivative satisfies*

$$-M \le g'(x) \le -m < 0$$

for all x, where M and m are given positive constants.

We have actually already seen the idea of Lipschitz continuity; this is the same condition that was required in §3.9 for uniqueness and convergence of fixed-point iterations.

---

[1] Rudolf Otto Sigismund Lipschitz (1832–1903) was born in Königsberg, Germany. He entered the University of Königsberg at the age of 15, and completed his studies at the University of Berlin, from which he was awarded a doctorate in 1853. By 1864, he was a full professor at the University of Bonn, where he remained the rest of his professional life. The Lipschitz condition first appeared in a paper on the existence of solutions to differential equations, which was published in book form as part of an 1877 treatise on analysis.

■ EXAMPLE 6.5

Just to illustrate, consider the two initial value problems

$$y' = 4y - e^{-t}, \quad y(0) = 1, \tag{6.6}$$

and

$$y' = -(1 + t^2)y + \sin t, \quad y(0) = 1. \tag{6.7}$$

For (6.6), we have $f(t, y) = 4y - e^{-t}$, so that

$$f(t, y_1) - f(t, y_2) = (4y_1 - e^{-t}) - (4y_2 - e^{-t}) = 4(y_1 - y_2);$$

hence,

$$|f(t, y_1) - f(t, y_2)| \le 4|y_1 - y_2|.$$

Hence, in this case f is Lipschitz continuous in y with constant $K = 4$. However, f is not smooth and uniformly monotone decreasing, because $f_y(t, y) = 4 > 0$ for all t and y. On the other hand, for (6.7), we have $f(t, y) = -(1 + t^2)y + \sin t$, so

$$f(t, y_1) - f(t, y_2) = (-(t^2 + 1)y_1 + \sin t) - (-(t^2 + 1)y_2 + \sin t) = -(t^2 + 1)(y_1 - y_2);$$

thus, for $0 \le t \le 1$,

$$|f(t, y_1) - f(t, y_2)| \le (t^2 + 1)|y_1 - y_2| \le 2|y_1 - y_2|.$$

Therefore, this f is Lipschitz continuous with constant $K = 2$. In addition, we have (again, for $0 \le t \le 1$),

$$f_y(t, y) = -(t^2 + 1) \Rightarrow -2 \le f_y(t, y) \le -1 < 0,$$

showing that this f is also smooth and uniformly monotone decreasing.

The significance of these definitions lies in how they affect what we know about the solution of the differential equation. The standard result for ordinary differential equations is the following.

Theorem 6.1 (Existence-Uniqueness of Solutions for IVPs, Version I) *Let $f(t, y)$ be continuous for all (t, y) in an open rectangle $R = \{(t, y) : a < t < b, c < y < d\}$, and Lipschitz continuous in y, with constant K. Then for all $(t_0, y_0) \in R$ there exists a unique solution to the initial value problem*

$$y' = f(t, y), \quad y(t_0) = y_0.$$

Moreover, if $z(t)$ is the solution to the same problem with initial data $z(t_0) = z_0$, then

$$|y(t) - z(t)| \le e^{K(t-t_0)}|y_0 - z_0|. \tag{6.8}$$

Proof: This result is proved in some texts on ordinary differential equations (see, for example, [6]). Interestingly, the original proof was done by constructing a discrete solution and showing that it converged, in the limit as $h \to 0$, to a differentiable function that solved the IVP. See Goldstine's book [10] for more details on the history behind this result. ●

If we impose only slightly more on the function f, then we get quite a different bound on the change in solutions due to a change in the initial data.

Theorem 6.2 (Existence-Uniqueness of Solutions for IVPs, Version II) *Let $f(t, y)$ be continuous for all (t, y) in a rectangle R, and smooth and uniformly monotone decreasing in y. Then, for all $(t_0, y_0) \in R$ there exists a unique solution to the initial value problem*

$$y' = f(t, y), \quad y(t_0) = y_0.$$

Moreover, if $z(t)$ is the solution to the same problem with initial data $z(t_0) = z_0$, then

$$|y(t) - z(t)| \le e^{-m(t-t_0)}|y_0 - z_0|. \tag{6.9}$$

Here m is the negative of the upper monotonicity constant from Definition 6.2.

Proof: Since f is smooth and uniformly monotone decreasing in y, we can apply the Mean Value Theorem to easily prove that it is Lipschitz in y, as well:

$$|f(t, y_1) - f(t, y_2)| = |f_y(t, \eta)(y_1 - y_2)| \le M|y_1 - y_2|.$$

Thus, we can apply Theorem 6.1 to get the existence and uniqueness of the solution. The inequality (6.9) takes a bit more work, and its proof is deferred to the Appendix. •

The inequalities (6.8) and (6.9) are known as *stability results* because they measure how stable the solutions to the initial value problem are, i.e., the extent to which small changes in the problem will lead to small changes in the solution. In the first case, if all we know is that f is Lipschitz, then small changes in the initial data can lead to very large changes in the solutions, if we consider t large enough. Note, however, that it still is the case that the differences in the two solutions can be bounded for all $t \le T$, where T is a fixed value, and that $z(t) \to y(t)$ on this interval, uniformly, as $z_0 \to y_0$. In the second instance, because we know that f is smooth and monotone decreasing, we know that the perturbation in the solution due to the initial error in the solution will *actually decay to zero as $t \to \infty$.* This will have similar implications for the error associated with numerical methods applied to the solution of the differential equation.

Exercises:

1. For each initial value problem below, verify (by direct substitution) that the given function y solves the problem.

   (a) $y' + 4y = 0$, $y(0) = 3$; $y(t) = 3e^{-4t}$;

   (b) $y' = t^2/y$, $y(0) = 1$; $y(t) = \sqrt{1 + \frac{2}{3}t^3}$;

   (c) $ty' - y = t^2$, $y(1) = 4$; $y(t) = 3t + t^2$.

2. For each initial value problem in Problem 1, write down the definition of f.

3. For each scalar equation below, write out explicitly the corresponding first-order system. What is the definition of f?

   (a) $y'' + 9y = e^{-t}$;

   (b) $y'''' + y = 1$;

   (c) $y'' + \sin y = 0$.

4. For each initial value problem below, verify (by direct substitution) that the given function y solves the problem.

(a) $y'' + 4y' + 4y = 0$, $y(0) = 0, y'(0) = 1$; $y(t) = te^{-2t}$;

(b) $t^2 y'' + 6ty' + 6y = 0$, $y(1) = 1, y'(1) = -3$; $y(t) = t^{-3}$;

(c) $y'' + 5y' + 6y = 0$, $y(0) = 1, y'(0) = -2$; $y(t) = e^{-2t}$.

5. For each initial value problem below, determine the Lipschitz constant, K, for the given rectangle.

(a) $y' = y(1 - y)$, $y(0) = \frac{1}{2}$, $R = (-1, 1) \times (0, 2)$;

(b) $y' = 1 - 3y$, $y(0) = 1$, $R = (-1, 1) \times (0, 2)$;

(c) $y' = y^2$, $y(0) = 1$, $R = (-1, 1) \times (0, 2)$.

6. Are any of the initial value problems in the previous exercise smooth and uniformly monotone decreasing over the given rectangle? If so, determine the values of M and m.

◁ ● ● ▷

6.2 EULER'S METHOD

Euler's method is the natural starting point for any discussion of numerical methods for IVPs. Although it is not the most accurate of the methods we study, it is by far the simplest, and much of what we learn from analyzing Euler's method in detail carries over to other methods without a lot of difficulty. Even though we treated Euler's method in Chapter 2, we are going to cover it again here, but this time more fully.

There are two main derivations; one is geometric, the other is analytic, and a third derivation (the one we used in Chapter 2) is given in §6.4. We start with the geometric derivation.

Consider Figure 6.1. This shows the graph of the solution $y(t)$ to the initial value problem

$$y' = f(t, y), \quad y(t_0) = y_0,$$

near the initial point t_0. All we know about y is that it passes through the point $(t_0, y_0) = (0, 0)$, and that it has slope $f(t_0, y_0)$ (why?) at that point. We therefore can approximate $y(t + h)$ for some small h by using the tangent line approximation:

$$y(t_0 + h) \approx y(t_0) + hf(t_0, y(t_0)). \tag{6.10}$$

For small values of h this will be a decent approximation, and if we define the next point in our grid by $t_1 = t_0 + h$, then it is very natural to extend (6.10) to get

$$y(t_1 + h) \approx y(t_1) + hf(t_1, y(t_1)), \tag{6.11}$$

or, more generally,

$$y(t_n + h) \approx y(t_n) + hf(t_n, y(t_n)). \tag{6.12}$$

The numerical method is then defined from (6.12) according to

$$y_{n+1} = y_n + hf(t_n, y_n),$$

with y_0 given and $y_n \approx y(t_n)$.

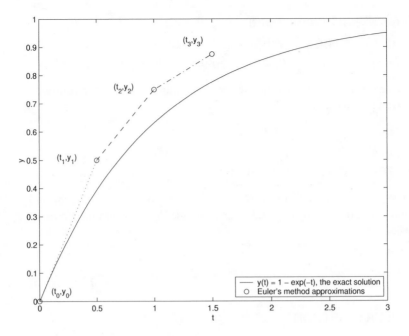

Figure 6.1   Geometric derivation of Euler's method.

This is a reasonable and easy-to-follow derivation, but it does not admit of much error analysis, and it also opens up a lot of questions. Perhaps the most important such question concerns the validity of extending the tangent line approximation from the first computed point, (t_1, y_1), to the next one. From the graph in Figure 6.1, it is clear that there is substantial error between y_1 and $y(t_1)$. How valid, then, is (6.11) in defining the next approximate point?

The analytic derivation is based, to no surprise, on Taylor's Theorem. We have that

$$y(t+h) = y(t) + hy'(t) + \frac{1}{2}h^2 y''(\theta),$$

for some θ between t and $t+h$. But the differential equation implies that $y'(t) = f(t, y(t))$, so we have

$$y(t+h) = y(t) + hf(t, y(t)) + \frac{1}{2}h^2 y''(\theta). \tag{6.13}$$

Now set $t = t_0$, define $t_{n+1} = t_n + h$, set y_n to be the approximate value, and drop the remainder to get the same expression that we had before:

$$y_{n+1} = y_n + hf(t_n, y_n). \tag{6.14}$$

The advantage of this derivation is that we have (6.13), which relates the exact solution to the numerical method with a very precise remainder. It is the key to getting an error estimate for Euler's method. To anticipate some more general terminology, the quantity

$$R(t, h) = \frac{1}{2}h^2 y''(\theta)$$

is called the *residual* for Euler's method; the quantity

$$\tau(t,h) = \frac{1}{h}R(t,h)$$

is called the *truncation error*.

Before getting into the analysis (see §6.3) some examples might be in order.

■ **EXAMPLE 6.6**

Consider the two differential equations

$$y' + y = 1, \quad y(0) = 0, \tag{6.15}$$

and

$$y' = y, \quad y(0) = 1. \tag{6.16}$$

The exact solutions can be found by typical ODE methods to be

$$y(t) = 1 - e^{-t}$$

for (6.15), and

$$y(t) = e^t$$

for (6.16). Table 6.1 shows the error made by Euler's method in solving each of these, using a sequence of decreasing meshes. The entries in the table give the maximum error over the interval $[0,1]$; that is,

$$E_n = \max_{k \le n} |y(t_k) - y_k|.$$

Table 6.1 Euler's method applied to (6.15) and (6.16).

$n = h^{-1}$	Error in approximating (6.15)	Error in approximating (6.16)
16	0.118053E–01	0.803533E–01
32	0.582415E–02	0.412917E–01
64	0.289292E–02	0.209369E–01
128	0.144173E–02	0.105428E–01
256	0.719686E–03	0.529020E–02
512	0.359550E–03	0.264983E–02
1024	0.179702E–03	0.132610E–02

The table shows that the errors in both cases are going down by about a factor of 2, from which we infer that Euler's method is $\mathcal{O}(h)$ accurate, only. Figures 6.2 and 6.3 show plots of the absolute error as a function of t for the $n = 1024$ cases; note that for (6.15) the worst error occurs near the beginning of the computation and that afterward the error actually decreases, whereas for (6.16) the worst error is always the most recent error. In fact, the error for (6.16) appears to be growing exponentially. We will touch on this issue in the next section, and try to explain it.

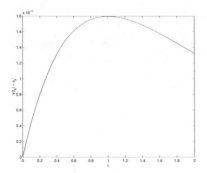

Figure 6.2   Error as a function of time for Eq'n. (6.15), using $h = 1/1024$.

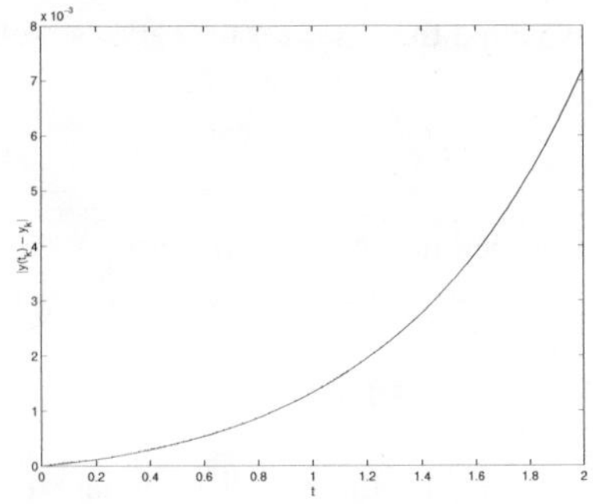

Figure 6.3   Error as a function of time for Eq'n. (6.16), using $h = 1/1024$.

Exercises:

1. Use Euler's method with $h = \frac{1}{4}$ to compute approximate values of $y(1)$ for each of the following initial value problems. Don't write a computer program, use a hand calculator to produce an orderly table of (t_k, y_k) pairs.

   (a) $y' = y(1 - y)$, $y(0) = \frac{1}{2}$;

   (b) $ty' = y(\sin t)$, $y(0) = 2$;

   (c) $y' = y(1 + e^{2t})$, $y(0) = 1$;

   (d) $y' + 2y = 1$, $y(0) = 2$.

2. For each initial value problem above, use the differential equation to produce approximate values of y' at each of the grid points, t_k, $k = 0, 1, 2, 3, 4$.

3. Write a computer program that solves each of the initial value problems in Problem 1, using Euler's method and $h = 1/16$.

4. For each initial value problem below, approximate the solution using Euler's method using a sequence of decreasing grids $h^{-1} = 2, 4, 8, \ldots$. For those problems where an exact solution is given, compare the accuracy achieved over the interval $[0, 1]$ with the theoretical accuracy.

   (a) $y' + 4y = 1$, $y(0) = 1$; $y(t) = \frac{1}{4}(3e^{-4t} + 1)$;

   (b) $y' = -y \ln y$, $y(0) = 3$; $y(t) = e^{(\ln 3)e^{-t}}$;

   (c) $y' + y = \sin 4\pi t$, $y(0) = \frac{1}{2}$;

   (d) $y' + \sin y = 0$, $y(0) = 1$.

◁ ● ● ● ▷

6.3 ANALYSIS OF EULER'S METHOD

In this section we will prove two results that establish the convergence and error estimate for Euler's method. We provide a fair amount of detail in this section, in order to avoid going into so much detail with more sophisticated methods that we will derive later. Throughout the section we are concerned with the approximate solution, via Euler's method, of the initial value problem

$$y' = f(t, y), \quad y(t_0) = y_0.$$

The first theorem shows that Euler's method is, indeed, first-order accurate.

Theorem 6.3 (Error Estimate for Euler's Method, version I) *Let f be Lipschitz continuous, with constant K, and assume that the solution $y \in C^2([t_0, T])$ for some $T > t_0$. Then*

$$\max_{t_k \leq T} |y(t_k) - y_k| \leq C_0|y(t_0) - y_0| + Ch\|y''\|_{\infty, [t_0, T]},$$

where

$$C_0 = e^{K(T-t_0)}$$

and

$$C = \frac{e^{K(T-t_0)} - 1}{2K}. \tag{6.17}$$

Proof: The key element in the proof of this result is the fact that the exact solution satisfies the same relationship as does the approximate solution, except for the addition of a remainder term. Thus we have (from (6.13) and (6.14)),

$$\begin{aligned} y(t_{n+1}) &= y(t_n) + hf(t_n, y(t_n)) + \frac{1}{2}h^2 y''(\theta_n), \\ y_{n+1} &= y_n + hf(t_n, y_n), \end{aligned}$$

which we subtract to get

$$y(t_{n+1}) - y_{n+1} = y(t_n) - y_n + hf(t_n, y(t_n)) - hf(t_n, y_n) + \frac{1}{2}h^2 y''(\theta_n).$$

Take absolute values and apply the Lipschitz continuity of f to get

$$|y(t_{n+1}) - y_{n+1}| \leq |y(t_n) - y_n| + Kh|y(t_n) - y_n| + \frac{1}{2}h^2|y''(\theta_n)|,$$

which we write as

$$e_{n+1} \leq \gamma e_n + R_n,$$

where $e_n = |y(t_n) - y_n|$, $\gamma = 1 + Kh$, and $R_n = \frac{1}{2}h^2|y''(\theta_n)|$, for notational simplicity. This is a simple recursive inequality, which we can "solve" as follows. We have

$$\begin{aligned} e_1 &\leq \gamma e_0 + R_0, \\ e_2 &\leq \gamma e_1 + R_1 \leq \gamma^2 e_0 + \gamma R_0 + R_1, \\ e_3 &\leq \gamma e_2 + R_2 \leq \gamma^3 e_0 + \gamma^2 R_0 + \gamma R_1 + R_2, \end{aligned}$$

and so on. An inductive argument can be applied to get the general result

$$e_n \leq \gamma^n e_0 + \frac{1}{2}h^2 \sum_{k=0}^{n-1} \gamma^k |y''(\theta_{n-1-k})|.$$

It remains only to simplify the summation term. We have

$$|y''(t)| \le \|y''\|_{\infty,[t_0,T]}$$

for all $t \in [t_0, T]$, so we then have

$$e_n \le \gamma^n e_0 + \left(\frac{1}{2}h^2\|y''\|_{\infty,[t_0,T]}\right)\left(\sum_{k=0}^{n-1} \gamma^k\right).$$

We can explicitly add up the sum now, to get

$$e_n \le \gamma^n e_0 + \frac{1}{2}h^2\|y''\|_{\infty,[t_0,T]}\frac{\gamma^n - 1}{\gamma - 1},$$

but $\gamma = 1 + Kh$ so that $\gamma - 1 = Kh$, thus we have

$$e_n \le \gamma^n e_0 + \frac{\gamma^n - 1}{2K}h\|y''\|_{\infty,[t_0,T]}.$$

Finally, we note that it follows from Taylor's Theorem (see Problem 1) that

$$(1 + x)^n \le e^{nx}, \quad x > -1;$$

so that

$$\gamma^n = (1 + Kh)^n \le e^{nKh} = e^{K(nh)} = e^{K(t_n - t_0)} \le e^{K(T - t_0)},$$

and we are done. ●

   This estimate shows that Euler's method is first order (i.e., the error is $\mathcal{O}(h)$), but it also shows that the constants multiplying the terms in the error estimate can become quite large. If we now assume that f is smooth and uniformly monotone decreasing, then we get an estimate that, while still first order, involves much smaller constants.

Theorem 6.4 (Error Estimate for Euler's Method, version II) *Let f be smooth and uniformly monotone decreasing in y, and assume that the solution $y \in C^2([t_0, T])$ for some $T > t_0$. Then, for h sufficiently small,*

$$\max_{t_k \le T} |y(t_k) - y_k| \le C_0|y(t_0) - y_0| + Ch\|y''\|_{\infty,[t_0,T]},$$

where $C_0 \le 1$, $C_0 \to 0$ as $k \to \infty$, and

$$C = \frac{1}{2m}.$$

Proof: The proof here is much akin to that in Theorem 6.3, at least in the beginning. We have

$$y(t_{n+1}) - y_{n+1} = y(t_n) - y_n + hf(t_n, y(t_n)) - hf(t_n, y_n) + \frac{1}{2}h^2y''(\theta_n). \quad (6.18)$$

Now, instead of taking absolute values and inequalities, we use the Mean Value Theorem to write

$$f(t_n, y(t_n)) - f(t_n, y_n) = \frac{\partial f}{\partial y}(t_n, \eta_n)(y(t_n) - y_n),$$

where η_n is a value between $y(t_n)$ and y_n. For simplicity we write this as

$$\frac{\partial f}{\partial y}(t_n, \eta_n)(y(t_n) - y_n) = k_n(y(t_n) - y_n),$$

where we note that the smooth and uniformly monotone decreasing hypothesis implies that $-M \le k_n \le -m < 0$. Then (6.18) becomes

$$y(t_{n+1}) - y_{n+1} = \gamma_n(y(t_n) - y_n) + \frac{1}{2}h^2 y''(\theta_n),$$

where $\gamma_n = 1 - k_n h$. *Now* take absolute values and use the triangle inequality to get

$$e_{n+1} \le |\gamma_n| e_n + R_n$$

where e_n and R_n are as in the proof of Theorem 6.3. Assume now that h is small enough that $1 - Mh > 0$; i.e., assume that $h < M^{-1}$. Then $\gamma > 0$ for all n, so that

$$|\gamma_n| = \gamma_n = 1 - k_n h \le \gamma = 1 - mh < 1.$$

This is the crucial difference between the two theorems and their proofs. By making a stronger assumption about the way f behaves, we are able to show that the rate constant in the error recursion is less than 1. We can then write

$$e_{n+1} \le \gamma e_n + R_n,$$

from which

$$e_n \le \gamma^n e_0 + \frac{1}{2}h^2 \|y''\|_{\infty, [t_0, T]} \frac{\gamma^n - 1}{\gamma - 1} \tag{6.19}$$

now follows as before. We now observe that

$$0 \le \frac{\gamma^n - 1}{\gamma - 1} = \frac{1 - \gamma^n}{1 - \gamma} \le \frac{1 - \gamma^n}{1 - (1 - mh)} = \frac{1 - \gamma^n}{mh} \le \frac{1}{mh},$$

so that (6.19) becomes

$$e_n \le \gamma^n e_0 + \frac{1}{2m}h\|y''\|_{\infty, [t_0, T]},$$

and we are done. ●

Both error theorems show that Euler's method is only first-order accurate; that is, the error is $\mathcal{O}(h)$. The difference between the theorems lies in the way that the estimates depend on the function f: If f is only Lipschitz continuous, then the constants multiplying the initial error and the mesh parameter can be quite large and rapidly growing; however, if f is smooth and uniformly monotone decreasing in y—which means that the differential equation is modeling a decay process, essentially—then the constants in the error estimate are bounded for all n. Note, in particular, how the initial error is affected by this. If f is monotone decreasing in y, then the initial error is multiplied by γ^n, where $0 \le \gamma < 1$; thus, the effect of any initial error decreases rapidly as the computation progresses. However, if f is only Lipschitz continuous, then (6.17) suggests that any initial error that is made could be amplified to something exponentially large. While the initial error is typically quite small—usually, it is $\mathcal{O}(\mathbf{u})$—if the term multiplying it grows exponentially, then the effects of this small initial error could dominate the calculation at a later point in time.

Now let's look back at Example 6.6 from §6.2. For one equation ($y' = y$), the error grew almost exponentially with time, and if we look at the definition of f for this case, we see that it is only Lipschitz continuous. For the second equation ($y' = 1 - y$), however, we have that f is smooth and uniformly monotone decreasing in y, thus the error growth with time is much more benign, as predicted by the theory and confirmed by the computation.

Exercises:

1. Use Taylor's Theorem to prove that

$$(1 + x)^n \leq e^{nx}$$

   for all $x > -1$. *Hint:* Expand e^x in a Taylor series, throw away the unnecessary terms, and then take powers of both sides.

2. For each initial value problem below, use the error theorems of this section to estimate the value of

$$E(h) = \frac{\max_{t_k \leq 1} |y(t_k) - y_k|}{\|y''\|_{\infty, [0,1]}}$$

   using $h = 1/16$, assuming that Euler's method was used to approximate the solution.

   (a) $y' = \sin y$, $y(0) = \frac{1}{2}\pi$;

   (b) $y' + 4y = 1$, $y(0) = 1$;

   (c) $y' = y(1 - y)$, $y(0) = 1/2$.

3. Consider the initial value problem

$$y' = e^{-t} - 16y, \quad y(0) = 1.$$

   (a) Confirm that this is smooth and uniformly monotone decreasing in y. What is M? What is m?

   (b) Approximate the solution using Euler's method and $h = \frac{1}{8}$. Do we get the expected behavior from the approximation? Explain.

6.4 VARIANTS OF EULER'S METHOD

Euler's method, of course, is not the only nor even the best scheme for approximating solutions to initial value problems, and what we need to do now is look at other methods that we might employ. Several ideas can be considered based on some simple extensions of one derivation of Euler's method.

Our third derivation of Euler's method, which also gives us a remainder term, is based on our difference methods for derivative approximation (from §2.2). We start with the differential equation

$$y'(t) = f(t, y(t))$$

and replace the derivative with the simple difference quotient derived in (2.1). This yields

$$\frac{y(t + h) - y(t)}{h} = f(t, y(t)) + \frac{1}{2}hy''(\theta_{t,h}),$$

where the subscripts on θ remind us that the value depends on both t and h. Euler's method is then obtained simply by dropping the remainder, and replacing t with t_n and $y(t)$ with y_n, and so on. This is the derivation we used in Chapter 2.

This begs an obvious question: What happens if we use other approximations to the derivative? For example, if we use

$$y'(t) = \frac{y(t) - y(t - h)}{h} - \frac{1}{2}hy''(\theta),$$

then we get the *backward* Euler method:

$$y_{n+1} = y_n + hf(t_{n+1}, y_{n+1}); \tag{6.20}$$

and if we use

$$y'(t) = \frac{y(t + h) - y(t - h)}{2h} - \frac{1}{6}h^2 y'''(\theta_{t,h})$$

we get what is commonly referred to as the *midpoint method*:

$$y_{n+1} = y_{n-1} + 2hf(t_n, y_n). \tag{6.21}$$

Or, we could use the derivative approximations based on interpolation (§4.5):

$$y'(t) \quad \approx \quad \frac{1}{2h}\left(-y(t + 2h) + 4y(t + h) - 3y(t)\right),$$

$$y'(t + 2h) \quad \approx \quad \frac{1}{2h}\left(3y(t + 2h) - 4y(t + h) + y(t)\right),$$

to get the two numerical methods

$$y_{n+1} = 4y_n - 3y_{n-1} - 2hf(t_{n-1}, y_{n-1}), \tag{6.22}$$

and

$$y_{n+1} = \frac{4}{3}y_n - \frac{1}{3}y_{n-1} + \frac{2}{3}hf(t_{n+1}, y_{n+1}). \tag{6.23}$$

Finally, we note that yet another set of methods can be derived by integrating the differential equation. We have that the exact solution satisfies

$$y(t + h) = y(t) + \int_t^{t+h} f(s, y(s))ds. \tag{6.24}$$

We can therefore apply the trapezoid rule to (6.24) to get

$$y(t + h) = y(t) + \frac{1}{2}h\left[f(t + h, y(t + h)) + f(t, y(t))\right] - \frac{1}{12}h^3 y'''(\theta_{t,h}), \tag{6.25}$$

where $\theta_{t,h} \in [t, t + h]$ and we remind the reader that

$$f(t, y(t)) = y'(t) \Rightarrow \frac{d^2}{dt^2}f(t, y(t)) = y'''(t).$$

Dropping the remainder from (6.25) leads to the numerical method (commonly called the trapezoid method, for obvious reasons)

$$y_{n+1} = y_n + \frac{1}{2}h\left(f(t_{n+1}, y_{n+1}) + f(t_n, y_n)\right). \tag{6.26}$$

Alternatively, we can use a midpoint rule approximation to integrating (6.24). This leads to

$$y(t+h) = y(t) + hf\left(t + \frac{1}{2}h, y\left(t + \frac{1}{2}h\right)\right) - \frac{1}{24}h^3 y'''(\theta_{t,h}),$$

which suggests the numerical method

$$y_{n+1} = y_n + hf(t_{n+1/2}, y_{n+1/2}), \tag{6.27}$$

where $t_{n+1/2} = t_n + \frac{1}{2}h$ and $y_{n+1/2} \approx y(t_n + \frac{1}{2}h)$. This is similar to (6.21).

What about these methods? Are any of them any good?

Several observations can be made immediately. The methods (6.21), (6.22), and (6.23) are all based on derivative approximations that are $\mathcal{O}(h^2)$, whereas Euler's method was based on a derivative approximation that is only $\mathcal{O}(h)$ (as is the backward Euler method (6.20)). This suggests to us (but does not, of course, prove) that (6.21), (6.22), and (6.23) should be more accurate than Euler (and backward Euler). Similarly, the methods (6.26) and (6.27) are based on integral approximations that are more accurate.

A second observation involves the midpoint method (6.21) and the two methods (6.22) and (6.23). Note that here we have formulas for y_{n+1} in terms of y_n *and* y_{n-1}. These are not *single-step methods*, they are *multistep methods*; that is, they depend on information from more than one previous approximate value of the unknown function. How do we actually implement these methods? The differential equation only gives us a single initial value, y_0; we need more to even start the recursion here.

A third observation concerns backward Euler and the methods (6.26) and (6.23). Note that all of these formulas involve $f(t_{n+1}, y_{n+1})$; we cannot explicitly solve for the new approximate value y_{n+1}, which is why these methods (and others like them) are called *implicit*, whereas methods like Euler, midpoint, and (6.22) are called *explicit*, because they define y_{n+1} *explicitly* in terms of information from previous steps.

We would like to address the issue of accuracy, at least experimentally, but we can't even implement several of the methods until we address the other problems. However, it will be useful, at this point, to introduce some terminology associated with the accuracy of the various methods.

6.4.1   The Residual and Truncation Error

We are, of course, interested in the accuracy of the methods we develop here. All of the numerical methods for IVPs that we will study can be written in the general format

$$y_{n+1} = \sum_{k=0}^{p} a_k y_{n-k} + hF(y_{n+1}, y_n, \ldots, y_{n-p}; f_{n+1}, f_n, \ldots, f_{n-p}), \tag{6.28}$$

where we have used $f_k = f(t_k, y_k)$ for simplicity.[2] Thus, for example, in Euler's method we have

$$y_{n+1} = y_n + hf(t_n, y_n),$$

so that

$$a_0 = 1,$$
$$a_k = 0, \quad \text{all } k \geq 1;$$

---

[2]We will return to this notational convention in §6.4.3.

and

$$F(y_{n+1}, y_n, \ldots, y_{n-p}; f_{n+1}, f_n, \ldots, f_{n-p}) = f_n = f(t_n, y_n).$$

For the trapezoid rule method (6.26) we have

$$
\begin{aligned}
a_0 &= 1, \\
a_k &= 0, \quad \text{all } k \geq 1;
\end{aligned}
$$

but

$$F(y_{n+1}, y_n, \ldots, y_{n-p}; f_{n+1}, f_n, \ldots, f_{n-p}) = \frac{1}{2}(f_n + f_{n+1}) = \frac{1}{2}(f(t_n, y_n) + f(t_{n+1}, y_{n+1})).$$

Crucial to the accuracy of these methods is the extent to which the exact solution $y(t)$ satisfies the numerical method (6.28). To this end we define the residual, the truncation error (sometimes called the local truncation error), and the concept of consistency.

Definition 6.3 (Residual, Truncation Error, and Consistency) *For a numerical method written as in (6.28), define the residual, R_n, as given by*

$$
\begin{aligned}
R_n &= y(t_{n+1}) - \sum_{k=0}^{p} a_k y(t_{n-k}) - h F(y(t_{n+1}), y(t_n), \ldots, y(t_{n-p}); f(t_{n+1}, y(t_{n+1})), \\
&\quad f(t_n, y(t_n)), \ldots, f(t_{n-p}, y(t_{n-p}))).
\end{aligned}
$$

The truncation error is defined as

$$\tau_n = \frac{1}{h} R_n,$$

and the method is said to be consistent if

$$\lim_{h \to 0} \max_{t_n \leq T} |\tau_n| = 0$$

for sufficiently smooth solutions y.

For any "reasonable" method, the error $\max_{t_k \leq T} |y(t_k) - y_k|$ is proportional to the truncation error, if the solution is smooth enough. Note that the residual and truncation error are both defined in terms of substituting the exact solution into the numerical method.[3]

For most methods that we develop, the residual and truncation error naturally come out of the construction process. This is the case, for example, with Euler's method and the several variants discussed above.

For example, consider some of the methods we have looked at so far.

■ **EXAMPLE 6.7**

Using either of the two analytic derivations for Euler's method, we can write

$$y(t_{n+1}) = y(t_n) + h f(t_n, y(t_n)) + \frac{1}{2} h^2 y''(\theta_n);$$

---

[3]There is a disturbing lack of consistency in the literature on the terminology used here. For example, Burden and Faires [5], Isaacson and Keller [12], and Allen and Isaacson [1], all define *truncation error* as we have here. Other books simply replace the word *truncation* with *discretization*. However, Atkinson [2] and Lambert [15], among others, use *truncation error* or *local truncation error* for what we have called the *residual*. So, when consulting references, be sure to check and see which definition is being used.

therefore, the residual is $\frac{1}{2}h^2 y''(\theta_n)$, and the truncation error is $\frac{1}{2}hy''(\theta_n)$. The method is consistent as long as the solution is C^2, because in this case we have

$$\lim_{h \to 0} \max_{t_n \le T} \frac{1}{2}h^2 |y''(\theta_n)| = \left(\lim_{h \to 0} \frac{1}{2}h^2\right)\left(\max_{t \le T} |y''(t)|\right) = \left(\lim_{h \to 0} \frac{1}{2}h^2\right)(Y_2) = 0,$$

where Y_2 is the maximum absolute value of y'' for $t \le T$.

■ EXAMPLE 6.8

Using what we know of the trapezoid rule for approximating integrals, we can write (see §2.5)

$$y(t_{n+1}) = y(t_n) + \frac{1}{2}h \left(f(t_n, y(t_n)) + f(t_{n+1}, y(t_{n+1}))\right) - \frac{1}{12}h^3 y'''(\theta_n);$$

therefore, the residual is $-\frac{1}{12}h^3 y'''(\theta_n)$, and the truncation error is $-\frac{1}{12}h^2 y'''(\theta_n)$. This time, the method is consistent if y is C^3.

Generally, the residual and truncation error are derived by expanding $y(t_{n+1})$ in a Taylor series about t_n and matching terms so they vanish, or by applying some approximate integration or differentiation method.

■ EXAMPLE 6.9

Consider the method defined by

$$y_{n+1} = y_n + \frac{1}{2}h(3f(t_n, y_n) - f(t_{n-1}, y_{n-1})),$$

which we encounter in §6.6 as one of the Adams–Bashforth methods. What are its residual and truncation error, and is it a consistent method?

We have

$$R_n = y(t_{n+1}) - y(t_n) - \frac{1}{2}h(3f(t_n, y(t_n)) - f(t_{n-1}, y(t_{n-1}))).$$

We write $y(t_{n+1})$ in a Taylor series as

$$y(t_{n+1}) = y(t_n) + hy'(t_n) + \frac{1}{2}h^2 y''(t_n) + \frac{1}{6}h^3 y'''(\theta),$$

and use the fact that $y'(t) = f(t, y(t))$ to write the residual as

$$\begin{aligned} R_n &= hy'(t_n) + \frac{1}{2}h^2 y''(t_n) + \frac{1}{6}h^3 y'''(\theta) - \frac{1}{2}h(3y'(t_n) - y'(t_{n-1})) \\ &= \frac{1}{2}h\left[hy''(t_n) - (y'(t_n) - y'(t_{n-1}))\right] + \frac{1}{6}h^3 y'''(\theta). \end{aligned}$$

Taylor's Theorem can again be applied, this time to y', to get

$$y'(t_{n-1}) = y'(t_n) - hy''(t_n) + \frac{1}{2}h^2 y'''(\eta),$$

so that we now have

$$R_n = \frac{1}{2}h\left[hy''(t_n) - hy''(t_n) + \frac{1}{2}h^2 y'''(\eta)\right] + \frac{1}{6}h^3 y'''(\theta) = h^3 \left(\frac{1}{2}y'''(\eta) + \frac{1}{6}y'''(\theta)\right).$$

Therefore, the residual is $\mathcal{O}(h^3)$ and the truncation error is $\mathcal{O}(h^2)$. Moreover, the method is consistent whenever the exact solution is C^3.

We close this section with a definition.

Definition 6.4 (Order of Accuracy) *If the truncation error for a numerical scheme for the solution of initial value problems is $\mathcal{O}(h^k)$, then we say that the method has* order of accuracy k.

Thus, the trapezoid rule method has order of accuracy $p = 2$, and Euler's method has order of accuracy $p = 1$. It is commonplace to say that the trapezoid rule is second-order accurate, and Euler's method is first-order accurate.

There is a potential for confusion here. Note that the definition of order of accuracy is based on what we defined as the *truncation error*, not the residual.

6.4.2 Implicit Methods and Predictor–Corrector Schemes

The methods (6.26), (6.27), and (6.23) are all examples of *implicit* methods, because in each instance it is not possible to solve for y_{n+1} in terms of y_n and other known values. Nonetheless it is possible to make use of these methods; the trapezoid method, (6.26), is in fact very useful.

One way to implement methods of this type is to view the equations (6.26), (6.27), or (6.23) as an instance of a single nonlinear equation in the single unknown y_{n+1}. That is, we *solve* for the unknown value of y_{n+1} using Newton's method or the secant method or a fixed-point iteration.

■ **EXAMPLE 6.10**

Consider the IVP

$$y' = -y \ln y, \quad y(0) = y_0. \tag{6.29}$$

Applying the trapezoid method (6.26) to this yields the computation

$$y_{n+1} = y_n - \frac{1}{2}h[y_{n+1} \ln y_{n+1} + y_n \ln y_n], \tag{6.30}$$

where we now view y_{n+1} as the (unknown) root of the function

$$F(y) = y - y_n + \frac{1}{2}h[y \ln y + y_n \ln y_n].$$

Alternatively, we can view (6.30) as defining y_{n+1} as the fixed point of the function

$$g(y) = y_n - \frac{1}{2}h[y \ln y + y_n \ln y_n].$$

Regardless of which view we take, we can apply the methods of Chapter 3 to solve for y_{n+1}. Denote the individual iterates as $y_{n+1,k}$, where k is the iteration counter.[4]

---

[4] At this point it is worthwhile to point out that some confusion can occur. We have two indices on y. One of them $(n + 1)$ refers to the time step in the numerical solution of the differential equation. The other one (k) refers to the iteration count for the solution—within each time step—of the nonlinear equation for the next approximate value of y.

Then Newton's method applied to $F(y)$ yields the iteration

$$y_{n+1,k+1} = y_{n+1,k} - \left(\frac{y_{n+1,k} - y_n - \frac{1}{2}h[f(t_{n+1}, y_{n+1,k}) + f(t_n, y_n)]}{1 - \frac{1}{2}hf_y(t_{n+1}, y_{n+1,k})} \right),$$

where

$$f_y(t, y) = \frac{\partial f}{\partial y}(t, y);$$

the secant method yields the iteration

$$y_{n+1,k+1} = y_{n+1,k} - \left(y_{n+1,k} - y_n - \frac{1}{2}h[f(t_{n+1}, y_{n+1,k}) + f(t_n, y_n)] \right)$$

$$\times \left(\frac{y_{n+1,k} - y_{n+1,k-1}}{y_{n+1,k} - \frac{1}{2}hf(t_{n+1}, y_{n+1,k}) - y_{n+1,k-1} + \frac{1}{2}hf(t_{n+1}, y_{n+1,k-1})} \right);$$

and a simple fixed-point iteration yields the iteration

$$y_{n+1,k+1} = y_n - \frac{1}{2}h[y_{n+1,k} \ln y_{n+1,k} + y_n \ln y_n]. \tag{6.31}$$

(Note that the fixed-point iteration is much the simplest of these three, although it will usually be the slowest to converge.) We know that all three of these will converge if h is sufficiently small and if we take the initial guess close enough to the actual value (for the fixed-point iteration this is a straightforward application of Theorem 3.6; see Problem 17); it is very reasonable to take $y_{n+1,0} = y_n$, i.e., to take the value from the previous step of the iteration as the initial guess for the next step. An alternate (and usually better) means of initializing the iteration is to use some explicit method—Euler's method, for example—to generate $y_{n+1,0}$.

Although there are circumstances in which it is necessary to carry out this type of iteration, for most problems it is, in fact, possible to use a much cruder means of estimating y_{n+1}. This leads us to the very important *predictor–corrector* idea.

Consider the trapezoid method, defined by (6.26); instead of actually solving for the exact value of y_{n+1} that satisfies this equation, we apply a simpler method to estimate y_{n+1}, and then use this estimated value on the right side of (6.26). For the trapezoid method, it is common to use Euler's method as the estimator, or *predictor*:

$$\bar{y}_{n+1} = y_n + hf(t_n, y_n); \tag{6.32}$$

and then the *corrector* step is accomplished by

$$y_{n+1} = y_n + \frac{1}{2}h\left[f(t_{n+1}, \bar{y}_{n+1}) + f(t_n, y_n)\right]. \tag{6.33}$$

This combination is generally referred to as the *trapezoid rule predictor–corrector method*. Note that it is the same as using Euler's method to initialize a fixed-point iteration that we arbitrarily stop after one step.

■ **EXAMPLE 6.11**

Consider the IVP

$$y' = -y \ln y, \quad y(0) = \frac{1}{2}.$$

We will use the trapezoid rule predictor–corrector to approximate the solution at $t = 1$ using the rather crude value $h = 0.25$. We have the following sequence of computations:

$$\bar{y} = y_0 + hf(t_0, y_0) = 0.5 - 0.25 \times 0.5 \ln 0.5 = 0.5866433976;$$
$$y_1 = y_0 + \frac{1}{2}h[f(t_0, y_0) + f(t_1, \bar{y})] = 0.58243161136465.$$

$$\bar{y} = y_1 + hf(t_1, y_1) = 0.66113901764113;$$
$$y_2 = y_1 + \frac{1}{2}h[f(t_1, y_1) + f(t_2, \bar{y})] = 0.65598199856663.$$

$$\bar{y} = y_2 + hf(t_2, y_2) = 0.72512609790319;$$
$$y_3 = y_2 + \frac{1}{2}h[f(t_2, y_2) + f(t_3, \bar{y})] = 0.71968686944048.$$

$$\bar{y} = y_3 + hf(t_3, y_3) = 0.77887015094459;$$
$$y_4 = y_3 + \frac{1}{2}h[f(t_3, y_3) + f(t_4, \bar{y})] = 0.77360953103925.$$

The exact value is $y(1) = 0.77492068450995$, so we are not far off, using this crude timestep.

■ **EXAMPLE 6.12**

To illustrate the difference between actually solving the nonlinear equation and using a predictor–corrector method, we continue with the same IVP (6.29) as used in Example 6.11. The exact solution here is $y = e^{(-\ln 2)e^{-t}}$. Table 6.2 shows the results of approximating the solution to this equation using the predictor–corrector method (6.32)–(6.33) (second column) and actually solving the equation (6.30) (using a fixed-point iteration) for the exact value of y_{n+1} (fourth column). Clearly, the two methods are not producing wildly different approximations. Note, also, that the calculation based on exactly solving the recursion is more accurate; in fact, the error in the iterated approximation (fifth column) starts out at about 61% of the error for the predictor–corrector approximation, and ends up at around 9%.

Generally speaking, unless the differential equation is *very* sensitive to changes in the data, a simple predictor–corrector method will be just as good as the more time-consuming process of solving for the exact value of y_{n+1}, which satisfies the implicit recursion.

Table 6.2   Implicit method applied to $y' = -y \ln y$.

n	Predictor–corrector		Fixed-point solution	
	y_n	Error	y_n	Error
1	0.521438685E+00	0.629350068E–05	0.521441082E+00	0.389659700E–05
2	0.542415401E+00	0.124265310E–04	0.542420677E+00	0.715031332E–05
3	0.562889543E+00	0.183888337E–04	0.562898148E+00	0.978351030E–05
4	0.582826911E+00	0.241665678E–04	0.582839249E+00	0.118283433E–04
5	0.602199375E+00	0.297435050E–04	0.602215794E+00	0.133243757E–04
6	0.620984504E+00	0.351021428E–04	0.621005290E+00	0.143164168E–04
7	0.639165163E+00	0.402246989E–04	0.639190535E+00	0.148526117E–04
8	0.656729097E+00	0.450939634E–04	0.656759208E+00	0.149827925E–04
9	0.673668505E+00	0.496939995E–04	0.673703441E+00	0.147570906E–04
10	0.689979615E+00	0.540106915E–04	0.690019401E+00	0.142247986E–04
11	0.705662271E+00	0.580321478E–04	0.705706870E+00	0.134334649E–04
12	0.720719531E+00	0.617489703E–04	0.720768874E+00	0.124059468E–04
13	0.735157278E+00	0.651544045E–04	0.735211228E+00	0.112040544E–04
14	0.748983859E+00	0.682443879E–04	0.749042237E+00	0.986672690E–05
15	0.762209745E+00	0.710175127E–04	0.762272333E+00	0.842946303E–05
16	0.774847210E+00	0.734749197E–04	0.774913760E+00	0.692422919E–05

Algorithm 6.1 *Trapezoid predictor–corrector Pseudocode*

```
input t0, y0, h, n
external f
for k = 1 to n do
!
!     First, predict:
!
      f0 = f(t0,y0)
      ybar = y0 + h*f0
!
!     Next, correct:
!
      y = y0 + 0.5*h*(f0 + f(t0+h,ybar))
!
!     Update for next pass through the loop
!
      y0 = y
      t0 = t
endfor
```

Programming Hint: Note how this code saves the function evaluation `f(t0,y0)` in order to re-use it in a later statement.

If the differential equation is linear, however, then we can entirely avoid the problem of implicitness, as we now illustrate. Write the general linear ODE

$$y' = a(t)y + b(t), \quad y(t_0) = y_0,$$

where a and b are known functions of t, only. If we apply the trapezoid method to this equation we initially have the implicit recursion

$$y_{n+1} = y_n + \frac{1}{2}h\left[a(t_n)y_n + b(t_n) + a(t_{n+1})y_{n+1} + b(t_{n+1})\right].$$

However, we can make this explicit, by using the linearity of the equation to solve for y_{n+1}:

$$y_{n+1} = \left(\frac{1 + \frac{1}{2}ha(t_n)}{1 - \frac{1}{2}ha(t_{n+1})}\right)y_n + \frac{1}{2}h\left(\frac{b(t_n) + b(t_{n+1})}{1 - \frac{1}{2}ha(t_{n+1})}\right). \qquad (6.34)$$

The significance of this will become apparent later, when we discuss stability and stiffness issues in more detail. For now, we content ourselves with an example.

■ **EXAMPLE 6.13**

Consider the IVP

$$y' = y, \quad y(0) = 1,$$

which we considered previously, using only Euler's method. This is a linear equation, so we can use the ideas of (6.34) to approximate this using the trapezoid rule. The

Table 6.3   Trapezoid rule applied to $y' = y$.

$n = h^{-1}$	$E(h)$
4	0.1432958E–01
8	0.3550064E–02
16	0.8855204E–03
32	0.2212558E–03
64	0.5530617E–04
128	0.1382606E–04
256	0.3456484E–05
512	0.8641191E–06
1024	0.2160295E–06

recursion becomes

$$y_{n+1} = \frac{1+h}{1-h}y_n, \quad y_0 = 1;$$

thus, for $h = \frac{1}{4}$ we very quickly get that

$$y_1 = \frac{5/4}{3/4}y_0 = 1.6667; \qquad y_2 = \frac{5}{3}y_1 = 2.7778;$$

$$y_3 = \frac{5}{3}y_2 = 4.630; \qquad y_4 = \frac{5}{3}y_3 = 7.7160.$$

If we continue with smaller and smaller values of h, then the error

$$E(h) = \max_{t_k \leq 1} |y(t_k) - y_k|$$

is as given in Table 6.3. Note that the error is going down by a factor of 4, as we expect from a second-order method.

The predictor–corrector idea can also be used to make (6.27) a practical method. If we use Euler's method as the predictor, then we have

$$\bar{y} = y_n + \frac{1}{2}hf(t_n, y_n) \tag{6.35}$$

$$, y_{n+1} = y_n + hf\left(t_n + \frac{1}{2}h, \bar{y}\right). \tag{6.36}$$

Note that we used the Euler predictor to take a "half step," and used this value in the corrector to approximate $y(t_n + h/2)$.

■ **EXAMPLE 6.14**

Let's compute approximate solutions to the IVP:

$$y' = -e^{-(t+y)}, \quad y(0) = 1,$$

which has exact solution $y = \ln(e - 1 + e^{-t})$. Using $h = \frac{1}{4}$, we compute as follows:

$$\bar{y} = y_0 + \frac{1}{2}h\left(-e^{-(t_0+y_0)}\right) = 1 - \frac{1}{8}e^{-1} = 0.9540150699,$$

so that then

$$y_1 = y_0 + h\left(-e^{-(t_0+h/2+\bar{y})}\right) = 1 - \frac{1}{4}e^{-\frac{1}{8}-0.9540150699} = 0.957508729.$$

The second value is computed similarly:

$$\bar{y} = y_1 + \frac{1}{2}h\left(-e^{-(t_1+y_1)}\right) = 0.920141092,$$

$$y_2 = y_1 + h\left(-e^{-(t_1+h/2+\bar{y})}\right) = 0.8890439235.$$

It can be shown (see the exercises) that this is also a second-order method.

6.4.3   Starting Values and Multistep Methods

Because they depend on more than just a single previous value, multistep methods such as (6.21), (6.22), and (6.23) require *starting values*. In essence, we must use some other method to compute approximations $y_1, y_2, \ldots, y_p$ before we can use the multistep method. For the methods presented so far, we need only a single additional value, y_1. For some of the methods presented in §6.6, we will need several such starting values. One reason for the development of the predictor–corrector methods in §6.4.2 is precisely to give us single-step methods that we can use to produce accurate starting values for the multistep methods, such as (6.21), (6.22), and (6.23).

Let's look at an example.

■ **EXAMPLE 6.15**

We take as our initial value problem the same one we looked at in the beginning of
§6.4.2,

$$y' = -y \ln y, \quad y(0) = \frac{1}{2}. \tag{6.37}$$

We want to apply, say, the midpoint method

$$y_{n+1} = y_{n-1} + 2hf(t_n, y_n)$$

with $h = \frac{1}{4}$ to this problem. We will first use Euler's method to provide the estimate
of $y(t_1)$ that is necessary to proceed with the computation. We get

$$y_1 = y_0 + hf(t_0, y_0) = \frac{1}{2} + \frac{1}{4}\left(-\frac{1}{2}\ln\frac{1}{2}\right) = 0.5866433976.$$

We can now compute using the midpoint method:

$$y_2 = y_0 + 2hf(t_1, y_1) = \frac{1}{2} + 2 \times \frac{1}{4}\left(-0.5866433976 \ln 0.5866433976\right) = 0.6564396503,$$

$$y_3 = y_1 + 2hf(t_2, y_2) = 0.7247991686,$$

and so on.

Table 6.4   Midpoint method applied to $y' = -y \ln y$, $h = 1/16$.

n	y_1 computed by predictor–corrector		y_1 computed by Euler's method	
	y_n	Error	y_n	Error
1	0.521438685E+00	0.629350068E–05	0.521660849E+00	–0.215870648E–03
2	0.542442735E+00	–0.149082970E–04	0.542433042E+00	–0.521501377E–05
3	0.562913366E+00	–0.543392852E–05	0.563136000E+00	–0.228068587E–03
4	0.582876067E+00	–0.249892345E–04	0.582854530E+00	–0.345267290E–05
5	0.602241522E+00	–0.124030017E–04	0.602465395E+00	–0.236276553E–03
6	0.621050404E+00	–0.307978504E–04	0.621015069E+00	0.453741300E–05
7	0.639220651E+00	–0.152631109E–04	0.639446837E+00	–0.241449481E–03
8	0.656807255E+00	–0.330637366E–04	0.656756294E+00	0.178973669E–04
9	0.673732972E+00	–0.147732929E–04	0.673962850E+00	–0.244651773E–03
10	0.690066203E+00	–0.325779206E–04	0.689997851E+00	0.357749024E–04
11	0.705732009E+00	–0.117057153E–04	0.705967262E+00	–0.246958289E–03
12	0.720811392E+00	–0.301122659E–04	0.720723877E+00	0.574032598E–04
13	0.735229212E+00	–0.678027590E–05	0.735471822E+00	–0.249390296E–03
14	0.749078472E+00	–0.263680981E–04	0.748969952E+00	0.821511474E–04
15	0.762281388E+00	–0.625682149E–06	0.762533643E+00	–0.252880568E–03
16	0.774942633E+00	–0.219485394E–04	0.774811136E+00	0.109548790E–03

Suppose now that we want to use the trapezoid rule predictor–corrector method
(6.32) and (6.33) to provide the estimate of $y(t_1)$. Again, using $h = \frac{1}{4}$ we compute

$$\bar{y} = y_0 + hf(t_0, y_0) = 0.5 - 0.25 \times 0.5 \ln 0.5 = 0.5866433976;$$

$$y_1 = y_0 + \frac{1}{2}h[f(t_0, y_0) + f(t_1, \bar{y})] = 0.5824316114.$$

so that

$$y_2 = y_0 + 2hf(t_1, y_1) = \frac{1}{2} + 2 \times \frac{1}{4}(-0.5824316114 \ln 0.5824316114) = 0.6574148126,$$

$$y_3 = y_1 + 2hf(t_2, y_2) = 0.7203046741,$$

and so on. Regardless of which method is used to produce y_1, we can now compute y_k values using only the midpoint method.

Now, by taking $h = 1/16$, and computing out to $t = 1$, we get the results shown in Table 6.4. Compare this to Table 6.2. Note that, as expected, we are getting better accuracy when we use the more accurate starting value, although it should be noted that the discrepancy is not large.

This example begs an interesting question. Note that we used Euler's method, which is only $\mathcal{O}(h)$, to generate the starting values for the midpoint method, which is $\mathcal{O}(h^2)$. Is the use of a lower-order method for the starting values going to affect the accuracy of the overall computation?

A complete explanation is beyond the scope of this book, but it can be shown that a $(p-1)$-order method can be used to generate the starting values for a p-order method, usually without affecting the overall order of convergence. It would of course be more accurate to use a p-order method than to use a $(p-1)$-order method for the starting values, but the net order of accuracy of the method should not be affected.

6.4.4   The Midpoint Method and Weak Stability

Let's consider the midpoint method (6.21) as applied to the very simple differential equation

$$y' = -y, \quad y(0) = 1, \tag{6.38}$$

which has exact solution $y(t) = e^{-t}$. To minimize the effects of any error in the starting values, let's use

$$y_1 = y(t_1) = e^{-h}, \tag{6.39}$$

where h is the mesh spacing; note that this is the *exact* value of $y(t_1)$; thus, there is no error in using this starting value. Figure 6.4 shows the results of applying (6.21) to (6.38), using (6.39) as the starting value, for a sequence of mesh values $h^{-1} = 4, 8, 16, \ldots, 128$. We have plotted the exact solution curve and each approximate curve, as generated by the numerical method.

We expect to get decent results, since the midpoint method is second-order; instead, we get terrible results for larger values of t. Taking a smaller value of h does appear to defer the onset of this problem, but it does not appear to eliminate it. As Figure 6.5 shows, even for $h^{-1} = 128$ we eventually get poor results.

What is going on here?

It is tempting to blame this problem—which is known as *weak stability*—on rounding error, but in fact the problem is inherent in the numerical method (6.21), and would occur in exact arithmetic. Because a good understanding of weak stability is necessary for the full discussion of stability for multistep methods (see §6.7.1), we will go into some detail here.

We first slightly generalize our model problem to

$$y' = \lambda y, \quad y(0) = 1,$$

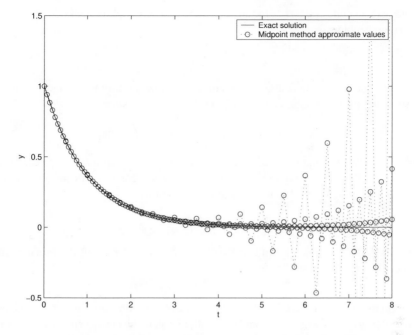

Figure 6.4   Illustration of Weak Stability, $h = 4^{-1}, 8^{-1}, 16^{-1}$.

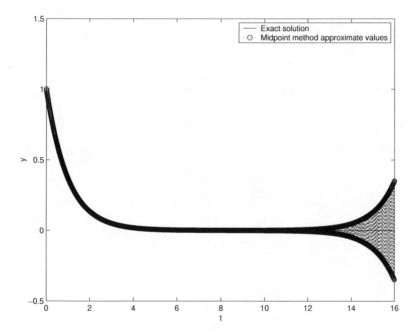

Figure 6.5   Illustration of Weak Stability, $h = 128^{-1}$

which has exact solution $y = e^{\lambda t}$ for any constant λ. The midpoint method applied to this problem then becomes

$$y_{n+1} = y_n + 2h\lambda y_n, \quad y_0 = 1, y_1 = e^{h\lambda}.$$

Note that the mesh spacing h and the parameter λ show up only when multiplying each other; so, to simplify the notation, we will write $h\lambda = \xi$ hereafter. Thus, we have

$$y_{n+1} = y_{n-1} + 2\xi y_n, \quad y_0 = 1, y_1 = e^{\xi}. \tag{6.40}$$

This is an example of what is called a *three-term recurrence relation*, and we can actually produce a formula to solve it, much as we would solve a second-order constant coefficient differential equation. We first look for solutions in the form

$$y_n = r^n \tag{6.41}$$

for some value r. We actually find *two* such values, r_1 and r_2, and the linearity of the recurrence then implies that

$$y_n = C_1 r_1^n + C_2 r_2^n$$

is a solution for any constants C_1 and C_2. The values of the constants are determined by the need to satisfy $y_0 = 1$ and $y_1 = e^{\xi}$, and that will complete the construction of the solution.

So much for the outline and preview. If we substitute (6.41) into (6.40), we get

$$r^{n+1} = r^{n-1} + 2\xi r^n.$$

Thus, dividing by r^{n-1} yields the quadratic equation

$$r^2 - 2\xi r - 1 = 0,$$

which defines r_1 and r_2, according to the quadratic formula:

$$r_1 = (\xi + \sqrt{\xi^2 + 1}), \quad r_2 = (\xi - \sqrt{\xi^2 + 1}). \tag{6.42}$$

The constants C_1 and C_2 are then found by solving the system

$$
\begin{aligned}
1 &= C_1 + C_2 \\
e^{\xi} &= C_1 r_1 + C_2 r_2
\end{aligned}
$$

to get

$$C_1 = \frac{r_2 - e^{\xi}}{r_2 - r_1} = \frac{(-\xi + \sqrt{\xi^2 + 1}) + e^{\xi}}{2\sqrt{\xi^2 + 1}}$$

and

$$C_2 = \frac{r_1 - e^{\xi}}{r_1 - r_2} = \frac{(\xi + \sqrt{\xi^2 + 1}) - e^{\xi}}{2\sqrt{\xi^2 + 1}}.$$

Hence, we have the solution

$$y_n = C_1 r_1^n + C_2 r_2^n \tag{6.43}$$

which the student can verify by direct substitution (see Problem 12) is the exact solution to the recurrence (6.40). We should repeat, for emphasis, that this solution formula is *exact*: This is what the computation will produce in the complete absence of rounding error.

Some modest work allows us to estimate the size of C_1 and C_2; thus (see Problem 14):

$$C_1 = 1 + \mathcal{O}(\xi^3), \quad C_2 = \mathcal{O}(\xi^3). \tag{6.44}$$

Consider, now, the behavior of the approximate solution as n grows, under the assumption that $\lambda < 0$. This assumption means that $\xi < 0$, and it also means that the exact solution will be a decaying exponential. However, the approximate solution satisfies

$$y_n = C_1 r_1^n + C_2 r_2^n,$$

where we have $0 < r_1 < 1$ and $r_2 < -1$ (see Problem 13). Hence, as $n \to \infty$, $r_1^n \to 0$, but $|r_2|^n \to \infty$, thus destroying the accuracy of the computation. Since $C_2 \neq 0$, this term will be present in the calculation, although the fact that $C_2 = \mathcal{O}(h^3)$ does mean that it will take more and more steps for the growing term to be noticed, for smaller and smaller values of h.

The fundamental issue here is that the exact solution to the recursion that defines the approximate solution contains solution components that do not correspond in any way to the exact solution of the ODE. The $C_1 r_1^n$ term, it can be shown, will converge to the exact solution and will be an $\mathcal{O}(h^2)$ approximation to it. But the other term, usually called a *parasite* or *parasitic solution*, will eventually dominate and corrupt the approximation.

Parasitic solutions are an inherent feature of multistep methods, and the stability theory for multistep methods is designed to minimize their ability to corrupt the solution. (See §6.7 for more details.)

Exercises:

1. Using the approximation

$$y'(t_n) \approx \frac{y(t_{n+1}) - y(t_{n-1})}{2h},$$

   derive the numerical method (6.21) for solving initial value problems. What is the residual? What is the truncation error? Is it a consistent method?

2. Using the approximation

$$y'(t_{n-1}) \approx \frac{-y(t_{n+1}) + 4y(t_n) - 3y(t_{n-1})}{2h},$$

   derive the numerical method (6.22) for solving initial value problems. What is the residual? What is the truncation error? Is it a consistent method?

3. Using the approximation

$$y'(t_{n+1}) \approx \frac{3y(t_{n+1}) - 4y(t_n) + y(t_{n-1})}{2h},$$

   derive the numerical method (6.23) for solving initial value problems. What is the residual? What is the truncation error? Is it a consistent method?

4. Use the trapezoid rule predictor–corrector with $h = \frac{1}{4}$ to compute approximate values of $y(1)$ for each of the following initial value problems. Don't write a computer program; use a hand calculator to produce an orderly table of (t_k, y_k) pairs.

    (a) $y' = y(1 + e^{2t})$, $y(0) = 1$;

    (b) $y' + 2y = 1$, $y(0) = 2$;

    (c) $y' = y(1 - y)$, $y(0) = \frac{1}{2}$;

    (d) $ty' = y(\sin t)$, $y(0) = 2$.

5. Repeat Problem 4, using the method (6.23) as a predictor–corrector, with Euler's method as the predictor. Also use Euler's method to produce the starting value, y_1.

6. Write a computer program that solves each of the initial value problems in Problem 4, using the trapezoid rule predictor–corrector and $h = 1/16$.

7. For each initial value problem below, approximate the solution using the trapezoid rule predictor–corrector with a sequence of decreasing grids $h^{-1} = 2, 4, 8, \ldots$. For those problems where an exact solution is given, compare the accuracy achieved over the interval $[0, 1]$ with the theoretical accuracy.

    (a) $y' + y = \sin 4\pi t$, $y(0) = \frac{1}{2}$;

    (b) $y' + \sin y = 0$, $y(0) = 1$;

    (c) $y' + 4y = 1$, $y(0) = 1$; $y(t) = \frac{1}{4}(3e^{-4t} + 1)$;

    (d) $y' = -y \ln y$, $y(0) = 3$; $y(t) = e^{(\ln 3)e^{-t}}$.

8. Here, we will consider a tumor growth model based on some work of H. P. Greenspan in *J. Theor. Biol.*, vol. 56, pp. 229-242, 1976. The differential equation is

$$R'(t) = -\frac{1}{3}S_i R + \frac{2\lambda\sigma}{\mu R + \sqrt{\mu^2 R^2 + 4\sigma}}, \qquad R(0) = a.$$

Here $R(t)$ is the radius of the tumor (assumed spherical); λ and μ are scale parameters, both $\mathcal{O}(1)$; S_i measures the rate at which cells at the core of the tumor die; and σ is a nutrient level. Take $\lambda = \mu = 1$, $a = 0.25$, $S_i = 0.8$, and $\sigma = 0.25$. Use the trapezoid rule predictor–corrector to solve the differential equation, using $h = 1/16$, and show that the tumor radius approaches a limiting value as $t \to \infty$.

9. Repeat Problem 8, but this time with $S_i = 0.90$, $\sigma = 0.05$, and $a = 0.50$. What happens now?

10. Now solve the differential equation using a variety of S_i, σ, and a values (your choices). What happens?

11. Now let's model some treatment for our mathematical tumor. In Problems 8–10, we assumed that the nutrient level was constant. Suppose that we are able to decrease the nutrient level according to the model

$$\sigma(t) = \sigma_\infty + (\sigma_0 - \sigma_\infty)e^{-qt}.$$

Here σ_0 is the initial nutrient level, σ_∞ is the asymptotic nutrient level, and q measures the rate at which the nutrient level drops. Investigate the effect of various choices of these parameters on the growth of the tumor, based on your observations from the earlier problems. Again, use the trapezoid rule predictor–corrector with $h = 1/16$ to solve the differential equation.

12. Verify by direct substitution that (6.43) satisfies the recursion (6.40) for all $n \geq 2$.

13. For the midpoint method (6.21), show that if $\lambda < 0$, then $0 < r_1 < 1$ and $r_2 < -1$.

14. Show that (6.44) is valid. *Hint:* First, show that $C_1 = 1 - C_2$. Next, use Taylor expansions to write

$$\frac{\xi - e^\xi}{\sqrt{\xi^2 + 1}} = \frac{-1 - \frac{1}{2}\xi^2 + \mathcal{O}(\xi^3)}{1 + \frac{1}{2}\xi^2 + \mathcal{O}(\xi^4)}.$$

15. Use the midpoint rule predictor–corrector method (6.35) and (6.36) to solve each of the IVPs given in Problem 4.

16. Show that the residual for the midpoint rule predictor–corrector is given by

$$R = y(t + h) - y(t) - hf(t + h/2, y(t) + (h/2)f(t, y(t))).$$

Then use Taylor's theorem to show that $R = \mathcal{O}(h^3)$, and hence that the midpoint rule predictor–corrector is a second-order method.

17. Assume that f is differentiable in y and that this derivative is bounded in absolute value for all t and y:

$$|f_y(t, y)| \leq F.$$

Show that using fixed-point iteration to solve for y_{n+1} in the trapezoid rule method will converge so long as h is sufficiently small. *Hint:* Recall Theorem 3.6.

18. Derive the numerical method based on using Simpson's rule to approximate the integral in

$$y(t + h) = y(t - h) + \int_{t-h}^{t+h} f(s, y(s))ds.$$

What is the order of accuracy of this method? What is the truncation error? Is it implicit or explicit? Is it a single-step or a multistep method?

$$\triangleleft \bullet \bullet \bullet \triangleright$$

6.5 SINGLE-STEP METHODS: RUNGE–KUTTA

The Runge–Kutta[5] family of methods is one of the most popular families of accurate solvers for initial value problems. The general derivation can become very involved; to avoid drowning in a sea of detail and notation, we will outline the basic ideas using the second-order case.

Recall the usual predictor–corrector formulation of the trapezoid method:

$$\begin{aligned}
\bar{y}_{n+1} &= y_n + hf(t_n, y_n), \\
y_{n+1} &= y_n + \frac{1}{2}h\left[f(t_{n+1}, \bar{y}_{n+1}) + f(t_n, y_n)\right].
\end{aligned}$$

---

[5]Martin Wilhelm Kutta (1867–1944) studied at Breslau and Munich, in addition to a year spent in Britain at Cambridge. Most of his professional career was spent in Stuttgart. Building on Runge's original idea (first presented in an 1894 article), Kutta published his version of the Runge–Kutta methods in 1901.

We can directly substitute the predictor into the corrector to write this as a single recursion:

$$y_{n+1} = y_n + \frac{1}{2}h\left[f(t_{n+1}, y_n + hf(t_n, y_n)) + f(t_n, y_n)\right]. \tag{6.45}$$

Since the differential equation implies that f gives values of y', it follows that we can view (6.45) as defining y_{n+1} from y_n by advancing along a straight line defined by the simple average of the two slopes $f(t_n, y_n)$ and $f(t_{n+1}, \bar{y})$, where $\bar{y} = y_n + hf(t_n, y_n)$. This raises the question, of course, of whether or not a different average of two slopes might yield a more accurate method. To this end, we consider the more general method

$$y_{n+1} = y_n + c_1 hf(t_n, y_n) + c_2 hf(t_n + \alpha h, y_n + \beta hf(t_n, y_n)), \tag{6.46}$$

where c_1, c_2, α, β are parameters as yet undetermined. We want to choose these so that the approximate solution defined by (6.46) is as accurate as possible; hence, we want to make the truncation error as small as possible, in terms of powers of h. Thus, we look at the (admittedly rather imposing) expression

$$R = y(t + h) - y(t) - c_1 hf(t, y(t)) - c_2 hf(t + \alpha h, y(t) + \beta hf(t, y(t))).$$

To reduce this so that we can infer the correct values of c_1, c_2, α, and β that will yield the smallest residual R, we will need to use Taylor's Theorem in two variables, which we state here without proof:

$$
\begin{aligned}
F(x + h, y + \eta) =\ & F(x,y) + hF_x(x,y) + \eta F_y(x,y) + \frac{1}{2}(h^2 F_{xx}(x,y) \\
& + h\eta F_{xy}(x,y) + \eta^2 F_{yy}(x,y)) + \mathcal{O}(h^3 + \eta^3).
\end{aligned}
$$

We can thus expand the last term in R as

$$
\begin{aligned}
f(t + \alpha h, y(t) + \beta hf(t,y(t))) =\ & f(t,y(t)) + \alpha h f_t(t,y(t)) + \beta h f(t,y(t)) f_y(t,y(t)) \\
& + \frac{1}{2}(\alpha^2 h^2 f_{tt}(t,y(t)) + \alpha\beta h^2 f(t,y(t)) f_{ty}(t,y(t)) \\
& + \beta^2 h^2 [f(t,y(t))]^2 f_{yy}(t,y(t))) + \mathcal{O}(h^3).
\end{aligned}
$$

We can also expand $y(t + h)$ in terms of $y(t)$ as follows:

$$y(t + h) = y(t) + hy'(t) + \frac{1}{2}h^2 y''(t) + \mathcal{O}(h^3).$$

But the differential equation implies that $y' = f(t, y(t))$ and, therefore, in addition,

$$y''(t) = \frac{d}{dt}f(t,y(t)) = f_t(t,y(t)) + f_y(t,y(t))y'(t) = f_t(t,y(t)) + f_y(t,y(t))f(t,y(t)).$$

Now let us substitute both of these expansions into our expression for R, where we get

$$
\begin{aligned}
R =\ & \left(y + hf + \frac{1}{2}h^2(f_t + f_y f) + \mathcal{O}(h^3)\right) - y - c_1 hf \\
& - c_2 h\left(f + \alpha h f_t + \beta h f f_y + \frac{1}{2}(\alpha^2 h^2 f_{tt} + \alpha\beta h^2 f f_{ty} + \beta^2 h^2 f^2 f_{yy}) + \mathcal{O}(h^3)\right).
\end{aligned}
$$

The reader should note that we have written this without explicit arguments to any of the f terms, including the partial derivatives, since all of the arguments are $(t, y(t))$. Remember that our goal here is to find the values of c_1, c_2, α, and β that make R as small (in terms of h) as possible. So, we rewrite our ponderous expression for R in terms of powers of h. We get

$$
\begin{aligned}
R &= h\left(f - c_1 f - c_2 f\right) + h^2\left(\frac{1}{2}(f_t + f_y f) - c_2 \alpha f_t - c_2 \beta f f_y\right) + \mathcal{O}(h^3) \\
&= h\left(f - c_1 f - c_2 f\right) + h^2\left[\left(\frac{1}{2} - c_2 \alpha\right) f_t + \left(\frac{1}{2} - c_2 \beta\right) f f_y\right] + \mathcal{O}(h^3),
\end{aligned}
$$

from which we can extract the equations

$$
1 - c_1 - c_2 = 0, \quad \frac{1}{2} - c_2 \alpha = 0, \quad \frac{1}{2} - c_2 \beta = 0.
$$

If these three equations are satisfied, then the residual satisfies $R = \mathcal{O}(h^3)$; thus, the truncation error is $\mathcal{O}(h^2)$. Since this is a set of three equations in four unknowns, there is more than one solution. Typically, c_1 is regarded as arbitrary, and then we have

$$
c_2 = 1 - c_1, \tag{6.47}
$$

$$
\alpha = \beta = \frac{1}{2c_2}. \tag{6.48}
$$

Thus, some possible solutions include:

$$
c_1 = c_2 = \frac{1}{2}, \quad \alpha = \beta = 1;
$$

this yields the trapezoid rule predictor–corrector method:

$$
y_{n+1} = y_n + \frac{1}{2}h\left[f(t_{n+1}, y_n + hf(t_n, y_n)) + f(t_n, y_n)\right];
$$

or,

$$
c_1 = 0, \quad c_2 = 1, \quad \alpha = \beta = \frac{1}{2};
$$

this yields the midpoint rule predictor–corrector (6.35) and (6.36):

$$
y_{n+1} = y_n + hf(t_n + \frac{1}{2}h, y_n + \frac{1}{2}hf(t_n, y_n)). \tag{6.49}
$$

We could also try

$$
c_1 = \frac{1}{4}, \quad c_2 = \frac{3}{4}, \quad \alpha = \beta = \frac{2}{3},
$$

which yields a method sometimes called the *method of Heun*:[6]

$$
y_{n+1} = y_n + \frac{1}{4}h\left[f(t_n, y_n) + 3f(t_n + \frac{2}{3}h, y_n + \frac{2}{3}hf(t_n, y_n))\right]. \tag{6.50}
$$

---

[6]This is perhaps a good time to point out that there is a lot of inconsistency in naming many of these methods. What we call the "trapezoid method" is called "improved Euler" or "modified Euler" by other authors, just as an example. We call the method in (6.50) Heun's method because Goldstine [10] says that Kutta used that attribution in his 1901 paper that introduced what we call Runge–Kutta method.

All three of these have second-order truncation errors, and it can be shown that all of them are second-order accurate in the sense that, for any $T > 0$,

$$\max_{t_k \leq T} |y(t_k) - y_k| \leq C_T h^2,$$

where C_T will depend on T, but not on h.

The most commonly used Runge–Kutta method is the fourth-order method, derived in a manner very similar to what we did here, but using four slope values instead of only two. The required manipulations are necessarily much more involved, so we skip them. The method is usually written as follows:

$$k_1 = hf(t_n, y_n), \tag{6.51}$$

$$k_2 = hf\left(t_n + \frac{1}{2}h, y_n + \frac{1}{2}k_1\right), \tag{6.52}$$

$$k_3 = hf\left(t_n + \frac{1}{2}h, y_n + \frac{1}{2}k_2\right), \tag{6.53}$$

$$k_4 = hf(t_n + h, y_n + k_3), \tag{6.54}$$

$$y_{n+1} = y_n + \frac{1}{6}(k_1 + 2k_2 + 2k_3 + k_4). \tag{6.55}$$

A pseudocode for implementing this is given at the end of the section.

■ **EXAMPLE 6.16**

Let's consider our usual example IVP:

$$y' = -y \ln y, \quad y(0) = \frac{1}{2},$$

which we will first solve using a very coarse grid, using both the second-order method (6.50) and the fourth-order method (6.51)–(6.55). We then show the results of solving the same equation using the same methods, but with finer grids.

Let $h = 1/2$; to compute out to $t = 1$ requires two steps with each method. For the method of Heun, we compute as follows:

Step 1:

$$\bar{f} = f(t_0, y_0) = f(0, 1/2) = -(1/2)\ln(1/2) = 0.3465735903;$$

$$\bar{y} = y_0 + (2/3)h\bar{f} = 0.5 + (1/3)\bar{f} = 0.6155245301;$$

$$y_1 = y_0 + \frac{1/2}{4}\left[\bar{f} + 3f(t_0 + (2/3)h, \bar{y})\right] = 0.5 + 0.125\left[0.3465735903 + 3(0.2987020395)\right]$$

$$= 0.6553349636.$$

Step 2:

$$\bar{f} = f(t_1, y_1) = f(0.5, 0.6553349636) = 0.276950309;$$

$$\bar{y} = y_1 + (2/3)h\bar{f} = 0.6553349636 + (1/3)(0.276950309) = 0.7476517333;$$

$$y_2 = y_1 + \frac{1/2}{4}\left[\bar{f} + f(t_1 + (2/3)h, \bar{y})\right]$$

$$= 0.6553349636 + 0.125\left[0.276950309 + 3(0.2174305868)\right]$$

$$= 0.7714902223.$$

For the fourth-order Runge–Kutta, the computation is longer, of course:
Step 1:

$$
\begin{aligned}
k_1 &= hf(t_0, y_0) = 0.17328679513999; \\
z_1 &= y_0 + (1/2)k_1 = 0.58664339756999; \\
k_2 &= hf(t_0 + (1/2)h, z_1) = 0.15643965031862; \\
z_2 &= y_0 + (1/2)k_2 = 0.57821982515931; \\
k_3 &= hf(t_0 + (1/2)h, z_2) = 0.15837474613445; \\
z_3 &= y_0 + k_3 = 0.65837474613445; \\
k_4 &= hf(t_0 + h, z_3) = 0.13759406302779; \\
y_1 &= y_0 + (1/6)\,(k_1 + 2k_2 + 2k_3 + k_4) = 0.65675160851232.
\end{aligned}
$$

Step 2:

$$
\begin{aligned}
k_1 &= hf(t_1, y_1) = 0.13806541027436; \\
z_1 &= y_1 + (1/2)k_1 = 0.72578431364950; \\
k_2 &= hf(t_1 + (1/2)h, z_1) = 0.11630780609087; \\
z_2 &= y_1 + (1/2)k_2 = 0.71490551155775; \\
k_3 &= hf(t_1 + (1/2)h, z_2) = 0.11996289517416; \\
z_3 &= y_1 + k_3 = 0.77671450368648; \\
k_4 &= hf(t_1 + h, z_3) = 9.8131054204024D - 02; \\
y_2 &= y_1 + (1/6)\,(k_1 + 2k_2 + 2k_3 + k_4) = 0.77487458634706.
\end{aligned}
$$

The exact value of the solution at $t = 1$ is $y(1) = 0.7749206846$, so we can see that the fourth-order method did very well even with this coarse grid, whereas the second-order method did less well.

Table 6.5   Runge–Kutta examples, $y' = -y \ln y$.

$n = h^{-1}$	RK2 Error	RK4 Error
4	0.767315E-03	0.269490E-05
8	0.180964E-03	0.162549E-06
16	0.439226E-04	0.997759E-08
32	0.108187E-04	0.617969E-09
64	0.268460E-05	0.384481E-10
128	0.668655E-06	0.239708E-11
256	0.166852E-06	0.149658E-12
512	0.416742E-07	0.976996E-14

If we apply the same two methods to the same problem, but with much finer grids, we of course will get better accuracy. Table 6.5 shows the maximum errors over the interval $[0, 1]$ for a sequence of values of the mesh parameter. Note that the errors do decrease at the expected rate, in both cases.

Figure 6.6 shows the exact solution along with the second-order approximation for $h = \frac{1}{4}$ (denoted by the small circles). Note that the approximate values are right on top of the exact solution curve, even for this coarse value of h. For smaller h, or with the higher-order method, the accuracy would only improve, of course.

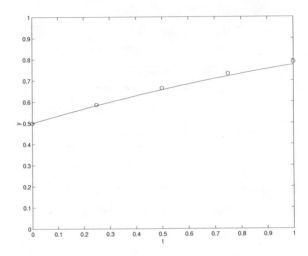

Figure 6.6 Solution of $y' = -y \ln y$, $y(0) = \frac{1}{2}$, along with second-order Runge–Kutta approximate values for $h = \frac{1}{4}$.

One major drawback of the Runge–Kutta methods is that they require more evaluations of the function f than do other methods. For simple scalar equations like our example, this is not a big deal, but for larger systems it is an issue. To get methods that are as accurate as the Runge–Kutta methods, but less expensive, we have to turn to the multistep methods.

Finally, we include here a pseudocode implementation of the fourth-order Runge–Kutta method, simply as an illustration. Note that four evaluations of f are required at each step.

Algorithm 6.2 *Pseudocode for Fourth-Order Runge–Kutta*

```
input t0, y0, h, n
external f
for k = 1 to n do
      t = t0+k*h
      t2 = t0 + 0.5*h
      v1 = f(t0,y0)
      v2 = f(t2,y0+0.5*h*v1)
      v3 = f(t2,y0+0.5*h*v2)
      v4 = f(t,y0+h*v3)
      y = (h/6.0)*(v1 + 2.0*(v2 + v3) + v4)
      y0 = y
      t0 = t
endfor
```

Exercises:

1. Use the method of Heun with $h = \frac{1}{4}$ to compute approximate values of $y(1)$ for each of the following initial value problems. Don't write a computer program, use a hand calculator to produce an orderly table of (t_k, y_k) pairs.

   (a) $y' = y(1 - y)$, $y(0) = \frac{1}{2}$;

   (b) $y' = y(1 + e^{2t})$, $y(0) = 1$;

   (c) $y' + 2y = 1$, $y(0) = 2$;

   (d) $ty' = y(\sin t)$, $y(0) = 2$.

2. Repeat Problem 1, using the fourth-order Runge–Kutta.

3. Write a computer program that solves each of the initial value problems in Problem 1, using the method of Heun, and $h = 1/16$.

4. Write a computer program that solves each of the initial value problems in Problem 1, using fourth-order Runge–Kutta, and $h = 1/16$.

5. For each initial value problem below, approximate the solution using the method of Heun with a sequence of decreasing grids $h^{-1} = 2, 4, 8, \ldots$. For those problems where an exact solution is given, compare the accuracy achieved over the interval $[0, 1]$ with the theoretical accuracy.

   (a) $y' + \sin y = 0$, $y(0) = 1$;

   (b) $y' + 4y = 1$, $y(0) = 1$; $y(t) = \frac{1}{4}(3e^{-4t} + 1)$;

   (c) $y' + y = \sin 4\pi t$, $y(0) = \frac{1}{2}$;

   (d) $y' = -y \ln y$, $y(0) = 3$; $y(t) = e^{(\ln 3)e^{-t}}$.

6. Repeat Problem 5, using fourth-order Runge–Kutta as the numerical method.

7. Repeat Problem 8 of §6.4, except this time use fourth-order Runge–Kutta to solve the differential equation, with $h = 1/8$.

8. Repeat Problem 9 of §6.4, except this time use fourth-order Runge–Kutta to solve the differential equation, with $h = 1/8$.

9. Repeat Problem 11 of §6.4, except this time use fourth-order Runge–Kutta to solve the differential equation, with $h = 1/8$.

10. Repeat Problem 10 of §6.4, except this time use fourth-order Runge–Kutta to solve the differential equation, with $h = 1/8$.

◁ • • ▷

6.6 MULTISTEP METHODS

6.6.1 The Adams Families

Two of the most popular families of multistep methods are the so-called *Adams families*, which are based on the exact integration of properly defined interpolating polynomials. One family (Adams–Bashforth) leads to explicit methods; the other (Adams–Moulton) leads to implicit methods.[7]

Recall the differential equation

$$y'(t) = f(t, y(t)), \quad y(t_0) = y_0.$$

Starting at any value $t_n \geq t_0$, we can solve this, formally, by simple integration to get the value at the next step:

$$y(t_{n+1}) = y(t_n) + \int_{t_n}^{t_{n+1}} f(s, y(s))ds.$$

Suppose now that we have values $y(t_{n-k}), k = 0, 1, 2, \ldots, p$, of the exact solution. We can construct a polynomial of degree p that interpolates to $F(s) = f(s, y(s))$ at these points:

$$q_p(s) = \sum_{k=0}^{p} L_k(s) f(t_{n-k}, y(t_{n-k})),$$

where the L_k are essentially the Lagrange functions from §4.1 (although the notation and indexing are slightly different):

$$L_k(s) = \prod_{\substack{i = 0 \\ i \neq k}}^{p} \frac{s - t_{n-i}}{t_{n-k} - t_{n-i}}. \tag{6.56}$$

---

[7]John Couch Adams (1819–1892) was born in Cornwall, England, and educated at St. John's College, Cambridge. Something of a child prodigy in mathematics, at Cambridge he compiled an extraordinary record and was awarded several prizes. While still an undergraduate he decided to study the irregularities in the orbit of the planet Uranus, to see if they could be explained by the gravitational attraction of an as-yet unknown eighth planet. Adams predicted the new planet's position, but, probably because of his youth, the Cambridge Observatory took no action so the credit for the discovery of Neptune went to Urbain Le Verrier, although the question of priority here is still controversial. Adams briefly held a position as professor of mathematics at St. Andrews College before being named Professor of Astronomy and director of the Cambridge Observatory.

Francis Bashforth (1819–1912) was born in Thurnscoe, England, the son of a farmer, and attended St. John's College of Cambridge at the same time as Adams. Although he had been ordained as an Anglican priest in 1851, upon graduation Bashforth worked first as a civil engineer and surveyor for a railroad company, and then (in 1864) obtained a position as professor of applied mathematics at what evolved into the Royal Artillery College. Although Bashforth made numerous important contributions to the study of ballistics, in 1872 an army reorganization left him with such a reduced position that he resigned and became a parish rector. The Adams–Bashforth method comes from a joint study of capillary action that the two men wrote in 1883.

Forest Ray Moulton (1872–1952), the youngest of eight children, was born on the family farm between Grand Rapids and Traverse City, Michigan. (The land had been given to his father as part of his bounty for serving in the Union army during the Civil War.) Forest—so named because he was born in a log cabin in the forest—was educated at Albion College in Michigan, and received a Ph.D. in astronomy from the University of Chicago, in 1899. He is credited, along with his colleague Thomas C. Chamberlain, with formulating the planetesimal hypothesis for the formation of the solar system. During World War I, Moulton did ballistics research for the U.S. Army at Aberdeen Proving Ground, Maryland, and it was during this period that he refined the original work of Adams and Bashforth into what we now know as the Adams–Moulton method for solving initial value problems.

We then have that

$$y(t_{n+1}) = y(t_n) + \int_{t_n}^{t_{n+1}} \sum_{k=0}^{p} L_k(s) f(t_{n-k}, y(t_{n-k})) ds + R_p(t_{n+1}), \qquad (6.57)$$

where (recall that $f(t, y(t)) = y'(t)$)

$$R_p(t_{n+1}) = \int_{t_n}^{t_{n+1}} \frac{1}{(p+1)!} (t - t_n)(t - t_{n-1}) \ldots (t - t_{n-p}) y^{(p+2)}(\theta_{h,t}) dt.$$

Since the polynomial part of the integrand does not change sign on $[t_n, t_{n+1}]$, there exists a value ξ_n such that

$$\begin{aligned}
R_p(t_{n+1}) &= y^{(p+2)}(\xi_n) \int_{t_n}^{t_{n+1}} \frac{1}{(p+1)!} (t - t_n)(t - t_{n-1}) \ldots (t - t_{n-p}) dt \\
&= \rho_p y^{(p+2)}(\xi_n),
\end{aligned}$$

for

$$\rho_p = \int_{t_n}^{t_{n+1}} \frac{1}{(p+1)!} (t - t_n)(t - t_{n-1}) \cdots (t - t_{n-p}) dt.$$

The expression (6.57) then simplifies to

$$y(t_{n+1}) = y(t_n) + \sum_{k=0}^{p} \lambda_k f(t_{n-k}, y(t_{n-k})) + \rho_p y^{(p+2)}(\xi_n);$$

where

$$\lambda_k = \int_{t_n}^{t_{n+1}} L_k(s) ds.$$

By dropping the residual and setting $y(t_k)$ to the approximate value y_k, we get the numerical method, known as the *Adams–Bashforth method* of order $p + 1$:

$$y_{n+1} = y_n + \sum_{k=0}^{p} \lambda_k f(t_{n-k}, y_{n-k}).$$

For simplicity's sake, we will hereafter write $f_{n-k} = f(t_{n-k}, y_{n-k})$.

If we assume a uniform grid with mesh spacing h, then the formulas for the λ_k and ρ_p simplify substantially, and they are routinely tabulated; Table 6.6 shows what they are for the common range of $p \leq 3$. Note that they are explicit methods, with order of accuracy $p + 1$, using $p + 1$ steps. For example, the third-order method would be written as

$$y_{n+1} = y_n + \frac{h}{12} \left(23 f_n - 16 f_{n-1} + 5 f_{n-2} \right),$$

where we again use the common notation

$$f_k = f(t_k, y_k).$$

Table 6.6 Adams–Bashforth coefficients and residual terms

Steps	p	λ_0	λ_1	λ_2	λ_3	R_p
1	0	h				$\frac{1}{2}h^2 y''(\xi_n)$
2	1	$\frac{3}{2}h$	$-\frac{1}{2}h$			$\frac{5}{12}h^3 y'''(\xi_n)$
3	2	$\frac{23}{12}h$	$-\frac{16}{12}h$	$\frac{5}{12}h$		$\frac{3}{8}h^4 y''''(\xi_n)$
4	3	$\frac{55}{24}h$	$-\frac{59}{24}h$	$\frac{37}{24}h$	$-\frac{9}{24}h$	$\frac{251}{720}h^5 y'''''(\xi_n)$

■ **EXAMPLE 6.17**

To illustrate their use, we return to our example problem,

$$y' = -y \ln y, \quad y(0) = \frac{1}{2},$$

the solution to which we will approximate using the second-order Adams–Bashforth scheme, commonly denoted "AB2," with $h = \frac{1}{8}$. We choose to use Euler's method to generate the starting value,[8] thus, our first computation is

$$y_1 = y_0 + hf(t_0, y_0) = \frac{1}{2} + \frac{1}{8}\left(-\frac{1}{2} \times \ln \frac{1}{2}\right) = 0.5433216988.$$

Now we can do the first AB2 step, for y_2:

$$
\begin{aligned}
y_2 &= y_1 + \frac{1}{2}h(3f(t_1, y_1) - f(t_0, y_0)) \\
&= 0.5433216988 + \frac{1}{2} \times \frac{1}{8}\left((-3(0.5433216988)\ln(0.5433216988) + \frac{1}{2}\ln\frac{1}{2}\right) \\
&= 0.583808738,
\end{aligned}
$$

followed by the second AB2 step, for y_3:

$$
\begin{aligned}
y_3 &= y_2 + \frac{1}{2}\frac{1}{8}(3f(t_2, y_2) - f(t_1, y_1)) \\
&= 0.583808738 + \frac{1}{2} \times \frac{1}{8}(-3(0.583808738)\ln(0.583808738) \\
&\quad + 0.5433216988\ln(0.5433216988)) \\
&= 0.622004388.
\end{aligned}
$$

Thus, $y(t_3) = y(0.375) \approx 0.622004388$. The exact solution is $y = e^{-(\ln 2)e^{-t}}$, thus $y(0.375) = 0.6210196063$, so our accuracy is not bad. We would have done better using a more accurate method for the starting value, of course.

The Adams–Bashforth methods are based on an interpolating polynomial that is defined using the nodes $t_n, t_{n-1}, \ldots, t_{n-p}$. If we use the same number of nodes, but include $t = t_{n+1}$, then (6.57) becomes

$$y(t_{n+1}) = y(t_n) + \int_{t_n}^{t_{n+1}} \sum_{k=-1}^{p-1} L_k(s)f(t_{n-k}, y(t_{n-k}))ds + R_p(t_{n+1}), \qquad (6.58)$$

---

[8]Recall from the discussion in §6.4.3, that we can use an order $p-1$ method to generate the starting values for a method of order p without a significant loss of accuracy,

where the Lagrange functions are now

$$L_k(s) = \prod_{\substack{i = -1 \\ i \neq k}}^{p-1} \frac{s - t_{n-i}}{t_{n-k} - t_{n-i}},$$ (6.59)

so the numerical method—known as the *Adams–Moulton method* of order $p + 1$—is

$$y_{n+1} = y_n + \sum_{k=-1}^{p-1} \gamma_k f(t_{n-k}, y_{n-k}),$$

where the constants are given by

$$\gamma_k = \int_{t_n}^{t_{n+1}} L_k(s)ds.$$

Table 6.7 summarizes these for the common range $p \leq 3$. Note that the Adams–Moulton methods are implicit, and that the $p + 1$ step method has order of accuracy $p + 1$. To be specific, the third-order method would be written in practice as

$$y_{n+1} = y_n + \frac{h}{12} \left(5f_{n+1} + 8f_n - f_{n-1}\right).$$

Table 6.7 Adams–Moulton coefficients and residual terms

steps	p	γ_{-1}	γ_0	γ_1	γ_2	R_p
1	0	h				$-\frac{1}{2}h^2 y''(\xi_n)$
2	1	$\frac{1}{2}h$	$\frac{1}{2}h$			$-\frac{1}{12}h^3 y'''(\xi_n)$
3	2	$\frac{5}{12}h$	$\frac{8}{12}h$	$-\frac{1}{12}h$		$-\frac{1}{24}h^4 y''''(\xi_n)$
4	3	$\frac{9}{24}h$	$\frac{19}{24}h$	$-\frac{5}{24}h$	$\frac{1}{24}h$	$-\frac{19}{720}h^5 y'''''(\xi_n)$

Note that three methods that we have already studied (Euler, backward Euler, and trapezoid rule) are included in these two tables. Whereas the Adams–Bashforth methods can be used by themselves (once starting values are generated), the Adams–Moulton methods require either the solution of a nonlinear equation or a predictor–corrector scheme. A very popular predictor–corrector scheme is to use the fourth-order Adams–Bashforth as the predictor and the fourth-order Adams–Moulton as the corrector.

■ **EXAMPLE 6.18**

Consider the example,

$$y' = -y \ln y, \quad y(0) = y_0,$$ (6.60)

which has exact solution $y = e^{(-\ln 2)e^{-t}}$ when $y_0 = \frac{1}{2}$. We solve this in two ways:

1. Via the second-order Adams–Bashforth method, with the trapezoid rule predictor–corrector used to generate the starting value;

Table 6.8   Adams family example for $y' = -y \ln y$.

$n = h^{-1}$	Second-order AB $\max_{k \le n} \lvert y(t_k) - y_k \rvert$	Fourth-order ABM PC $\max_{k \le n} \lvert y(t_k) - y_k \rvert$
8	0.28474335730512E–02	0.19742002576040E–05
16	0.67170745938105E–03	0.12099205870530E–06
32	0.16212367000479E–03	0.72652883709168E–08
64	0.39755173993794E–04	0.44197434601045E–09
128	0.98386454340238E–05	0.27204238861600E–10
256	0.24469431793017E–05	0.16866508190105E–11
512	0.61013351659867E–06	0.10536016503693E–12
1024	0.15233231198675E–06	0.76605388699136E–14

2. Via the fourth-order Adams–Bashforth–Moulton predictor–corrector, using the fourth-order Runge–Kutta method to generate the starting values.

The error results are summarized in Table 6.8.

Since they are such high-order methods, we need accurate techniques for getting the starting values for the Adams methods. Generally, the Runge–Kutta methods (§6.5) are used for this purpose.

6.6.2   The BDF Family

The Adams families are relatively old as numerical methods, with the original work going back to 1883. More recently, because of an interest in *stiff* differential equations, another family of methods has become popular. Known formally as the *Backward differentiation formula* family (or BDF family), they are derived from a different use of interpolating polynomials. See Table 6.9 for a summary of the coefficient values for these methods.

Again, we let the differential equation be

$$y'(t) = f(t, y(t)),$$

and this time we construct an interpolating polynomial to $y(t)$ (instead of $f(t, y(t)) = y'(t)$) at the nodes $t_{n-p+1}, t_{n-p+2}, \ldots, t_{n+1}$. Thus, we have

$$q(t) = \sum_{k=-1}^{p-1} L_k(t) y(t_{n-k}),$$

where the L_k are defined as in the case of Adams–Moulton. Now we use this polynomial to approximate $y'(t_{n+1})$, using the scheme outlined in §4.5. (This choice of argument to y' means that the BDF methods are all implicit.) Thus, we have

$$y'(t_{n+1}) = q'(t_{n+1}) + \frac{1}{(p+1)!} w'_p(t_{n+1}) y^{(p+1)}(\xi_p),$$

where

$$w_p(t) = \prod_{k=-1}^{p-1} (t - t_k).$$

Table 6.9 Backward difference family coefficients and residual terms.

Step	p	ν	μ_0	μ_1	μ_2	μ_3	R_p
1	1	1	1				$-\frac{1}{2}h^2y''(\xi_n)$
2	2	$\frac{2}{3}h$	$\frac{4}{3}$	$-\frac{1}{3}$			$-\frac{2}{9}h^3y'''(\xi_n)$
3	3	$\frac{6}{11}h$	$\frac{18}{11}$	$-\frac{9}{11}$	$\frac{2}{11}$		$-\frac{3}{22}h^4y''''(\xi_n)$
4	4	$\frac{12}{25}h$	$\frac{48}{25}$	$-\frac{36}{25}$	$\frac{16}{25}$	$-\frac{3}{25}$	$-\frac{12}{125}h^5y'''''(\xi_n)$

We substitute this into the differential equation to get

$$\sum_{k=-1}^{p-1} L_k'(t_{n+1})y(t_{n-k}) = f(t_{n+1}, y(t_{n+1})) - \frac{1}{(p+1)!}w_p'(t_{n+1})y^{(p+1)}(\xi_p),$$

which we can solve for $y(t_{n+1})$ to get

$$y(t_{n+1}) = \sum_{k=0}^{p-1} \mu_k y(t_{n-k}) + \nu f(t_{n+1}, y(t_{n+1})) - \frac{\nu}{(p+1)!}w_p'(t_{n+1})y^{(p+1)}(\xi_p),$$

where the constants are defined as

$$\mu_k = -\frac{L_k'(t_{n+1})}{L_{n+1}'(t_{n+1})}, \quad \nu = \frac{1}{L_{n+1}'(t_{n+1})}.$$

Table 6.9 summarizes these values for $p \leq 4$, under the assumption that the grid spacing is uniform. Note that these methods have order of accuracy p. To illustrate, the third-order method is given by

$$y_{n+1} = \frac{1}{11}(18y_n - 9y_{n-1} + 2y_{n-2}) + \frac{6h}{11}f_{n+1}.$$

A complete explanation of why the BDF family is particularly useful requires a little more discussion of the stability problem and the notion of *stiffness* for a differential equation, which we defer to §6.8.2. In general, they have better stability properties than those of the Adams methods. For now it is worth noting that these methods can be used either as the corrector in a predictor–corrector scheme, in conjunction with a nonlinear equation solver at each step, or to explicitly solve linear equations as was done in §6.4.2.

Exercises:

1. Use second-order Adams–Bashforth with $h = \frac{1}{4}$ to compute approximate values of $y(1)$ for each of the following initial value problems. Don't write a computer program, use a hand calculator to produce an orderly table of (t_k, y_k) pairs. Use Euler's method to generate the needed starting value.

   (a) $y' + 2y = 1, y(0) = 2$;
   (b) $y' = y(1 + e^{2t}), y(0) = 1$;
   (c) $ty' = y(\sin t), y(0) = 2$;
   (d) $y' = y(1 - y), y(0) = \frac{1}{2}$.

2. Verify that the λ_k values and ρ_p are correct for second-order Adams–Bashforth.

3. Write a computer program that solves each of the initial value problems in Problem 1, using second-order Adams–Bashforth and $h = 1/16$. Use Euler's method to generate the starting value.

4. Write a computer program that solves each of the initial value problems in Problem 1, using fourth-order Adams–Bashforth, with fourth-order Runge–Kutta to generate the starting values.

5. For each initial value problem below, approximate the solution using the fourth-order Adams–Bashforth–Moulton predictor–corrector, with fourth-order Runge–Kutta to generate the starting values. Use a sequence of decreasing grids $h^{-1} = 2, 4, 8, \ldots$. For those problems where an exact solution is given, compare the accuracy achieved over the interval $[0, 1]$ with the theoretical accuracy.

   (a) $y' + \sin y = 0$, $y(0) = 1$;

   (b) $y' + 4y = 1$, $y(0) = 1$; $y(t) = \frac{1}{4}(3e^{-4t} + 1)$;

   (c) $y' + y = \sin 4\pi t$, $y(0) = \frac{1}{2}$;

   (d) $y' = -y \ln y$, $y(0) = 3$; $y(t) = e^{(\ln 3)e^{-t}}$.

6. Derive the second-order Adams–Bashforth method under the assumption that the grid is *not* uniform. Assume that $t_{n+1} - t_n = h$, and $t_n - t_{n-1} = \eta$, with $\eta = \theta h$. What is the truncation error in this instance?

7. Repeat Problem 8 from §6.4, this time using second-order Adams–Bashforth with $h = 1/16$. Use simple Euler to generate the starting value y_1. If you did the earlier problems of this type, compare your results now to what you got before.

8. Repeat Problem 9 from §6.4, this time using second-order Adams–Bashforth with $h = 1/16$. Use simple Euler to generate the starting value y_1. If you did the earlier problems of this type, compare your results now to what you got before.

9. Repeat Problem 11 from §6.4, this time using second-order Adams–Bashforth with $h = 1/16$. Use simple Euler to generate the starting value y_1. If you did the earlier problems of this type, compare your results now to what you got before.

10. Repeat Problem 10 from §6.4, this time using second-order Adams–Bashforth with $h = 1/16$. Use simple Euler to generate the starting value y_1. If you did the earlier problems of this type, compare your results now to what you got before.

◁ • • • ▷

6.7   STABILITY ISSUES

6.7.1   Stability Theory for Multistep Methods

We have already seen an example of stability problems in the solution of IVPs, in §6.4.4 on the weak stability of the midpoint method. In this section, we will outline the issues involved in a more general study of stability.

It is important to understand, at the outset, the distinction between the stability of the solution to the differential equation, and the stability of the numerical method used to approximate that solution. To illustrate, consider the example initial value problem

$$y' = 100y + e^{-t}, y(0) = -\frac{1}{101}.$$

The exact solution here is $y = -\frac{1}{101}e^{-t}$, which is a bounded, continuous, very well-behaved function. Yet, as a practical matter, almost any numerical method applied to this problem will result in a solution that grows rapidly and eventually blows up. The reason is that the *general* solution to the differential equation is

$$y = Ce^{100t} - \frac{1}{101}e^{-t}.$$

Note the presence of the e^{100t} term; this is missing from the exact solution to the IVP because the initial value is carefully chosen to make the constant C in the general solution 0. However, the computed value y_1 will not be the exact value $y(t_1)$; thus, considering the new initial value problem

$$Y' = 100Y + e^{-t}, Y(h) = y_1,$$

the exact solution here is

$$Y(t) = -\frac{1}{101}e^{-t} + e^{100(t-h)}\left(y_1 + \frac{1}{101}e^{-h}\right);$$

thus, the e^{100t} term is multiplied by the error in the initial value y_1. Since this error will *not* be zero, any approximate solution will almost surely depend on the e^{100t} term; thus, the approximate values for $t > h$ will be affected by this term, and will generate a huge growth of the approximate solution. Although this is indeed a kind of instability—large growth of the solution due to a small change in the problem—it is an instability that is inherent in the differential equation itself, *not* in the numerical methods that we apply to compute solutions. We are interested in stability problems that come out of the numerical algorithms.

To motivate the larger discussion of stability, consider applying the $p = 1$ case of the BDF family to our standard initial value problem,

$$y' = f(t, y), \quad y(0) = y_0.$$

Thus, we have that the exact solution satisfies

$$y(t_{n+1}) = \frac{4}{3}y(t_n) - \frac{1}{3}y(t_{n-1}) + \frac{2}{3}hf(t_{n+1}, y(t_{n+1})) - \frac{2}{9}h^3y'''(\xi_n),$$

and the approximate solution satisfies

$$y_{n+1} = \frac{4}{3}y_n - \frac{1}{3}y_{n-1} + \frac{2}{3}hf(t_{n+1}, y_{n+1}).$$

Subtracting these two, we get that the error satisfies

$$e_{n+1} = \frac{4}{3}e_n - \frac{1}{3}e_{n-1} + \frac{2}{3}hf(t_{n+1}, y(t_{n+1})) - \frac{2}{3}hf(t_{n+1}, y_{n+1}) - \frac{2}{9}h^3y'''(\xi_n).$$

The error, of course, depends on the mesh spacing h, as well as the index n, so we can write the error as $e_n(h)$ to make this clear. Note, however, that if we consider the limiting case $h = 0$, then we find that

$$e_{n+1}(0) = \frac{4}{3}e_n(0) - \frac{1}{3}e_{n-1}(0). \tag{6.61}$$

Now, the fundamental issue—which we will not prove in any sort of formal sense—is the following:

> If the error blows up as $n \to \infty$ when $h = 0$, then elementary considerations of continuity imply that it will still blow up for h small but nonzero. On the other hand, if the error for $h = 0$ does not blow up as $n \to \infty$, then the same continuity considerations imply that it will be bounded for at least some range of nonzero values of h, as $n \to \infty$.

We thus look at the recursion (6.61) and solve it using the same ideas as in §6.4.4: We look for solutions in the form

$$e_n(0) = r^n.$$

This leads to a second-degree polynomial in r,

$$\sigma(r) = r^2 - \frac{4}{3}r + \frac{1}{3},$$

which has roots $r_1 = 1$ and $r_2 = \frac{1}{3}$. Hence, the $h = 0$ part of the error satisfies

$$e_n(0) = C_1 r_1^n + C_2 r_2^n,$$

where the coefficients depend on the values $e_0(0)$ and $e_1(0)$. From this we conclude that this particular BDF method is stable, because as $n \to \infty$ the $h = 0$ error does not blow up; quite to the contrary, one term goes to zero and the other is bounded. For this method, the parasitic solutions do not dominate the approximation for n large.

On the other hand, consider the method

$$y_{n+1} = 4y_n - 3y_{n-1} - 2hf(t_{n-1}, y_{n-1}), \tag{6.62}$$

which we derived in §6.4. This has a truncation error that is second-order in the mesh, comparable to the trapezoid method and the second-order Runge–Kutta methods, so we can call it an accurate method (it is consistent). However, the polynomial is, in this case,

$$\sigma(r) = r^2 - 4r + 3,$$

which has roots $r_1 = 1$ and $r_2 = 3$. Now, as $n \to \infty$ we see that one component of the exact solution to the approximate problem will blow up. Thus, we say that this method is unstable, and we expect that it will not perform well, as can be seen in some of the exercises.

The stability of a multistep method,[9] then, is determined by the roots of the polynomial corresponding to σ. A formal statement is based on the following definitions.

---

[9] Stability in this sense is not as much of an issue for single-step methods since they do not have parasitic solutions, although it is an issue when considering *systems* of equations.

Definition 6.5 (Stability Polynomial) *Consider a multistep method in the form*

$$y_{n+1} = \sum_{k=0}^{p} a_k y_{n-k} + \sum_{k=-1}^{p} b_k f(t_{n-k}, y_{n-k}).$$

Then the stability polynomial for this method is given by

$$\sigma(r) = r^{p+1} - \sum_{k=0}^{p} a_k r^{p-k}. \tag{6.63}$$

Definition 6.6 (Root Conditions) *Consider a multistep method with stability polynomial σ. Denote the roots of σ by $r_0, r_1, \ldots, r_p$. Define the following conditions.*

Root Condition: *If*

$$|r_k| \leq 1, \quad 0 \leq k \leq p$$

and all roots that satisfy $|r_j| = 1$ are simple roots, then we say that the stability polynomial satisfies the root condition.

Strong Root Condition: *If*

$$r_0 = 1, |r_k| < 1, \quad 1 \leq k \leq p$$

then we say that the stability polynomial satisfies the strong root condition.

We have the following:

1. If the stability polynomial satisfies the root condition, then the method is stable, in the sense that for h sufficiently small it will deliver accurate results over a closed interval $[0, T]$.

2. If the stability polynomial satisfies the strong root condition, then the method is *relatively stable*, meaning that, for h sufficiently small, the parasitic solution components will go to zero as $n \to \infty$.

3. A method that is stable, but not relatively stable, is called *weakly stable*, and will exhibit the type of behavior seen in §6.4.4.

■ **EXAMPLE 6.19**

Consider the $p + 1$-step Adams–Bashforth method:

$$y_{n+1} = y_n + \sum_{k=0}^{p} \lambda_k f(t_{n-k}, y_{n-k}).$$

The stability polynomial, based on the formal definition (6.63), will be

$$\sigma(r) = r^{p+1} - r^p,$$

thus, the roots are $r = 1$ and $r = 0$ (multiplicity p). This satisfies the strong root condition, so we know that the Adams–Bashforth methods are relatively stable. The same analysis holds for the Adams–Moulton methods, so we know that they are relatively stable as well.

6.7.2 Stability Regions

The definition of stability that we developed above was based on the behavior of the method in the limit, as $h \to 0$. Although perfectly valid, it is not immediately clear what it means for the practical implementation of methods, since we have to actually compute with $h > 0$. All we know, based on the material in §6.7.1, is that the Adams–Bashforth methods (for example) will be stable "for h sufficiently small."

To address this need for specific information about when a multistep method will work well, we have the notion of "stability regions." Consider the model problem

$$y' = \lambda y, \quad y(0) = 1, \tag{6.64}$$

where we assume that λ is *complex*. If we apply, say, the second-order Adams–Bashforth method to this problem, then we have the recursion

$$y_{n+1} = \left(1 + \frac{3}{2}\lambda h\right) y_n - \frac{1}{2}\lambda h y_{n-1}.$$

If we look for the exact solution of this recursion as we have before, we are led to the polynomial

$$p(r) = r^2 - \left(1 + \frac{3}{2}\xi\right) r + \frac{1}{2}\xi$$

for $\xi = \lambda h$. The stability region is then defined as follows:

> The stability region for a multistep method is that part of the complex plane where the method, when applied to the model problem (6.64), is absolutely stable; i.e., all components of the approximate solution decay to zero as $n \to \infty$.

If we denote the roots of the polynomial as

$$r_0(\xi), r_1(\xi), \dots, r_p(\xi),$$

then the stability region is that part of the complex plane[10] where $|r_k(\xi)| < 1$, for all k.

Figures 6.7 to 6.9 show the stability regions for three families of methods:

1. the Adams–Bashforth family (Fig. 6.7);

2. the Adams–Moulton family (Fig. 6.8);

3. the Adams–Bashforth–Moulton predictor–corrector family (Fig. 6.9; in this family we use the AB method as a predictor, and the AM method of the same accuracy as the corrector).

Note that, in general, as the accuracy of the method increases, the size of the stability region decreases. Note also that the implicit Adams–Moulton methods have larger stability regions than those of the Adams–Bashforth–Moulton predictor–corrector methods, which have larger stability regions than the Adams–Bashforth methods do. The reason that only two of the Adams–Moulton methods (orders 3 and 4) have their stability regions plotted is that for the order 1 (backward Euler) and order 2 (trapezoid rule) schemes, the stability

---

[10]The reason that we have to consider the complex plane is that when we consider *systems* of differential equations, the quantities corresponding to the λ in the differential equation are *eigenvalues* of a real matrix, and these can of course be complex.

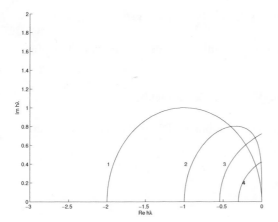

Figure 6.7 Regions of absolute stability for the Adams–Bashforth methods of orders 1 to 4.

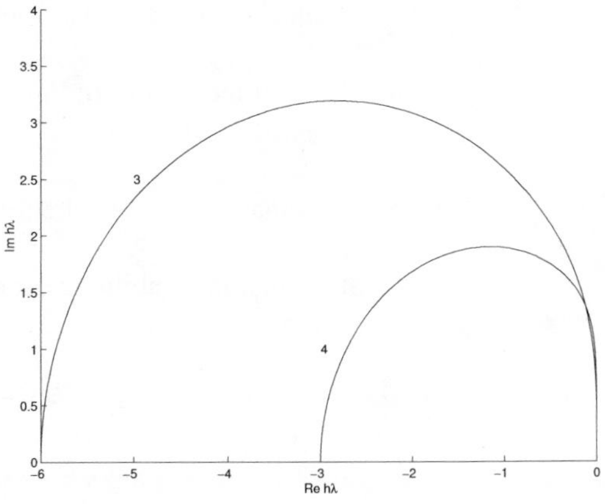

Figure 6.8 Regions of absolute stability for the Adams–Moulton methods of orders 3 and 4.

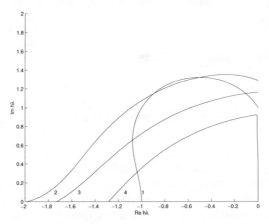

Figure 6.9 Regions of absolute stability for the Adams–Bashforth–Moulton predictor–corrector methods of orders 1 to 4.

region is the entire left half-plane (see Problem 4). Finally, note that the regions are symmetric about the horizontal axis, but only the top half is shown.

Exercises:

1. Determine the residual and the truncation error for the method defined by $y_{n+1} = 4y_n - 3y_{n-1} - 2hf(t_{n-1}, y_{n-1})$. Try to use it to approximate solutions of the IVP $y' = -y \ln y$, $y(0) = 2$. Comment on your results.

2. Show that the method defined by $y_{n+1} = 4y_n - 3y_{n-1} - 2hf(t_{n-1}, y_{n-1})$ is unstable.

3. Consider the method defined by

$$y_{n+1} = y_{n-1} + \frac{1}{3}h\left[f(t_{n-1}, y_{n-1}) + 4f(t_n, y_n) + f(t_{n+1}, y_{n+1})\right],$$

which is known as *Milne's method.* Is this method stable or strongly stable?

4. Show for the backward Euler and trapezoid rule methods, that the stability region is the entire left half-plane.

5. Show that all four members of the BDF family are stable.

6. Show that the stability region for the fourth-order BDF method contains the entire negative real axis.

◁ ● ● ● ▷

6.8 APPLICATION TO SYSTEMS OF EQUATIONS

So far in this chapter all of our examples of implementing methods have been for ordinary scalar equations. In this section we will look a bit at applying some of our methods to *systems* of equations.

6.8.1 Implementation Issues and Examples

Consider the system

$$y_1' = y_1 - y_1 y_2, \quad y_1(0) = y_{10}, \tag{6.65}$$
$$y_2' = y_1 y_2 - y_2, \quad y_2(0) = y_{20}. \tag{6.66}$$

This is generally known as the *Lotka–Volterra population model,* since it was first used in the 1920s to model the population dynamics of two species, one of which feeds off the other. To apply any of the methods that we have studied so far to this problem, it helps to formally write it out in terms of components and functions f_i, as follows:

$$y_1' = f_1(t, y_1, y_2), \quad y_1(0) = y_{10}, \tag{6.67}$$
$$y_2' = f_2(t, y_1, y_2), \quad y_2(0) = y_{20}, \tag{6.68}$$

where

$$f_1(t, y_1, y_2) = y_1 - y_1 y_2, \tag{6.69}$$
$$f_2(t, y_1, y_2) = y_1 y_2 - y_2. \tag{6.70}$$

Then, for example, an Euler's method approximation would require that we compute using the recursions (note the plural)[11]

$$y_{1,n+1} = y_{1,n} + h f_1(t_n, y_{1,n}, y_{2,n}),$$
$$y_{2,n+1} = y_{2,n} + h f_2(t_n, y_{1,n}, y_{2,n}),$$

---

[11]Note also the use of double subscripts. The first one counts the element in the solution vector y_n, and the second one indicates the time step number.

whereas the second-order Runge–Kutta method (6.49) would require that we compute using the recursions

$$\bar{y}_{1,n+1} = y_{1,n} + \frac{1}{2}hf_1(t_n, y_{1,n}, y_{2,n}),$$

$$\bar{y}_{2,n+1} = y_{2,n} + \frac{1}{2}hf_2(t_n, y_{1,n}, y_{2,n}),$$

$$y_{1,n+1} = y_{1,n} + hf_1\left(t_n + \frac{1}{2}h, \bar{y}_{1,n+1}, \bar{y}_{2,n+1}\right),$$

$$y_{2,n+1} = y_{2,n} + hf_2\left(t_n + \frac{1}{2}h, \bar{y}_{1,n+1}, \bar{y}_{2,n+1}\right),$$

and the second-order Adams–Bashforth method would use

$$y_{1,n+1} = y_{1,n} + \frac{1}{2}h\left(3f_1(t_n, y_{1,n}, y_{2,n}) - f_1(t_n, y_{1,n}, y_{2,n})\right),$$

$$y_{2,n+1} = y_{2,n} + \frac{1}{2}h\left(3f_2(t_n, y_{1,n}, y_{2,n}) - f_2(t_n, y_{1,n}, y_{2,n})\right).$$

■ EXAMPLE 6.20

To illustrate, let's do several example computations with the system

$$\begin{aligned} y_1' &= y_1 - y_1 y_2 + \sin \pi t, \quad y_1(0) = 2, \\ y_2' &= y_1 y_2 - y_2, \quad y_2(0) = 1. \end{aligned}$$

This is an instance of the Lotka–Volterra model in which the population of one of the species is driven by an external sinusoid. We have, in this case,

$$\begin{aligned} f_1(t, y_1, y_2) &= y_1 - y_1 y_2 + \sin \pi t, \\ f_2(t, y_1, y_2) &= y_1 y_2 - y_2. \end{aligned}$$

A single step of Euler's method, using $h = \frac{1}{4}$, would have us do the following computations:

$$y_{11} = y_{10} + hf_1(0, y_{10}, y_{20}) = 2 + \frac{1}{4} \times (2 - (2)(1)) + \sin 0 = 2,$$

$$y_{21} = y_{20} + hf_2(0, y_{10}, y_{20}) = 1 + \frac{1}{4} \times ((2)(1) - 1) = \frac{5}{4}.$$

Now we can continue the computation using second-order Adams–Bashforth, using these Euler's method values as the starting values. In this case we get

$$\begin{aligned} y_{12} &= y_{11} + \frac{1}{2}h\left(3f_1(t_1, y_{11}, y_{2,1}) - f_1(t_0, y_{10}, y_{20})\right) \\ &= 2 + \frac{1}{2} \times \frac{1}{4}(3(2 - (5/4)2 + \sin \frac{1}{4}\pi) - (2 - (2)(1) + \sin 0)) \\ &= 2.078, \end{aligned}$$

$$\begin{aligned} y_{22} &= y_{21} + \frac{1}{2}h\left(3f_2(t_1, y_{11}, y_{21}) - f_2(t_0, y_{10}, y_{20})\right), \\ &= \frac{5}{4} + \frac{1}{2} \times \frac{1}{4}(3((2)(5/4) - (5/4)) - ((2)(1) - 1)), \\ &= 1.625. \end{aligned}$$

Generalizations to more accurate and more involved schemes simply involve a little more care and work.

An important special case, as with scalar equations, occurs when the system is a linear system. In this case we can write it in a matrix–vector form as

$$y' = Ay + g, \quad y(0) = y_0,$$

where now A is an $N \times N$ matrix, and y, g, and y_0 are vectors of N components. It now becomes possible to (once again) employ implicit methods by the device of solving the linear equations explicitly. The difference between this case and what we had in §6.4.2 is that now we have a linear *system* to solve. For example, a direct solution of our general example via the trapezoid rule method involves the solution of the linear system

$$\left(I - \frac{1}{2}hA\right) y_{n+1} = \left(I + \frac{1}{2}hA\right) y_n + \frac{1}{2}h \left(g(t_n) + g(t_{n+1})\right)$$

at each step of the computation.

■ **EXAMPLE 6.21**

Consider the system

$$
\begin{aligned}
y_1' &= -4y_1 + y_2 + \sin \pi t, \quad y_1(0) = 1, \\
y_2' &= y_1 - 4y_2, \quad y_2(0) = 2,
\end{aligned}
$$

which we can write in matrix–vector form as

$$y' = Ay + g, \quad y(0) = y_0.$$

Here

$$
A = \begin{bmatrix} -4 & 1 \\ 1 & -4 \end{bmatrix}, \quad
y = \begin{bmatrix} y_1 \\ y_2 \end{bmatrix}, \quad
g = \begin{bmatrix} \sin \pi t \\ 0 \end{bmatrix}, \quad
y_0 = \begin{bmatrix} y_{10} \\ y_{20} \end{bmatrix}.
$$

If we now do two steps of the trapezoid rule method for this system, using $h = \frac{1}{4}$, we get the following:

$$
\begin{aligned}
y_1 &= \left(I - \frac{h}{2}A\right)^{-1}\left(I + \frac{h}{2}A\right) y_0 + \frac{1}{2}h \left(I - \frac{h}{2}A\right)^{-1}(g(t_0) + g(t_1)) \\
&= \begin{bmatrix} 0.6258 \\ 0.8021 \end{bmatrix},
\end{aligned}
$$

$$
\begin{aligned}
y_2 &= \left(I - \frac{h}{2}A\right)^{-1}\left(I + \frac{h}{2}A\right) y_1 + \frac{1}{2}h \left(I - \frac{h}{2}A\right)^{-1}(g(t_1) + g(t_2)) \\
&= \begin{bmatrix} 0.7690 \\ 0.8141 \end{bmatrix}.
\end{aligned}
$$

We will end this section by giving an algorithm in pseudocode form for approximating solutions of the system

$$y' = f(t, y), \quad y(0) = y_0$$

using the trapezoid rule predictor–corrector method. The code assumes that the individual vector functions $f_j(t, y)$ can be called using an index passed into a single function name, thus: f(j,t,y). The number of unknowns in the code is taken to be m.

Algorithm 6.3 *Trapezoid Rule Predictor–Corrector for Systems.*

```
input h,n,m,y0,t0
external f
!
t = t0
for j = 1 to m do
     y(j) = y0(j)
endfor
for k = 1 to n do
!
!      Euler predictor
!
       for j=1 to m do
            yb(j) = y(j) + h*f(j,t,y)
       endfor
!
!      Trapezoid rule corrector
!
       for j=1 to m do
            yc(j) = y(j) + (h/2)*(f(j,t,y) + f(j,t+h,yb))
       endfor
!
!      Update
!
       for j=1 to m do
            y(j) = yc(j)
       endfor
       t = k*h
endfor
```

6.8.2 Stiff Equations

In some applications, we must deal with a large system of coupled differential equations, modeling processes that are occurring at substantially different rates. This is the essence of the notion of *stiffness* of a differential equation. Stiff systems can be *very* difficult to solve numerically.

To illustrate the phenomenon, consider the linear system

$$y_1' \;=\; 198y_1 + 199y_2, \quad y_1(0) = 1, \qquad\qquad (6.71)$$

$$y_2' \;=\; -398y_1 - 399y_2, \quad y_2(0) = -1, \qquad\qquad (6.72)$$

which we write in matrix–vector form as

$$y' = Ay, \quad y(0) = y_0,$$

where

$$A = \begin{bmatrix} 198 & 199 \\ -398 & -399 \end{bmatrix}, \quad y = \begin{pmatrix} y_1 \\ y_2 \end{pmatrix}, \quad y_0 = \begin{pmatrix} y_1(0) \\ y_2(0) \end{pmatrix}.$$

The exact solution is

$$y_1(t) = e^{-t}, \quad y_2(t) = -e^{-t},$$

and we will compute approximations using second order Adams–Bashforth, with Euler's method to compute the starting value. The results are given in Table 6.10; note that we get decent accuracy at the outset, but then horrible accuracy until we have $h < 1/128$. Why is this?

Table 6.10   Example of stiff approximation (Adams–Bashforth)

$n = h^{-1}$	Error in y_1	Error in y_2
4	0.313335E–02	0.313335E–02
8	0.653059E–03	0.653059E–03
16	0.142067E–03	0.142069E–03
32	0.634997E+07	0.126999E+08
64	0.976963E+19	0.195393E+20
128	0.778921E+15	0.155784E+16
256	0.473265E–06	0.473265E–06
512	0.117631E–06	0.117631E–06
1024	0.293221E–07	0.293221E–07
2048	0.731982E–08	0.731982E–08

The answer lies in the *eigenvalues* of the matrix A, which are $\lambda_1 = -1, \lambda_2 = -200$. This means that the general solution of the system will involve components of the form e^{-200t}, in addition to e^{-t}. The initial condition was cleverly chosen to make the e^{-200t} components vanish. However, even though the e^{-200t} term is not present in the solution, we still have to use a value of h that is small enough to accurately resolve it. (To be precise, we have to take h small enough so that $-200h$ is in the stability region for second-order Adams–Bashforth; note that this is consistent with the results in Table 6.10.) A complete development of how this comes about is deferred to Appendix A. For our part, we will call a linear system of differential equations *stiff* if the eigenvalues of the coefficient matrix A all have negative real parts, and if the ratio of the largest eigenvalue to the smallest, in absolute value, is "large."

One of the best ways to deal with a stiff system is to use an implicit method to compute the approximation. If the system is linear, the implicitness can be dealt with by solving a linear system; if the system is nonlinear, then some kind of root-finding scheme has to be used to solve the nonlinear equations, or a predictor–corrector method has to be used.

If we apply the trapezoid rule method to the solution of this example, we get the results shown in Table 6.11 (next page). Note that the error is much smaller, and does not show the erratic and undesirable behavior that we saw with the Adams–Bashforth method. The material in Appendix A gives some explanation for this.

6.8.3   A-Stability

Because of the phenomenon of stiffness, we are interested in numerical methods for ODEs that will always produce decaying approximate solutions whenever applied to a differential

Table 6.11   Example of stiff approximation (trapezoid method)

$n = h^{-1}$	Error in y_1	Error in y_2
4	0.192913E–02	0.192913E–02
8	0.479822E–03	0.479822E–03
16	0.119803E–03	0.119803E–03
32	0.299413E–04	0.299413E–04
64	0.748472E–05	0.748472E–05
128	0.187114E–05	0.187114E–05
256	0.467784E–06	0.467784E–06
512	0.116946E–06	0.116946E–06
1024	0.292364E–07	0.292364E–07
2048	0.730911E–08	0.730911E–08

equation that has decaying solutions, regardless of the mesh size being used. This is the concept of A-stability. (The "A" stands for "asymptotic.") The formal definition is as follows:

Definition 6.7 (A-Stability) *Let y_k be the approximate solution to*

$$y' = \lambda y, \quad y(0) = 1$$

for λ complex, as computed by some numerical method. If

$$\lim_{n \to \infty} y_n = 0$$

for all h and all λ such that $\Re(\lambda) < 0$, then we say that the numerical method is A-stable.

The trapezoid method and the backward Euler method are examples of A-stable methods. (See Problem 3.) The advantage of an A-stable method is that the parasitic solutions will always decay, regardless of the mesh size; hence, they can be used to solve stiff systems. Unfortunately, there are no A-stable multistep methods of greater than $\mathcal{O}(h^2)$ accuracy.

Exercises:

1. Using a hand calculator and $h = \frac{1}{4}$, compute approximate solutions to the initial value problem

$$
\begin{aligned}
y_1' &= -4y_1 + y_2, \quad y_1(0) = 1; \\
y_2' &= y_1 - 4y_2, \quad y_2(0) = -1.
\end{aligned}
$$

Compute out to $t = 1$, using the following methods:

   (a) Euler's method;

   (b) RK4;

   (c) AB2 with Euler's method for the starting values;

   (d) Trapezoid rule predictor–corrector.

Organize your computations and results neatly to minimize mistakes.

2. Consider the second-order equation

$$y'' + \sin y = 0, \quad y(0) = \frac{1}{8}\pi, y'(0) = 0;$$

write this as a first-order system and compute approximate solutions using a sequence of grids and the following methods:

   (a) Second-order Adams–Bashforth;

   (b) Fourth-order Runge–Kutta;

   (c) The method of Heun.

   The solution will be periodic; try to determine, experimentally, the period of the motion.

3. Show that backward Euler and the trapezoid rule methods are both A-stable.

4. Show that the second-order BDF method is A-stable.

5. Apply the third-order BDF method, using the trapezoid method to obtain the starting values, to the stiff example in the text. Compare your results to the exact solution and those obtained in the text.

6. Consider the following system of differential equations:

$$
\begin{aligned}
y_1'(t) &= a y_1(t) - b y_2(t) y_1(t), y_1(0) = y_{10} \\
y_2'(t) &= -c y_2(t) + d y_1(t) y_2(t), y_2(0) = y_{20}
\end{aligned}
$$

   Solve this system for a sequence of mesh sizes, using the data $a = 4, b = 1, c = 2, d = 1$, with initial values $y_{10} = 3/2$ and $y_{20} = 4$. Use (a) second-order Adams–Bashforth, with Euler's method providing the starting value; and, (b) fourth-order Runge–Kutta. Plot the solutions in two ways: as a single plot showing y_1 and y_2 versus t; and as a *phase plot* showing y_1 versus y_2. The solutions should be periodic. Try to determine, experimentally, the period. (*Note:* This problem is an example of a *predator–prey* model, in which y_1 represents the population of a prey species and y_2 represents the population of a predator that uses the prey as its only food source. See Braun [4] for an excellent discussion of the dynamics of such systems, as well as the history behind the problem.)

7. Consider now a situation in which two species, denoted x_1 and x_2, *compete* for a common food supply. A standard model for this is

$$
\begin{aligned}
x_1'(t) &= a x_1(t) - (b x_1(t) + c x_2(t)) x_1(t) \\
x_2'(t) &= A x_2(t) - (B x_1(t) + C x_2(t)) x_2(t)
\end{aligned}
$$

   with initial conditions $x_1(0) = x_1^0$, $x_2(0) = x_2^0$. Using any of the methods in this chapter, solve this system for $x_1^0 = x_2^0 = 10,000$, using $a = 4, b = 0.0003, c = 0.004, A = 2, B = 0.0002$, and $C = 0.0001$. Vary the parameters of the problem slightly and observe what happens to the solution.

8. An interesting and somewhat unusual application of systems of differential equations is to *combat modeling*. Here, we denote the force levels of the two sides at war by

x_1 and x_2, and make various hypotheses about how one side's force level affects the losses suffered by the other side. For example, if we hypothesize that losses are proportional to the size of the opposing force, then (in the absence of reinforcements) we get the model

$$\begin{aligned} x_1' &= -ax_2, \\ x_2' &= -bx_1. \end{aligned}$$

If we hypothesize that losses are proportional to the *square* of the opposing force, we get the model

$$\begin{aligned} x_1' &= -\alpha x_2^2, \\ x_2' &= -\beta x_1^2. \end{aligned}$$

The constants α, β, a, and b represent the military efficiency of one side's forces.

Using any of the methods of this section, consider both models with

$$a = 1, \alpha = 1, x_1(0) = 40, x_2(0) = 120$$

and observe how changing b and β affects the long-term trend of the solutions. In particular, can you find values of b and β such that the smaller force annihilates the larger one? Combat models such as this are known as *Lanchester models* after the British mathematician F. W. Lanchester, who introduced them.

9. Consider the following system of ODEs:

$$\begin{aligned} x' &= \sigma(y - x), \\ y' &= x(\rho - z) - y, \\ z' &= xy - \beta z. \end{aligned}$$

   Use two different second-order methods (your choice) to approximate solutions to this system, for $\sigma = 10$, $\beta = 8/3$, and $\rho = 28$. Take $x(0) = y(0) = z(0) = 1$ and compute out to $t = 65$. Plot each component.

10. Consider the second-order equation

$$x'' - \mu(1 - x^2)x' + x = 0, \quad x(0) = 0.5, \quad x'(0) = 0.$$

    First, write this as a system. Then, as in Problem 9, choose two different second-order methods and use each to solve the system for $\mu = 0.1$ and $\mu = 10$. Compute out to $t = 20$.

11. One simple, but important, application of systems of differential equations is to the spread of epidemics. Perhaps the simplest model is the so-called *SIR model*, which models an infectious disease (such as measles) that imparts immunity to those who have had it and recovered. We divide the population into three categories:

    - $S(t)$: These are the people who are susceptible; that is, they are well at time t but might get sick at some future time;
    - $I(t)$: These are the people who are infected, i.e., sick;

- $R(t)$: These are the people who have recovered and are therefore protected from getting the disease by being immune.

Under a number of simplifying assumptions,[12] the SIR model involves the following system, sometimes known as the continuous Kermack–McKendrick model.

$$\frac{dS}{dt} = -rSI,$$
$$\frac{dI}{dt} = rSI - aI,$$
$$\frac{dR}{dt} = aI.$$

The parameters r and a measure the rate at which susceptible people become sick upon contact with infectives, and the rate at which infected people are cured.

(a) For the case $I_0 = 1$, $S_0 = 762$, $R_0 = 0$, $a = 0.44$, and $r = 2.18 \times 10^{-3}$, solve the system using the numerical method and step size of your choice, and plot the solutions. These data are taken from a study of an influenza epidemic at an English boys' school; the time scale here is in days.

(b) Now consider the case where $S_0 = 10^6$, $R_0 = 0$, $a = 1.74$, $r = 0.3 \times 10^{-5}$. Solve the system (again, using the method of your choice) for a range of values of $I_0 > 0$. Is there a critical value of I_0 beyond which the disease spreads to almost all of the population? These data are taken, roughly, from a study of a plague epidemic in Bombay; the time scale is weeks.

(c) In part (b), take $I_0 = 10^4$ and vary the value of r. Try to find the critical value of r such that, for $r < r_{\text{crit}}$, the number of infected drops monotonically, and for $r > r_{\text{crit}}$, the number of infected rises to a peak before falling off.

◁ ● ● ▷

6.9 ADAPTIVE SOLVERS

Just as we did with numerical quadrature, we can devise methods for initial value problems that implement some kind of variable step size and at the same time some automatic error control to minimize the work while obtaining a user-defined accuracy. The basic principles are the same as those that we used in §5.8.3 for numerical integration: We do two computations of the same value, and from these derive a computable estimate of the error. If this computable estimate of the error is small enough, we accept the computed value and go on to the next point. If the error estimate is not small enough, then we go to a smaller mesh size and try again.

The "professional" codes use various schemes. One of the first such codes, DIFSUB [9], used a variable-order Adams–Bashforth–Moulton scheme. Others (the MATLAB routine ode45) use a pair of Runge–Kutta methods, an idea first published by Erwin Fehlberg in 1969, so these schemes are often called Runge–Kutta–Fehlberg methods. What we will

---

[12]See J. D. Murray, *Mathematical Biology*, Springer-Verlag, Berlin, 1989, pp. 611ff.

describe here is a simpler but less accurate scheme that nonetheless contains the essential elements of the others, followed by some experiments with several schemes from MATLAB, which contains several ODE solvers, most based on the article by Shampine [17].

Consider the initial value problem

$$y' = f(t, y), \quad y(0) = y_0,$$

and assume that we have computed accurate approximate solutions out as far as some t_n. Thus, we have $y_n \approx y(t_n)$. We have a user-defined error criterion τ. We define $u_n(t)$ as the exact solution of the IVP

$$y' = f(t, y), \quad y(t_n) = y_n;$$

thus, $u_n(t)$ defines the solution of the ODE that passes through our most recent approximate value. We will design our algorithm to control the *local error*, i.e., the error between y_{n+1} and $u_n(t_{n+1})$. The *global error* $y(t_{n+1}) - y_{n+1}$ can then be estimated from the local error and the differential equation.

Our algorithm, which we will develop and justify in more detail, is based on using two approximate values generated from the trapezoid rule method, which of course is one instance of second-order Runge–Kutta. We require one preliminary lemma before proceeding with the design of the algorithm.

Lemma 6.1 *If the solution y is sufficiently smooth, then the truncation error for the second-order Runge–Kutta scheme*

$$y_{n+1} = y_n + \frac{1}{2}h \left(f(t_n, y_n) + f(t_{n+1}, y_n + hf(t_n, y_n)) \right)$$

satisfies

$$\tau(t, h) = -h^2 \left[\frac{1}{12} y'''(t) - \frac{1}{4} y''(t) f_y(t, y(t)) \right] + \mathcal{O}(h^3). \tag{6.73}$$

Proof: We have

$$\tau(t, h) = h^{-1} \left[y(t + h) - y(t) - \frac{1}{2}h \left(f(t, y(t)) + f(t, y(t) + hf(t, y(t))) \right) \right].$$

We proceed essentially as we did in deriving the Runge–Kutta methods, by expanding $y(t + h)$ and $f(t, y(t) + hf(t, y(t)))$ using Taylor's Theorem:

$$y(t + h) = y(t) + hy'(t) + \frac{1}{2}h^2 y''(t) + \frac{1}{6}h^3 y'''(t) + \mathcal{O}(h^4)$$

and

$$\begin{aligned} f(t, y(t) + hf(t, y(t))) &= f(t, y(t)) + hf_t(t, y(t)) + hf(t, y(t))f_y(t, y(t)) \\ &+ \frac{1}{2} \left(h^2 f_{tt}(t, y(t)) + 2h^2 f(t, y(t))f_{ty}(t, y(t)) \right. \\ &\left. + h^2 f^2(t, y(t))f_{yy}(t, y(t)) \right) + \mathcal{O}(h^3). \end{aligned}$$

Putting this all together gives us

$$\begin{aligned} \tau(t, h) &= \left(\frac{1}{6} y'''(t) - \frac{1}{4} \left(f_{tt}(t, y(t)) + 2f(t, y(t))f_{ty}(t, y(t)) + f^2(t, y(t))f_{yy}(t, y(t)) \right) \right) h^2 \\ &+ \mathcal{O}(h^3), \end{aligned}$$

and the conclusion follows by computing $y'''(t)$ in terms of f and its derivatives. •

This lemma is the key to constructing our adaptive algorithm. Note that (6.73) implies that—if the solution is smooth enough—we can use Taylor's Theorem to show that, for any real scalar a,

$$\tau(t_n + ah, h) = \tau(t_n, h) + \mathcal{O}(h^3).$$
(6.74)

Our algorithm is then the following.

Algorithm 6.4 *Simple Adaptive Solver for Initial Value Problems.*
Given t_n, y_n, a desired error tolerance, ϵ, the current value of the mesh, h, and a value $y_n \approx y(t_n)$, do the following:

1. *Compute*

$$Y_1 = y_n + \frac{1}{2}h\left(f(t_n, y_n) + f(t_{n+1}, y_n + hf(t_n, y_n))\right);$$

   note that this is a single-step $\mathcal{O}(h^2)$ approximation to $y(t_{n+1})$.

2. *Compute*

$$Y_2 = W + \frac{1}{4}h\left(f\left(t_n + \frac{1}{2}h, W\right) + f\left(t_{n+1}, W + \frac{1}{2}hf\left(t_n + \frac{1}{2}h, W\right)\right)\right),$$

   where

$$W = y_n + \frac{1}{4}h\left(f(t_n, y_n) + f\left(t_n + \frac{1}{2}h, y_n + \frac{1}{2}hf(t_n, y_n)\right)\right);$$

   note that Y_2 is a two-step $\mathcal{O}(h^2)$ approximation to $y(t_{n+1})$.

3. *Compute $Y_3 = (4Y_2 - Y_1)/3$; note that this is the Richardson extrapolated value for Y_1 and Y_2.*

4. *Compute $E = (Y_2 - Y_1)/3$; note that this is the Richardson estimated error.*

5. *If $\frac{1}{4}h\epsilon \leq |E| \leq h\epsilon$, then we set $y_{n+1} = Y_3$ and start all over with the next point.*

6. *If $|E| > h\epsilon$, then the step size is too large, so we cut h in half ($h \leftarrow h/2$) and repeat the computation.*

7. *If $|E| < \frac{1}{4}h\epsilon$, then the step size is too small, so we set $y_{n+1} = Y_3$ but we double the step size ($h \leftarrow 2h$) for the next step.*

The reason for the last step is that we don't want the step size to become too small, because that would result in our doing too much work by taking many tiny steps where fewer, larger, steps would be acceptably accurate.

The last two steps are a very simple way to control the step size; more sophisticated methods actually use the information from one step to predict what the next value of h ought to be. These same ideas can be used to predict an initial value of h, as well.

Because we do not accept an approximate value y_{n+1} unless the estimated truncation error is less than $h\epsilon$, we are confident that the local error satisfies

$$|u_n(t_{n+1}) - y_{n+1}| \le h\epsilon. \qquad (6.75)$$

This can be made precise in the following theorem.

Theorem 6.5 *For Y_1, Y_2, Y_3, and E as defined above, we have (assuming that f is continuously differentiable in y)*

$$u_n(t_{n+1}) - Y_3 = \mathcal{O}(h^4)$$

and

$$u_n(t_{n+1}) - Y_2 = E + \mathcal{O}(h^3).$$

Proof: We prove this by simply going through the details of the error analysis, using the results of the lemma proved above. We have (since $u_n(t_n) = y_n$)

$$u_n(t_{n+1}) = y_n + \frac{1}{2}h\left(f(t_n, y_n) + f(t_{n+1}, y_n + hf(t_n, y_n))\right) + h\tau(t_n, h)$$

and

$$Y_1 = y_n + \frac{1}{2}h\left(f(t_n, y_n) + f(t_{n+1}, y_n + hf(t_n, y_n))\right),$$

so that

$$u_n(t_{n+1}) - Y_1 = h\tau(t_n, h).$$

Similarly, we can show that the error in the half-step is given by ($t_{n+\frac{1}{2}} = t_n + \frac{1}{2}h$)

$$u_n(t_{n+\frac{1}{2}}) - W = \frac{1}{2}h\tau\left(t_n, \frac{1}{2}h\right).$$

We then can get $u_n(t_{n+1}) - Y_2$ by looking at the next half-step:

$$
\begin{aligned}
u_n(t_{n+1}) &= u_n(t_{n+\frac{1}{2}}) + \frac{1}{4}h\left(f(t_{n+\frac{1}{2}}, u_n(t_{n+\frac{1}{2}})) + f(t_{n+1}, u_n(t_{n+\frac{1}{2}})\right.\\
&\left. + \frac{1}{2}hf(t_{n+\frac{1}{2}}, u_n(t_{n+\frac{1}{2}})))\right) + \frac{1}{2}h\tau\left(t_{n+\frac{1}{2}}, \frac{1}{2}h\right)\\
Y_2 &= W + \frac{1}{4}h\left(f(t_{n+\frac{1}{2}}, W) + f\left(t_{n+1}, W + \frac{1}{2}hf(t_n, W)\right)\right).
\end{aligned}
$$

Subtract these two and use the Mean Value Theorem to write

$$f(t_{n+\frac{1}{2}}, u_n(t_{n+\frac{1}{2}})) - f(t_{n+\frac{1}{2}}, W) = f_y(t_{n+\frac{1}{2}}, v)(u_n(t_{n+\frac{1}{2}}) - W),$$

where v is a value between W and $u_n(t_{n+\frac{1}{2}})$. This yields

$$u_n(t_{n+1}) - Y_2 = \frac{1}{2}h\tau(t_n, \frac{1}{2}h)\left[1 + \frac{1}{2}hf_y + \frac{1}{8}h^2 f_y^2\right] + \frac{1}{2}h\tau\left(t_{n+\frac{1}{2}}, \frac{1}{2}h\right),$$

where $f_y = f_y(t_{n+\frac{1}{2}}, v)$. Using (6.74), we can now simplify this to

$$u_n(t_{n+1}) - Y_2 = h\tau\left(t_n, \frac{1}{2}h\right) + \mathcal{O}(h^4).$$

Therefore,

$$
\begin{aligned}
u_n(t_{n+1}) - Y_3 &= u_n(t_{n+1}) - \frac{1}{3}(4Y_2 - Y_1) \\
&= \frac{4}{3}(u_n(t_{n+1}) - Y_2) - \frac{1}{3}(u_n(t_{n+1}) - Y_1) \\
&= \frac{4}{3}\left(h\tau\left(t_n, \frac{1}{2}h\right) + \mathcal{O}(h^4)\right) - \frac{1}{3}h\tau(t_n, h) \\
&= \mathcal{O}(h^4),
\end{aligned}
$$

and

$$
\begin{aligned}
E &= \frac{1}{3}(Y_2 - Y_1) \\
&= \frac{1}{3}\left(Y_2 - u_n(t_{n+1}) + u_n(t_{n+1}) - Y_1\right) \\
&= \frac{1}{3}\left(-h\tau\left(t_n, \frac{1}{2}h\right) + \mathcal{O}(h^4) + h\tau(t_n, h)\right) \\
&= -\frac{1}{4}h\tau\left(t_n, \frac{1}{2}h\right) + \mathcal{O}(h^3) \\
&= u_n(t_{n+1}) - Y_2 + \mathcal{O}(h^3),
\end{aligned}
$$

and we are done. •

Thus, we can use E as a computable estimate to the local error in Y_2; if this estimate is small enough, we use the (more accurate) value Y_3 as the approximation; this is known as *local extrapolation*.

The use of $\frac{1}{4}h\epsilon$ as the lower bound for the error in a single step is somewhat arbitrary. Also, it is typical to have an absolute minimum step size; if the computation attempts to take a step size smaller than $h_{\min}$, then the computation terminates. Similarly, it is a good idea to have an absolute maximum step size, so that the code does not take too large a step.

This particular algorithm has a number of drawbacks compared to others, but it is simple enough to illustrate the ideas behind adaptive methods. The most glaring flaw—beyond the fact that it is a relatively low order method—in this algorithm is that it requires a total of five different function evaluations in each step.

■ EXAMPLE 6.22

We can show how the algorithm works by using it to approximate, with a crude specified error, the solution to the initial value problem

$$
y' = 16y(1 - y), \quad y(0) = 1/1024,
$$

whose exact solution is $y(t) = 1/(1 + 1023e^{-16t})$. We take $\epsilon = 0.001$ and use $h = 0.125$ as the initial step size. We get

$$
\begin{aligned}
Y_1 &= 0.0048714, \\
W &= 0.0024385, \\
Y_2 &= 0.0060786,
\end{aligned}
$$

so the estimated error is

$$
E = 0.00040239,
$$

which is larger than the desired tolerance of $0.001h = 1.25 \times 10^{-4}$. Thus, we cut h in half to $h = 0.0625$ and repeat the computation. This time we get

$$
\begin{aligned}
Y_1 &= 0.0024385, \\
W &= 0.0015860, \\
Y_2 &= 0.0025749,
\end{aligned}
$$

and the estimated error is now $E = 0.000045460 < \epsilon h = 6.25 \times 10^{-5}$. Therefore, we accepted the *extrapolated* value

$$
Y_3 = (4Y_2 - Y_1)/3 = 0.0026204
$$

as the value of $y_1 \approx y(h)$; since the error is *not* less than $\frac{1}{4}\epsilon h$, we do not change h before continuing the computation.

We will show one more step. To compute y_2, we go through essentially the same computations as for y_1; we get

$$
\begin{aligned}
Y_1 &= 0.0065304, \\
W &= 0.0042517, \\
Y_2 &= 0.0068921, \\
E &= 1.2056 \times 10^{-4}.
\end{aligned}
$$

This estimated error is again too large, so we again cut h in half to $h = 0.03125$ and again repeat the computation. This time we get

$$
\begin{aligned}
Y_1 &= 0.0042517, \\
W &= 0.0033550, \\
Y_2 &= 0.0042946, \\
E &= 1.4296 \times 10^{-5}.
\end{aligned}
$$

This time the estimated error is less than $\epsilon h = 3.125 \times 10^{-5}$, so we accept the extrapolated value $Y_3 = 0.0043089$ as the value for $y_2 \approx y(2h)$. Again, the estimated error is not *too* small, so we leave h alone and go on to the next step.

■ **EXAMPLE 6.23**

We will now attack this problem with several different routines:

- odeje, an implementation of Algorithm 6.4 written by the author;

- ode113, a variable-order Adams–Bashforth–Moulton method;

- ode23, a low-order Runge–Kutta method, based on [3];

- ode45, a medium-order Runge–Kutta method, based on [7];

- BV78, a Runge–Kutta method based on a seventh- and eighth-order pair; details may be found in [18];

- odevr7, a Runge–Kutta method based on a different seventh- and eighth-order pair; details may be found in [19].

The routines ode113, ode23, and ode45 are part of the usual MATLAB distribution; BV78 and odevr7 are more recent developments, so are not yet part of MATLAB[13] but may be downloaded from http://faculty.smu.edu/shampine/current.html.

We will perform two tests with each routine on the same problem as that in Example 6.22: one with the local error tolerance $\tau = 10^{-3}$, and one with $\tau = 10^{-6}$. We expect all six routines to produce acceptably accurate solutions; the evaluation of performance will be based on two measures: (1) the length of the output arrays, N, which indicates how many steps the method took—obviously, a method that takes more steps is doing more work and therefore is being less efficient; and (2) the maximum absolute error over the points used, $E_{\max}$.

The results for the six routines are summarized in Tables 6.12 and 6.13.

Table 6.12   Comparison of adaptive ODE methods for Example 6.23.

	odeje		ode23		ode45	
τ	N	$E_{\max}$	N	$E_{\max}$	N	$E_{\max}$
10^{-3}	82	8.4604e–4	32	0.0039	53	3.2667e–4
10^{-6}	2470	1.0409e–7	133	1.9882e–4	121	1.5205e–4

Table 6.13   Comparison of adaptive ODE methods for Example 6.23.

	ode113		odevr7		BV78	
τ	N	$E_{\max}$	N	$E_{\max}$	N	$E_{\max}$
10^{-3}	29	0.0011	71	1.5349e–4	61	2.6109e–5
10^{-6}	62	9.5061e–5	134	5.8655e–6	91	1.3374e–5

The first observation is that there is no "clear winner" among the six routines. The author's routine produced the smallest error, but used by far the most points; ode113 used the fewest points and gave tolerable error when $\tau = 10^{-6}$, but very poor error for $\tau = 10^{-3}$. The point is that adaptive routines are very sensitive to the specifics of a problem. One tries to write a robust routine that will perform well across a broad spectrum of problems. It is worth noting that the "new" routines (BV78, odevr7) produced less error than the others did.

Figure 6.10 shows the exact solution, with the points used by ode113 (for $\tau = 10^{-6}$) marked with circles.

One thing about our results that might be worrisome: In several instances, the actual error was greater than the imposed tolerance. Does this mean that something is going wrong? No, it does not. First, remember that the adaptivity is based on using computable *estimates* of the error, and it should not surprise us that some of the estimates are inexact. Also, recall that we are controlling the *local error*, i.e., the error in each step. The *global error* is something else, and while it can be bounded in terms of the local error, it quite likely will be bigger. The precise theorems are the following.

Theorem 6.6 (Global Error for Adaptive Methods, Version I) *If f is Lipschitz continuous in y, if the local error at each step satisfies (6.75), and if $h_{\max}/h_{\min} < c$, then, for*

---

[13] As of early 2013; they may well be part of later MATLAB releases.

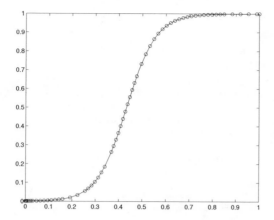

Figure 6.10   Plot of exact solution for adaptive algorithm example.

$h_{\max}$ *sufficiently small, the global error is bounded by*

$$|y(t_n) - y_n| \leq C_0|y(t_0) - y_0| + C_1\epsilon,$$

where

$$C_0 = e^{cKt_n}, \quad C_1 = \frac{e^{cKt_n} - 1}{K}$$

and K is the Lipschitz constant for f.

Proof: We start with the triangle inequality, which gives us

$$|y(t_{n+1}) - y_{n+1}| \leq |y(t_{n+1}) - u_n(t_{n+1})| + |u_n(t_{n+1}) - y_{n+1}|. \tag{6.76}$$

Note that the last term is bounded by the local error hypothesis (6.75). From the original IVP and the definition of u_n we can get

$$y(t) = y(t_n) + \int_{t_n}^{t} f(s, y(s))ds,$$

$$u_n(t) = y_n + \int_{t_n}^{t} f(s, u_n(s))ds.$$

Subtraction then yields

$$y(t) - u_n(t) = y(t_n) - y_n + \int_{t_n}^{t} (f(s, y(s)) - f(s, u_n(s)))\,ds. \tag{6.77}$$

Take absolute values, upper bounds, and use the Lipschitz condition to get

$$|y(t) - u_n(t)| \leq |y(t_n) - y_n| + K(t - t_n) \max_{t_n \leq s \leq t} |y(s) - u_n(s)|. \tag{6.78}$$

Since this is true for all $t \in [t_n, t_{n+1}]$, it follows that

$$\max_{t_n \leq s \leq t_{n+1}} |y(s) - u_n(s)| \leq |y(t_n) - y_n| + K(t_{n+1} - t_n) \max_{t_n \leq s \leq t_{n+1}} |y(s) - u_n(s)|, \tag{6.79}$$

so that (assuming that $1 - K(t_{n+1} - t_n) > 0$),

$$\max_{t_n \leq s \leq t_{n+1}} |y(s) - u_n(s)| \leq \frac{1}{1 - K(t_{n+1} - t_n)} |y(t_n) - y_n|. \tag{6.80}$$

Therefore,

$$|y(t_{n+1}) - u_n(t_{n+1})| \leq \max_{t_n \leq s \leq t_{n+1}} |y(s) - u_n(s)| \leq \frac{1}{1 - K(t_{n+1} - t_n)} |y(t_n) - y_n|. \tag{6.81}$$

Hence, if $h_{n+1} = t_{n+1} - t_n$ satisfies $h_{n+1} < 1/K$, then

$$|y(t_{n+1}) - y_{n+1}| \leq \frac{1}{1 - Kh_{n+1}} |y(t_n) - y_n| + |u_n(t_{n+1}) - y_{n+1}|. \tag{6.82}$$

The local error hypothesis allows us to write this as

$$|y(t_{n+1}) - y_{n+1}| \leq \frac{1}{1 - Kh_{n+1}} |y(t_n) - y_n| + h_{n+1}\epsilon. \tag{6.83}$$

If we assume that h is small enough at each step (this amounts to making sure that h_{max} is small enough, that is, $h_{max} < 1/K$) and the the local error hypothesis is valid at each step, then we can solve this recursion to get

$$|y(t_n) - y_n| \leq \gamma^n |y(t_0) - y_0| + \epsilon \sum_{k=0}^{n-1} \gamma^{n-k} h_k,$$

where

$$\gamma = \frac{1}{1 - Kh_{max}} \leq 1 + Kh_{max} \leq e^{Kh_{max}}.$$

Thus,

$$\begin{aligned}
|y(t_n) - y_n| &\leq \gamma^n |y(t_0) - y_0| + \epsilon h_{max} \sum_{k=0}^{n-1} \gamma^{n-k} \\
&= \gamma^n |y(t_0) - y_0| + \epsilon h_{max} \frac{1 - \gamma^n}{1 - \gamma} \\
&= \gamma^n |y(t_0) - y_0| + \epsilon h_{max} \frac{1 - \gamma^n}{Kh_{max}} (1 - Kh_{max}) \\
&\leq \gamma^n |y(t_0) - y_0| + \epsilon \frac{\gamma^n - 1}{K} \\
&\leq e^{Knh_{max}} |y(t_0) - y_0| + \epsilon \frac{e^{Knh_{max}} - 1}{K}.
\end{aligned}$$

Now, since we assume that $h_{max}/h_{min} < c$, then it follows that $nh_{max} < ct_n$; therefore,

$$|y(t_n) - y_n| \leq e^{cKt_n} |y(t_0) - y_0| + \epsilon \frac{e^{cKt_n} - 1}{K},$$

and we are done. •

If we assume that f is smooth and uniformly monotone decreasing in y, we can get a better result, and we actually get it more easily.

Theorem 6.7 (Global Error for Adaptive Methods, Version II) *If f is smooth and uniformly monotone decreasing in y, and if the local error at each step satisfies (6.75), then*

$$|y(t_n) - y_n| \le |y(t_0) - y_0| + \epsilon t_n.$$

Proof: We begin as we did for Theorem 6.6, writing the global error in terms of the local error, using the triangle inequality:

$$|y(t_{n+1}) - y_{n+1}| \le |y(t_{n+1}) - u_n(t_{n+1})| + |u_n(t_{n+1}) - y_{n+1}|. \qquad (6.84)$$

The local error hypothesis tells us that

$$|u_n(t_{n+1}) - y_{n+1}| \le h_{n+1}\epsilon,$$

where $h_{n+1} = t_{n+1} - t_n$ is the step used in the most recent computation, so we need only worry about the first term on the right. Recall now the definition of u_n as the solution of the ODE with initial value $u_n(t_n) = y_n$. In other words, u_n satisfies

$$u_n' = f(t, u_n), \quad u_n(t_n) = y_n.$$

But we also have, of course, that y satisfies

$$y' = f(t, y), \quad y(t_n) = y(t_n).$$

We can therefore use the stability estimate for smooth and uniformly monotone equations (6.9) to get that

$$|y(t) - u_n(t)| \le \left(e^{-m(t-t_n)}\right)|y(t_n) - y_n|;$$

therefore,

$$|y(t_{n+1}) - u_n(t_{n+1})| \le \left(e^{-m(t_{n+1}-t_n)}\right)|y(t_n) - y_n| = e^{-mh_{n+1}}|y(t_n) - y_n| = \gamma|y(t_n) - y_n|,$$

where $\gamma < 1$. We therefore have that

$$|y(t_{n+1}) - y_{n+1}| \le \gamma|y(t_n) - y_n| + h_{n+1}\epsilon, \qquad (6.85)$$

which is a recursive inequality in $e_k = |y(t_k) - y_k|$:

$$e_{n+1} \le \gamma e_n + h_{n+1}\epsilon.$$

We can solve this recursion to get

$$e_n = |y(t_n) - y_n| \le C_n|y(t_n) - y_n| + \epsilon\left(\sum_{k=1}^{n} \gamma^{k-1} h_n\right),$$

where

$$C_n = \gamma^n \le 1$$

and

$$\epsilon\left(\sum_{k=1}^{n} \gamma^{k-1} h_n\right) \le \epsilon t_n,$$

and we are done. ●

Note that this result is much better than the more general one that used only Lipschitz continuity of f. In either case, we see that the global error is thus controlled when the local error is controlled, but not in the same way or to the same degree.

We will conclude this section by looking, in the same way as we did in Example 6.23, at how our several routines perform on two other examples.

■ **EXAMPLE 6.24**

Consider now the IVP

$$y' = y \cos t, \quad y(0) = 1.$$

The exact solution is $y = e^{\sin t}$. We will look at how our various routines solve this problem for $0 \le t \le 6.25$. Tables 6.14 and 6.15 give the results. Fig. 6.11 shows the solution produced by ode45 using $\tau = 10^{-6}$.

Table 6.14 Comparison of adaptive ODE methods for Example 6.24.

	odeje		ode23		ode45	
τ	N	$E_{\max}$	N	$E_{\max}$	N	$E_{\max}$
10^{-3}	50	7.6422e–4	23	0.0021	45	1.9687e–4
10^{-6}	1450	3.7617e–7	188	3.5508e–6	97	3.3921e–6

Table 6.15 Comparison of adaptive ODE methods for Example 6.24.

	ode113		odevr7		BV78	
τ	N	$E_{\max}$	N	$E_{\max}$	N	$E_{\max}$
10^{-3}	28	0.0012	71	1.2346e–6	67	1.0923e–7
10^{-6}	56	3.4993e–6	78	5.2362e–7	67	1.2023e–7

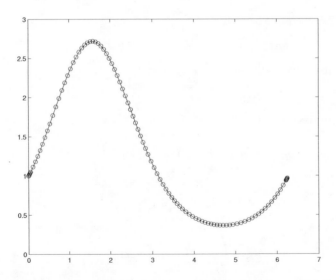

Figure 6.11 Solution to Example 6.24 using ode45; computed points are marked with circles.

Note that BV78 and odevr7 are again the most accurate, especially for the lower value of τ, but ode113 continues to be, in some sense, the most efficient of the solvers, delivering very good accuracy while using a modest number of points.

■ **EXAMPLE 6.25**

This time we look at the IVP

$$y' = \frac{1}{4}y(1 - y/20), \quad y(0) = 1.$$

The exact solution is

$$y = \frac{20}{1 + 19e^{-t/4}}.$$

We will look at how our various routines solve this problem for $0 \le t \le 25$. Tables 6.16 and 6.17 give the results. Fig. 6.12 shows the solution produced by ode113, with the approximation points marked by plus signs.

Table 6.16   Comparison of adaptive ODE methods for Example 6.25.

	odeje		ode23		ode45	
τ	N	$E_{\max}$	N	$E_{\max}$	N	$E_{\max}$
10^{-3}	34	8.0988e–4	15	0.0102	45	1.446e–4
10^{-6}	1020	2.9652e–8	110	2.4368e–5	77	2.2664e–5

Table 6.17   Comparison of adaptive ODE methods for Example 6.25.

	ode113		odevr7		BV78	
τ	N	$E_{\max}$	N	$E_{\max}$	N	$E_{\max}$
10^{-3}	19	0.0062	71	9.467e–7	67	3.5139e–8
10^{-6}	41	8.4995e–6	78	7.7524e–7	67	4.7566e–8

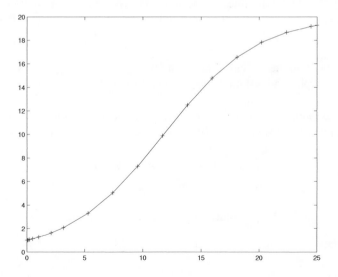

Figure 6.12   Solution for Example 6.25 using ode113.

Both of these examples looked at computing over longer time intervals, to see how efficiently the various routines would handle this. One issue that is present is that all of the routines (except the author's) define a maximum Δt as `MaxStep = (tf - t0)/10`, in other words, the maximum step is one-tenth of the total computational interval. The MATLAB routines allow this to be reset, and it might be interesting to see how the MATLAB routines performed with a smaller maximum step. (Note that in Fig. 6.12 it is evident that several of the time steps are rather large, clearly $\gg 1$, yet we still achieve acceptable accuracy.

Exercises:

1. Use Algorithm 6.4 to solve the following IVP using the local error tolerance of $\epsilon = 0.001$:

$$y' = y^2, \quad y(0) = 1.$$

   Do this as a hand calculation. Compute four or five steps. Compare the computed solution to the exact solution $y = (1 - t)^{-1}$.

2. Apply one of the MATLAB routines to each of the following ODEs:

   (a) $y' + 4y = 1, y(0) = 1$;

   (b) $y' + \sin y = 0, y(0) = 1$;

   (c) $y' + y = \sin 4\pi t, y(0) = \frac{1}{2}$;

   (d) $y' = \frac{y-t}{y+t}, y(0) = 4$.

   Use both $\tau = 1.e - 3$ and $\tau = 1.e - 6$, and compute out to $t = 10$. Plot your solutions. You may need to adjust `MaxStep`.

3. The MATLAB ODE routines can be applied to systems. Use each of `ode23`, `ode45`, and `ode113` to solve the system in Problem 9 of §6.8, using $\tau = 1.e - 3$ and also $\tau = 1.e - 6$.

4. Apply `ode45` to the system in Problem 10 of §6.8, using $\mu = 10$. Compute out to $t = 60$, and plot both solution components. Comment.

5. The three routines we used in this section are not designed for stiff ODEs, but that does not mean that we can't try it! Apply `ode23` to the system

$$\begin{aligned} y_1' &= 198y_1 + 199y_2, \quad y_1(0) = 1, \\ y_2' &= -398y_1 - 399y_2, \quad y_2(0) = -1, \end{aligned}$$

   using $\tau = 10^{-6}$. Plot the solution components over the interval $(0, 5)$.

◁ ● ● ▷

6.10 BOUNDARY VALUE PROBLEMS

We have already briefly encountered boundary value problems, in §§2.7 and 4.9. In this
section we will take a more in-depth look at the problem of approximating BVPs. We will
confine ourselves to two broad classes of problems:

1. Linear BVPs in the form

$$\begin{aligned} -u'' + p(x)u' + q(x)u &= f(x), \quad 0 \le x \le 1, \\ u(0) &= g_0, \\ u(1) &= g_1, \end{aligned}$$

where p, q, and f are known functions;

2. Nonlinear BVPs that can be written in the form

$$\begin{aligned} -u'' &= F(x, u, u'), \quad 0 \le x \le 1, & (6.86) \\ u(0) &= g_0, & (6.87) \\ u(1) &= g_1. & (6.88) \end{aligned}$$

The first class of problems can be attacked by a modest extension of the ideas from §2.7 or
§4.9. The second class we attack by adapting our IVP methods to apply to the boundary
value problem. In both cases we assume that a unique solution to the original BVP exists,
and that this solution is smooth. Precisely how smooth will be apparent from the hypotheses
to each result that we obtain.

6.10.1 Simple Difference Methods

The basic ideas are the same as in §2.7: We first define a grid

$$0 = x_0 < x_1 < x_2 < \cdots < x_{n-1} < x_n = 1$$

with $x_k - x_{k-1} = h$ (therefore $x_k = kh$), and then use the derivative approximations
derived from

$$\begin{aligned} u'(x) - \frac{u(x+h) - u(x-h)}{2h} &= \frac{1}{6}h^2 u'''(\xi_{x,h}), \\ u''(x) - \frac{u(x-h) - 2u(x) + u(x+h)}{h^2} &= \frac{1}{12}h^2 u^{(4)}(\eta_{x,h}), \end{aligned}$$

to write the BVP as the difference equation

$$-\left(1 + \frac{1}{2}p(x)h\right)u(x-h) + \left(2 + q(x)h^2\right)u(x) - \left(1 - \frac{1}{2}p(x)h\right)u(x+h)$$
$$= h^2 f(x) + R(x, h),$$

where the remainder term is given by

$$R(x, h) = \frac{1}{12}h^4 u^{(4)}(\eta_{x,h}) + \frac{1}{6}p(x)h^4 u'''(\xi_{x,h}). \qquad (6.89)$$

Note that $R = \mathcal{O}(h^4)$. Denote the approximate function as $U_h(x) \approx u(x)$, and use $U_i = U_h(x_i)$ for the approximation at the grid points. Then the vector of U_i values is defined by the system of (tridiagonal) linear equations

$$\left(2 + q_1 h^2\right) U_1 - \left(1 - \frac{1}{2} p_1 h\right) U_2 = \left(1 + \frac{1}{2} p_1 h\right) g_0 + h^2 f_1, \tag{6.90}$$

$$-\left(1 + \frac{1}{2} p_i h\right) U_{i-1} + \left(2 + q_i h^2\right) U_i - \left(1 - \frac{1}{2} p_i h\right) U_{i+1} = h^2 f_i, \tag{6.91}$$

$$-\left(1 + \frac{1}{2} p_{n-1} h\right) U_{n-2} + \left(2 + q_{n-1} h^2\right) U_{n-1} = \left(1 - \frac{1}{2} p_{n-1} h\right) g_1 + h^2 f_{n-1}, \tag{6.92}$$

where we have used $p_i = p(x_i)$ and similarly for q_i and f_i, and, in the middle equation, we have $2 \leq i \leq n - 2$.

At this point it might be useful to summarize all this as an algorithm in outline form.

Algorithm 6.5 *Finite Difference Method for Linear Boundary Value Problems.*

1. *Given $h = (b - a)/n$, form the $(n - 1) \times (n - 1)$ tridiagonal matrix T, where*

$$t_{ii} = \left(2 + q_i h^2\right), \quad t_{i,i+1} = -\left(1 - \frac{1}{2} p_i h\right), \quad t_{i,i-1} = -\left(1 + \frac{1}{2} p_i h\right);$$

2. *Form the vector $F \in \mathbb{R}^{n-1}$ given by*

$$F_i = \begin{cases} \left(1 + \frac{1}{2} p_1 h\right) g_0 + h^2 f_1, & i = 1 \\ h^2 f_i, & 2 \leq i \leq n - 2, \\ \left(1 - \frac{1}{2} p_{n-1} h\right) g_1 + h^2 f_{n-1}, & i = n - 1; \end{cases}$$

3. *Solve the system $TU = F$;*

4. *Then the approximate values are $U_i \approx u(x_i)$, $1 \leq i \leq n - 1$.*

■ **EXAMPLE 6.26**

Consider the BVP

$$\begin{aligned} -u'' + u' + u &= \sin \pi x, \quad 0 \leq x \leq 1, \\ u(0) &= 1, \\ u(1) &= 0. \end{aligned}$$

Therefore $p(x) = 1$, $q(x) = 1$, and $f(x) = \sin \pi x$. If we take $h = \frac{1}{4}$, then $n = 4$ and the linear system is therefore 3×3. We have

$$T = \begin{bmatrix} 2.0625 & -0.875 & 0 \\ -1.125 & 2.0625 & -0.875 \\ 0 & -1.125 & 2.0625 \end{bmatrix}, \quad F = \begin{bmatrix} 1.1692 \\ 0.0625 \\ 0.04419 \end{bmatrix},$$

and the solution to the linear system is

$$U = \begin{bmatrix} 0.8423 \\ 0.6489 \\ 0.3754 \end{bmatrix},$$

thus giving us our approximate solution,

$$u(1/4) \approx 0.8423, \quad u(1/2) \approx 0.6489, \quad u(3/4) \approx 0.3754.$$

The following theorem guarantees that the tridiagonal linear system can be solved using Algorithm 2.6 from §2.6.

Theorem 6.8 *Let p and q both be continuous on $[0, 1]$, with $|p(x)| \leq P_M$ and $0 < Q_m \leq q(x)$ for all $x \in [0, 1]$. Then, for h such that*

$$h \leq 2/P_M, \tag{6.93}$$

the discrete system (6.90)–(6.92) has a unique solution that can be computed by Algorithm 2.6.

Proof: The condition on h implies that

$$1 - \frac{1}{2}P_M h \geq 0;$$

therefore, $1 \pm \frac{1}{2}p_i h \geq 0$ regardless of the sign of p_i. Hence, we have

$$
\begin{aligned}
\left|1 + \frac{1}{2}p_i h\right| + \left|1 - \frac{1}{2}p_i h\right| &\leq 2 \\
&< 2 + h^2 Q_m \\
&\leq 2 + h^2 q_i
\end{aligned}
$$

for all i (the argument for the first and last equations is a little different); thus, the matrix is diagonally dominant and we can invoke Algorithm 2.6. ●

Now that we know that the approximate solution can always be computed, we can ask how accurate it is.

Theorem 6.9 *Under the same hypotheses as in Theorem 6.8, there exists a constant $C > 0$, independent of h, such that*

$$\max_{1 \leq i \leq n-1} |u(x_i) - U_i| \leq Ch^2 \left(\|u^{(4)}\|_\infty + \|u'''\|_\infty \right).$$

Proof: Define $e_i = u(x_i) - U_i$, with $e_0 = e_n = 0$. Subtracting (6.89) and (6.91) shows that the e_i satisfy the system of equations

$$
\begin{aligned}
\left(2 + q_1 h^2\right) e_1 - \left(1 - \frac{1}{2}p_1 h\right) e_2 &= R_1, \\
-\left(1 + \frac{1}{2}p_i h\right) e_{i-1} + \left(2 + q_i h^2\right) e_i - \left(1 - \frac{1}{2}p_i h\right) e_{i+1} &= R_i, \quad 2 \leq i \leq n-2, \\
-\left(1 + \frac{1}{2}p_{n-1}h\right) e_{n-2} + \left(2 + q_{n-1}h^2\right) e_{n-1} &= R_{n-1},
\end{aligned}
$$

where we have used $R_k = R(x_k, h)$ for the remainder as given in (6.89). Let J be the index for which e_i is largest in absolute value: $|e_J| \geq |e_i|$ for all i; assume for simplicity that $J \neq 1$ and $J \neq n - 1$. We then have

$$
\begin{aligned}
(2 + q_J h^2)|e_J| &\leq \left|1 + \frac{1}{2}p_J h\right| |e_{J-1}| + \left|1 - \frac{1}{2}p_J h\right| |e_{J+1}| + |R_J| \\
&\leq \left(\left|1 + \frac{1}{2}p_J h\right| + \left|1 - \frac{1}{2}p_J h\right|\right) |e_J| + |R_J| \\
&= 2|e_J| + |R_J|.
\end{aligned}
$$

Now subtract $2|e_J|$ from both sides of the equation to get

$$
q_J h^2 |e_J| \leq |R_J| \Rightarrow |e_J| \leq \frac{1}{h^2 Q_m} \max_{x \in [0,1]} |R(x, h)|.
$$

But,

$$
\max_{x \in [0,1]} |R(x, h)| \leq \frac{1}{6} h^4 \left(\|u^{(4)}\|_\infty + \|u'''\|_\infty\right),
$$

so we are done, with $C = (6Q_m)^{-1}$, since $|e_J| = \max_{1 \leq i \leq N-1} |u(x_i) - U_i|$ is the error we are trying to bound. ●

This result is an excellent example of what mathematicians would call a "not sharp" result. It seems to suggest that the accuracy of the approximation would depend adversely on Q_m, when in fact we will retain second-order accuracy even if $q(x) \equiv 0$, i.e., $Q_m = 0$. This requires a different and more subtle mode of proof, though, which we choose to omit. It should be noted that the effect of P_M that is implied in the two theorems is very real; unless h is small compared to $p(x)$, we will not get the expected accuracy. Some of the exercises are designed to demonstrate this.

■ EXAMPLE 6.27

To illustrate, consider the example

$$-u'' + 25u' + 26u = 1, \tag{6.94}$$

$$u(0) = 0, u(1) = 1, \tag{6.95}$$

for which the the exact solution is

$$u(x) = \frac{1}{26}\left(1 - \frac{e^{26} + 25}{e^{26} - e^{-1}}e^{-x} + \frac{25 + e^{-1}}{e^{26} - e^{-1}}e^{26x}\right).$$

We solve this with the above ideas, using a sequence of grids with mesh size $h = 2^{-k}$, $2 \leq k \leq 10$. Table 6.18 shows the maximum error over the entire interval for each value of h. Note that we do not begin to get the appropriate rate of convergence until $h \leq 1/16$, as we should expect. Figure 6.13 shows the exact solution, with the approximate solution for $h = 1/16$ marked by the circles.

■ EXAMPLE 6.28

A second example,

$$-u'' + u' = 2, \tag{6.96}$$

$$u(0) = 0, u(1) = 1, \tag{6.97}$$

Table 6.18   Finite Difference Approximation to Examples 6.27 and 6.28.

$n = h^{-1}$	Error in Ex. 6.27	Error in Ex. 6.28
4	0.410152632559E+00	0.617591875069D–03
8	0.226940305241E+00	0.156383538506D–03
16	0.795659554206E–01	0.392871138047D–04
32	0.185637947707E–01	0.982751528289D–05
64	0.437975179182E–02	0.245793638298D–05
128	0.110156098833E–02	0.614467541404D–06
256	0.274526544131E–03	0.153625683064D–06
512	0.685779319546E–04	0.384063436609D–07
1024	0.171423695873E–04	0.960190649213D–08

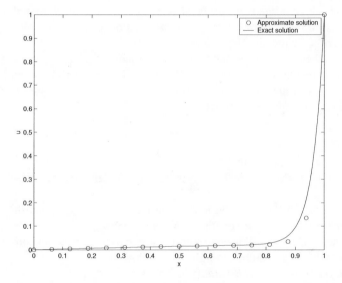

Figure 6.13   Solution for Example 6.27, together with the approximate solution for $h = 1/16$.

shows that the condition $q > 0$ is not required for second-order accuracy. Here the exact solution is

$$u(x) = 2x + \frac{1}{e - 1}\left(1 - e^x\right),$$

and the problem was solved using the same sequence of grids as in Example 6.27. Table 6.18 summarizes the results.

6.10.2 Shooting Methods

The finite difference methods of §6.10.1 can be applied to nonlinear problems such as (6.86)–(6.88), but the result is a nonlinear system of equations, rather than a linear one, and we are not yet prepared to approximate the solutions to such a problem. (But see §7.9.)

What we do here is reformulate the problem so that we can apply our initial value problem methods.

Recall the problem of interest:

$$
\begin{aligned}
-u'' &= F(x, u, u'), \quad 0 \le x \le 1, \\
u(0) &= g_0, \\
u(1) &= g_1.
\end{aligned}
$$

Consider now the new problem

$$
\begin{aligned}
-y'' &= F(x, y, y'), \quad 0 \le x \le 1, & (6.98) \\
y(0) &= g_0, & (6.99) \\
y'(0) &= p, & (6.100)
\end{aligned}
$$

and note that this is an *initial value problem*, although we are using x instead of t to denote the independent variable. We will view y as depending on the choice of initial value for y', and write $y(x; p)$ to denote this dependence.

The question then is: Can we find a value of p such that $y(1, p) = g_1 = u(1)$? If we can, then the fact that both the boundary and initial value problems have unique solutions implies that we have found the solution of the BVP (6.86)–(6.88). Essentially, if we can find the value of p that makes the function

$$
f(p) = y(1; p) - g_1 \tag{6.101}
$$

equal to zero, we will have found the correct value of p, and therefore have solved the BVP. But (6.101) is nothing more than a root-finding problem, just like those we solved in Chapter 3!

The *shooting method* is based on making an initial guess for p, and then applying a root-finding method to an approximation to f. Note that (6.98)–(6.100) is a second-order differential equation, so we will have to recast it as a first-order system in order to proceed.

A naive approach to the problem might be as follows:

Shooting Method for Nonlinear Boundary Value Problem.

1. Define $y_h(p)$ to be the (approximate) value of $y(1)$ that is found using a mesh spacing of h and an initial value of p for y', together with a reasonable initial value problem method applied to (6.98)–(6.100).

2. Use, for example, the secant method to update values of p:

$$
p_{k+1} = p_k - y_h(p_k) \left(\frac{p_k - p_{k-1}}{y_h(p_k) - y_h(p_{k-1})} \right).
$$

Note that each time we iterate on this, we have to solve the initial value problem for a new value of $p = p_k$; this is an example of a function whose root we are seeking, and which might take a substantial amount of time to compute.

3. Once the iteration converges, the approximate solution of the initial value problem is also an approximate solution of the boundary value problem.

Because of the high cost that is potentially involved in solving the IVP several times, it is usually more economical to initially iterate for p using a very coarse mesh, and then refine the value of p obtained in this way using the desired, finer, mesh.

■ **EXAMPLE 6.29**

To illustrate, consider the nonlinear BVP defined by

$$-u'' = (u')^2, \tag{6.102}$$
$$u(0) = 0, u(1) = 1, \tag{6.103}$$

which has exact solution $u = \ln((e-1)x+1)$. We used the trapezoid rule predictor–corrector method to solve the initial value problems, and used the secant method with $p_0 = 1$ and $p_1 = 0.75$ to solve (iteratively) the root-finding problem. For the first several iterations we used the very crude mesh size $h = \frac{1}{4}$, but on subsequent iterations we used the much smaller value $h = 1/128$. Table 6.19 shows the progress of the iteration for p, and Figure 6.14 shows a plot of the final approximate solution.

Let's look at the first iteration in more detail. The first-order system for this second-order equation is $w' = f(t,w)$, $w(0) = w_0$, where

$$w(t) = \begin{bmatrix} w_1(t) \\ w_2(t) \end{bmatrix} = \begin{bmatrix} u(t) \\ u'(t) \end{bmatrix}, \quad f(t,w) = \begin{bmatrix} w_2(t) \\ -w_2^2(t) \end{bmatrix}, \quad w_0 = \begin{bmatrix} 0 \\ p \end{bmatrix},$$

and we take the initial value of p to be 1. Then the first solution over the interval $[0,1]$, using $h = 1/4$ and the trapezoid rule predictor–corrector, goes like this:

$$\bar{w} = w_0 + hf(0, w_0) = \begin{bmatrix} w_{0,1} + hw_{0,2} \\ w_{0,2} - hw_{0,2}^2 \end{bmatrix} = \begin{bmatrix} 0.25 \\ 0.75 \end{bmatrix},$$

$$w_1 = w_0 + (h/2)(f(h, \bar{w}) + f(0, w_0)) = \begin{bmatrix} w_{0,1} + (h/2)(\bar{w}_2 + w_{0,2}) \\ w_{0,2} - (h/2)(\bar{w}_2^2 + w_{0,2}^2) \end{bmatrix} = \begin{bmatrix} 0.21875 \\ 0.8046875 \end{bmatrix};$$

$$\bar{w} = w_0 + hf(0, w_0) = \begin{bmatrix} w_{1,1} + hw_{1,2} \\ w_{1,2} - hw_{1,2}^2 \end{bmatrix} = \begin{bmatrix} 0.419921875 \\ 0.64280700683594 \end{bmatrix},$$

$$w_2 = w_1 + (h/2)(f(2h, \bar{w}) + f(h, w_1)) = \begin{bmatrix} w_{1,1} + (h/2)(\bar{w}_2 + w_{1,2}) \\ w_{1,2} - (h/2)(\bar{w}_2^2 + w_{1,2}^2) \end{bmatrix} = \begin{bmatrix} 0.39968681335449 \\ 0.67209714741330 \end{bmatrix};$$

$$\bar{w} = w_2 + hf(h, w_1) = \begin{bmatrix} w_{2,1} + hw_{2,2} \\ w_{2,2} - hw_{2,2}^2 \end{bmatrix} = \begin{bmatrix} 0.56771110020782 \\ 0.55916850352302 \end{bmatrix},$$

$$w_3 = w_2 + (h/2)(f(3h, \bar{w}) + f(2h, w_2)) = \begin{bmatrix} w_{2,1} + (h/2)(\bar{w}_2 + w_{2,2}) \\ w_{2,2} - (h/2)(\bar{w}_2^2 + w_{2,2}^2) \end{bmatrix} = \begin{bmatrix} 0.55359501972153 \\ 0.57654914855164 \end{bmatrix};$$

$$\bar{w} = w_3 + hf(3h, w_3) = \begin{bmatrix} w_{3,1} + hw_{3,2} \\ w_{3,2} - hw_{3,2}^2 \end{bmatrix} = \begin{bmatrix} 0.69773230685944 \\ 0.49344691837773 \end{bmatrix},$$

$$w_4 = w_3 + (h/2)(f(4h, \bar{w}) + f(3h, w_3)) = \begin{bmatrix} w_{3,1} + (h/2)(\bar{w}_2 + w_{3,2}) \\ w_{3,2} - (h/2)(\bar{w}_2^2 + w_{3,2}^2) \end{bmatrix} = \begin{bmatrix} 0.68734452808770 \\ 0.50456180080763 \end{bmatrix}.$$

Now, if we had used the correct value of $p = u'(0)$ to start the computation, we would have $w_{4,1} \approx u(1) = 1$. But clearly, $p = 1$ was not a good enough approximation, so we want to do a secant iteration on the function $F(p) = w_{4,1} - 1$; this requires that we take a second initial guess and compute $w_{4,1}$ based on that guess. We take

$p = 0.75$ as our second guess and repeat the computation outlined above, getting -0.4434700177 as our new value of $w_{4,1} - 1$. The secant step is then

$$
\begin{aligned}
p_2 &= p_1 - F(p_1)\frac{p_1 - p_0}{F(p_1) - F(p_0)} \\
&= 0.75 - (-0.4434700177)\left(\frac{0.75 - 1.00}{-0.4434700177 - (-0.3126554719)}\right) \\
&= 1.597516641,
\end{aligned}
$$

and we would continue the computation using p_2 as the initial value for the first derivative; i.e., we would set $w_{0,2} = p_2$.

Table 6.19 summarizes the results we get by continuing this computation. Note that we use the crude mesh spacing $h = 1/4$ until we get a converged value for p, then switch to the smaller $h = 1/128$ mesh size. This is a more economical approach than doing the entire computation with the finer grid.

Table 6.19   Shooting approximation to (6.102) and (6.103).

k	$n = h^{-1}$	p_k	$u_h(1) - 1$
0	4	0.1000000000E+01	−0.3126554719E+00
1	4	0.7500000000E+00	−0.4434700177E+00
2	4	0.1597516641E+01	−0.6162711822E−01
3	4	0.1734300664E+01	−0.1368453961E−01
4	4	0.1773343754E+01	−0.5578476751E−03
5	4	0.1775002976E+01	−0.5342620750E−05
6	4	0.1775019020E+01	−0.2114179276E−08
7	128	0.1717716619E+01	−0.2161118864E−03
8	128	0.1718310128E+01	0.2246518221E−05
9	128	0.1718304022E+01	0.2427305024E−09
10	128	0.1718304021E+01	−0.5551115123E−15

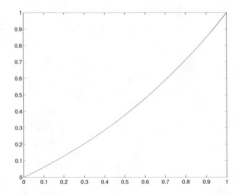

Figure 6.14   Approximate solution via shooting method for Example 6.29, using $h = 1/128$.

6.10.3   Finite Element Methods for BVPs

There are much more sophisticated methods for attacking boundary value problems, and in this subsection we discuss one of them, the finite element method.

The term *finite element method* is only one of a number of names used for the body of ideas we will introduce here. Others include the *Galerkin method*, and the *Ritz-Galerkin method*[14].

To begin the discussion, we treat the following BVP:

$$
\begin{aligned}
-u'' + q(x)u &= f(x), \quad 0 \le x \le 1, \\
u(0) &= 0, \\
u(1) &= 0,
\end{aligned}
$$

and we assume a uniform grid of points $x_i = ih$, although one of the advantages of the finite element approach is that it can easily handle a nonuniform grid.

The basic idea in the finite element method is to find an approximate solution u_h, in the form

$$
u_h(x) = \sum_{i=1}^{n-1} u_i \phi_i^h(x), \tag{6.104}
$$

where the u_i values are coefficients to be found, and the ϕ_i^h functions are an especially simple set of basis functions, defined formally by

$$
\phi_i^h(x) = \begin{cases}
(x - x_{i-1})/h, & x \in (x_{i-1}, x_i), \\
(x_i - x)/h, & x \in (x_i, x_{i+1}), \\
0, & \text{otherwise.}
\end{cases}
$$

Graphs of one of these functions (using $h = 0.1$) and its derivative are shown in Figs. 6.15 and 6.16, from which we see why these functions are sometimes known as "tent functions." Higher-order approximations, such as the B-spline basis used in §4.9, are of course possible. The central idea is to have a set of simple, *locally defined* basis functions.

We would like to proceed as we did in §4.9, and substitute our expansion (6.104) into the BVP and manipulate to get some kind of (hopefully, tridiagonal) linear system satisfied by the coefficients u_i, but we can't do that, because the tent functions, being only piecewise linear, are not twice differentiable.

---

[14]Walther Ritz (1878–1909) was a Swiss theoretical physicist who studied in Zurich and Göttingen. In 1908 he wrote a paper ("Über eine neue Methode zur Lösung gewisser Variationsprobleme der mathemtischen Physik," *J. Reine Angew. Math.*, 1908) which introduced his method for problems involving elastic plates, based on a minimization principle. Tragically, Ritz received little credit for his insight before he died at a young age, probably of tuberculosis, although some sources say it was pleurisy.

Boris Grigorievich Galerkin (1871–1945) came from a poor family in the Belarus region of Imperial Russia. He attended secondary school in Minsk, then began studying at the St. Petersburg Technological Institute in 1893. He graduated in 1899 and began working as an engineer. Politically active, he was arrested by the Czarist government in 1905, but released after 18 months. In 1909 his academic career began when he took a teaching position at the St. Petersburg Institute, and continued until his death shortly after the end of World War II. Galerkin may have met Ritz during a tour of Western Europe which Galerkin made in 1909. Galerkin's method was published in Russian in 1915, and was basically what Ritz had proposed earlier, although Galerkin was able to show, following I. G. Bubnov, that a minimization principle was not necessary. Galerkin called it the "Ritz method" in his own work.

A very thorough treatment of the history of this method may be found in a recent article by Martin Gander and Gerhard Wanner, "From Euler, Ritz, and Galerkin to Modern Computing," *SIAM Rev.*, vol. 54, pp. 627–666, 2012.

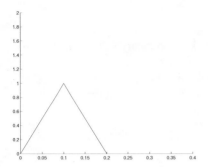

Figure 6.15   Plot of a "tent function."   **Figure 6.16**   Plot of the derivative of a "tent function."

If we multiply the differential equation by an arbitrary *smooth* function v (which also vanishes at the boundary points), we get

$$-vu'' + q(x)vu = vf(x).$$

Now integrate this over the interval of interest to get

$$\int_0^1 \left\{ -vu'' + q(x)vu \right\} dx = \int_0^1 vf(x)dx.$$

This doesn't look like we have accomplished much, but now comes the key step. Consider the first integral on the left side, and apply ordinary integration by parts to it:

$$\int_0^1 (-vu'')dx = \int_0^1 (v'u')dx + (v(1)u'(1) - v(0)u'(0)).$$

The boundary terms from the integration by parts vanish because of the assumption we made about v at the boundary, so we therefore have the new equation

$$\int_0^1 \left\{ v'u' + q(x)vu \right\} dx = \int_0^1 vf(x)dx. \tag{6.105}$$

This is known as the *weak form* of the original differential equation. (It is called "weak" because we assume less differentiability of the solution and because the equation is posed in an integral form, rather than in the usual pointwise form.) We can substitute our approximation (6.104) into this form of the problem, because the tent functions (and therefore u_h) are indeed first-order differentiable. To construct our approximation, we require that (6.105) hold for u replaced by u_h, and for v replaced by each of the basis functions ϕ_i^h:

$$\int_0^1 \left\{ (\phi_i^h)' \left(\sum_{j=1}^{n-1} u_j \phi_j^h(x) \right)' + q(x)\phi_i^h \left(\sum_{j=1}^{n-1} u_j \phi_j^h(x) \right) \right\} dx = \int_0^1 \phi_i^h(x)f(x)dx, \tag{6.106}$$

for all $i = 1, 2, \ldots, n - 1$. This is sometimes known as the *discrete weak form* of the problem; it looks imposing, but a little manipulation shows that it reduces to the system of equations

$$Ku^h = f^h,$$

where $K \in \mathbb{R}^{(n-1) \times (n-1)}$ is a matrix whose elements are defined by

$$k_{ij} = \int_0^1 \left\{ \left(\frac{d}{dx} \phi_i^h(x) \right) \left(\frac{d}{dx} \phi_j^h(x) \right) + q(x) \phi_i^h(x) \phi_j^h(x) \right\} dx;$$

$u^h \in \mathbb{R}^{(n-1)}$ is the vector of (unknown) coeficients u_i, and $f^h \in \mathbb{R}^{(n-1)}$ is the right-hand-side vector defined by

$$f_i^h = \int_{x_{i-1}}^{x_{i+1}} f(x) \phi_i^h(x) dx.$$

At this point the *local* nature of the tent functions becomes important. Because $\phi_i^h(x) = 0$ for $x \notin (x_{i-1}, x_{i+1})$, most of the matrix elements vanish. In fact, it is not difficult to show that K is, indeed, tridiagonal, and is given by

$$K = \begin{bmatrix} \frac{2}{h} + \kappa_1 & -\frac{1}{h} + r_1 & 0 & \cdots & \cdots & 0 \\ -\frac{1}{h} + r_1 & \frac{2}{h} + \kappa_2 & -\frac{1}{h} + r_2 & 0 & \cdots & 0 \\ 0 & -\frac{1}{h} + r_2 & \frac{2}{h} + \kappa_3 & -\frac{1}{h} + r_3 & 0 & \vdots \\ \vdots & & \ddots & \ddots & \ddots & 0 \\ \vdots & & & \ddots & \ddots & -\frac{1}{h} + r_{n-2} \\ 0 & \cdots & \cdots & 0 & -\frac{1}{h} + r_{n-2} & \frac{2}{h} + \kappa_{n-1} \end{bmatrix},$$

where κ_i is defined by

$$\kappa_i = \int_{x_{i-1}}^{x_{i+1}} q(x) \left(\phi_i^h(x) \right)^2 dx$$

and r_i by

$$r_i = \int_{x_i}^{x_{i+1}} q(x) \phi_i^h(x) \phi_{i+1}^h(x) dx.$$

Using this method seems very involved, and it can be, which begs the question: Why bother? The answer is that what makes this method involved also makes it applicable to a wide variety of problems that are more general than simple BVPs.

■ **EXAMPLE 6.30**

Let's look, first, at an example we used in Chapter 2; i.e., $q(x) = f(x) = 1$:

$$\begin{aligned} -u'' + u &= 1, \quad 0 \le x \le 1, \\ u(0) &= 0, \\ u(1) &= 0. \end{aligned}$$

In this case the integrals defining q_i, r_i and f_i^h are easy to do exactly, and we get

$$\kappa_i = \int_{x_{i-1}}^{x_{i+1}} \left(\phi_i^h(x) \right)^2 dx = \int_{x_{i-1}}^{x_i} \left(\frac{x - x_{i-1}}{h} \right)^2 dx + \int_{x_i}^{x_{i+1}} \left(\frac{x_{i+1} - x}{h} \right)^2 dx = \frac{2}{3} h,$$

$$r_i = \int_{x_i}^{x_{i+1}} \phi_i^h(x) \phi_{i+1}^h(x) dx = \int_{x_i}^{x_{i+1}} \left(\frac{x_{i+1} - x}{h} \right) \left(\frac{x - x_i}{h} \right) dx = \frac{1}{6} h,$$

$$f_i^h = \int_{x_{i-1}}^{x_{i+1}} \phi_i^h(x)dx = h,$$

so the matrix in the tridiagonal system becomes

$$
\begin{bmatrix}
\frac{2}{h} + \frac{2}{3}h & -\frac{1}{h} + \frac{1}{6}h & 0 & \cdots & \cdots & 0 \\
-\frac{1}{h} + \frac{1}{6}h & \frac{2}{h} + \frac{2}{3}h & -\frac{1}{h} + \frac{1}{6}h & 0 & \cdots & 0 \\
0 & -\frac{1}{h} + \frac{1}{6}h & \frac{2}{h} + \frac{2}{3}h & -\frac{1}{h} + \frac{1}{6}h & 0 & \vdots \\
\vdots & & \ddots & \ddots & \ddots & 0 \\
\vdots & & & \ddots & \ddots & -\frac{1}{h} + \frac{1}{6}h \\
0 & \cdots & \cdots & 0 & -\frac{1}{h} + \frac{1}{6}h & \frac{2}{h} + \frac{2}{3}h
\end{bmatrix}.
$$

This is very slightly different from the system in §2.7, reflecting the very different ideas underlying the method. How does it perform? Table 6.20 gives the error for the same sequence of uniform mesh sizes as in §2.7. The errors are very similar to those in Table 2.10, and it appears that they are decreasing by a factor of 4, which suggests that the error is $\mathcal{O}(h^2)$.

Table 6.20 Finite element example: $-u'' + u = 1$

h^{-1}	$\max |u(x_k) - U_k|$
4	5.378273070081180e–4
8	1.336649851621518e–4
16	3.336711303851547e–5
32	8.338712962463468e–6
64	2.084486755454806e–6
128	5.211097669488574e–7
256	1.302768320582404e–7
512	3.256977841592512e–8
1024	8.145399799097674e–9

We haven't invested a lot of discussion in the theoretical basis of this method, and there is a reason for that: A complete analysis requires a heavy dose of some very sophisticated mathematical tools and machinery. We will mainly content ourselves with a few more examples that illustrate the method, but we will say that the $\mathcal{O}(h^2)$ tendency illustrated in this first example can be made theoretically sound. In fact, using higher-degree approximating functions will often result in a higher order of accuracy. (It also needs to be said that pointwise norms are not the "natural" ones for use in the finite element method, but we defer to their simplicity for our examples here.)

■ **EXAMPLE 6.31**

What about a problem with nonconstant $q(x)$ or $f(x)$? How do we compute the matrix elements? To illustrate, consider the problem

$$-u'' + \left(\frac{1}{1+x}\right)u = f(x), \quad 0 \leq x \leq 1,$$
$$u(0) = 0,$$
$$u(1) = 0.$$

For $f(x) = 13x^2 - x^3 - 2$, the exact solution is $u(x) = x^2(1 - x^2)$ (the reader should verify this). How do we implement the method here?

The only real issue is the fact that the coefficient $q(x)$ and forcing function $f(x)$ are not constant; thus, we need to do some integration to get the values κ_i, r_i, and f_i^h. We could perhaps do the integrations exactly, but it is simpler to use a quadrature, although this does introduce some additional error. We will use the midpoint rule over the intervals (x_{i-1}, x_i) and (x_i, x_{i+1}). This gives us ($x_{i\pm1/2} = x_i \pm \frac{1}{2}h$):

$$\kappa_i = \int_{x_{i-1}}^{x_{i+1}} \left(\phi_i^h(x)\right)^2 q(x)dx = \int_{x_{i-1}}^{x_{i+1}} \left(\phi_i^h(x)\right)^2 \left(\frac{1}{1+x}\right)dx$$
$$\approx \frac{1}{4}h\left(\frac{1}{1+x_{i-1/2}} + \frac{1}{1+x_{i+1/2}}\right)$$
$$r_i = \int_{x_i}^{x_{i+1}} q(x)\phi_i^h(x)\phi_{i+1}^h(x)dx \approx \frac{1}{4}h\left(\frac{1}{1+x_{i+1/2}}\right)$$
$$f_i^h = \int_{x_{i-1}}^{x_{i+1}} f(x)\phi_i^h(x)dx = \int_{x_{i-1}}^{x_{i+1}} (11x^3 + x^2 - 2)\phi_i^h(x)dx$$
$$\approx \frac{1}{2}h\left(\left(13x_{i-1/2}^3 - x_{i-1/2}^3 - 2\right) + \left(13x_{i+1/2}^3 - x_{i+1/2}^3 - 2\right)\right)$$

Table 6.21 shows the errors, and they continue to look like the error is $\mathcal{O}(h^2)$. If we were using higher-order basis functions, we would need to use higher-order quadrature to maintain this accuracy.

Table 6.21   Errors in Example 6.31.

h^{-1}	$\max |u(x_k) - u_h(x_k)|$
4	0.00811313662254
8	0.00201198958946
16	5.038365178675064e–4
32	1.261547324883650e–4
64	3.153473944955687e–5
128	7.885241891991690e–6
256	1.971295138653018e–6
512	4.928228100331555e–7
1024	1.232058168265660e–7

How do we treat more general boundary conditions? If the problem imposes nonzero function value boundary conditions, it is a simple modification to handle them.

■ **EXAMPLE 6.32**

Consider the problem given below.

$$
\begin{aligned}
-u'' + u &= 1, \quad 0 \le x \le 1, \\
u(0) &= 1, \\
u(1) &= 0,
\end{aligned}
$$

The exact solution is

$$
u(x) = 1 - \frac{e}{e^2 - 1}e^x + \frac{e}{e^2 - 1}e^{-x}.
$$

(the reader should verify this). We use the same expansion as before, but we add to it a function that satisfies the boundary conditions:

$$
u_h(x) = (1 - x) + \sum_{i=1}^{n-1} u_i \phi_i^h(x).
$$

We substitute this into the discrete weak form (6.106) and proceed as before. The $1 - x$ term enters into the approximation via the right-hand side of the linear algebra problem. We have almost the same linear algebra problem as in Example 6.30:

$$
Ku^h = \hat{f}^h,
$$

where

$$
K = \text{tridiag}\left(-\frac{1}{h} + \frac{1}{6}h, \quad \frac{2}{h} + \frac{2}{3}h, \quad -\frac{1}{h} + \frac{1}{6}h \right)
$$

and

$$
\hat{f}_i^h = \int_{x_{i-1}}^{x_{i+1}} x\phi_i^h(x)dx + \int_{x_{i-1}}^{x_{i+1}} (\phi_i^h(x))'dx.
$$

(In one of the exercises we ask the student to fill in the details here.) We use a midpoint rule quadrature on the first term in $\hat{f}_i^h$ (the derivative term integrates to zero) to get

$$
\hat{f}_i^h = hx_i + \mathcal{O}(h^3).
$$

Table 6.22 shows the errors in our approximation as h decreases, and Fig. 6.17 shows the solution plot for $h = 1/32$. Given that the exact solution is nearly a straight line, the accuracy is perhaps not surprising.

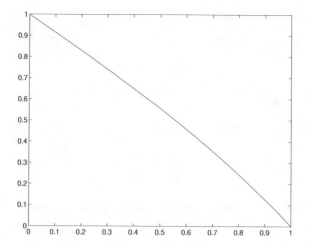

Figure 6.17   Solution for Example 6.32 for $h = 1/32$.

Table 6.22   Error in Example 6.32.

h^{-1}	$\max |u(x_k) - u_h(x_k)|$
4	2.689136535040104e–004
8	6.884707685311797e–005
16	1.722228151940230e–005
32	4.318665783487052e–006
64	1.079566930628495e–006
128	2.698855354199558e–007
256	6.747435599141482e–008
512	1.686886730301040e–008
1024	4.218667526423303e–009

There is much more to the finite element method than has been discussed in this brief treatment. We will do a little more with these ideas in Chapter 9.

Exercises:

1. Use a mesh size of $h = \frac{1}{4}$ ($n = 4$) to solve the boundary value problem (via finite differences)

$$-u'' + u = \sin x, 0 < x < 1;$$
$$u(0) = u(1) = 0.$$

Don't bother to write a computer program; do this as a hand calculation.

2. Repeat the above for the BVP

$$-u'' + (2 - x)u = x, 0 < x < 1;$$
$$u(0) = u(1) = 0.$$

Don't bother to write a computer program; do this as a hand calculation.

3. Repeat the above for the BVP

$$-u'' + u = e^{-x}, 0 < x < 1;$$
$$u(0) = u(1) = 0.$$

4. Consider the linear boundary value problem

$$-u'' + (10\cos 2x)u = 1,$$
$$u(0) = u(\pi) = 0.$$

Solve this using finite difference methods and a decreasing sequence of grids, starting with $h = \pi/16, \pi/32, \ldots$. Do the approximate solutions appear to be converging to a solution?

5. Consider the nonlinear boundary value problem

$$-u'' + e^u = 1,$$
$$u(0) = u(1) = 0.$$

Use shooting methods combined with the trapezoid rule predictor–corrector to construct solutions to this equation, then use a fourth-order Runge–Kutta scheme and compare the two approximate solutions. Are they nearly the same?

6. Write a program to use the finite element method to solve the BVP

$$-u'' + u = e^{-x}, 0 < x < 1;$$
$$u(0) = u(1) = 0,$$

using a sequence of grids, $h^{-1} = 4, 8, 16, \ldots, 1024$.

7. Repeat Problem 6, using the different boundary conditions $u(0) = 0, \quad u(1) = 1$.

8. No exact solution was provided for either of Problems 6 or 7 (although anyone having completed a sophomore ODE course—or with access to a symbolic algebra program such as Maple or Mathematica—ought to be able to produce a solution). Write an essay addressing the following question: On what basis are you confident that your codes are producing the correct solution?

6.11 LITERATURE AND SOFTWARE DISCUSSION

The numerical solution of initial value problems for ordinary differential equations is one of the most important areas of scientific computation. Good overviews of the development of the subject in the last 40 years or so may be found in the monographs [11], [9], [15], and [16]; more recent treatments are in [8] and [13]. A selection of easily available automatic software—as well as a rich summary of recent results—is given by Shampine

[17]. Additional, more recent, codes are discussed in [18] and [19]. Other codes can be found by searching the Netlib repository at http://www.netlib.org.

Less is available about two-point boundary value problems. Much of what we did here is based on Chapter 8 (Sec. 7) of the book by Isaacson and Keller [12]; a more comprehensive survey is in [14]. One reason less is written about BVPs for ordinary differential equations is that they can in almost all respects be considered as special cases of the Poisson partial differential equation, which we treat in §9.3. However, the ordinary differential equation case is *so* much simpler that it really deserves its own exposition, which can serve as an introduction to the PDE case.

REFERENCES

1. Allen, Myron B., and Isaacson, Eli L.; *Numerical Analysis for Applied Science*, John Wiley & Sons, Inc., New York, 1998.

2. Atkinson, Kendall, *An Introduction to Numerical Analysis*, John Wiley & Sons, Inc., New York, 1987 (2nd edition).

3. Bogacki, P., and Shampine, L. F., "A 3(2) pair of Runge–Kutta formulas," *Appl. Math. Lett.*, vol. 2, pp. 1-9, 1989.

4. Braun, Martin, *Differential Equations and Their Applications*, Springer-Verlag, New York, 1993 (4th Edition).

5. Burden, Richard, and Faires, J. Douglas, *Numerical Analysis*, Brooks/Cole, Pacific Grove, CA, 1997 (6th Edition).

6. Coddington, E. A., and Levinson, N., *Theory of Ordinary Differential Equations*, McGraw-Hill, New York, 1955.

7. Dormand, J. R., and Prince, P.J., "A family of embedded Runga–Kutta formulae," J. Comp. Appl. Math., vol. 6, no. 1, pp. 19–26, 1980.

8. Eriksson, K., Estep, D., Hansbo, P., and Johnson, C. *Computational Differential Equations*, Cambridge University Press, Cambridge, UK, 1996.

9. Gear, C. William, *Numerical Initial Value Problems in Ordinary Differential Equations*, Prentice-Hall, Englewood Cliffs, NJ, 1971.

10. Goldstine, Herman H., *A History of Numerical Analysis from the 16th Through the 19th Century*, Springer-Verlag, New York, 1977.

11. Henrici, Peter; *Discrete Variable Methods in Ordinary Differential Equations*, John Wiley & Sons, Inc., New York, 1962.

12. Isaacson, E., and Keller, H. B., *Analysis of Numerical Methods*, Dover, New York, 1994; originally published by John Wiley & Sons, Inc., 1966.

13. Iserles, Arieh, *A First Course in the Numerical Analysis of Differential Equations*, Cambridge University Press, Cambridge, UK, 1996.

14. Keller, H. B., *Numerical Solution of Two Point Boundary Value Problems*, SIAM, Philadelphia, 1976.

15. Lambert, J. D., *Computational Methods for Ordinary Differential Equations*, John Wiley & Sons, Inc., London, 1973.

16. Shampine, L. F., and Gordon, M. K., *Computer Solution of Ordinary Differential Equations*, W.H. Freeman and Company, San Francisco, 1975.

17. Shampine, L. F., "The MATLAB ODE Suite," *SIAM J. Sci. Comput.*, vol. 18, pp. 1–22, 1997.

18. Shampine, L. F., "Vectorized Solution of ODEs in MATLAB," available online at `http://faculty.smu.edu/shampine/current.html`.

19. Shampine, L. F., "Vectorized Solution of ODEs in MATLAB with Control of Residual and Error," available online at `http://faculty.smu.edu/shampine/current.html`.

CHAPTER 7

NUMERICAL METHODS FOR THE SOLUTION OF SYSTEMS OF EQUATIONS

Here we will look at numerical methods for solving two important problems from linear algebra:

1. *The linear systems problem*: Given a matrix A and a vector b, both known, find the vector x such that

$$Ax = b.$$

2. *The nonlinear systems problem*: Given a vector-valued function F, find the vector x such that

$$F(x) = 0.$$

A third important problem—the algebraic eigenvalue problem—is deferred to Chapter 8. These two chapters—7 and 8—are most heavily affected by the use of MATLAB, which in many ways was originally designed to be an easy-to-use interface for the FORTRAN packages LINPACK (linear systems) and EISPACK (eigenvalue problems). It is fair to ask why we are going to spend time describing in detail algorithms that can be executed with a single line of MATLAB code. Part of the answer lies in a bit of philosophy: The author believes very strongly that students need some exposure to the details of an algorithm in order to understand the material, but we will not go into deep detail on some of the more complicated algorithms, relying on the appropriate MATLAB constructs.

We begin with a review of the linear algebra concepts and notation that are especially necessary for this chapter.

An Introduction to Numerical Methods and Analysis, Second Edition. By James F. Epperson
Copyright © 2013 John Wiley & Sons, Inc.

7.1 LINEAR ALGEBRA REVIEW

A *vector* $x \in \mathbb{R}^n$ is defined to be an ordered n-tuple of real numbers:

$$x = (x_1, x_2, \ldots, x_n)^T,$$

where the superscript (denoting transpose) indicates that the vector is considered to be a column vector. A *matrix* $A \in \mathbb{R}^{m \times n}$ is a rectangular array of m rows and n columns:

$$A = \begin{bmatrix} a_{11} & a_{12} & a_{13} & \cdots & a_{1n} \\ a_{21} & a_{22} & a_{23} & \cdots & a_{2n} \\ a_{31} & a_{32} & a_{33} & \cdots & a_{3n} \\ \vdots & \vdots & \vdots & \ddots & \vdots \\ a_{m1} & a_{m2} & a_{m3} & \cdots & a_{mn} \end{bmatrix}.$$

We will assume that the student is familiar with the basic operations of addition and multiplication of matrices.

We will use no diacritical marks to distinguish vectors from scalars. In general, matrices will be denoted by upper case Roman letters, and vectors by lower case Roman letters. The notational correspondence between the vector or matrix and its components will almost always be as in the above examples.

Our assumption that vectors are column vectors means that we can regard them as matrices in $\mathbb{R}^{n \times 1}$, and this allows us to write the ordinary vector dot product as

$$(x, y) = x \cdot y = x^T y = \sum_{i=1}^{n} x_i y_i.$$

Note that reversing the order of multiplication results in a matrix, not a scalar:

$$xy^T = \begin{bmatrix} x_1 y_1 & x_1 y_2 & x_1 y_3 & \cdots & x_1 y_n \\ x_2 y_1 & x_2 y_2 & x_2 y_3 & \cdots & x_2 y_n \\ x_3 y_1 & x_3 y_2 & x_3 y_3 & \cdots & x_3 y_n \\ \vdots & \vdots & \vdots & \ddots & \vdots \\ x_n y_1 & x_n y_2 & x_n y_3 & \cdots & x_n y_n \end{bmatrix}.$$

Note also that this notation allows us to write

$$(x, Ay) = x^T Ay = (A^T x)^T y = (A^T x, y),$$

which is one of the important properties of the transpose of a matrix.

Given a square matrix $A \in \mathbb{R}^{n \times n}$, if there exists a second square matrix $B \in \mathbb{R}^{n \times n}$ such that $AB = BA = I$, then we say that B is the inverse of A and we write $B = A^{-1}$. *Not all square matrices have an inverse!* If a matrix $A \in \mathbb{R}^{n \times n}$ has an inverse, we say that A is *nonsingular*; otherwise, we say that A is *singular*.

The following theorem summarizes the conditions under which a matrix is nonsingular, and also connects them to the solvability of the linear systems problem.

Theorem 7.1 *Given a matrix* $A \in \mathbb{R}^{n \times n}$, *the following statements are equivalent:*

1. *A is nonsingular;*

2. *The columns of A form an independent set of vectors;*

3. *The rows of A form an independent set of vectors;*

4. *The linear system* $Ax = b$ *has a unique solution for all vectors* $b \in \mathbb{R}^n$;

5. *The homogeneous system* $Ax = 0$ *has only the trivial solution* $x = 0$;

6. *The determinant is nonzero:* $\det A \neq 0$.

An important corollary to this theorem is the following.

Corollary 7.1 *If* $A \in \mathbb{R}^{n \times n}$ *is singular, then there exist infinitely many vectors* $x \in \mathbb{R}^n$, $x \neq 0$, *such that* $Ax = 0$.

There are a number of special classes of matrices. In §2.6 we looked at *tridiagonal* matrices, and later in this chapter we will look at *symmetric, positive definite* matrices. One class of special matrix that the student ought to be familiar with is the *triangular* matrix: A *square matrix* is *lower (upper) triangular* if all the elements above (below) the main diagonal are zero. Thus,

$$U = \begin{bmatrix} 1 & 2 & 3 \\ 0 & 4 & 5 \\ 0 & 0 & 6 \end{bmatrix}$$

is upper triangular, and

$$L = \begin{bmatrix} 1 & 0 & 0 \\ 2 & 3 & 0 \\ 4 & 5 & 6 \end{bmatrix}$$

is lower triangular.

The student who is not familiar with the concepts of *independence/dependence, spanning, basis, vector space/subspace, dimension*, and *orthogonal/orthonormal* should review the appropriate sections of a linear algebra text. Students should also be familiar with the basic properties of determinants, although we do not use determinants as much as might be expected. (They are used somewhat in the exercises in some of the early sections, especially in §7.2.) While we defer our treatment of the computation of eigenvalues and eigenvectors to Chapter 8, the student should still be familiar with the basic properties and definition of these quantities, as they are mentioned in several places in this chapter.

Exercises:

1. Assume Theorem 7.1 and use it to prove Corollary 7.1.

2. Use Theorem 7.1 to prove that a triangular matrix is nonsingular if and only if the diagonal elements are all nonzero.

3. Suppose that we can write $A \in \mathbb{R}^{n \times n}$ as the product of two triangular matrices $L \in \mathbb{R}^{n \times n}$ and $U \in \mathbb{R}^{n \times n}$ where the diagonal elements of L and U are all nonzero. Prove that A is nonsingular.

◁ ● ● ▷

7.2   LINEAR SYSTEMS AND GAUSSIAN ELIMINATION

In §2.6 we constructed an algorithm that solved tridiagonal linear systems by first reducing them to triangular form, and then solving the triangular system. In this section, we will construct a general version of that algorithm. We begin by writing the linear system as a single augmented matrix:

$$A' = [A \mid b],$$

where the vertical bar is supposed to separate the coefficient matrix from the right-side vector. The solution algorithm is then applied to A'.

The algorithm is the same one that is taught in a standard linear algebra course, and which we first saw in §2.6: Gaussian elimination. It works by systematically eliminating nonzero elements below the main diagonal of the coefficient matrix. This is accomplished by using only those operations that preserve the solution set of the system, known as *elementary row operations*:

1. Multiply a row by a nonzero scalar, c;

2. Interchange two rows;

3. Multiply a row by a nonzero scalar, c, and add the result to another row.

If we can manipulate from one matrix to another using only elementary row operations, then the two matrices are said to be *row equivalent*.

The important theorem that connects the elementary row operations to the solution of linear systems is the following.

Theorem 7.2 *Let A' be the augmented matrix corresponding to the linear system $Ax = b$, and suppose that A' is row equivalent to $A'' = [T \ \ c]$. Then the two linear systems have precisely the same solution sets.*

Our goal, then, is to use elementary row operations to reduce the augmented matrix A' to the new augmented matrix $A'' = [U \mid c]$, where U is *upper triangular*. This will mean that the new system $Ux = c$ will be easy to solve.

■ EXAMPLE 7.1

To illustrate the process, let's look at a concrete example that we will work through in detail. Consider the system of equations

$$\begin{aligned}
4x_1 + 2x_2 - x_3 &= 5 \\
x_1 + 4x_2 + x_3 &= 12 \\
2x_1 - x_2 + 4x_3 &= 12
\end{aligned}$$

which can be written in matrix–vector form as

$$\begin{bmatrix} 4 & 2 & -1 \\ 1 & 4 & 1 \\ 2 & -1 & 4 \end{bmatrix} \begin{bmatrix} x_1 \\ x_2 \\ x_3 \end{bmatrix} = \begin{bmatrix} 5 \\ 12 \\ 12 \end{bmatrix}.$$

We write this as an augmented matrix:

$$A' = \begin{bmatrix} 4 & 2 & -1 & 5 \\ 1 & 4 & 1 & 12 \\ 2 & -1 & 4 & 12 \end{bmatrix}.$$

Then the elimination algorithm proceeds as follows:

$$A' = \begin{bmatrix} 4 & 2 & -1 & 5 \\ 1 & 4 & 1 & 12 \\ 2 & -1 & 4 & 12 \end{bmatrix} \sim \begin{bmatrix} 4 & 2 & -1 & 5 \\ 0 & \frac{7}{2} & \frac{5}{4} & \frac{43}{4} \\ 2 & -1 & 4 & 12 \end{bmatrix} \sim \begin{bmatrix} 4 & 2 & -1 & 5 \\ 0 & \frac{7}{2} & \frac{5}{4} & \frac{43}{4} \\ 0 & -2 & \frac{9}{2} & \frac{19}{2} \end{bmatrix}.$$

The first step was accomplished by multiplying the first row by $\frac{1}{4}$ and subtracting the result from the second row; the second step was accomplished by multiplying the first row by $\frac{1}{2}$ and subtracting the result from the third row. To finish the job, we have (by multiplying the second row by $-\frac{4}{7}$ and subtracting from the third row)

$$A' \sim \begin{bmatrix} 4 & 2 & -1 & 5 \\ 0 & \frac{7}{2} & \frac{5}{4} & \frac{43}{4} \\ 0 & -2 & \frac{9}{2} & \frac{19}{2} \end{bmatrix} \sim \begin{bmatrix} 4 & 2 & -1 & 5 \\ 0 & \frac{7}{2} & \frac{5}{4} & \frac{43}{4} \\ 0 & 0 & \frac{73}{14} & \frac{219}{14} \end{bmatrix} = A''.$$

This augmented matrix represents a triangular system—meaning that the coefficient matrix is triangular—as follows:

$$A'' = [U \mid c] \Rightarrow Ux = c,$$

that is,

$$\begin{bmatrix} 4 & 2 & -1 \\ 0 & \frac{7}{2} & \frac{5}{4} \\ 0 & 0 & \frac{73}{14} \end{bmatrix} \begin{bmatrix} x_1 \\ x_2 \\ x_3 \end{bmatrix} = \begin{bmatrix} 5 \\ \frac{43}{4} \\ \frac{219}{14} \end{bmatrix};$$

and we can now solve by interpreting each row as follows:

Third Row: $\frac{73}{14}x_3 = \frac{219}{14} \Rightarrow x_3 = 3$;

Second Row: $\frac{7}{2}x_2 + \frac{5}{4}x_3 = \frac{43}{4} \Rightarrow x_2 = 2$;

First Row: $4x_1 + 2x_2 - x_3 = 5 \Rightarrow x_1 = 1$.

To render this example process into a general algorithm, we note the essential features:

> Work down each column, eliminating (i.e., converting to a zero) each component below the main diagonal, and modifying the rest of the corresponding row appropriately.

We start by presenting what we call "naive" Gaussian elimination, a version of the algorithm that is easy to understand yet, is not completely general. We then move quickly to a more robust and complete version.

Algorithm 7.1 *Naive Gaussian Elimination Algorithm for Ax = b (Pseudocode)*

```
for i=1 to n-1
    for j=i+1 to n
        m = a(j,i)/a(i,i)
        for k=i+1 to n
            a(j,k) = a(j,k) - m*a(i,k)
        endfor
        b(j) = b(j) - m*b(i)
    endfor
endfor
```

Programming Notes: Let's talk about what this algorithm does and how it implements what was done in the example.

The outermost loop (the i loop) ranges over the columns of the matrix; the last column is skipped because we do not need to perform any eliminations there, since there are no elements below the diagonal element. (If we were doing elimination on a nonsquare matrix with more rows than columns, then we would have to include the last column in this loop.)

The middle loop (the j loop) ranges down the ith column, below the diagonal (hence j ranges only from $i+1$ to n). We first compute the *multiplier*, m, for each row. This is the constant that we use to multiply the ith row by in order to eliminate the a_{ji} element. Note that we *overwrite* the previous values with the new ones, and we do not actually carry out the computation that makes a_{ji} zero. Note also that this loop is where the right-hand-side vector is modified to reflect the elimination step.

The innermost loop (the k loop) ranges across the jth row, starting after the ith column, modifying each element appropriately to reflect the elimination of a_{ji}.

Finally, we must be aware that the algorithm does not actually create the zeros in the lower triangular half of A; this would be wasteful of computer time since we don't need to have these zeros in place for the algorithm to work. If we were to apply our algorithm to Example 7.1, then the computer storage for A would look like this when the process was done:

$$
B = \begin{bmatrix} 4 & 2 & -1 \\ 1 & \frac{7}{2} & \frac{5}{4} \\ 2 & -2 & \frac{73}{14} \end{bmatrix}.
$$

The process works because we only work with the upper triangular part of the matrix from this point forward, so the lower triangular elements need never be referenced. This will change somewhat in §7.4.

To finish the solution process, we apply what is known as the backward solution or *backsolve algorithm* to the augmented matrix that results from the elimination step. It's called the *backward solution step* because we proceed backwards up the diagonal from the bottom to the top.

Algorithm 7.2 *Backward Solution Algorithm for Ax = b (Pseudocode)*

```
x(n) = b(n)/a(n,n)
for i=n-1 to 1
    sum = 0
    for j=i+1 to n
        sum = sum + a(i,j)*x(j)
    endfor
    x(i) = (b(i) - sum)/a(i,i)
endfor
```

Programming Notes: This algorithm simply marches backwards up the diagonal, computing each x_i in turn. Formally, we are computing

$$
x_i = \frac{1}{a_{ii}} \left(b_i - \sum_{j=i+1}^{n} a_{ij} x_j \right),
$$

which is what is necessary to solve a triangular system. The j loop is simply accumulating the summation term in this formula.

To summarize, then, the algorithm is nothing more than a systematic application of the ordinary Gaussian elimination algorithm from basic linear algebra. There is one problem, however. In the computation of the multiplier, we divide by the current (note that the values of the matrix elements change throughout the elimination process) value of the diagonal element. Clearly, if this value is zero, then we cannot proceed further and the algorithm breaks down. Note that this does not mean that the matrix is singular, since

$$A = \begin{bmatrix} 0 & 1 \\ 1 & 0 \end{bmatrix}$$

is an example of a nonsingular matrix for which the algorithm cannot even get started. What do we do to fix this?

Recall that one of our allowed elementary row operations is the interchange of two rows. In the example above, if we switch the two rows, we get the new matrix

$$A' = \begin{bmatrix} 1 & 0 \\ 0 & 1 \end{bmatrix}$$

and from this point the elimination algorithm can obviously proceed to completion. (In fact, it is already finished!)

The pivotal issue in this discussion is whether or not the diagonal elements become zero at any point in the process. Because they are so important in this regard, the diagonals are called *pivots* or *pivot elements*, and the process of swapping rows to avoid a zero element on the diagonal is called *pivoting*. There are actually several kinds of pivoting. *Partial pivoting*, in which only entries in the same column below the current diagonal are examined, is the commonest and is all that we will discuss here in any depth. *Complete pivoting*, which searches not only on the current column, but also on all subsequent columns, is known to be more stable, but is also much more expensive to implement. Finally, it is sometimes necessary to scale the rows before pivoting, and this is called *scaled pivoting*.

The partial pivoting algorithm is relatively easy to describe:

Algorithm 7.3 *Partial Pivoting (Outline).*

1. *Suppose that we are about to work on the ith column of the matrix. Then we search that portion of the ith column below and including the diagonal, and find the element that has the largest absolute value. Let p denote the index of the row that contains this element.*

2. *Interchange rows i and p.*

3. *Proceed with the elimination.*

Note, by the way, that this algorithm does not merely search for the first nonzero element in the column to make that the pivot, but searches for the *largest* element in the column to use as the pivot. We will address this issue momentarily. It is not difficult to update our algorithm to include pivoting.

Although the actual problem with naive Gaussian elimination is the potential division by a zero pivot, the entire algorithm will be less susceptible to rounding error if we choose to use the largest possible pivot element.

■ **EXAMPLE 7.2**

To gain some insight into why this is so, consider the following example system:

$$\begin{bmatrix} \epsilon & 1 \\ 1 & 1 \end{bmatrix} \begin{bmatrix} x_1 \\ x_2 \end{bmatrix} = \begin{bmatrix} 1 \\ 2 \end{bmatrix}.$$

For any $\epsilon \neq 1$, the matrix is nonsingular and so a unique solution exists. The exact solution, expressed in terms of ϵ (see Problem 7), is

$$x_1 = \frac{1}{1-\epsilon} = 1 + \mathcal{O}(\epsilon), \quad x_2 = \frac{1-2\epsilon}{1-\epsilon} = 1 + \mathcal{O}(\epsilon), \tag{7.1}$$

but what happens when we solve this system—under the assumption that ϵ is very small—using the naive versus the pivoting algorithms?

If we don't pivot, then the computation goes like this:

$$\left[\begin{array}{cc|c} \epsilon & 1 & 1 \\ 1 & 1 & 2 \end{array}\right] \sim \left[\begin{array}{cc|c} \epsilon & 1 & 1 \\ 0 & 1-\frac{1}{\epsilon} & 2-\frac{1}{\epsilon} \end{array}\right].$$

Suppose now that ϵ is so small that, within the machine arithmetic, $1 - \frac{1}{\epsilon} = -\epsilon^{-1}$ and $2 - \frac{1}{\epsilon} = -\epsilon^{-1}$. Then we have

$$\left[\begin{array}{cc|c} \epsilon & 1 & 1 \\ 0 & 1-\frac{1}{\epsilon} & 2-\frac{1}{\epsilon} \end{array}\right] \approx \left[\begin{array}{cc|c} \epsilon & 1 & 1 \\ 0 & -\frac{1}{\epsilon} & -\frac{1}{\epsilon} \end{array}\right] \Rightarrow x_2 \approx 1, \quad x_1 \approx 0.$$

Although the value of x_2 is accurate, the value of x_1 is not. This error is caused by the rounding error associated with the large number $1/\epsilon$. On the other hand, if we first pivot by switching the rows, then the computation becomes

$$\left[\begin{array}{cc|c} \epsilon & 1 & 1 \\ 1 & 1 & 2 \end{array}\right] \sim \left[\begin{array}{cc|c} 1 & 1 & 2 \\ \epsilon & 1 & 1 \end{array}\right] \sim \left[\begin{array}{cc|c} 1 & 1 & 2 \\ 0 & 1-\epsilon & 1-2\epsilon \end{array}\right] \approx \left[\begin{array}{cc|c} 1 & 1 & 2 \\ 0 & 1 & 1 \end{array}\right] \Rightarrow x_2 \approx 1, \quad x_1 \approx 1.$$

This example illustrates (but, of course, does not *prove*) the utility of partial pivoting even if we do not have a zero pivot. This is especially important when we consider that we cannot predict what the sizes of the pivot elements will be.

The appropriate MATLAB routine to solve a system is the backslash command: x = A\b. Some of the exercises ask you to write you own code, some ask you to use the MATLAB commands and even compare the results.

We end this section with a pseudocode for a Gaussian elimination routine that does partial pivoting. It might be appropriate, at this time, to remind students that it is important to understand the methods and algorithms well enough to write your own working code. A degree of facility with MATLAB is a good thing, but true understanding will only come from writing your own codes.

Algorithm 7.4 *Gaussian Elimination with Partial Pivoting (Pseudocode)*

```
for i=1 to n-1
     am = abs(a(i,i))
     p = i
     for j=i+1 to n
          if abs(a(j,i)) > am then
               am = abs(a(j,i))
               p = j
          endif
     endfor
     if p > i then
          for k = i to n
               hold = a(i,k)
               a(i,k) = a(p,k)
               a(p,k) = hold
          endfor
          hold = b(i)
          b(i) = b(p)
          b(p) = hold
     endif
     for j=i+1 to n
          m = a(j,i)/a(i,i)
          for k=i+1 to n
               a(j,k) = a(j,k) - m*a(i,k)
          endfor
          b(j) = b(j) - m*b(i)
     endfor
endfor
```

Exercises:

For the sake of simplicity here, we will define at the outset several families of matrices, parameterized by their dimension. These are referred to in several of the exercises throughout the chapter.

$$H_n = [h_{ij}], \quad h_{ij} = \frac{1}{i+j-1}.$$

$$K_n = [k_{ij}], \quad k_{i,j} = \begin{cases} 2, & i = j; \\ -1, & |i-j| = 1; \\ 0, & \text{otherwise.} \end{cases}$$

$$T_n = [t_{ij}], \quad t_{i,j} = \begin{cases} 4, & i = j; \\ 1, & |i-j| = 1; \\ 0, & \text{otherwise.} \end{cases}$$

$$A_n = [a_{ij}], \quad a_{i,j} = \begin{cases} 1, & i = j; \\ 4, & i - j = 1; \\ -4, & i - j = -1; \\ 0, & \text{otherwise.} \end{cases}$$

Even if we do not know the solution to a linear system, we can check the accuracy of a computed solution x_c by means of the residual $r = b - Ax_c$. If x_c is the exact solution, then each component of r will be zero; in floating-point arithmetic, there might be a small amount of rounding error, unless the matrix is "nearly singular," a concept that we will discuss in detail in §7.5.

1. Write a naive Gaussian elimination code and use it to solve the system of equations $Ax = b$, where

$$A = \begin{bmatrix} 14 & 14 & -9 & 3 & -5 \\ 14 & 52 & -15 & 2 & -32 \\ -9 & -15 & 36 & -5 & 16 \\ 3 & 2 & -5 & 47 & 49 \\ -5 & -32 & 16 & 49 & 79 \end{bmatrix}$$

and $b = [-15, -100, 106, 329, 463]^T$. The correct answer is $x = [0, 1, 2, 3, 4]^T$.

2. Write a naive Gaussian elimination code and use it to solve the system of equations

$$T_5 x = b,$$

where $b = [1, 6, 12, 18, 19]^T$. The correct answer is $x = [0, 1, 2, 3, 4]^T$.

3. Write a naive Gaussian elimination code and use it to solve the system of equations

$$H_5 x = b,$$

where $b = [5.0, 3.550, 2.81428571428571, 2.34642857142857, 2.01746031746032]^T$. The correct answer is $x = [1, 2, 3, 4, 5]^T$.

4. Repeat Problem 3, except now use $b_1 = 5.0001$; how much does the answer change?

5. Write your own naive Gaussian elimination code, based on the material in this chapter, and test it on the indicated families, over the range of $4 \le n \le 20$. Take b to be the vector of appropriate size, each of whose entries is 1.

   (a) H_n;

   (b) T_n;

   (c) K_n.

   For each value of n, compute the value of $\max_{1 \le i \le n} |r_i|$, where $r = b - Ax$.

6. Modify the Gaussian elimination algorithm to handle more than a single right-hand side. Test it on a 5×5 example of your own design, using at least three right-hand-side vectors.

7. Use the naive Gaussian elimination algorithm to solve (by hand) the following system. You should get the same results as in (7.1).

$$\begin{bmatrix} \epsilon & 1 \\ 1 & 1 \end{bmatrix} \begin{bmatrix} x_1 \\ x_2 \end{bmatrix} = \begin{bmatrix} 1 \\ 2 \end{bmatrix}.$$

8. Write a Gaussian elimination code that does partial pivoting, and use it to solve the system of equations $Ax = b$, where

$$A = \begin{bmatrix} 9 & 3 & 2 & 0 & 7 \\ 7 & 6 & 9 & 6 & 4 \\ 2 & 7 & 7 & 8 & 2 \\ 0 & 9 & 7 & 2 & 2 \\ 7 & 3 & 6 & 4 & 3 \end{bmatrix}$$

and $b = [35, 58, 53, 37, 39]^T$. The correct answer is $x = [0, 1, 2, 3, 4]^T$.

9. Write a naive Gaussian elimination code and use it to solve the system of equations $Ax = b$, where

$$A = \begin{bmatrix} 1 & 1/2 & 1/3 \\ 1/2 & 1/3 & 1/4 \\ 1/3 & 1/4 & 1/5 \end{bmatrix}$$

and $b = [7/6, 5/6, 13/20]^T$. The correct answer is $x = [0, 1, 2]^T$.

10. Use the naive Gaussian elimination algorithm to solve (by hand) the following system, using only three-digit decimal arithmetic. Repeat, using Gaussian elimination with partial pivoting. Comment on your results.

$$\begin{bmatrix} 0.0001 & 1 \\ 1 & 1 \end{bmatrix} \begin{bmatrix} x_1 \\ x_2 \end{bmatrix} \begin{bmatrix} 1 \\ 2 \end{bmatrix}$$

11. Write a code to do Gaussian elimination with partial pivoting, and apply it to the system $A_5 x = b$, where $b = [-4, -7, -6, -5, 16]^T$ and the solution is $x = [0, 1, 2, 3, 4]^T$.

12. Use MATLAB's rand function to generate A, a random 10×10 matrix, and a random vector $b \in \mathbb{R}^{10}$; solve the system $Ax = b$ (1) Using your own code; and (2) Using MATLAB's backslash command: x = A\b. Obviously, you should get the same results both times.

13. Repeat Problem 12, this time using a 20×20 random matrix and appropriate random right-hand side.

◁ • • ▷

7.3 OPERATION COUNTS

Since many practical problems in numerical linear algebra involve very large matrices, an important issue is the number of operations that a specific algorithm requires in order to operate on a matrix of a given size. In this section we will go through the details of deriving the operation counts for Gaussian elimination and the backsolve step; in subsequent sections we will give the operation count without going into the details.

We first have to decide what constitutes an "operation." The historical convention is to count only the multiplications and divisions, since most multiplications (and divisions) are associated with a subsequent addition (or subtraction).[1] This is how we will count things, but the reader should be aware that other texts will occasionally count *all* operations, which will lead to different results (generally by a factor of 2) than what we get here. MATLAB has a very useful command, `flops`, which can be used to estimate the number of operations in a computation (Be aware that `flops` counts *all* operations.) Several of the exercises in this section involve comparisons of our estimates with what `flops` produces.

There is little subtlety and no grace at all in the way we count the operations. For the naive Gaussian elimination algorithm, we simply write out the loop structure as a series of summations; inside each summation we put the number of operations that occur inside that loop. We thus have

$$C = \sum_{i=1}^{n-1} \sum_{j=i+1}^{n} \left(2 + \sum_{k=i+1}^{n} 1\right).$$

We now proceed to add all this up. An exact answer will be more than a little bit ugly; fortunately, we are more interested in the gross size of the operation count than its exact value. What we really want to know is how the largest term in the count depends on the size of the matrix. We thus write

$$\sum_{j=i+1}^{n} \left(2 + \sum_{k=i+1}^{n} 1\right) = \sum_{j=i+1}^{n} \left(2 + \sum_{m=1}^{n-i} 1\right) = \sum_{j=i+1}^{n} (n-i+2) = (n-i)(n-i+2),$$

hence,[2]

$$C = \sum_{i=1}^{n-1}(n-i)(n-i+2) = \sum_{m=1}^{n-1}(m^2+2m) = \frac{1}{6}n(n-1)(2n-1)+n(n-1) = \frac{1}{3}n^3+\mathcal{O}(n^2).$$

So we get an operation count for (naive) Gaussian elimination of

$$C = \frac{1}{3}n^3 + \mathcal{O}(n^2).$$

If we do the same steps (Problem 10) for the backward solution algorithm we get

$$C = \frac{1}{2}n^2 + \mathcal{O}(n)$$

so that the total cost of solving a linear system can be estimated by

$$\underbrace{C_T}_{\text{Total operations}} = \underbrace{\frac{1}{3}n^3 + \frac{1}{2}n^2 + \mathcal{O}(n^2)}_{\text{Cost of elimination}} + \underbrace{\frac{1}{2}n^2 + \mathcal{O}(n)}_{\text{Cost of backward solve}}$$
$$= \frac{1}{3}n^3 + n^2 + \mathcal{O}(n)$$
$$= \frac{1}{3}n^3 + \mathcal{O}(n^2).$$

---

[1] Traditionally, it was also true that multiplications and divisions were more costly than additions and subtractions, and thus it was more important to keep track of the more expensive operations. Divisions are still more costly than the other operations, but it is no longer the case that multiplications are substantially more expensive than additions and subtractions.

[2] The student should be aware that we have used the formulas

$$\sum_{k=1}^{n} k = \frac{1}{2}n(n+1) \quad \sum_{k=1}^{n} k^2 = \frac{1}{6}n(n+1)(2n+1)$$

here. Sometimes these are part of the calculus discussion of Riemann sums.

The significant thing here is that the solution step is a full power of n cheaper than the elimination step.

Exercises:

1. Determine the operation count for the tridiagonal solution algorithm of §2.6.

2. What is the operation count for computing the dot product $x \cdot y$ of two vectors?

3. Create a pair of random vectors in $\mathbb{R}^{10}$ and compute their dot product using the MATLAB command `dot`. What is the estimated operations count, according to `flops`? Repeat for a pair of vectors in $\mathbb{R}^{20}$ and a pair in $\mathbb{R}^{100}$. Comment on your results, compared to your answer in the previous problem.

4. What is the operation count for computing the matrix–vector product Ax?

5. Repeat Problem 4, assuming that A is tridiagonal.

6. What is the operation count for a matrix-matrix product, AB?

7. Repeat Problem 6, assuming that A is tridiagonal.

8. Repeat Problem 6, assuming now that both A and B are tridiagonal.

9. What is the operation count for the *outer product* xy^T?

10. Determine the operation count for the backward solution algorithm.

11. Repeat Problem 10 for the Gaussian elimination code you wrote in §7.2.

12. Repeat Problem 10 for the tridiagonal solver you wrote back in §2.6.

13. Use the `rand` command to create a sequence of linear system problems of increasing size, say $4 \leq n \leq 100$. Use the backslash operator to solve each problem, and estimate the operations count using `flops`. Plot the estimated cost as a function of n.

14. Use the `diag` command to form a sequence of tridiagonal systems, similar to what you did in Problem 13, and solve these using the backslash operator. What is the estimated cost of solving these systems, according to `flops`?

15. Assume that you are working on a computer that does one operation every 10^{-9} seconds. How long, roughly, would it take such a computer to solve a linear system for $n = 100,000$, using the cost estimates derived in this section for Gaussian elimination? What is the time estimate if the computer only does one operation every 10^{-6} seconds?

◁ ● ● ▷

7.4 THE *LU* FACTORIZATION

The Gaussian elimination algorithm developed in §7.2 is deficient in one respect: There is no way to "save" the work done in reducing A to upper triangular form. If we want to solve several linear systems having the same coefficient matrix, we can only do it (see Problem 6 of §7.2) if all the right-side vectors are known ahead of time. However, we saw in §6.8.2 that it is possible to have a sequence of linear systems problems in which the right-side vector for one problem depends on the solution to the preceding problem. Thus, unless we can somehow "save" the elimination step, we will have to repeat all that work, which is, in fact, the bulk of the work involved in the solution of the problem.

Our goal in this section is to develop a *matrix factorization* that allows us to save the work from the elimination step. Before proceeding with that effort, though, perhaps we should explain why an obvious way to proceed is not the best way to proceed. Why don't we just compute A^{-1}?

The answer is that it is not cost-effective to do so. The standard (and, with a little care, the best) way to compute the inverse is to solve the matrix–matrix system

$$AX = I, \tag{7.2}$$

where I is the identity matrix. The solution X is the desired inverse: $X = A^{-1}$. Let's compute the cost of finding the inverse and of using it to solve linear systems.

The matrix equation (7.2) is a collection of n simple linear systems problems. We can easily modify the Gaussian elimination algorithm to find X by adding a single extra loop to account for having more than one right-side vector. The elimination computation now takes $\frac{5}{6}n^3 + \mathcal{O}(n^2)$ operations, and the triangular backsolves take $\frac{1}{2}n^3 + \mathcal{O}(n^2)$; thus, the total cost is $\frac{4}{3}n^3 + \mathcal{O}(n^2)$. (In Problem 7 we ask you to provide the details for all this.) A few operations can be saved by taking into account the fact that I is all 1's and 0's. When this is done the total cost of producing the inverse is $n^3 + \mathcal{O}(n^2)$. Compare this with the cost of Gaussian elimination, alone, which is $\frac{1}{3}n^3 + \mathcal{O}(n^2)$. If we can find some way to "save" the Gaussian elimination work, we will have a more efficient means of solving linear systems than computing and using the inverse would give us.

What we will do is show that we can *factor* the matrix A into the product of a lower triangular and an upper triangular matrix:

$$A = LU.$$

This allows us to solve linear systems by, instead, solving two triangular systems:

$$Ax = b \Rightarrow Ux = y, \quad \text{where} \quad Ly = b.$$

Thus, we first solve $Ly = b$ and then $Ux = y$ to get the solution:

$$Ux = y, \quad Ly = b \Rightarrow Ux = L^{-1}b \Leftrightarrow LUx = b \Leftrightarrow Ax = b.$$

This shows that we can solve linear systems by (a) computing the LU factorization of A; (b) then solving appropriate lower and upper triangular systems. An example might be useful at this point.

■ **EXAMPLE 7.3**

Consider the matrices

$$L = \begin{bmatrix} 1 & 0 & 0 \\ \frac{1}{2} & 1 & 0 \\ 0 & \frac{1}{2} & 1 \end{bmatrix}$$

and

$$U = \begin{bmatrix} 4 & 2 & 0 \\ 0 & 2 & 1 \\ 0 & 0 & 2 \end{bmatrix}.$$

These are lower and upper triangular, respectively, and their product is

$$LU = A = \begin{bmatrix} 4 & 2 & 0 \\ 2 & 3 & 1 \\ 0 & 1 & \frac{5}{2} \end{bmatrix}.$$

Consider now the linear systems problem

$$Ax = b,$$

where $b = (2, 5, 6)^T$. Since we have an LU factorization of A, we can go ahead and use it to compute the solution. We first solve

$$Ly = b$$

using a *forward* solution algorithm, which is perfectly analogous to the backward solution algorithm except that it works by going forward *down* the diagonal. We have

$$\begin{bmatrix} 1 & 0 & 0 \\ \frac{1}{2} & 1 & 0 \\ 0 & 0 & 1 \end{bmatrix} \begin{bmatrix} y_1 \\ y_2 \\ y_3 \end{bmatrix} = \begin{bmatrix} 2 \\ 5 \\ 6 \end{bmatrix}$$

and we get $y = (2, 4, 4)^T$. Now solve

$$Ux = y$$

using our backward solution algorithm; we have

$$U = \begin{bmatrix} 4 & 2 & 0 \\ 0 & 2 & 1 \\ 0 & 0 & 2 \end{bmatrix} \begin{bmatrix} x_1 \\ x_2 \\ x_3 \end{bmatrix} = \begin{bmatrix} 2 \\ 4 \\ 4 \end{bmatrix}$$

and we get $x = (0, 1, 2)^T$, which is exactly the solution to the original linear system.

Since it turns out (see below) that the LU factorization is nothing more than a properly organized Gaussian elimination, the total cost of this process is

$$C = \underbrace{\frac{1}{3}n^3 + \mathcal{O}(n^2)}_{\text{Cost of elimination}} + \underbrace{\frac{1}{2}n^2 + \mathcal{O}(n)}_{\text{Cost of backsolve}} + \underbrace{\frac{1}{2}n^2 + \mathcal{O}(n)}_{\text{Cost of forward solve}} = \underbrace{\frac{1}{3}n^3 + \mathcal{O}(n^2)}_{\text{Total cost}}.$$

If we already have done the factorization, then the cost of the two solution steps is simply

$$C = \frac{1}{2}n^2 + \mathcal{O}(n) + \frac{1}{2}n^2 + \mathcal{O}(n) = n^2 + \mathcal{O}(n).$$

Constructing the LU factorization is surprisingly easy. In fact, we have already done it; we just don't know it!

We start by considering the matrix

$$E_1 = I - \begin{bmatrix} 0 & 0 & \cdots & 0 \\ m_{21} & 0 & \cdots & 0 \\ m_{31} & 0 & \cdots & 0 \\ \vdots & 0 & \cdots & 0 \\ m_{n1} & 0 & \cdots & 0 \end{bmatrix},$$

where $m_{i1} = a_{i1}/a_{11}$. It is not difficult to show that

$$E_1 A = \begin{bmatrix} \begin{array}{c|ccc} a_{11} & a_{12} & \cdots & a_{1n} \\ \hline 0 & & & \\ \vdots & & A' & \\ 0 & & & \end{array} \end{bmatrix},$$

where A' is an $n-1 \times n-1$ matrix. Thus, we see that multiplication by E_1 accomplishes the same thing as the first pass through the i loop in the naive Gaussian elimination algorithm. Then define E_2 as

$$E_2 = I - \begin{bmatrix} 0 & 0 & 0 & \cdots & 0 \\ 0 & 0 & 0 & \cdots & 0 \\ 0 & m_{32} & 0 & \cdots & 0 \\ 0 & m_{42} & 0 & \cdots & 0 \\ \vdots & \vdots & 0 & \cdots & 0 \\ 0 & m_{n2} & 0 & \cdots & 0 \end{bmatrix},$$

where the m_{i2} values are defined from the elements of A' so that

$$E_2 E_1 A = \begin{bmatrix} \begin{array}{cc|ccc} a_{11} & a_{12} & a_{13} & \cdots & a_{1n} \\ 0 & a'_{12} & a'_{13} & \cdots & a'_{1n} \\ \hline 0 & 0 & & & \\ \vdots & \vdots & & A'' & \\ 0 & 0 & & & \end{array} \end{bmatrix}.$$

The general trend ought to be clear, but a little notation helps. Let's write $A^{(0)} = A$, $A^{(1)} = E_1 A^{(0)}$, $A^{(2)} = E_2 A^{(1)} = E_2 E_1 A$, and so on, with a similar superscript being placed on the individual matrix components: $a_{ij}^{(k)}$. Then we can define the E_k matrices rather easily as

$$E_k = I - R_k,$$

where R_k is almost entirely zeros:

$$r_{ij}^{(k)} = \begin{cases} a_{ji}^{(k)}/a_{ii}^{(k)}, & k = i \text{ and } j > i; \\ 0, & \text{otherwise.} \end{cases}$$

Note that the nonzero elements here are *precisely* the multipliers for the Gaussian elimination step. It follows, then, that the matrix

$$U = E_{n-1} E_{n-2} \cdots E_2 E_1 A$$

will be upper triangular. In fact, it will be the same upper triangular matrix that is produced by naive Gaussian elimination. Moreover, each of the E_i matrices is lower triangular and nonsingular; therefore, their inverses all exist and are lower triangular. We can therefore write

$$A = (E_{n-1}E_{n-2}\cdots E_2E_1)^{-1}U = LU,$$

where L is defined by

$$L = (E_{n-1}E_{n-2}\cdots E_2E_1)^{-1} = E_1^{-1}E_2^{-1}\cdots E_{n-1}^{-1}.$$

But the simplicity of the E_k matrices allows us to write their inverses explicitly (Problem 8):

$$E_k^{-1} = I + R_k; \tag{7.3}$$

moreover, it is easy to show that the sequential products of these matrices are very simple (Problem 8):

$$E_k^{-1}E_{k-1}^{-1} = I + R_k + R_{k-1}. \tag{7.4}$$

Thus, when all is said and done, we have (again, Problem 8)

$$L = I + R_1 + R_2 + \cdots + R_{n-1}.$$

But, since the nonzero elements of each R_k are nothing more than the multipliers, the lower triangular matrix is nothing more than the matrix of multipliers! Thus, the *LU* factorization is nothing more than a very slight re-organization of the same Gaussian elimination algorithm that we studied earlier in this chapter.

This allows for a very simple and compact algorithm for computing the *LU* decomposition, which we have split into two parts.

Algorithm 7.5 *LU decomposition algorithm (no pivoting)*

Part A (Factorization)

```
!
!       Compute decomposition
!
for i=1 to n-1
      for j=i+1 to n
            a(j,i) = a(j,i)/a(i,i)
            for k=i+1 to n
                  a(j,k) = a(j,k) - m*a(i,k)
            endfor
            b(j) = b(j) - m*b(i)
      endfor
endfor
```

Algorithm 7.6 *LU decomposition algorithm (no pivoting)*

Part B (Solution)

```
!
!      Solve Ly = b
!
x(1) = b(1)
for i=2 to n
     sum = 0.0
     for j=1 to i-1
          sum = sum + a(i,j)*x(j)
     endfor
     x(i) = b(i) - sum
endfor
!
!      Solve Ux = y
!
x(n) = x(n)/a(n,n)
for i=n-1 to 1
     sum = 0.0
     for j=i+1 to n
          sum = sum + a(i,j)*x(j)
     endfor
     x(i) = (x(i) - sum)/a(i,i)
endfor
```

Note that the first half of the solution algorithm (the forward solve) made explicit use of the fact that L was unit lower triangular; thus, the diagonal elements are all 1's.

It is important to understand how these algorithms manipulate the actual computer storage. Note that we use the lower triangular part of A to store L, and the upper triangular part to store U; separate arrays are not needed (and would be wasteful). Let's see how this would work in a particular case.

■ **EXAMPLE 7.4**

Take

$$A = \begin{bmatrix} 4 & 1 & 0 \\ 1 & 4 & 1 \\ 0 & 1 & 4 \end{bmatrix};$$

then the LU decomposition algorithm produces the following steps:

$$\begin{bmatrix} 4 & 1 & 0 \\ 1 & 4 & 1 \\ 0 & 1 & 4 \end{bmatrix} \sim \begin{bmatrix} 4 & 1 & 0 \\ (1/4) & 15/4 & 1 \\ (0) & 1 & 4 \end{bmatrix} \sim \begin{bmatrix} 4 & 1 & 0 \\ (1/4) & 15/4 & 1 \\ (0) & (4/15) & 56/15 \end{bmatrix},$$

where we have indicated the multipliers by putting parentheses around them. Thus, we have

$$L = \begin{bmatrix} 1 & 0 & 0 \\ 1/4 & 1 & 0 \\ 0 & 4/15 & 1 \end{bmatrix}; \quad U = \begin{bmatrix} 4 & 1 & 0 \\ 0 & 15/4 & 1 \\ 0 & 0 & 56/15 \end{bmatrix}.$$

Solving the system

$$\begin{bmatrix} 4 & 1 & 0 \\ 1 & 4 & 1 \\ 0 & 1 & 4 \end{bmatrix} \begin{bmatrix} x_1 \\ x_2 \\ x_3 \end{bmatrix} = \begin{bmatrix} 6 \\ 12 \\ 14 \end{bmatrix}$$

therefore requires that we solve the two triangular systems

$$\begin{bmatrix} 1 & 0 & 0 \\ 1/4 & 1 & 0 \\ 0 & 4/14 & 1 \end{bmatrix} \begin{bmatrix} y_1 \\ y_2 \\ y_3 \end{bmatrix} = \begin{bmatrix} 6 \\ 12 \\ 14 \end{bmatrix}$$

and

$$\begin{bmatrix} 4 & 1 & 0 \\ 0 & 15/4 & 1 \\ 0 & 0 & 56/15 \end{bmatrix} \begin{bmatrix} x_1 \\ x_2 \\ x_3 \end{bmatrix} = \begin{bmatrix} y_1 \\ y_2 \\ y_3 \end{bmatrix}.$$

The reader should check that we get the solution $x_1 = 1$, $x_2 = 2$, $x_3 = 3$.

Pivoting and the LU Decomposition   The previous discussion and resulting algorithm both assumed that no pivoting was done. However, we have seen that pivoting is sometimes necessary and often desirable. Can we use pivoting in the *LU* decomposition without destroying the algorithm? If so, how?

The answer is easier to explain than it is to prove. Because of the triangular structure of the *LU* factors, we can implement pivoting almost exactly as we did before. The difference is that we must keep track of how the rows are interchanged in order to properly apply the forward and backward solution steps.

The best way to see this is to look at an example.

■ **EXAMPLE 7.5**

Consider the 4×4 matrix

$$A = \begin{bmatrix} 4 & 0 & 1 & 1 \\ 3 & 1 & 3 & 1 \\ 0 & 1 & 2 & 0 \\ 3 & 2 & 4 & 1 \end{bmatrix}.$$

We will apply the *LU* decomposition to this, showing the values as stored in the computer. As in Example 7.4, to distinguish the multipliers that are stored in the lower triangular positions, we enclose them in parentheses, e.g., $(3/4)$. Eliminating in the first column is very routine:

$$\begin{bmatrix} 4 & 0 & 1 & 1 \\ 3 & 1 & 3 & 1 \\ 0 & 1 & 2 & 0 \\ 3 & 2 & 4 & 1 \end{bmatrix} \sim \begin{bmatrix} 4 & 0 & 1 & 1 \\ (3/4) & 1 & 9/4 & 1/4 \\ (0) & 1 & 2 & 0 \\ (3/4) & 2 & 13/4 & 1/4 \end{bmatrix}.$$

To carry out the next step of elimination, we see that we need to pivot. Let's go ahead and swap the rows, just as we did before, and see what happens.

$$
\begin{bmatrix}
4 & 0 & 1 & 1 \\
(3/4) & 1 & 9/4 & 1/4 \\
(0) & 1 & 2 & 0 \\
(3/4) & 2 & 13/4 & 1/4
\end{bmatrix}
\sim
\begin{bmatrix}
4 & 0 & 1 & 1 \\
(3/4) & 2 & 13/4 & 1/4 \\
(0) & (1/2) & 3/8 & -1/8 \\
(3/4) & (1/2) & 5/8 & 1/8
\end{bmatrix}.
$$

We have to pivot again (since $3/8 < 5/8$), so we end up with

$$
\begin{bmatrix}
4 & 0 & 1 & 1 \\
(3/4) & 2 & 13/4 & 1/4 \\
(3/4) & (1/2) & 5/8 & 1/8 \\
(0) & (1/2) & 3/8 & -1/8
\end{bmatrix}
\sim
\begin{bmatrix}
4 & 0 & 1 & 1 \\
(3/4) & 2 & 13/4 & 1/4 \\
(3/4) & (1/2) & 5/8 & 1/8 \\
(0) & (1/2) & (3/5) & -1/5
\end{bmatrix}.
$$

This suggests, then, that the L and U matrices are

$$
L =
\begin{bmatrix}
1 & 0 & 0 & 0 \\
3/4 & 1 & 0 & 0 \\
3/4 & 1/2 & 1 & 0 \\
0 & 1/2 & 3/5 & 1
\end{bmatrix},
\quad
U =
\begin{bmatrix}
4 & 0 & 1 & 1 \\
0 & 2 & 13/4 & 1/4 \\
0 & 0 & 5/8 & 1/8 \\
0 & 0 & 0 & -1/5
\end{bmatrix}.
$$

If we multiply these out, we get

$$
\begin{bmatrix}
1 & 0 & 0 & 0 \\
3/4 & 1 & 0 & 0 \\
3/4 & 1/2 & 1 & 0 \\
0 & 1/2 & 3/5 & 1
\end{bmatrix}
\begin{bmatrix}
4 & 0 & 1 & 1 \\
0 & 2 & 13/4 & 1/4 \\
0 & 0 & 5/8 & 1/8 \\
0 & 0 & 0 & -1/5
\end{bmatrix}
=
\begin{bmatrix}
4 & 0 & 1 & 1 \\
3 & 2 & 4 & 1 \\
3 & 1 & 3 & 1 \\
0 & 1 & 2 & 0
\end{bmatrix}
\neq A.
$$

However, we do note that the product LU is the same as A, *once we have imposed the same row interchanges on A as we did in the pivoting process.* Thus, if we keep track of how the rows were swapped, we can then impose this reordering on the right-side vector and carry on with the solution process.

So, how do we keep track of the row interchanges? One way is to use an *index array*, which is simply an integer-valued vector. It is initialized to the natural row ordering,

$$
J =
\begin{bmatrix}
1 \\
2 \\
3 \\
\vdots \\
n
\end{bmatrix},
$$

and then every time two rows in the matrix are swapped, the corresponding two elements of the index array J are swapped. Thus, in our example, the final version of J is

$$
J =
\begin{bmatrix}
1 \\
4 \\
2 \\
3
\end{bmatrix}.
$$

The student should check that this is correct, of course. We interpret this as follows:

- The first row of A is the first row of the product matrix LU.

- The fourth row of A is the second row of the product matrix LU.

- The second row of A is the third row of the product matrix LU.

- The third row of A is the fourth row of the product matrix LU.

- In general, if we denote the product matrix LU by A', then

$$a'_{i,j} = a_{J_i, j}.$$

How do we use this information in the solution process? Since the product matrix $A' = LU$ does not equal A, we have to reorder the b vector accordingly. This is accomplished by the simple expedient of copying b into b', where

$$b'_i = b_{J_i}.$$

The factorization and solution algorithm (again, in two parts) now becomes:

Algorithm 7.7 *LU Decomposition with Partial Pivoting (Pseudocode)*

Part A (Factorization with Pivoting)

```
for i=1 to n-1
!
!     This begins the pivoting code
!
      am = abs(a(i,i))
      p = i
      for j=i+1 to n
            if abs(a(j,i)) > am then
                  am = abs(a(j,i))
                  p = j
            endif
      endfor
      if p > i then
            for k = 1 to n
                  hold = a(i,k);  a(i,k) = a(p,k);   a(p,k) = hold
            endfor
            ihold = indx(i);  indx(i) = indx(p);  indx(p) = ihold
      endif
!
!     This ends the pivoting code
!
!
!     Now do the elimination step
!
      for j=i+1 to n
            a(j,i) = a(j,i)/a(i,i)
            for k=i+1 to n
                  a(j,k) = a(j,k) - a(j,i)*a(i,k)
            endfor
      endfor
endfor
```

---

and the forward–backward solve algorithm becomes:

Algorithm 7.8 *Forward–Backward Solution, Using LU Decomposition (Pseudocode).*
Part B (Solution Steps with Re-ordering)

```
!
!     First, re-order the b vector
!
for k=1 to n
      x(k) = b(indx(k))
endfor
for k=1 to n
      b(k) = x(k)
endfor
!
!     Solve Ly = b
!
y(1) = b(1)
for i=2 to n
      sum = 0.0
      for j=1 to i-1
            sum = sum + a(i,j)*y(j)
      endfor
      y(i) = b(i) - sum
endfor
!
!     Next, solve Ux = y
!
x(n) = y(n)/a(n,n)
for i=n-1 to 1
      sum = 0.0
      for j=i+1 to n
            sum = sum + a(i,j)*x(j)
      endfor
      x(i) = (y(i) - sum)/a(i,i)
endfor
```

Programming Notes: Some comments on these algorithms might be in order at this point. First, note that we used indx as the variable name for the index vector J. Second, note that the loop where the rows are interchanged now runs across the entire matrix (in other words, the loop starts at k = 1 instead of k = i). This is because the LU decomposition requires that we interchange the elements of L (the multipliers), as well as the elements of U. Finally, note that we used three separate vectors (y, b, and x) in the solution step. This really is not necessary but does make it easier to follow the algorithm.

The MATLAB command for doing an LU decomposition is, of course, lu:

$$[\text{L, U, P}] = \text{lu(A)}$$

This produces a unit lower triangular L, upper triangular U, and a permutation matrix[3] such that $PA = LU$. There is also the command linsolve, which uses an LU decomposition

---

[3] A permutation matrix is a re-ordering of the identity that accomplishes the re-ordering of the rows of A.

to solve the system $Ax = b$ (lu just produces the factorization). If you want to produce the inverse of a matrix, use inv(A).

Exercises:

1. Do, by hand, an LU factorization of the matrix

$$A = \begin{bmatrix} 2 & 1 \\ 1 & 2 \end{bmatrix}$$

   and use it to solve the system $Ax = b$, where $b = (1,2)^T$. The exact solution is $x = (0,1)^T$. Verify that $LU = A$.

2. Repeat Problem 1 for

$$A = \begin{bmatrix} 4 & 1 \\ 1 & 5 \end{bmatrix}$$

   and $b = (2, 10)^T$; here the exact solution is $x = (0, 2)^T$.

3. Write an LU factorization code and use it to solve the system of equations $Ax = b$, where

$$A = \begin{bmatrix} 14 & 14 & -9 & 3 & -5 \\ 14 & 52 & -15 & 2 & -32 \\ -9 & -15 & 36 & -5 & 16 \\ 3 & 2 & -5 & 47 & 49 \\ -5 & -32 & 16 & 49 & 79 \end{bmatrix}$$

   and $b = [-15, -100, 106, 329, 463]^T$. The correct answer is $x = [0, 1, 2, 3, 4]^T$.

4. Write an LU factorization code and use it to solve the system of equations

$$T_5 x = b,$$

   where $b = [1, 6, 12, 18, 19]^T$. The correct answer is $x = [0, 1, 2, 3, 4]^T$.

5. Write an LU factorization code and use it to solve the system of equations

$$H_5 x = b,$$

   where $b = [5.0, 3.550, 2.81428571428571, 2.34642857142857, 2.01746031746032]^T$. The correct answer is $x = [1, 2, 3, 4, 5]^T$.

6. Write up your own LU factorization code, based on the material in this chapter, and test it on the following examples. In each case, have your code multiply out the L and U factors to check that the routine is working.

   (a) $K_5 x = b$, $b = [-1, 0, 0, 0, 5]^T$; the solution is $x = [0, 1, 2, 3, 4]^T$;

   (b) $A_5 x = b$, $b = [-4, -7, -6, -5, 16]^T$; the solution is $x = [0, 1, 2, 3, 4]^T$.

7. Determine the operation count for computing the inverse of a matrix, as outlined in this section.

8. Show that

   (a)
   $$E_k^{-1} = I + R_k \qquad (7.5)$$

   for all k;

   (b)
   $$E_k^{-1} E_{k-1}^{-1} = I + R_k + R_{k-1} \qquad (7.6)$$

   for all k;

   (c) $L = I + R_1 + R_2 + \cdots + R_{n-1}$.

9. Modify the tridiagonal solution algorithm from Chapter 2 to produce an LU decomposition. Be sure to maintain the simple storage of the matrix that was used in Chapter 2, and assume that no pivoting is required.

10. Write an LU factorization code with partial pivoting, and apply it to the system $A_5 x = b$, where $b = [-4, -7, -6, -5, 16]^T$ and the solution is $x = [0, 1, 2, 3, 4]^T$.

11. Write an LU factorization code that does partial pivoting and use it to solve the system of equations $Ax = b$, where

$$A = \begin{bmatrix} 9 & 3 & 2 & 0 & 7 \\ 7 & 6 & 9 & 6 & 4 \\ 2 & 7 & 7 & 8 & 2 \\ 0 & 9 & 7 & 2 & 2 \\ 7 & 3 & 6 & 4 & 3 \end{bmatrix}$$

and $b = [35, 58, 53, 37, 39]^T$. The correct answer is $x = [0, 1, 2, 3, 4]^T$.

12. Compare your LU factorization-and-solution code to MATLAB'S `linsolve` command by creating a random 10×10 system and solving with both routines. (They should produce *exactly* the same solution.) Use `flops` to estimate the operation count for each.

13. Again, generate a random 10×10 linear system (matrix and right-hand-side vector). Then solve this system four ways:

    (a) Using your LU factorization-and-solution code;

    (b) Using MATLAB's `linsolve` command;

    (c) Using the MATLAB backslash operation;

    (d) Using MATLAB's `inv` command to compute the inverse of the matrix, and then multiplying this by the right-hand-side vector.

    Use `flops` to estimate the cost of each solution technique, and rank the methods for their efficiency in this regard.

14. Repeat Problem 13 for the matrix K_{20} defined at the end of §7.2, using a random right-hand-side vector, and get the `flops` estimate. Then apply the tridiagonal solver from Chapter 2 to this problem, and again get the flops estimate. Comment on your results.

◁ ● ● ▷

7.5 PERTURBATION, CONDITIONING, AND STABILITY

In this section we will study how the matrix solution process as outlined §7.4 is affected by changes to the problem. The main reason for doing this is to attain an understanding of why some linear system problems are difficult to solve. Usually, this has to do with a concept known as *conditioning*.

To motivate our discussion, let's consider a simple example.

■ **EXAMPLE 7.6**

Let A be given by

$$A = \begin{bmatrix} 1.002 & 1 \\ 1 & 0.998 \end{bmatrix}$$

and let b be given by

$$b = \begin{bmatrix} 2.002 \\ 1.998 \end{bmatrix}.$$

Then a direct computation shows that the solution to $Ax = b$ is given by $x = (1,1)^T$. Suppose now that we perturb the right-side vector a little bit, and define

$$b_c = \begin{bmatrix} 2.0021 \\ 1.998 \end{bmatrix}.$$

Note that this is a change of less than one-half of a percent in one component of the vector b. Now a direct computation shows that the solution to $Ax = b_c$ is

$$x_c = \begin{bmatrix} -23.95 \\ 26.00 \end{bmatrix}.$$

Most of us would probably agree that this is a large change in the solution for such a small change in the problem. (And, we should emphasize, this is an *exact* computation.) Why did this happen?

This is a serious issue in mathematics and computation; we like to know that small changes in the problem will produce correspondingly small changes in the solution, and we like to understand why the exceptions occur. And we especially want to know if the large change due to a small perturbation is an artifact of the problem or of the computational scheme.

Before we can directly tackle the issues of this section, we need to introduce some notation and terminology that will allow us to measure the size of vectors and matrices. For the most part, this is an extension of the function norm ideas introduced in §4.3, as well as a natural continuation of the idea of length of a vector that is part of an ordinary linear algebra course.

7.5.1 Vector and Matrix Norms

Because we will be concerned with measuring errors in vectors and matrices, it is necessary to introduce the concepts of a *vector norm* and a *matrix norm*. To be informal, these norms are simply a generalization of the function norm concept that we introduced in §4.3. They are means of measuring the *size* of a vector or matrix. The formal definition of a vector norm is as follows.

Definition 7.1 (Vector Norm) *A vector norm on $\mathbb{R}^n$ is any mapping $\| \cdot \|$, defined on $\mathbb{R}^n$ with values in $[0, \infty)$, which satisfies the three conditions:*

1. *$\|x\| > 0$ for any vector $x \neq 0$;*

2. *$\|ax\| = |a|\|x\|$ for any scalar a.*

3. *$\|x + y\| \leq \|x\| + \|y\|$ for any two vectors x and y.*

The most common examples of vector norms are the *infinity norm* and the *Euclidian 2-norm*, defined as

$$\|x\|_\infty = \max_{1 \leq i \leq n} |x_i|,$$

and

$$\|x\|_2 = \left(\sum_{i=1}^{n} x_i^2 \right)^{1/2}.$$

We could use the same norm definition for matrix norms, but this would not give us a couple of important properties that we want to have. So we use a special type of matrix norm, known as an *operator norm*.

Definition 7.2 (Matrix Norm) *Let $\| \cdot \|$ be a given vector norm defined on $\mathbb{R}^n$. Define the corresponding matrix norm, for matrices $A \in \mathbb{R}^{n \times n}$, by*

$$\|A\| = \max_{x \neq 0} \frac{\|Ax\|}{\|x\|}. \tag{7.7}$$

One important consequence of this definition of matrix norm is that the norm of a product is always less than or equal to the product of the norms:

$$\|AB\| \leq \|A\|\|B\| \tag{7.8}$$

and

$$\|Ax\| \leq \|A\|\|x\|. \tag{7.9}$$

See Problem 15.

As we have defined it here, matrix norms are always associated with a particular vector norm. Strictly speaking, this is not necessary; one can define a norm on a matrix that does not require any particular vector norm. However, such norms will not usually satisfy (7.8) and (7.9), and these conditions are necessary for what we want to do throughout this section.

The matrix infinity norm, $\|A\|_\infty$, can be shown to be equivalent to the following computation:

$$\|A\|_\infty = \max_{1 \leq i \leq n} \sum_{j=1}^{n} |a_{ij}|.$$

Thus, $\|A\|_\infty$ is defined as the maximum row sum. The matrix 2-norm is more subtle. It can be shown that we have

$$\|A\|_2 = \sqrt{\Lambda(A^T A)},$$

where $\Lambda(B)$ is the largest (in absolute value) eigenvalue of the matrix B. Since this makes it much more difficult to compute with the matrix 2-norm, we will almost always use the infinity norm.

■ **EXAMPLE 7.7**

Let A be the simple matrix given by

$$A = \begin{bmatrix} 4 & -6 & 2 \\ 0 & 4 & 1 \\ 1 & 2 & 3 \end{bmatrix}.$$

Then $\|A\|_\infty = \max\{12, 5, 6\} = 12$. Furthermore,

$$A^T A = \begin{bmatrix} 17 & -22 & 11 \\ -22 & 56 & -2 \\ 11 & -2 & 14 \end{bmatrix},$$

which has eigenvalues $\lambda = 66.6816, 19.8065, 0.5118$, so that $\|A\|_2 = \sqrt{66.6816} = 8.1659$.

7.5.2 The Condition Number and Perturbations

Suppose that we want to look at the linear system

$$Ax = b,$$

and suppose further that we are most interested in understanding how the solution x changes as the right-side vector b changes. Thus we might look at the two specific systems

$$Ax_1 = b_1$$

and

$$Ax_2 = b_2.$$

We have (assuming that A is nonsingular)

$$x_1 - x_2 = A^{-1}(b_1 - b_2),$$

so that the relative error in x_2 as an approximation to x_1 is given by

$$\frac{\|x_1 - x_2\|}{\|x_1\|} \le \|A^{-1}\| \frac{\|b_1 - b_2\|}{\|x_1\|}.$$

We would like to bound the change in solution by something that did not depend on the solution; thus, we want to get rid of the x_1 in the denominator on the right. To do this, we note that $\|A\|\|x_1\| \ge \|b_1\|$, so that

$$\frac{1}{\|x_1\|} \le \frac{\|A\|}{\|b_1\|};$$

hence, we have

$$\frac{\|x_1 - x_2\|}{\|x_1\|} \le \|A\|\|A^{-1}\| \frac{\|b_1 - b_2\|}{\|b_1\|}.$$

The multiplying coefficient $\|A\|\|A^{-1}\|$ is interesting. It depends entirely on the matrix in the problem, not the right-side vector, yet it shows up as an amplifier to the relative change in the right-hand-side vector. We will see this quantity often enough in the rest of this section that it is worth giving it a name and special notation. We call it the *condition number*.

Definition 7.3 (Condition Number) *For a given matrix $A \in \mathbb{R}^{n \times n}$ and a given matrix norm $\| \cdot \|$, the condition number with respect to the given norm is defined by*

$$\kappa(A) = \|A\| \|A^{-1}\|. \tag{7.10}$$

If A is singular, then we take $\kappa(A) = \infty$.

The justification for taking $\kappa(A) = \infty$ if A is singular is the following theorem.

Theorem 7.3 *Let $A \in \mathbb{R}^{n \times n}$ be given, nonsingular. Then, for any singular matrix $B \in \mathbb{R}^{n \times n}$, we have*

$$\frac{1}{\kappa(A)} \leq \frac{\|A - B\|}{\|A\|}.$$

Proof: We have

$$\frac{1}{\kappa(A)} = \frac{1}{\|A\| \|A^{-1}\|}$$

$$= \frac{1}{\|A\|} \left(\frac{1}{\max_{x \neq 0} \frac{\|A^{-1}x\|}{\|x\|}} \right)$$

$$\leq \frac{1}{\|A\|} \left(\frac{1}{\frac{\|A^{-1}x\|}{\|x\|}} \right)$$

$$\leq \frac{1}{\|A\|} \left(\frac{\|Ay\|}{\|y\|} \right),$$

where $y \in \mathbb{R}^n$ is arbitrary. Now let y be any nonzero vector such that $By = 0$; since B is singular, we know that such things exist. Then we can write

$$\frac{1}{\kappa(A)} \leq \frac{\|(A - B)y\|}{\|A\| \|y\|} \leq \frac{\|(A - B)\| \|y\|}{\|A\| \|y\|} = \frac{\|(A - B)\|}{\|A\|},$$

and we are done. •

The importance of this result is that it tells us that if A is close to a singular matrix, then the reciprocal of the condition number will be near zero; i.e., $\kappa(A)$ itself will be "large." Thus, the condition number measures how close the matrix is to being singular; if $\kappa(A)$ is "large," then we know that A is close to being singular. One of the points of this section is that we will learn that solving systems that are nearly singular can produce large errors.

Suppose that we want to solve the system

$$Ax = b,$$

where $A \in \mathbb{R}^{n \times n}$ and $b \in \mathbb{R}^n$ are given. However, we know that the solution process will be affected by rounding error. Although each instance of rounding error will no doubt be small, if n is large, then there will be an enormous number of individual rounding errors to add up. Is it possible that these will combine to dominate the computation and render the results meaningless? For example, if $n = 200$, which is not considered a large system, then ordinary Gaussian elimination will do about 2.7×10^6 operations to produce the solution. If each operation has error that is $\mathcal{O}(10^{-7})$, and if they all accumulate, then the net error will be about 0.27, hardly small.

There are several main results in this section, which tell us how the solution of a linear system is affected by errors in the matrix or the right-hand-side vector. Since some of the proofs can be rather involved, some of them are deferred to the Appendix or are omitted entirely.

Theorem 7.4 (Effects of Perturbation in b) *Let $A \in \mathbb{R}^{n \times n}$ (nonsingular) and $b \in \mathbb{R}^n$ be given, and define $x \in \mathbb{R}^n$ as the solution of the linear system $Ax = b$. Let $\delta b \in \mathbb{R}^n$ be a small perturbation of b, and define $x + \delta x \in \mathbb{R}^n$ as the solution of the system $A(x + \delta x) = b + \delta b$. Then,*

$$\frac{\|\delta x\|}{\|x\|} \leq \kappa(A) \frac{\|\delta b\|}{\|b\|}.$$

Proof: This is a straightforward computation. We have $Ax = b$ and $A(x + \delta x) = b + \delta b$, so that $A\delta x = \delta b$. Thus,

$$\delta x = A^{-1} \delta b \Rightarrow \|\delta x\| \leq \|A^{-1}\| \|\delta b\|$$

so we have

$$\frac{\|\delta x\|}{\|x\|} \leq \frac{\|A\| \|A^{-1}\| \|\delta b\|}{\|A\| \|x\|}.$$

But $\|A\| \|x\| \geq \|b\|$ and the definition of condition number imply that

$$\frac{\|A\| \|A^{-1}\| \|\delta b\|}{\|A\| \|x\|} \leq \kappa(A) \frac{\|\delta b\|}{\|b\|},$$

and we are done. •

The importance of this theorem is to show us that perturbations in the problem affect the final solution in a form that is amplified by the condition number of the matrix. Thus, if the matrix is ill-conditioned—meaning that the condition number, $\kappa(A)$, is large—then a small change in the data could lead to a large change in the solution. A similar result involves the *residual* of a solution: Let x_c be a "computed" solution to $Ax = b$; then the residual is the vector $r = b - Ax_c$; i.e., r is the amount by which x_c fails to solve the system. If $r = 0$, then x_c is exact; one might think, then, that if r is small, then x_c is close to the exact solution. This isn't always so, as the following theorem shows.

Theorem 7.5 (Effects of Residual on Accuracy) *Let $A \in \mathbb{R}^{n \times n}$ (nonsingular) and $b \in \mathbb{R}^n$ be given, and define $x_c \in \mathbb{R}^n$ as a computed solution of the linear system $Ax = b$. Let $r \in \mathbb{R}^n$ be the residual $r = b - Ax_c$. Then,*

$$\frac{\|x - x_c\|}{\|x\|} \leq \kappa(A) \frac{\|r\|}{\|b\|}.$$

Proof: This is a small change from the preceding proof, and so is left to the student. See Problem 18. •

The gist of these two results is that, in some sense, we cannot trust the solution to a problem involving an ill-conditioned matrix unless we have taken special care in the solution. A similar but more involved result applies if we consider perturbations in the matrix itself.

Theorem 7.6 (Effects of Perturbation in A**)** *Let* $A \in \mathbb{R}^{n \times n}$ *be given, nonsingular, and let* $E \in \mathbb{R}^{n \times n}$ *be a perturbation of* A. *Let* $x \in \mathbb{R}^n$ *be the unique solution of* $Ax = b$. *If* $\kappa(A)\|E\| < \|A\|$, *then the perturbed system* $(A + E)x_c = b$ *has a unique solution and*

$$\frac{\|x - x_c\|}{\|x\|} \le \frac{\theta}{1 - \theta},$$

where

$$\theta = \kappa(A)\frac{\|E\|}{\|A\|}.$$

Proof: See the Appendix. •

The importance of these perturbation results is that we can establish that any computed (i.e., approximate) solution to the linear system problem is the *exact* solution to some "nearby" problem. The next result gives a sense of what is meant by this.

Theorem 7.7 *Let* $A \in \mathbb{R}^{n \times n}$ *(nonsingular) and* $b \in \mathbb{R}^n$ *be given, and define* $x \in \mathbb{R}^n$ *as the* exact *solution of the linear system* $Ax = b$ *and* $x_c \in \mathbb{R}^n$ *as a* computed *(that is, approximate) solution. Then there exists* $\delta b \in \mathbb{R}^n$ *such that* x_c *is the* exact *solution of the perturbed system*

$$Ax_c = b + \delta b,$$

where $\delta b = Ax_c - b = -r$ *is the (negative) residual for* x_c.

Proof: This is a straightforward computation:

$$Ax_c = A(x + (x_c - x)) = b + A(x_c - x) = b + \delta b.$$

On the other hand, we have

$$\delta b = A(x_c - x) = Ax_c - b = -r,$$

and we are done. •

Thus, we can consider any computed solution to a linear system as the exact solution to a problem using slightly perturbed data. The idea of using this approach to estimating the error between two quantities—by establishing that they are both exact solutions to "nearby" problems—is called *backward error analysis* and is due to the British mathematician James H. Wilkinson,[4] widely regarded as the founder of modern numerical analysis.

A similar result shows that we can regard the computed solution as the exact solution to a problem involving the same right-side vector, but a perturbed matrix. Considering how much more complicated the proof of Theorem 7.6 is compared to that of Theorem 7.4, it is surprising how easy this argument is. Note, however, that this result is restricted to the 2-norm, whereas our previous results have been independent of the choice of norm.

---

[4]James Hardy Wilkinson (1919–1986) was born in Kent, England, and at the age of 16 won a scholarship to Trinity College, Cambridge, where he was widely regarded as one of the top students of his time. During World War II he did ballistics work, and soon after the end of the war he began work at Great Britain's National Physical Laboratory, on early computer designs. Out of this work came his studies of numerical algorithms, most notably the effects of rounding errors on Gaussian elimination and methods for the algebraic eigenvalue problem, works that remain classics to this day. In 1969 he was elected to the Royal Society of London.

Theorem 7.8 *Let $A \in \mathbb{R}^{n \times n}$ and $b \in \mathbb{R}^n$ be given, with A nonsingular; define $x \in \mathbb{R}^n$ as the solution of $Ax = b$. Let x_c be an approximate solution to this system with residual $r = b - Ax_c$ and define*

$$E = r x_c^T / \|x_c\|_2^2. \tag{7.11}$$

Then, x_c is the exact solution to $(A + E)x_c = b$ and

$$\frac{\|E\|_2}{\|A\|_2} \leq \frac{\|r\|_2}{\|A\|_2 \|x_c\|_2}. \tag{7.12}$$

Proof: We have

$$(A + E)x_c = Ax_c + \frac{1}{\|x_c\|_2^2}(b - Ax_c)x_c^T x_c = Ax_c + (b - Ax_c)\frac{\|x_c\|_2^2}{\|x_c\|_2^2} = b.$$

Thus, x_c is the exact solution of the perturbed system. Next, for any $y \in \mathbb{R}^n$,

$$Ey = \frac{1}{\|x_c\|_2^2} r x_c^T y = \alpha_y r,$$

where

$$\alpha_y = \frac{x_c^T y}{\|x_c\|_2^2}.$$

Note that we have

$$|\alpha_y| \leq \frac{\|x_c\|_2 \|y\|_2}{\|x_c\|_2^2} = \frac{\|y\|_2}{\|x_c\|_2}.$$

Therefore,

$$\frac{\|Ey\|_2}{\|y\|_2} = \frac{|\alpha_y| \|r\|_2}{\|y\|_2} \leq \frac{\|r\|}{\|x_c\|_2},$$

from which

$$\|E\|_2 \leq \frac{\|r\|}{\|x_c\|_2}$$

follows, and we are done. ●

This last result allows us to estimate, via an upper bound, the size of the perturbation in A that our computed solution corresponds to, using the easily computable residual.

To use these perturbation results, we need to have one final set of results which estimates the effects of machine arithmetic—rounding error—on the Gaussian elimination process. In this connection we need one more definition.

Definition 7.4 (Gaussian Elimination Growth Factor) *Let $A^{(k)}$ be the matrix after k columns have been eliminated—thus, $A^{(0)} = A$ and $A^{(n-1)} = U$—and let $a_{i,j}^{(k)}$ denote the individual elements of $A^{(k)}$. Then the growth factor for Gaussian elimination applied to A is defined by*

$$\rho_n = \max_{i,j,k} \frac{|a_{i,j}^{(k)}|}{\|A\|_\infty}.$$

To illustrate this, let's go back to one of our computational examples from earlier in this chapter.

■ **EXAMPLE 7.8**

Consider the matrix from Example 7.5:

$$A = \begin{bmatrix} 4 & 0 & 1 & 1 \\ 3 & 1 & 3 & 1 \\ 0 & 1 & 2 & 0 \\ 3 & 2 & 4 & 1 \end{bmatrix}.$$

After eliminating in the first column, we get $A^{(1)}$:

$$A^{(1)} = \begin{bmatrix} 4 & 0 & 1 & 1 \\ 0 & 1 & 9/4 & 1/4 \\ 0 & 1 & 2 & 0 \\ 0 & 2 & 13/4 & 1/4 \end{bmatrix}.$$

Proceeding further, we get

$$A^{(2)} = \begin{bmatrix} 4 & 0 & 1 & 1 \\ 0 & 2 & 13/4 & 1/4 \\ 0 & 0 & 3/8 & -1/8 \\ 0 & 0 & 5/8 & 1/8 \end{bmatrix},$$

and

$$A^{(3)} = \begin{bmatrix} 4 & 0 & 1 & 1 \\ 0 & 2 & 13/4 & 1/4 \\ 0 & 0 & 3/8 & -1/8 \\ 0 & 0 & 0 & -1/5 \end{bmatrix}.$$

Now, it is easy to compute $\|A\|_\infty = 10$. Then, to compute the growth factor, we simply look throughout all the values in these four matrices to find the one that is largest in absolute value. In this case, we pick out $a_{1,1}^{(0)} = 4$, so, for this matrix, the growth factor is

$$\rho_4 = 4/10.$$

The final piece of the puzzle, then, is the following theorem, taken from the classic study by Forsythe and Moler [3].

Theorem 7.9 (Rounding Error Effects) *Let L_c and U_c be the lower and upper triangular factors of A as computed on a machine with rounding unit $\mathbf{u}$, using Gaussian elimination with either partial or complete pivoting.*

1. Then there is a matrix E_1 such that $L_c U_c = A + E_1$ and

$$\|E_1\|_\infty \le n^2 \rho_n \mathbf{u} \|A\|_\infty.$$

2. If x_c is the solution to the system $Ax = b$, computed using L_c and U_c, then there is a matrix E_2 such that $(A + E_2)x_c = b$ and

$$\|E_2\|_\infty \le 1.01 n^2 (n+3) \rho_n \mathbf{u} \|A\|_\infty.$$

Thus, we see that Gaussian elimination produces the exact solution to a "nearby" problem, and we can combine Theorem 7.9 with Theorem 7.6 to get a bound on the error between the computed solution x_c and the exact solution, x:

$$\frac{\|x - x_c\|_\infty}{\|x\|_\infty} \leq \frac{1.01n^2(n+3)\rho_n\mathbf{u}\kappa_\infty(A)}{1 - 1.01n^2(n+3)\rho_n\mathbf{u}\kappa_\infty(A)}. \tag{7.13}$$

Or can we? Just how big is ρ_n going to be? Note that we don't get the bound (7.13) unless $1.01n^2(n+3)\rho_n\mathbf{u}\kappa_\infty(A)$ is "sufficiently small," and we don't (yet) know how large ρ_n is going to be.

The bounds on ρ_n depend on what type of pivoting we do. If we do complete pivoting—which is rather costly in terms of computer time—then it can be shown that

$$\rho_n \leq 1.8(n^{0.25\ln n}),$$

which grows rather slowly: For $n = 100$, $\rho_n = \mathcal{O}(400)$. However, if we only do partial pivoting, then it can be shown that there are matrices for which

$$\rho_n = 2^{n-1};$$

in other words, ρ_n grows exponentially. For comparison, we now have $\rho_{100} = \mathcal{O}(10^{30})$.

So it is fair to ask: Is Gaussian elimination with partial pivoting a stable process?

Strictly speaking, (7.13) says that it is: For a sufficiently accurate computer (i.e., $\mathbf{u}$ small enough) and a sufficiently small problem (n small enough), then Gaussian elimination with partial pivoting will produce solutions that are stable and accurate. The problem is that we don't want to have to limit ourselves to computer architectures that use unrealistic word lengths to achieve extraordinary accuracy, nor do we want to restrict ourselves to very small problems in order to keep ρ_n small. So, as a practical matter, it appears that Gaussian elimination with partial pivoting is perhaps not stable—except that the kinds of matrices that lead to large values of ρ_n do not appear to occur in practice. It can be shown that ρ_n is directly related to the growth of the elements in the upper triangular factor, U_c (this should be evident from the example we did above), and this growth is not observed in problems of interest. To quote from page 166 of *Numerical Linear Algebra*, by Trefethen and Bau [10]: "In fifty years of computing, no matrix problems that excite an explosive instability are known to have arisen under natural circumstances." This book also gives an interesting probabilistic argument to explain why this might be the case. Moreover, we should point out that ρ_n is relatively easy to compute as part of the elimination process, so if in any particular case we are worried about the size of ρ_n, we can compute it.

So, in the end, we conclude that, as a practical matter—meaning, for problems of interest that actually occur—Gaussian elimination with partial pivoting *is* a stable process.

It is possible to combine some of the other results to get a more meaningful *a posteriori* (after the fact) result. Note, again, that this result is specific to the 2-norm, because it uses Theorem 7.8.

Theorem 7.10 *Let x_c be a computed solution to the linear system $Ax = b$, and let $r = b - Ax_c$ be the corresponding residual. If $\kappa_2(A)\|r\|_2$ is sufficiently small, then, we have*

$$\frac{\|x - x_c\|_2}{\|x\|_2} \leq \frac{\kappa_2(A)\frac{\|r\|_2}{\|b-r\|_2}}{1 - \kappa_2(A)\frac{\|r\|_2}{\|b-r\|_2}}.$$

Proof: This follows by cleverly combining several of the earlier results in this section, and is left as an exercise. (See Problem 16.) •

For practical purposes, the following "rule of thumb" is very useful.

Rule of Thumb: Let E and δb be the perturbations such that the computed solution exactly solves $(A + E)x_c = b + \delta b$. If we can assume the following:

1. $\mathbf{u} \leq C_1 \times 10^{-s}$;

2. $\|E\| \leq C_2\mathbf{u}\|A\|$;

3. $\|\delta b\| \leq C_3\mathbf{u}\|A\|$;

4. $\kappa(A) \leq C_4 \times 10^t$;

then, for $t - s$ sufficiently negative, there is a constant $C > 0$ such that

$$\frac{\|x - x_c\|}{\|x\|} \leq C \times 10^{t-s}.$$

What this says is that, in practice, we observe that the error in a linear system computation goes like the condition number times the machine rounding unit.

7.5.3   Estimating the Condition Number

As the terminology implies, singular matrices, are perhaps something of a rarity. This does not mean that we don't have to worry about them, however. It can be shown, in a very mathematically precise sense, that all singular matrices are arbitrarily close to a nonsingular matrix. Once we begin doing numerical computations, then, the odds are very high that the inherent rounding error will perturb the matrix away from being singular. The resulting nonsingular matrix, however, will be very ill-conditioned. This is why we have to be careful when doing matrix computations. The algorithms will fail (in the sense of having a zero pivot) if the matrix is singular, but this is only reliable in infinite-precision arithmetic. In floating-point arithmetic, we have to recognize that small pivots might mean that the matrix is *nearly singular*, or ill-conditioned. Generally speaking,

> *if the solution to a linear system changes a great deal when the problem changes only very slightly, then we suspect that the matrix is ill-conditioned.*

Since the condition number is such an important indicator of how well we can compute with the matrix, it would be useful if it could be easily computed. Alas, this is asking too much, but we can *estimate* $\kappa_\infty(A)$ without too much more effort than is needed to solve the system $Ax = b$. A condition number estimator is a fairly standard feature of modern linear systems software.

Recall the definition:

$$\kappa = \|A\|_\infty\|A^{-1}\|_\infty.$$

Computing $\|A\|_\infty$ is not difficult; computing $\|A^{-1}\|_\infty$ is.[5] However, the norm definition tells us that

$$\|A^{-1}\|_\infty = \max_{x \neq 0} \frac{\|A^{-1}x\|_\infty}{\|x\|_\infty},$$

---

[5]In any case, if A is ill-conditioned, any computation of A^{-1} will be unreliable.

so that, for any fixed $w \in \mathbb{R}^n$, $w \neq 0$, we have

$$\|A^{-1}\|_\infty \geq \frac{\|A^{-1}w\|_\infty}{\|w\|_\infty}.$$

Now set $y = A^{-1}w$ and substitute into the condition number definition to get

$$\kappa_\infty(A) \geq \|A\|_\infty \frac{\|y\|_\infty}{\|w\|_\infty}, \qquad (7.14)$$

and this holds for any $y \in \mathbb{R}^n$, $y \neq 0$, $w = Ay$. The trick, now, is to find the choice of y or w that maximizes the right-hand side of (7.14). Since the inequality holds for any $y \in \mathbb{R}^n$, $w = Ay$, the larger right-hand side is, the more accurate the estimate of $\kappa_\infty(A)$.

Recall that a singular matrix has at least one eigenvalue equal to zero. It therefore follows that a matrix that is "close to singular" will have at least one eigenvalue "close to zero." If we choose y to be the eigenvector of A corresponding to the smallest (in absolute value) eigenvalue of A, then

$$\frac{\|y\|_\infty}{\|w\|_\infty} = \frac{\|y\|_\infty}{\|Ay\|_\infty} = \frac{\|y\|_\infty}{\|\lambda y\|_\infty} = |\lambda|^{-1},$$

and this will tend to maximize the estimate of the condition number. In §8.3 we will see that the recursion

$$y^{(i+1)} = A^{-1}y^{(i)}/\|y^{(i)}\|_\infty, \quad i = 1, 2, \ldots$$

will produce a sequence of vectors that tend to be in the direction of the eigenvector corresponding to the smallest eigenvalue. Note that we would not compute this sequence by actually inverting A; rather, we would solve the sequence of linear system problems

$$Ay^{(i+1)} = y^{(i)}/\|y^{(i)}\|_\infty$$

using an *existing LU* decomposition. Note also that if we are estimating the condition number as part of an *LU* decomposition routine, that the factorization will already have been done, and all that has to be done are the relatively inexpensive triangular solves. The number of steps to take in the recursion is an issue; taking more steps is more costly but will lead to a more accurate estimate. The public domain package LAPACK uses five iterations in the corresponding section of its condition number estimator, and that seems to be a reasonable choice.

In any event, with the recursion computed, we have

$$\kappa_\infty(A) \geq \|A\|_\infty \frac{\|y^{(5)}\|_\infty}{\|Ay^{(5)}\|_\infty} = \frac{\alpha v}{\omega}, \qquad (7.15)$$

where

$$\alpha = \|A\|_\infty, \quad v = \|y^{(5)}\|_\infty, \quad \omega = \|Ay^{(5)}\|_\infty.$$

We could thus use the estimate

$$\kappa^* = \frac{\alpha v}{\omega}$$

for the condition number. However, the way we have set things up, we have

$$\omega = \|Ay^{(5)}\|_\infty = \left\|\frac{y^{(4)}}{\|y^{(4)}\|_\infty}\right\|_\infty = 1.$$

So, we actually have

$$\kappa(A) \approx \kappa^* = \alpha v.$$

A complete algorithm then is the following.

Algorithm 7.9 *Condition Number Estimation*
Given an LU factorization of A, compute as follows:

1. *Compute $\alpha = \|A\|_\infty$;*

2. *Take a random initial guess $y^{(0)}$;*

3. *Compute $y^{(5)}$ in the sequence defined by*

$$y^{(i+1)} = A^{-1}y^{(i)}/\|y^{(i)}\|_\infty, \quad i = 0, 1, \ldots, 4 \qquad (7.16)$$

by solving the systems using the extant factorization of A, and set $v = \|y^{(5)}\|_\infty$;

4. *Set $\kappa^* = \alpha v$.*

A more sophisticated process for choosing y is slightly better, but justifying it requires more linear algebra background than we have here, and this will work reasonably well. MATLAB has a couple of commands that are relevant here:

- cond(A) computes the *exact* condition number in the 2-norm, κ_2;

- rcond computes an estimate of the *reciprocal* of the condition number in the 1-norm, κ_1.

■ **EXAMPLE 7.9**

To illustrate, consider the matrix

$$A = \begin{bmatrix} 1 & \frac{1}{2} & \frac{1}{3} \\ \frac{1}{2} & \frac{1}{3} & \frac{1}{4} \\ \frac{1}{3} & \frac{1}{4} & \frac{1}{5} \end{bmatrix}.$$

Since this is only 3×3 we can compute the exact condition number without too much effort, and we get $\kappa_\infty(A) = 748$. Applying the estimation algorithm here goes as follows. First, we compute an LU decomposition for A, getting

$$L = \begin{bmatrix} 1.0000 & 0 & 0 \\ 0.5000 & 1.0000 & 1.0000 \\ 0.3333 & 1.0000 & 0 \end{bmatrix}, U = \begin{bmatrix} 1.0000 & 0.5000 & 0.3333 \\ 0 & 0.0833 & 0.0889 \\ 0 & 0 & -0.0056 \end{bmatrix}.$$

Choose a random vector for $y^{(0)}$; for example,

$$y^{(0)} = \begin{bmatrix} 0.2190 \\ 0.0470 \\ 0.6789 \end{bmatrix}.$$

Computing in the recursion (7.16), we get

$$y^{(5)} = \begin{bmatrix} 66.5558 \\ -372.1152 \\ 359.0420 \end{bmatrix}.$$

Then, $\alpha = \|A\|_\infty = 1.833$ and $\upsilon = \|y^{(5)}\|_\infty = 372.1$, so $\kappa^* = 682.2$, which is not a bad estimate. For comparison's sake, we note that cond applied to this matrix shows that $\kappa_2(A) = 524.0568$, and rcond produces the estimate $\kappa_1(A) \approx 680.8093$.

7.5.4 Iterative Refinement

Since Gaussian elimination can be adversely affected by rounding error, especially if the matrix is ill-conditioned, it is sometimes worthwhile to improve the accuracy of a computed solution. This leads to the algorithm known as *iterative refinement* or *iterative improvement*.

We first observe that if x_c is a computed solution to $Ax = b$, then the error satisfies a very similar problem.

Theorem 7.11 *Let $A \in \mathbb{R}^{n \times n}$ be given, nonsingular, and let $x_c \in \mathbb{R}^n$ be a computed solution to the linear system $Ax = b$. Then the error $e = x - x_c$ satisfies $Ae = r$, where $r = b - Ax_c$ is the residual for x_c.*

Proof: This is a simple, direct computation:

$$Ae = A(x - x_c) = Ax - Ax_c = b - Ax_c = r. \qquad \bullet$$

Thus, given a computed solution $x_c \approx x$ and the LU factorization from which it was computed, we can then quickly solve the new problem

$$Ae = r$$

and set $x = x_c + e$.

If all the computations were carried out in exact arithmetic, then the "improved" value x computed above would be the exact solution. However, since the computations are almost surely not done exactly, the improved value will quite likely not be the exact solution. So it is worthwhile to continue the process iteratively. Describing this requires some better notation, though.

Let $x^{(0)}$ be the initial computed solution, and let $r^{(0)} = b - Ax^{(0)}$ be the corresponding residual. Then we compute the sequence of improved values $x^{(k)}$ according to the following algorithm.

Iterative refinement. For k from 0 until sufficient accuracy is achieved, do:

1. Compute $r^{(k)} = b - Ax^{(k)}$;

2. Solve $Ae = r^{(k)}$ using the existing LU factorization of A;

3. Define $x^{(k+1)} = x^{(k)} + e$.

Programming Hints: (1) Be sure to compute the residual using the original matrix A and not the product LU. (2) Because of the potential for subtractive cancellation, refinement works best when the residuals are computed in a higher precision than the rest of the calculation, if that is possible. (3) If $\kappa(A)\mathbf{u}$ is too large, then refinement may not work.

Note that the only additional work in the refinement algorithm is the computation of the residual and the two solution steps needed to complete the solution of the systems $Ae = r^{(k)}$; the factorization has already been done.

■ **EXAMPLE 7.10**

Let's consider the simple linear system

$$\begin{bmatrix} 1 & \frac{1}{2} & \frac{1}{3} \\ \frac{1}{2} & \frac{1}{3} & \frac{1}{4} \\ \frac{1}{3} & \frac{1}{4} & \frac{1}{5} \end{bmatrix} \begin{bmatrix} x_1 \\ x_2 \\ x_3 \end{bmatrix} = \begin{bmatrix} 3 \\ 23/12 \\ 43/30 \end{bmatrix},$$

which has exact solution $x = (1, 2, 3)^T$. The correct LU factorization of the coefficient matrix is

$$L = \begin{bmatrix} 1 & 0 & 0 \\ \frac{1}{2} & 1 & 0 \\ \frac{1}{3} & 1 & 1 \end{bmatrix}, \quad U = \begin{bmatrix} 1 & \frac{1}{2} & \frac{1}{3} \\ 0 & \frac{1}{12} & \frac{1}{12} \\ 0 & 0 & \frac{1}{180} \end{bmatrix},$$

but let's introduce some error by using only six digits to represent these:

$$L_c = \begin{bmatrix} 1 & 0 & 0 \\ 0.5 & 1 & 0 \\ 0.333333 & 1 & 1 \end{bmatrix}, \quad U_c = \begin{bmatrix} 1 & 0.5 & 0.333333 \\ 0 & 0.0833333 & 0.0833333 \\ 0 & 0 & 0.555556 \times 10^{-2} \end{bmatrix}.$$

Thus, the solution we compute from the forward–backward solution steps will be inexact, however slightly. In fact, we get

$$x^{(0)} = \begin{bmatrix} 1.00002960003512 \\ 1.99982440014289 \\ 3.00017759985791 \end{bmatrix}.$$

Now apply the refinement algorithm. Our initial residual is

$$r^{(0)} = 10^{-5} \times \begin{bmatrix} -0.10000591998960 \\ -0.06666963332513 \\ -0.14866856774542 \end{bmatrix}.$$

We solve the system

$$L_c U_c e = r^{(0)}$$

to get

$$e = 10^{-3} \times \begin{bmatrix} -0.02960008024619 \\ 0.17559977427749 \\ -0.17759977587713 \end{bmatrix}$$

so the first improved solution is

$$x^{(1)} = x^{(0)} + e = \begin{bmatrix} 0.99999999995487 \\ 1.99999999991716 \\ 3.00000000008204 \end{bmatrix},$$

which is a substantial improvement over the original computed solution. Another step of refinement would continue to improve the solution, of course.

Exercises:

1. Let

$$A = \begin{bmatrix} 1 & 2 & -7 \\ 4 & -8 & 0 \\ 2 & 1 & 0 \end{bmatrix}$$

Compute $\|A\|_\infty$.

2. Let

$$A = \begin{bmatrix} 5 & 6 & -9 \\ 1 & 2 & 3 \\ 0 & 7 & 2 \end{bmatrix}$$

Compute $\|A\|_\infty$.

3. Let

$$A = \begin{bmatrix} -8 & 0 & -1 \\ 3 & 12 & 0 \\ 1 & 2 & 3 \end{bmatrix}$$

Compute $\|A\|_\infty$.

4. Show that

$$\|A\|_\star = \max_{i,j} |a_{ij}|$$

does not define a matrix norm, according to our defintion. *Hint:* Show that one of the conditions fails to hold by finding a specific case where it fails.

5. Let

$$A = \begin{bmatrix} 1 & \frac{1}{2} & 0 \\ \frac{1}{2} & \frac{1}{3} & \frac{1}{4} \\ 0 & \frac{1}{4} & \frac{1}{5} \end{bmatrix}.$$

Compute, directly from the definition, $\kappa_\infty(A)$. You should get $\kappa_\infty(A) = 18$.

6. Repeat Problem 5 for

$$A = \begin{bmatrix} 4 & 1 & 0 \\ 1 & 4 & 1 \\ 0 & 1 & 4 \end{bmatrix},$$

for which $\kappa_\infty(A) = 2.5714$.

7. Consider the linear system

$$\begin{bmatrix} 1.002 & 1 \\ 1 & 0.998 \end{bmatrix} \begin{bmatrix} x_1 \\ x_2 \end{bmatrix} = \begin{bmatrix} 0.002 \\ 0.002 \end{bmatrix},$$

which has exact solution $x = (1, -1)^T$ (verify this). What is the residual $b - Ax_c$ for the "approximate" solution $x_c = (0.29360067817338, -0.29218646673249)^T$? Explain.

8. Consider the linear system problem

$$Ax = b,$$

where

$$A = \begin{bmatrix} 4 & 2 & 0 \\ 1 & 4 & 1 \\ 0 & 2 & 4 \end{bmatrix}, \quad b = \begin{bmatrix} 8 \\ 12 \\ 16 \end{bmatrix},$$

for which the exact solution is $x = (1, 2, 3)^T$. (Check this.) Let $x_c = x + (0.002, 0.01, 0.001)^T$ be a computed (i.e., approximate) solution to this system. Use Theorem 7.8 to find the perturbation matrix E such that x_c is the exact solution to $A x_c = b$.

9. Compute the growth factor for Gaussian elimination for the matrix in Problem 8.

10. Let

$$A = \begin{bmatrix} 2 & 1 & 0 \\ 1 & 2 & 1 \\ 0 & 1 & 2 \end{bmatrix}$$

This has exact condition number $\kappa_\infty(A) = 8$. Use the condition number estimator in this section to approximate the condition number.

11. Repeat Problem 10 for

$$A = \begin{bmatrix} 1 & \frac{1}{2} & 0 \\ \frac{1}{2} & \frac{1}{3} & \frac{1}{4} \\ 0 & \frac{1}{4} & \frac{1}{5} \end{bmatrix}$$

for which $\kappa_\infty(A) = 18$.

12. Use the condition number estimator to produce a plot of κ^* versus n for each of the following matrix families (you may want to consider a semilog or log-log scale for some of the plots):

    (a) $T_n, 4 \le n \le 20$;

    (b) $K_n, 4 \le n \le 20$;

    (c) $H_n, 4 \le n \le 20$;

    (d) $A_n, 4 \le n \le 20$.

Compare your estimates with the exact values from cond and the estimates from rcond.

13. Produce a plot of the growth factor for Gaussian elimination for each matrix family in Problem 12, as a function of n.

14. Given a matrix $A \in \mathbb{R}^{n \times n}$, show that

$$\mu(A) = \frac{\max_{x \ne 0} \frac{\|Ax\|}{\|x\|}}{\min_{x \ne 0} \frac{\|Ax\|}{\|x\|}}$$

is equivalent to the condition number as defined in (7.10).

15. Prove that (7.8) and (7.9) follow from the definition of matrix norm.

16. Prove Theorem 7.10.

17. Give an argument to support the validity of the "rule of thumb" following Theorem 7.10.

18. Prove Theorem 7.5.

19. Consider the linear system problem $Ax = b$ for

$$A = \begin{bmatrix} 1 & \frac{1}{2} \\ \frac{1}{2} & \frac{1}{3} \end{bmatrix}, \quad b = \begin{bmatrix} 2 \\ 7/6 \end{bmatrix}.$$

Note that the exact solution is $x = (1, 2)^T$.

(a) Represent the exact factorization of A using only four decimal digits, and solve the system.

(b) Do two steps of iterative refinement to improve your solution.

20. Do two more steps of refinement for the problem in Example 7.10.

◁ • • • ▷

7.6 SPD MATRICES AND THE CHOLESKY DECOMPOSITION

There exist a number of classes of special matrices whose properties make the solution of linear systems easier. One important such class consists of *symmetric positive definite* matrices.

Definition 7.5 (SPD Matrices) *Let $A \in \mathbb{R}^{n \times n}$ be given. If A satisfies*

$$A = A^T$$

and

$$x^T A x > 0, \quad all \ x \neq 0,$$

then we say that A is symmetric positive definite, *abbreviated by writing $A \in$ SPD.*

Like diagonally dominant matrices, SPD matrices appear often in applications involving approximation of functions and the solution of differential equations. What makes them important is that a special factorization and solution algorithm exists for them, which is about half as costly as ordinary Gaussian elimination. The result is due to the French mapmaker Cholesky.[6]

Theorem 7.12 (Cholesky Theorem) *Let $A \in \mathbb{R}^{n \times n}$ be given, with $A \in$ SPD. Then:*

1. A is nonsingular;

---

[6]André-Louis Cholesky (1875–1918), a French military officer who was killed in the final year of World War I, was born in Montguyon, educated at École Polytechnique, and served as a mapmaker for the French military, earning some distinction for his efforts in mapping Crete and some of the French colonies in North Africa. His method for the solution of positive definite systems of equations, which he developed for the solution of the so-called "normal equations" that arise in least squares problems, was published posthumously by a fellow officer, in 1924, with due credit being given to Cholesky.

2. *There exists a lower triangular matrix $G \in \mathbb{R}^{n \times n}$, with $g_{ii} > 0$, such that $GG^T = A$.*

Proof: The nonsingularity is a straightforward consequence of the condition $x^T A x > 0$; if A is singular, there is a nonzero vector x, such that $Ax = 0$, hence $x^T A x = 0$, therefore a singular matrix cannot be SPD.

For the construction of G, we use an argument that is based on induction on the size of the matrix. Let $n = 2$; then

$$A = \begin{bmatrix} a_{11} & a_{21} \\ a_{21} & a_{22} \end{bmatrix}$$

and the condition $x^T A x > 0$ implies that both diagonals a_{ii} must be positive (Problem 1). Then it can easily be verified (by direct computation) that

$$G = \begin{bmatrix} \sqrt{a_{11}} & 0 \\ a_{21}/\sqrt{a_{11}} & \sqrt{a_{22} - a_{21}^2/a_{11}} \end{bmatrix}$$

is such that $A = GG^T$. We know that both of the square roots are defined because $A \in$ SPD implies (Problems 1 and 2) that the diagonals must be positive and that $a_{22} - a_{21}^2/a_{11} > 0$. Therefore, G exists, at least for the special case of $n = 2$.

Assume now that G exists for $n = k - 1$; we will use this to show that it must exist for $n = k$, and this will complete the induction. We write A (which is $k \times k$) in the partitioned form

$$A = \begin{bmatrix} A_{11} & a \\ a^T & a_{kk} \end{bmatrix},$$

where A_{11} is $(k-1) \times (k-1)$ and a is a vector in $\mathbb{R}^{k-1}$. The fact that $A \in$ SPD forces (Problem 3) $A_{11} \in$ SPD. Therefore, by the inductive hypothesis, there exists G_{11}, lower triangular, such that $A_{11} = G_{11}G_{11}^T$. Construct

$$G = \begin{bmatrix} G_{11} & 0 \\ g^T & g_{kk} \end{bmatrix},$$

where $g \in \mathbb{R}^{k-1}$ and $g_{kk} \in \mathbb{R}$ are yet to be determined. Note that if we can determine these two quantities, we will have finished the proof because we will have found the lower triangular matrix for the $k \times k$ case and the induction will be complete.

To see what g and g_{kk} have to be, we multiply out GG^T and set this equal to A:

$$GG^T = \begin{bmatrix} G_{11}G_{11}^T & G_{11}g \\ g^T G_{11}^T & g_{kk}^2 + g^T g \end{bmatrix} = \begin{bmatrix} A_{11} & a \\ a^T & a_{kk} \end{bmatrix}.$$

Note that the upper-left element is correct: $G_{11}G_{11}^T$ does indeed equal A_{11} by the inductive hypothesis. To finish the job we have to have

$$G_{11}g = a \tag{7.17}$$

and

$$g_{kk}^2 + g^T g = a_{kk}.$$

Since the diagonal elements of G_{11} (which is upper triangular) are all positive, it is necessarily nonsingular; thus, a unique solution exists to the system (7.17), and g is defined. With g defined, we simply set

$$g_{kk} = \sqrt{a_{kk} - g^T g}.$$

The argument to the square root has to be positive; otherwise, there will be an x such that $x^T A x \leq 0$, which violates the positive definite condition (Problem 4). The induction is complete, hence the theorem is proved. •

There are a number of different ways of actually constructing the Cholesky decomposition, one of which can be deduced from the proof we gave above. The MATLAB command is $G = \texttt{chol(A)}$. It can be shown that the Cholesky factorization is unique, so all of these different constructions are just different ways or organizing the computation. One common scheme uses the following formulas:

$$
g_{ij} = \frac{1}{g_{jj}} \left[a_{ij} - \sum_{k=1}^{j-1} g_{ik} g_{jk} \right], \quad 1 \leq j < 1;
$$

$$
g_{ii} = \left[a_{ii} - \sum_{k=1}^{i-1} g_{ik}^2 \right]^{1/2}.
$$

Cholesky is a very efficient algorithm; the operation count (Problem 5) is $\frac{1}{6} n^3 + \mathcal{O}(n^2)$. Round-off error for a nearly singular SPD matrix can sometimes cause the algorithm to try and take a negative square root. A variant exists that computes a factorization of the form $A = LDL^T$, where L is unit lower triangular and D is diagonal; this can be computed without using any square roots.

We will defer a detailed discussion of the Cholesky method to §9.3.2, where it is used to solve the large systems that arise in the discretization of partial differential equations.

Banded Systems   In many important applications (see Chapter 9 for examples) the coefficient matrix is very sparse (meaning most of the elements are zero), with the nonzero elements concentrated in a relatively narrow band around the main diagonal of the matrix. This is the case, for example, for the matrices that arise in the discretization of partial differential equations. Note, also, that the tridiagonal matrices that were studied in Chapter 2 are a particularly simple example of banded matrices.

Banded systems can be factored "within the band;" that is, we do not need to store nor do we need to compute with the large number of zeros that are located away from the diagonal ("outside the band"). For large problems this can result in a considerable savings in storage and execution time. Since the precise algorithm to use is almost always very specific to the problem being solved, we will defer a detailed discussion to Chapter 9, where we discuss the simple discretization of some PDEs.

Exercises:

1. Show that a matrix with one or more non-positive diagonal elements cannot be SPD. *Hint:* Try to find a vector x such that $x^T A x = a_{ii}$.

2. For a 2×2 SPD matrix, show that $a_{22} - a_{21}^2/a_{11} > 0$. *Hint:* Consider the positive definite condition with $x = (a_{21}, -a_{11})^T$.

3. Let $A \in \mathbb{R}^{n \times n}$ be partitioned as

$$
A = \left[\begin{array}{cc} A_{11} & A_{21} \\ A_{21} & A_{22} \end{array} \right]'
$$

where A_{11} and A_{22} are square matrices. Show that both A_{11} and A_{22} must be SPD if $A \in \text{SPD}$.

4. In the proof of the Cholesky theorem, show that $a_{kk} - g^T g > 0$ must hold, or else there is a vector x such that $x^T A x \leq 0$. *Hint:* Use Problem 2 as a starting point.

5. Derive the operation count given in the text for the Cholesky decomposition.

6. Prove the following: If $A \in$ SPD, then there exists a unit lower triangular matrix L, and a diagonal matrix D with positive elements, such that $A = LDL^T$.

7. Derive an algorithm for the LDL^T factorization that does not require square roots. *Hint:* Look at the 3×3 case, explicitly. Multiply out LDL^T and set the result equal to A, and from this deduce the relationships that are necessary. In other words, if

$$
\begin{bmatrix} 1 & 0 & 0 \\ l_{21} & 1 & 0 \\ l_{31} & l_{32} & 1 \end{bmatrix}
\begin{bmatrix} d_{11} & 0 & 0 \\ 0 & d_{22} & 0 \\ 0 & 0 & d_{33} \end{bmatrix}
\begin{bmatrix} 1 & l_{21} & l_{31} \\ 0 & 1 & l_{32} \\ 0 & 0 & 1 \end{bmatrix}
=
\begin{bmatrix} a_{11} & a_{12} & a_{13} \\ a_{21} & a_{22} & a_{23} \\ a_{31} & a_{32} & a_{33} \end{bmatrix},
$$

what is the relationship between the components of L, D, and A? Then generalize to a problem of arbitrary size. Don't forget that A is symmetric!

8. The T_n and K_n families of matrices are both positive definite. Use the `chol` command to produce the triangular factor, then write a routine `cholsol` which takes the triangular factor and applies appropriate forward and backward solution steps to solve linear systems $T_n x = b$ and $K_n x = b$, over the range $4 \leq n \leq 20$, using random right-hand sides. Confirm the accuracy of you solutions by computing the norm of the residual.

9. If A is tridiagonal *and* SPD, then the Cholesky factorization can be modified to work only with the *two* distinct nonzero diagonals in the matrix. Construct this version of the algorithm and test it on the T_n and K_n families, as above.

10. The H_n family of matrices is also SPD, but ill-conditioned. Try to apply `chol` over the range $4 \leq n \leq 20$. What happens? Can you explain this?

$$\triangleleft \bullet \bullet \bullet \triangleright$$

7.7   ITERATIVE METHODS FOR LINEAR SYSTEMS: A BRIEF SURVEY

If the coefficient matrix is *very* large and sparse—meaning that the number of nonzero elements is a small fraction of the total number of elements in the matrix—then Gaussian elimination may not be the best way to solve the linear system problem. In these circumstances, elimination techniques will be drastically slowed by the memory management issues associated with handling the very large matrix,[7] and they will also destroy the sparsity of the original matrix—even though $A = LU$ is sparse, the individual factors L and U may not be *as* sparse as A was.

We can illustrate this last point with a simple example.

---

[7]This is because most modern architectures use *virtual memory* concepts to enable a large amount of disk storage to substitute for a large amount of RAM storage. However, there is a time cost associated with writing the current RAM contents to disk and bringing new material into the RAM, and this time cost can be more than the cost of doing the actual computation.

■ **EXAMPLE 7.11**

Let

$$
A = \left[\begin{array}{cccc|cccc|cccc|cccc}
-4 & 1 & 0 & 0 & 1 & 0 & 0 & 0 & 0 & 0 & 0 & 0 & 0 & 0 & 0 & 0 \\
1 & -4 & 1 & 0 & 0 & 1 & 0 & 0 & 0 & 0 & 0 & 0 & 0 & 0 & 0 & 0 \\
0 & 1 & -4 & 1 & 0 & 0 & 1 & 0 & 0 & 0 & 0 & 0 & 0 & 0 & 0 & 0 \\
0 & 0 & 1 & -4 & 0 & 0 & 0 & 1 & 0 & 0 & 0 & 0 & 0 & 0 & 0 & 0 \\
\hline
1 & 0 & 0 & 0 & -4 & 1 & 0 & 0 & 1 & 0 & 0 & 0 & 0 & 0 & 0 & 0 \\
0 & 1 & 0 & 0 & 1 & -4 & 1 & 0 & 0 & 1 & 0 & 0 & 0 & 0 & 0 & 0 \\
0 & 0 & 1 & 0 & 0 & 1 & -4 & 1 & 0 & 0 & 1 & 0 & 0 & 0 & 0 & 0 \\
0 & 0 & 0 & 1 & 0 & 0 & 1 & -4 & 0 & 0 & 0 & 1 & 0 & 0 & 0 & 0 \\
\hline
0 & 0 & 0 & 0 & 1 & 0 & 0 & 0 & -4 & 1 & 0 & 0 & 1 & 0 & 0 & 0 \\
0 & 0 & 0 & 0 & 0 & 1 & 0 & 0 & 1 & -4 & 1 & 0 & 0 & 1 & 0 & 0 \\
0 & 0 & 0 & 0 & 0 & 0 & 1 & 0 & 0 & 1 & -4 & 1 & 0 & 0 & 1 & 0 \\
0 & 0 & 0 & 0 & 0 & 0 & 0 & 1 & 0 & 0 & 1 & -4 & 0 & 0 & 0 & 1 \\
\hline
0 & 0 & 0 & 0 & 0 & 0 & 0 & 0 & 1 & 0 & 0 & 0 & -4 & 1 & 0 & 0 \\
0 & 0 & 0 & 0 & 0 & 0 & 0 & 0 & 0 & 1 & 0 & 0 & 1 & -4 & 1 & 0 \\
0 & 0 & 0 & 0 & 0 & 0 & 0 & 0 & 0 & 0 & 1 & 0 & 0 & 1 & -4 & 1 \\
0 & 0 & 0 & 0 & 0 & 0 & 0 & 0 & 0 & 0 & 0 & 1 & 0 & 0 & 1 & -4
\end{array}\right].
$$

This is a "small" example of the kind of sparse matrix we are talking about here; it has a total of 256 elements but only 64 are nonzero. If we factor this using the LU factorization, then there will be a lot of "fill-in:" Many of the elements that are zero in A would be nonzero in L or U. We can see this—without drowning in a sea of numbers—by creating a *sparsity plot* of the matrix and its LU factorization, using MATLAB's spy command; spy(A) produces a plot that puts a single mark at the location of every nonzero element, and leaves a blank at the zero elements. Figure 7.1 shows the sparsity plots for A, the individual L and U factors, and the matrix A' which denotes the sparsity pattern of $A = LU$ as it would actually appear in the computer storage after the elimination process had run to completion. Note that the total number of nonzero elements has nearly doubled due to fill-in during the elimination step. (The captions give the number of nonzero elements in the matrix.)

Now, it was possible to store the original matrix using only five vectors (to represent the diagonals), each of length ≤ 16, and fewer if we took advantage of the symmetry of the matrix. On the other hand, storing the L and U factors will require much more storage. Moreover, we can easily write a code that will carry out matrix–vector multiplications of the form $z = Au$ by storing only the five diagonal vectors plus the u and z vectors.

It is because of these issues that Gaussian elimination and factorization methods are sometimes not the best choice of solution technique. Instead, we are driven to consider *iterative* methods for solving the linear system. This might seem counterintuitive at first, since we are giving up a method (Gaussian elimination) that is exact in the absence of rounding error for a method that is, by definition, mathematically inexact. Nonetheless, the iterative methods are cost-effective for large, sparse, problems, especially when A is symmetric and exhibits a strong structure to the placement of its zero and nonzero elements.

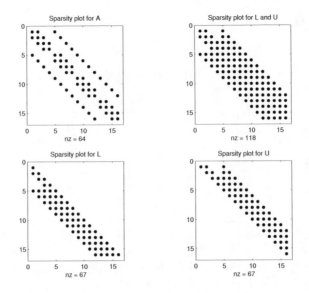

Figure 7.1   Sparsity plot for matrix A and its LU factors.

Unfortunately, a complete understanding of the mathematics underlying most iterative methods is beyond the intended level of this book. In addition, the most common realm of application for iterative methods is to solve the large, sparse, structured systems that come out of discretizing partial differential equations (PDEs). For this reason, we defer much of our discussion of iterative methods to Chapter 9; in this section we will outline some of the simpler ideas and show how they work on a simple 4×4 example. The most powerful technique for iteratively solving symmetric systems, the method of conjugate gradients, will be presented in §9.3.3. Still, we will be able to outline some significant ideas in this section.

An important class of matrix iterative methods come under the heading of what are called *splitting methods*, because they are based on the notion of *splitting* the coefficient matrix into two parts.

Suppose that we want to solve the linear system

$$Ax = b,$$

where $A \in \mathbb{R}^{n \times n}$ is assumed to be nonsingular. We split A into a difference $M - N$:

$$A = M - N,$$

where M is such that systems of the form $Mz = f$ are "easy" to solve. Then we have

$$Ax = b \Rightarrow (M - N)x = b \Rightarrow Mx = Nx + b.$$

Since $Mz = f$ is assumed to be "easy" to solve, it follows that $M^{-1}f$ is easy to compute[8] for any f. Thus, we can write

$$x = M^{-1}Nx + M^{-1}b.$$

---

[8]Note that this does not mean that we actually compute the inverse.

This suggests the following iteration: Given an initial guess,[9] $x^{(0)}$, compute the sequence of vectors according to

$$x^{(k+1)} = M^{-1}Nx^{(k)} + M^{-1}b, \tag{7.18}$$

until the sequence converges.

A very rich literature exists for this general scheme of iterations, showing that the iteration converges if and only if the *iteration matrix* $T = M^{-1}N$ is less than 1 in some sense. Note, incidentally, that we can write the iteration matrix as

$$T = M^{-1}N = M^{-1}(M - A) = I - M^{-1}A,$$

showing, more clearly perhaps, that the method will work best when $M^{-1} \approx A^{-1}$ or $M \approx A$.

The formal result, which we shall not prove, is the following.

Theorem 7.13 *Let $A \in \mathbb{R}^{n \times n}$ be given. Define $T = M^{-1}N$, where $A = M - N$. Then the iteration (7.18) converges for all initial guesses $x^{(0)}$ if and only if there exists a norm $\| \cdot \|$ such that $\|T\| < 1$.*

A second, very important result, relates the existence of such a norm to a fairly quantitative condition on the eigenvalues of the matrix T. First, a definition.

Definition 7.6 (Spectral Radius of a Matrix) *Let $A \in \mathbb{R}^{n \times n}$ be given. Then the spectral radius of A, denoted $\rho(A)$, is the largest (in magnitude) of all the eigenvalues of A. Formally,*

$$\rho(A) = \max |\lambda|,$$

where the maximum is taken over all the eigenvalues of A.

■ **EXAMPLE 7.12**

If A is the matrix

$$A = \begin{bmatrix} 4 & -3 \\ 2 & 1 \end{bmatrix},$$

then A has eigenvalues $2.5000 + 1.9365i$ and $2.5000 - 1.9365i$; thus, the spectral radius of A is

$$\rho(A) = 3.1623 = |2.5000 \pm 1.9365i|.$$

Then the theorem connecting the spectral radius to the convergence of iterative methods is the following.

Theorem 7.14 *Let $A \in \mathbb{R}^{n \times n}$ be given. Then there exists a norm, $\| \cdot \|$, such that $\|A\| < 1$ if and only if the spectral radius satisfies $\rho(A) < 1$.*

From this we conclude that

> The iteration (7.18) converges for all initial guesses $x^{(0)}$ if and only if $\rho(T) < 1$.

---

[9]We will, as a matter of convention, use parenthesized superscripts as iteration counters for vector quantities. This will also be an issue in §7.8, where we study nonlinear systems, and in Chapter 8 when we study eigenvalue–eigenvector approximations.

Having outlined the basic theory, let's now look at some specific methods. The simplest—and therefore, the slowest to converge—is known as the *Jacobi iteration*.[10] It might be more descriptive to call it "diagonal inversion," since it involves inversion of the diagonal part of A, only, at each step of the iteration. We use the splitting

$$A = D - (D - A),$$

where D is the diagonal part of A. Then the iteration is

$$x^{(k+1)} = (I - D^{-1}A)x^{(k)} + D^{-1}b.$$

The most efficient implementation of the method—and this is the case for all the splitting methods—depends greatly on the structure of the coefficient matrix, so it is difficult to discuss in general. If we have a specialized routine for forming matrix–vector products involving A (using the structure of A to maximize efficiency), then we can write the iteration to take advantage of that, thus:

$$x^{(k+1)} = x^{(k)} - D^{-1}(Ax^{(k)}) + D^{-1}b. \tag{7.19}$$

In Chapter 9 we will show how it would be done for the kinds of systems that arise from discretizing partial differential equations. Right now, let's look at a simple example.

■ **EXAMPLE 7.13**

Suppose that we want to solve the system $Ax = b$, where

$$A = \begin{bmatrix} 4 & 1 & 0 & 0 \\ 1 & 5 & 1 & 0 \\ 0 & 1 & 6 & 1 \\ 1 & 0 & 1 & 4 \end{bmatrix}$$

and $b = (1, 7, 16, 14)^T$, which means that the exact solution is $x = (0, 1, 2, 3)^T$. Take $x^{(0)} = (0, 0, 0, 0)^T$; then the first three iterations of the Jacobi iteration for this example go as follows:

$$x^{(1)} = x^{(0)} - D^{-1}Ax^{(0)} + D^{-1}b,$$

where $D = \text{diag}(4, 5, 6, 4)$, meaning the diagonal matrix having elements (in order down the diagonal) $4, 5, 6, 4$. This yields

$$x^{(1)} = \begin{bmatrix} 0.2500 \\ 1.4000 \\ 2.6667 \\ 3.5000 \end{bmatrix};$$

$$x^{(2)} = x^{(1)} - D^{-1}Ax^{(1)} + D^{-1}b,$$

---

[10]Karl Gustav Jacob Jacobi (1804–1851) was born in Potsdam, Prussia (now part of Germany) and was educated at the University of Berlin. He earned his doctorate in 1825 and had a position at the University of Königsberg from 1826 until 1844, after which he returned to Berlin where he remained until his death. Jacobi founded what we know as elliptic function theory and studied the determinant of functions known as the Jacobian. His iterative method for solving linear systems was first published in an astronomical journal in 1845.

so that

$$x^{(2)} = \begin{bmatrix} -0.1000 \\ 0.8167 \\ 1.8500 \\ 2.7708 \end{bmatrix};$$

and

$$x^{(3)} = x^{(2)} - D^{-1}Ax^{(2)} + D^{-1}b,$$

with

$$x^{(3)} = \begin{bmatrix} 0.0458 \\ 1.0500 \\ 2.0688 \\ 3.0625 \end{bmatrix}.$$

Even after only three iterations, it is becoming evident that the iteration is converging, and in fact we do converge, in the sense that $\|x^{(k)} - x^{(k-1)}\|_\infty < 10^{-6}$, in 19 iterations.

The student should bear in mind that this example is to illustrate the barest workings of the method. There is very little reason to use an iterative method for solving a 4×4 linear system. For work with more realistic examples, see §9.3.

An obvious extension of the Jacobi iteration is to invert the entire lower triangular part of A, not just the diagonal. This is what is known as the *Gauss–Seidel*[11] *iteration.* We split A into[12]

$$A = L - (L - A),$$

so that the iteration becomes

$$x^{(k+1)} = (I - L^{-1}A)x^{(k)} + L^{-1}b,$$

or

$$x^{(k+1)} = x^{(k)} - L^{-1}(Ax^{(k)}) + L^{-1}b. \tag{7.20}$$

Again, we will defer most of the implementation details until Chapter 9, when we can work with realistic examples. A simple example will illustrate and allow some comparison to the Jacobi method.

■ **EXAMPLE 7.14**

We use the same matrix A and the same initial vector as in Example 7.13. Then

$$x^{(1)} = x^{(0)} - L^{-1}(Ax^{(0)}) + L^{-1}b,$$

where

$$L = \begin{bmatrix} 4 & 0 & 0 & 0 \\ 1 & 5 & 0 & 0 \\ 0 & 1 & 6 & 0 \\ 1 & 0 & 1 & 4 \end{bmatrix}.$$

---

[11]Ludwig Philipp von Seidel (1821–1896) was born in Zweibrücken, Bavaria, and studied at the University of Berlin, beginning in 1840. Later he moved to Königsberg, where he studied under Jacobi. He received his doctorate from yet a third university, Munich, where he would eventually serve as a professor for many years. The bulk of his work was in optics and mathematical analysis. What we know as the Gauss–Seidel method was the result of a collaboration with Jacobi; Seidel published a memoir on the subject in 1874, but Gauss had described a similar method to a friend, who had published it in 1845.

[12]Note that L here is not the lower-triangular matrix from the LU decomposition of A, but is simply the lower triangular part of A.

Of course, we do not actually compute the inverse; rather, we solve the system

$$Lz = Ax^{(k)}$$

for each index k. Since this is a lower triangular system, it is quite easy to solve. For this example, we get

$$x^{(1)} = \begin{bmatrix} 0.2500 \\ 1.3500 \\ 2.4417 \\ 2.8271 \end{bmatrix} ; \quad x^{(2)} = \begin{bmatrix} -0.0875 \\ 0.9292 \\ 2.0406 \\ 3.0117 \end{bmatrix} ; \quad x^{(3)} = \begin{bmatrix} 0.0177 \\ 0.9883 \\ 2.0000 \\ 2.9956 \end{bmatrix} .$$

Once again, the example displays evidence of convergence, and in fact the convergence appears to be faster than for the Jacobi iteration. If we carry on the computation, we converge (in the same sense as in Example 7.13) in 11 iterations.[13]

The two examples seem to indicate that Gauss–Seidel is converging faster, and this is consistent with our intuition, since Gauss–Seidel is based on inverting "more" of the matrix at each step. The formal theoretical results are the following.

Theorem 7.15 (Jacobi and Gauss–Seidel Convergence Theorem)

1. If A is diagonally dominant, then both Jacobi and Gauss–Seidel converge, and Gauss–Seidel converges faster in the sense that $\rho(T_{GS}) < \rho(T_J)$, where T_J is the iteration matrix for the Jacobi iteration, and T_{GS} is the iteration matrix for Gauss–Seidel.

2. If $A \in$ SPD, then both Jacobi and Gauss–Seidel will converge.

While Gauss–Seidel is convergent for SPD matrices, it is still somewhat slow to converge. In the 1930s and 1940s, a number of schemes were proposed and developed to improve the convergence of Gauss–Seidel, usually by changing very slightly (or *relaxing*) the condition that defines the iteration. In the late 1940s and early 1950s this effort culminated with the development of the *SOR* (successive over-relaxation) *iteration*, which can be a substantial improvement over Gauss–Seidel. The problem is that this improved performance depends on the proper choice of a *relaxation parameter*.[14]

The splitting for SOR is somewhat complicated, and it is simpler to derive SOR as an extension of Gauss–Seidel. Let $\overline{x_i}$ be the value of the ith component as produced by Gauss–Seidel. Then the next SOR value is produced from the Gauss–Seidel value by averaging with the previous iterate:

$$x_i^{(k+1)} = \omega \overline{x_i} + (1 - \omega)x_i^{(k)}, \tag{7.21}$$

where ω is a real-valued parameter whose effect on the iteration needs to be studied. Putting this simple calculation in the form of a matrix splitting looks to be a bit of a challenge, but it can be done.[15]

---

[13] We could easily "cheat" and compute the exact solution in any one of a number of easy ways, but in a real application we wouldn't be able to do that, and would therefore have to use something like the difference in successive iterates—or the size of the residual—as our measure of convergence.

[14] A considerable amount of the history behind the SOR iteration can be found in David Young's doctoral dissertation, which is available online (a link will be on the text website).

[15] Writing SOR as a matrix splitting is absolutely necessary for any analysis that is going to be done, but for the purposes of calculation, (7.21) is the way to go.

Let D be the diagonal of A, and let L and U be the *strictly* lower and upper triangular parts of A; that is, L and U have zeros on the diagonal. Then, given ω, we split A as follows:

$$A = \left(\frac{1}{\omega}D + L\right) - \left(\left(\frac{1}{\omega} - 1\right)D - U\right),$$

leading to the iteration

$$x^{(k+1)} = x^{(k)} - Q_\omega^{-1}Ax^{(k)} + Q_\omega^{-1}b, \tag{7.22}$$

where

$$Q_\omega = \left(\frac{1}{\omega}D - L\right) = \omega^{-1}(D - \omega L),$$

which is lower triangular. What is interesting is that the best performance occurs when we take $\omega \in [1, 2]$, which means that it is not a "true" average.

Let's look at an example, using the arbitrary value of $\omega = 1.05$.

■ **EXAMPLE 7.15**

The iteration, in general, is

$$x^{(k+1)} = x^{(k)} - Q_\omega^{-1}Ax^{(k)} + Q_\omega^{-1}b,$$

where

$$Q_\omega = \begin{bmatrix} 3.8095 & 0 & 0 & 0 \\ 1.0000 & 4.7619 & 0 & 0 \\ 0 & 1.0000 & 5.7143 & 0 \\ 1.0000 & 0 & 1.0000 & 3.8095 \end{bmatrix}.$$

For the individual iterates we get

$$x^{(1)} = \begin{bmatrix} 0.2625 \\ 1.4149 \\ 2.5524 \\ 2.9361 \end{bmatrix}; \quad x^{(2)} = \begin{bmatrix} -0.1220 \\ 0.8889 \\ 2.0030 \\ 3.0344 \end{bmatrix}; \quad x^{(3)} = \begin{bmatrix} 0.0353 \\ 0.9975 \\ 1.9943 \\ 2.9905 \end{bmatrix}.$$

Again, it is apparent that the iteration is converging, and it is also apparent that it is converging faster than for Jacobi or Gauss–Seidel. If we run the iteration to completion, we get convergence (in the sense used in the two preceding examples) in nine iterations, slightly faster than for Gauss–Seidel.

As might be imagined, the convergence theory for SOR depends heavily on the choice of the parameter ω—for example, taking $\omega = 1.75$ results in much *slower* convergence than for Gauss–Seidel. There exists a substantial theory for choosing the "best" value of ω, and this theory applies in a surprisingly wide range of important cases. Still, it and most of the theory for SOR are beyond the intended level of this book, and so the reader is referred to the appropriate references at the end of the chapter. The most general theorem that can be succinctly stated is the following.

Theorem 7.16 (SOR Convergence Theorem) *Let $A \in \mathbb{R}^{n \times n}$, SPD be given. Then, for any $b \in \mathbb{R}^n$, any initial guess $x^{(0)} \in \mathbb{R}^n$, and any value $\omega \in (0, 2)$ the SOR iteration will converge to the exact solution of the linear system $Ax = b$. If $\omega < 0$ or $\omega > 2$, then the iteration will not converge.*

It is worth noting that the methods outlined in this section have been, to a great extent, supplanted by more powerful iterative techniques, which we will outline in §9.3.3. But the ideas contained in these iterations are still an important part of numerical methods and analysis.

Since the important applications of these methods are to the solution of the types of linear systems that occur when partial differential equations are solved approximately, we defer any detailed treatment of them to the appropriate sections of Chapter 9. In the exercises we look at some very simple examples.

Exercises:

1. Let

$$A = \begin{bmatrix} 4 & -1 & 0 & 0 \\ -1 & 4 & -1 & 0 \\ 0 & -1 & 4 & -1 \\ -1 & 0 & -1 & 4 \end{bmatrix}$$

and $b = (-4, 2, 4, 10)^T$.

(a) Verify that the solution to $Ax = b$ is $x = (0, 1, 2, 3)^T$.

(b) Do three iterations (by hand) of the Jacobi iteration for this matrix, using $x^{(0)} = (0, 0, 0, 0)^T$.

(c) Do three iterations (by hand) of the Gauss–Seidel iteration for this problem, using the same initial guess.

2. Do three iterations of SOR for Problem 1, using $\omega = 1.05$.

3. Solve the same system using SOR for a wide set of values of $\omega \in (0, 2)$. For each ω, compute $r(\omega) = \|b - Ax^{(k)}\|_\infty$. Graph $r(\omega)$ for $k = 1, 3, 5, 10$.

4. Write a computer code that does Jacobi for Problem 1, for a specified number of iterations. How many iterations does it take to get convergence, in the sense that the consecutive iterates differ by less than 10^{-6}?

5. Repeat Problem 4 for Gauss–Seidel.

6. Repeat Problem 4 for SOR. Make sure that your code can accept different values of ω as an input parameter.

7. Let A be the 16×16 matrix given at the beginning of this section. Take

$$b = (5, 11, 18, 21, 29, 40, 48, 48, 57, 72, 80, 76, 69, 87, 94, 85)^T.$$

Write a computer code to do Jacobi, Gauss–Seidel, and SOR on this system of equations. Write the code to only store the nonzero diagonals of A, and make the code as efficient as possible.

8. Prove that the spectral radius of A, $\rho(A)$, is bounded above by $\|A\|$ for any norm such that $\|Ax\| \leq \|A\|\|x\|$.

◁ • • ▷

7.8 NONLINEAR SYSTEMS: NEWTON'S METHOD AND RELATED IDEAS

In Chapter 3 we spent considerable time and effort developing algorithms for the solution of a *single* nonlinear equation. In this section we extend some of these ideas to tackle *systems* of nonlinear equations. Because we are talking about systems, we will use a lot of linear algebra to give structure and simplicity to the discussion.

Let f be a given function from a domain in $\mathbb{R}^k$ to a range also in $\mathbb{R}^k$, i.e., $f : \mathbb{R}^k \to \mathbb{R}^k$. We want to find the vector(s) $x \in \mathbb{R}^k$ such that $f(x) = 0$, where 0 here is understood to be the zero vector in $\mathbb{R}^k$.

For example, we might have

$$2x_1 - x_2 + \frac{1}{9}e^{-x_1} = 1, \tag{7.23}$$

$$-x_1 + 2x_2 + \frac{1}{9}e^{-x_2} = 0. \tag{7.24}$$

This implicitly defines $f = (f_1(x_1, x_2), f_2(x_1, x_2))^T$ via

$$f_1(x_1, x_2) = 2x_1 - x_2 + \frac{1}{9}e^{-x_1} - 1, \tag{7.25}$$

$$f_2(x_1, x_2) = -x_1 + 2x_2 + \frac{1}{9}e^{-x_2}. \tag{7.26}$$

How can we approximate the solution to this system?

This is a very active area of research, and many of the best ideas are relatively new. We will discuss how to derive and apply Newton's method for this kind of problem, and then show how certain variants might actually be superior to Newton.

One issue that is new here, compared to the work we did in Chapter 3, is the *cost* of certain operations. This is entirely due to the "curse of dimensionality:" the simple fact that we are now working in $\mathbb{R}^k$ makes the size of the problem a *lot* bigger, and drives the cost up.

Because we are again doing iterations with a vector variable, we once again will use a parenthesized superscript for the iteration counter, to avoid confusion with the subscript denoting the element of the vector. Thus, $x^{(k)}$ will refer to the vector at iteration k, and $x_j^{(k)}$ will refer to the jth component of that vector.

7.8.1 Newton's Method

To derive Newton's method for systems, take $x^{(0)} = (x_1^{(0)}, x_2^{(0)})$ as an initial approximation to the solution and expand both component functions in a Taylor's series about that point:

$$f_1(x_1, x_2) = f_1(x_1^{(0)}, x_2^{(0)}) + (x_1 - x_1^{(0)})\frac{\partial f_1}{\partial x_1}(x_1^{(0)}, x_2^{(0)})$$
$$+ (x_2 - x_2^{(0)})\frac{\partial f_1}{\partial x_2}(x_1^{(0)}, x_2^{(0)}) + R_1, \tag{7.27}$$

$$f_2(x_1, x_2) = f_2(x_1^{(0)}, x_2^{(0)}) + (x_1 - x_1^{(0)})\frac{\partial f_2}{\partial x_1}(x_1^{(0)}, x_2^{(0)})$$
$$+ (x_2 - x_2^{(0)})\frac{\partial f_2}{\partial x_2}(x_1^{(0)}, x_2^{(0)}) + R_2. \tag{7.28}$$

Now, set $f_1(x_1, x_2) = f_2(x_1, x_2) = 0$, and drop the remainders; this yields

$$
\begin{aligned}
0 &= f_1(x_1^{(0)}, x_2^{(0)}) + (x_1 - x_1^{(0)}) \frac{\partial f_1}{\partial x_1}(x_1^{(0)}, x_2^{(0)}) \\
&\quad + (x_2 - x_2^{(0)}) \frac{\partial f_1}{\partial x_2}(x_1^{(0)}, x_2^{(0)}),
\end{aligned} \tag{7.29}
$$

$$
\begin{aligned}
0 &= f_2(x_1^{(0)}, x_2^{(0)}) + (x_1 - x_1^{(0)}) \frac{\partial f_2}{\partial x_1}(x_1^{(0)}, x_2^{(0)}) \\
&\quad + (x_2 - x_2^{(0)}) \frac{\partial f_2}{\partial x_2}(x_1^{(0)}, x_2^{(0)}).
\end{aligned} \tag{7.30}
$$

We want to solve this for the new values of x_1 and x_2, so we re-organize this as a matrix–vector equation, thus:

$$
0 = f(x^{(0)}) + A_f(x^{(0)})(x - x^{(0)}),
$$

where

$$
x^{(0)} = (x_1^{(0)}, x_2^{(0)})^T, x = (x_1, x_2)^T, A_f(x^{(0)}) = \begin{bmatrix} \frac{\partial f_1}{\partial x_1}(x_1^{(0)}, x_2^{(0)}) & \frac{\partial f_1}{\partial x_2}(x_1^{(0)}, x_2^{(0)}) \\ \frac{\partial f_2}{\partial x_1}(x_1^{(0)}, x_2^{(0)}) & \frac{\partial f_2}{\partial x_2}(x_1^{(0)}, x_2^{(0)}) \end{bmatrix}.
$$

We can solve for x by multiplying by A_f^{-1} (assuming it exists):

$$
x = x^{(0)} - A_f^{-1}(x^{(0)}) f(x^{(0)}).
$$

Generalizing the notation, then, we have

$$
x^{(n+1)} = x^{(n)} - A_f^{-1}(x^{(n)}) f(x^{(n)}),
$$

where now $x^{(n)} = (x_1^{(n)}, x_2^{(n)})^T$ is the nth approximate (vector) value. This, of course, is almost exactly the classical Newton's method from §3.2, especially when we notice that the matrix A_f is the Jacobian matrix of first partial derivatives; essentially, this plays the role that the ordinary derivative did in the single-variable case. For this reason, we will introduce the notation of the derivative for this matrix.

Definition 7.7 (Gradient of a Function) *If f is a differentiable function from $\mathbb{R}^k$ to $\mathbb{R}^k$, then we will denote by f' the $k \times k$ matrix of first partial derivatives, also known as the gradient of f:*

$$
f'(x) = \left\{ \frac{\partial f_i}{\partial x_j}(x) \right\}.
$$

An alternate notation for the gradient is ∇f. Sometimes $\operatorname{grad} f$ is also used.

We will not go into detail on the theory of Newton's method for systems; suffice it to say that the basic ideas of Chapter 3 carry over: If the initial guess $x^{(0)}$ is close enough to the actual solution, then convergence will occur and the convergence will be quadratic. However, because the implementation of Newton's method for systems is more problematic, we will spend some time on those issues.

For example, while in Chapter 3 we had to worry about the possibility of dividing by a zero derivative in Newton's method, here we have to worry about a matrix being singular; moreover, we have to factor that matrix every step of the iteration. For the simple 2×2 example we gave above, this is not going to be important, but for a larger system it is an issue. The scalar version of Newton's method cost two function evaluations in each step; the k-dimensional version of Newton's method costs $k + k^2$ function evaluations, in addition to the roughly $\frac{1}{3}k^3$ cost of factoring and solving the linear system. Even for quadratic convergence, this can be a steep price to pay.

Table 7.1 shows the result of applying Newton's method to our example system, with the initial guess $x^{(0)} = (1, 1)$.[16] Note the rapid convergence, which we expect with Newton's method when the initial guess is close to the actual solution. (The fact that this example is *nearly* linear helps.)

Table 7.1   Two dimensional Newton's method example

n	$x_1^{(n)}$	$x_2^{(n)}$	Estimated error
0	1.000000000000	1.000000000000	N/A
1	0.605042638147	0.267104846820	0.732895E+00
2	0.597220155135	0.255587267973	0.115176E–01
3	0.597216751762	0.255582530497	0.473748E–05
4	0.597216751761	0.255582530496	0.821260E–12

For comparison's sake, consider the chord method, which we introduced in §3.11.2. Here we compute according to

$$x^{(n+1)} = x^{(n)} - \left[f'(x^{(0)}) \right]^{-1} f(x^{(n)}),   \tag{7.31}$$

which means that we do not need to re-evaluate nor re-factor the gradient matrix at each step. Thus, each step of the chord iteration is substantially cheaper than Newton's method. As indicated in §3.11.2, we could periodically update the point on which the gradient is based, thus doing something like

$$x^{(n+1)} = x^{(n)} - \left[f'(x^*) \right]^{-1} f(x^{(n)}),   \tag{7.32}$$

with x^* being updated every, say, p iterations. This reduces the cost compared to Newton's method, but does speed up the convergence some compared to the chord method. Tables 7.2 and 7.3 give the results of applying two chord iterations to our example problem. Table 7.2 uses the pure chord method (7.31); Table 7.3 uses (7.32), with $p = 3$. Compare the total cost to convergence with that for Newton's method, as given in Table 7.1; even though Newton's method took fewer iterations than either version of the chord method, the overall cost was cheaper for the chord methods. For Newton's method, the cost per iteration will be $k + k^2$ function evaluations plus $\frac{1}{3}k^3 + k^2$ operations in solving the linear system. This works out to a total of about 24 function evaluations plus roughly 46 operations in solving the linear system. For (7.31), the pure chord method, the cost per iteration is only k function evaluations plus k^2 operations in solving the linear system; there is also an overhead cost of k^2 function evaluations and about $\frac{1}{3}k^3$ operations in factoring the linear system *once*. This works out to a total of about 22 function evaluations plus about 39 operations with the linear system. Finally, for the updated chord method (7.32) the cost per iteration is the same as for the pure chord method, with an additional k^2 function evaluations plus $\frac{1}{3}k^3$

---

[16]In this and all the other nonlinear system examples, the "Estimated error" refers to the difference between consecutive iterates:

$$E_n = \max_{i=1,2} |x_i^{(n)} - x_i^{(n-1)}|.$$

operations for factoring the new gradient each time it is formed. This works out to about 18 function evaluations and 26 operations with the linear system. These differences are not gigantic, but they do show how the chord method can be cost-effective compared to Newton's method even though it is slower in terms of iterations to convergence.

Table 7.2 Illustration of two-dimensional chord method (7.31).

n	$x_1^{(n)}$	$x_2^{(n)}$	Estimated error
0	1.000000000000	1.000000000000	N/A
1	0.605042638147	0.267104846820	0.732895E+00
2	0.597506380968	0.255993159449	0.111117E−01
3	0.597227336747	0.255597398559	0.395761E−03
4	0.597217136546	0.255583069746	0.143288E−04
5	0.597216765729	0.255582550060	0.519685E−06
6	0.597216752268	0.255582531206	0.188547E−07
7	0.597216751779	0.255582530522	0.684125E−09
8	0.597216751762	0.255582530497	0.248232E−10
9	0.597216751761	0.255582530496	0.900727E−12

Table 7.3 Illustration of modified two-dimensional chord method (7.32), using $p = 3$.

n	$x_1^{(n)}$	$x_2^{(n)}$	Estimated error
0	1.000000000000	1.000000000000	N/A
1	0.605042638147	0.267104846820	0.732895E+00
2	0.597506380968	0.255993159449	0.111117E−01
3	0.597216756247	0.255582536629	0.410623E−03
4	0.597216751761	0.255582530496	0.613292E−08
5	0.597216751761	0.255582530496	0.184247E−12

7.8.2 Fixed-Point Methods

The problem with many of the Newton-like methods is that they require that a linear system be solved at each iteration. Sometimes, this requirement is more of a computational burden than the need to evaluate the functions. Thus, a method that took more iterations (and therefore more function evaluations) but was significantly cheaper in cost for each iteration might well be more efficient. This brings us back to the notion of fixed-point iterations, which we first saw in §3.9.

Given the vector function f, mapping from $\mathbb{R}^k$ to $\mathbb{R}^k$, let us assume that we have associated with f a second vector function, say g, such that

$$\alpha = g(\alpha) \Leftrightarrow f(\alpha) = 0.$$

In other words, α is a root of f if and only if α is a fixed point of g. We can then think about employing the fixed-point iteration

$$x^{(n+1)} = g(x^{(n)}) \tag{7.33}$$

as a means of approximating α. We have a substantial theory for fixed-point methods in one dimension, developed in Chapter 3, and most of this carries over to the multidimensional case.

Theorem 7.17 (Multidimensional Fixed-Point Theorem) *Let g be a vector function defined on a domain $D \subset \mathbb{R}^k$, continuous, with the property that $g(x) \in D$ for all $x \in D$. Assume that there exists $\gamma \in [0, 1)$ such that*

$$\|g(x) - g(y)\| \leq \gamma \|x - y\|$$

for all $x, y \in D$. Then there exists a unique fixed point $\alpha \in D$ and the iteration $x^{(n+1)} = g(x^{(n)})$ will converge to α for any initial value $x^{(0)} \in D$; furthermore, we have the error estimate

$$\|\alpha - x^{(n)}\| \leq \frac{\gamma^n}{1 - \gamma} \|x^{(1)} - x^{(0)}\|.$$

Proof: This is essentially Theorem 3.5 from §3.9, and the proof follows almost exactly as was done there. •

What this tells us is that if we can construct a function g from f such that the hypotheses of the theorem hold, then the iteration (7.33) can be used to compute approximate values of the solution to $f(x) = 0$. Although there is no "general theory" for deriving g from f—it is something that is usually best done on a case-by-case basis—there are some general guidelines for broad classes of problems. We might note, for example, that one way to construct g is to use

$$g(x) = x - Af(x),$$

where A is a $k \times k$ matrix. To mimic the Newton scheme, we want $A \approx [f'(x)]^{-1}$. One such choice, very simplistic, would be to invert only the diagonal part of f'—thus doing a kind of Jacobi iteration for a nonlinear problem; this requires no matrix factorizations, and only k additional function evaluations. If we apply this idea to our example system, we find that it takes 43 iterations to reach convergence. This is slow, in terms of the number of iterations, compared to all our other methods. More sophisticated choices, however, work very well, as we shall see in the next section. The student might want to consider, for example, how to apply the Gauss–Seidel ideas to this setting.

Exercises:

1. Consider the nonlinear system

$$2x_1 - x_2 + \frac{1}{9}e^{-x_1} = -1, \qquad (7.34)$$

$$-x_1 + 2x_2 + \frac{1}{9}e^{-x_2} = 1. \qquad (7.35)$$

Take $x^{(0)} = (1, 1)^T$ and do two iterations of Newton's method; you should get

$$x^{(2)} = (-0.48309783661427, 0.21361449746996)^T.$$

2. Write a computer code to solve the system in Problem 1, using

   (a) Newton's method;

   (b) The chord method;

    (c) The chord method, updating every three iterations.

3. Rewrite the system in Problem 1 as

$$Kx + \phi(x) = b,$$

where K is the 2×2 matrix

$$K = \begin{bmatrix} 2 & -1 \\ -1 & 2 \end{bmatrix},$$

$\phi(x)$ is defined by

$$\phi(x) = \begin{pmatrix} \frac{1}{9}e^{-x_1} \\ \frac{1}{9}e^{-x_2} \end{pmatrix},$$

and $b = (-1, 1)^T$.

    (a) Do two iterations (by hand) of the fixed-point iteration

$$x^{(k+1)} = \frac{1}{2}(b - \phi(x^{(k)} - Kx^{(k)} + 2x^{(k)})$$

       for this system, using $x^{(0)} = (1,1)^T$.

    (b) Do two iterations (by hand) of the fixed-point iteration

$$x^{(k+1)} = K^{-1}(b - \phi(x^{(k)}))$$

       for this system, using $x^{(0)} = (1,1)^T$.

    (c) Which one do you think is going to converge faster?

4. Write a computer code to implement the fixed-point iterations outlined in Problem 3. Compare the total "cost to convergence" with your results for Newton's method and the chord iterations.

7.9 APPLICATION: NUMERICAL SOLUTION OF NONLINEAR BOUNDARY VALUE PROBLEMS

When we studied the approximate solution of boundary value problems in §6.10, we were unable to apply the difference techniques of §6.10.1 to a nonlinear differential equation because we did not know how to solve the resulting system of (nonlinear) equations. Now we do, based on the material in §7.8.

    Consider the problem

$$-u'' = (u')^2, \tag{7.36}$$

$$u(0) = 0, u(1) = 1, \tag{7.37}$$

which we first looked at in Chapter 6. The exact solution is $u = \log((e - 1)x + 1)$. In Chapter 6 we had to use shooting methods to solve this problem, because the finite difference techniques would lead to a system of nonlinear equations, thus:

$$2u_1 - u_2 - \frac{1}{4}u_2^2 = 0, \tag{7.38}$$

$$-u_{j-1} + 2u_j - u_{j+1} - \frac{1}{4}(u_{j+1} - u_{j-1})^2 = 0, \quad 2 \le j \le N - 2, \tag{7.39}$$

$$-u_{N-2} + 2u_{N-1} - 1 - \frac{1}{4}(1 - u_{N-2})^2 = 0. \tag{7.40}$$

The student should verify (Problem 1) that this is the correct system, assuming a grid using $h = 1/N$.

We can solve this nonlinear system using any one of the methods discussed in §7.8. Rather than go through all of those in detail, we elect to illustrate the choice of a fixed-point iteration, and leave the others to the exercises. (See Problems 2 – 4.)

Recall that a common fixed-point iteration to solve the nonlinear equation $f(u) = 0$ would be

$$u^{(n+1)} = u^{(n)} - Af(u^{(n)}),$$

where, ideally, the matrix A would be chosen to approximate the gradient of f. We can write our nonlinear system in the form

$$Ku + \phi(u) = b,$$

where K is the tridiagonal matrix with components given by

$$k_{ij} = \begin{cases} 2, & i = j; \\ -1, & |i - j| = 1; \\ 0, & \text{otherwise}; \end{cases}$$

ϕ is the nonlinear function given by

$$\phi(u) = -\begin{pmatrix} \frac{1}{4}u_2^2 \\ \frac{1}{4}(u_3 - u_1)^2 \\ \vdots \\ \frac{1}{4}(u_N - u_{N-2})^2 \\ \frac{1}{4}(1 - u_{N-1})^2 \end{pmatrix};$$

and b is the constant vector

$$b = (0, 0, \ldots, 0, 1)^T.$$

Since $f(u) = Ku + \phi(u) - b$, it is not hard to show (Problem 5) that the gradient is given by $f'(u) = K + \phi'(u)$, where

$$\phi'(u) = -\begin{pmatrix} 0 & \frac{1}{2}u_2 & 0 & \cdots & & 0 \\ \frac{1}{2}(u_1 - u_3) & 0 & \frac{1}{2}(u_3 - u_1) & \cdots & & 0 \\ \vdots & \ddots & \ddots & \ddots & & \vdots \\ 0 & \cdots & \frac{1}{2}(u_{N-2} - u_N) & 0 & (u_N - u_{N-2}) \\ 0 & \cdots & & 0 & \frac{1}{2}(u_{N-1} - 1) & 0 \end{pmatrix} \tag{7.41}$$

is the gradient of ϕ. Thus, a very efficient choice of A would be $A = K$; not only is K an important part of the gradient of f, but it is also independent of the iteration. Thus, we can factor it once and then need only do the forward and backward solution steps each iteration. Table 7.4 shows the results that we get when we do this computation, using a sequence of meshes $h = \frac{1}{4}, \frac{1}{8}, \dots, \frac{1}{1024}$; for each mesh we printed out the number of iterations needed for convergence, and the error between the computed solution and the exact solution, as measured in the vector infinity norm. For the first case ($h = \frac{1}{4}$), we took as our initial guess the straight line connecting the boundary values. For subsequent cases, we took as our initial guess a vector based on the final solution from the previous case; thus, by starting off very close to the solution, we took fewer iterations to find it. For comparison's sake, if we had started out to do the $h = 1/1024$ case with the straight-line initial guess, it would have taken 15 iterations to converge.

Table 7.4 Number of iterations and maximum error for approximate solutions to the nonlinear BVP (7.36–7.37).

n	Iterations	Error
4	13	0.154547E–02
8	12	0.383951E–03
16	12	0.958279E–04
32	10	0.239854E–04
64	9	0.600085E–05
128	8	0.149813E–05
256	7	0.374157E–06
512	6	0.947443E–07
1024	5	0.242657E–07

Exercises:

1. Set up the nonlinear system for the example (7.36–7.37), and verify that the result given in (7.38)–(7.40) is correct.

2. Apply Newton's method and the chord method to the approximate solution of the nonlinear BVP (7.36–7.37). Compare the number of iterations to converge and the overall cost of convergence. Use the sequence of grids $h^{-1} = 4, 8, \dots, 1024$.

3. Consider the nonlinear BVP

$$-u'' + e^{-u} = 1,$$
$$u(0) = u(1) = 1.$$

   Use finite difference techniques to reduce this (approximately) to a system of nonlinear algebraic equations, and solve this system using several of the methods discussed in this chapter. Test the program on the sequence of grids $h^{-1} = 4, 8, \dots, 1024$ (and further, if practical on your system). Compare the cost of convergence for each method in terms of the number of iterations and in terms of the number of operations.

4. Now consider the nonlinear BVP

$$-u'' = \frac{u}{u+1},$$
$$u(0) = 0, \quad u(1) = 1.$$

Repeat the kind of study required in Problem 3.

5. Verify that (7.41) gives the correct gradient for (7.38)–(7.40).

◁ ● ● ● ▷

7.10 LITERATURE AND SOFTWARE DISCUSSION

The subject of numerical linear algebra has seen, in the last 25 years, an explosion of references and monographs. The standard references are the books by Stewart [9], Golub and van Loan [4], Watkins [11], Datta [1], and the more recent works by Trefethen and Bau [10] and Demmel [2]. Another classic, although dated, work is that of Householder [5]. For nonlinear systems the classic reference is still Ortega and Rheinboldt [7]; recent books by Kelley [6] and Saad [8] offer some more modern insight. The standard references on rounding-error issues are those of Forsythe and Moler [3] and, of course, Wilkinson [12].

In terms of software, the most important development has undoubtedly been the widespread use of MATLAB, which has made it very easy for beginning students to use very sophisticated computational linear algebra techniques.

REFERENCES

1. Datta, Biswa Nath, *Numerical Linear Algebra and Applications*, Brooks/Cole, Pacific Grove, CA, 1995.

2. Demmel, James, *Applied Numerical Linear Algebra*, SIAM, Philadelphia, 1997.

3. Forsythe, George, and Moler, Cleve, *Computer Solution of Linear Algebraic Systems*, Prentice-Hall, Englewood Cliffs, NJ, 1967.

4. Golub, G., and van Loan, C., *Matrix Computations*, Johns Hopkins Press, Baltimore, 2012 (Fourth edition).

5. Householder, Alston, *The Theory of Matrices in Numerical Analysis*, Dover, 1975 (originally published by Blaisdell Publishing Co., 1964).

6. Kelley, C. T., *Iterative Methods for Linear and Nonlinear Equations*, SIAM, Philadelphia, 1995.

7. Ortega, J. M., and Rheinboldt, W. C., *Iterative Solution of Nonlinear Equations in Several Variables*, Academic Press, New York, 1970.

8. Saad, Yousef, *Iterative Methods for Sparse Linear Systems*, PWS Publishing, Boston, 1996.

9. Stewart, G. W., *Introduction to Matrix Computations*, Academic Press, Orlando, FL, 1973.

10. Trefethen, L. N., and Bau, David, *Numerical Linear Algebra*, SIAM, Philadelphia, 1997.

11. Watkins, David, *Fundamentals of Matrix Computations*, John Wiley & Sons, Inc., New York, Third Edition, 2010.

12. Wilkinson, J.H., *Rounding Errors in Algebraic Processes*, Prentice-Hall, Englewood Cliffs, NJ, 1963.

CHAPTER 8

APPROXIMATE SOLUTION OF THE ALGEBRAIC EIGENVALUE PROBLEM

In this chapter we will discuss, in some detail, some iterative methods for finding single eigenvalue–eigenvector pairs (*eigenpairs* is a common term) of a given real matrix A; we will also give an overview of more powerful and general methods that are commonly used to find *all* the eigenpairs of a given real A. As in Chapter 7, our discussion here will depend a fair amount on MATLAB, although we will look at some algorithms in detail.

8.1 EIGENVALUE REVIEW

The algebraic eigenvalue problem is as follows: Given a matrix $A \in \mathbb{R}^{n \times n}$, find a nonzero vector $x \in \mathbb{R}^n$ and the scalar λ such that

$$Ax = \lambda x.$$

Note that this says that the vector Ax is parallel to x, with λ being an amplification factor, or *gain*. Note also that the above implies that

$$(A - \lambda I)x = 0,$$

showing (by Theorem 7.1) that $A - \lambda I$ is a singular matrix. Hence, $\det(A - \lambda I) = 0$; it is easy to show that this determinant is a polynomial (of degree n) in λ, known as the *characteristic polynomial* of A, $p(\lambda)$, so that the eigenvalues are the roots of a polynomial. Although this is *not* a good way to compute the eigenvalues, it does give us some insight

An Introduction to Numerical Methods and Analysis, Second Edition. By James F. Epperson
Copyright © 2013 John Wiley & Sons, Inc.

into their properties. Thus, we know that an $n \times n$ matrix has n eigenvalues, that the eigenvalues can be repeated, and that a real matrix can have complex eigenvalues, but these must occur in conjugate pairs. We summarize these and a number of other basic eigenvalue properties in the following theorem, presented without proof.

Theorem 8.1 (Basic Eigenvalue Properties) *Let $A \in \mathbb{R}^{n \times n}$ be given. Then we have the following:*

1. *There are exactly n eigenvalues, counting multiplicities; complex eigenvalues will occur in conjugate pairs.*

2. *Eigenvectors corresponding to distinct eigenvalues are independent.*

3. *If an $n \times n$ matrix A has n independent eigenvectors, then there exists a nonsingular matrix P such that $P^{-1}AP = D$ is diagonal and A is called* diagonalizable. *Moreover, the columns of P are the eigenvectors of A and the elements $d_{ii} = \lambda_i$ are the eigenvalues of A.*

4. *If A is symmetric $(A = A^T)$, then the eigenvalues are real and we can choose the eigenvectors to be real and orthogonal.*

5. *If A is symmetric, then there is an orthogonal matrix[1] Q such that $Q^T AQ = D$ is diagonal, where the elements $d_{ii} = \lambda_i$ are the eigenvalues of A.*

6. *If A is triangular, then the eigenvalues are the diagonal elements, $\lambda_i = a_{ii}$.*

■ **EXAMPLE 8.1**

Let

$$A = \begin{bmatrix} 4 & 1 & 0 \\ 1 & 4 & 1 \\ 0 & 1 & 4 \end{bmatrix}.$$

The characteristic polynomial for this matrix is

$$p(\lambda) = \lambda^3 - 12\lambda^2 + 46\lambda - 56,$$

and the eigenvalues are

$$\lambda_1 = 5.4142..., \lambda_2 = 4.000, \lambda_3 = 2.5858... \, .$$

Since the matrix is symmetric, we can find a set of orthonormal eigenvectors:

$$x_1 = \begin{bmatrix} 0.5000 \\ 0.7071 \\ 0.5000 \end{bmatrix}, x_2 = \begin{bmatrix} -0.7071 \\ 0.0000 \\ 0.7071 \end{bmatrix}, x_3 = \begin{bmatrix} -0.5000 \\ 0.7071 \\ -0.5000 \end{bmatrix}.$$

Since these vectors are orthonormal, when we arrange them as the columns of a matrix, we get an orthogonal matrix:

$$Q = [x_1 | x_2 | x_3] = \begin{bmatrix} 0.5000 & -0.7071 & -0.5000 \\ 0.7071 & 0.0000 & 0.7071 \\ 0.5000 & 0.7071 & -0.5000 \end{bmatrix}.$$

---

[1]Recall that an orthogonal matrix is one such that $Q^T = Q^{-1}$.

Moreover, we have $Q^{-1}AQ = Q^T AQ = D$, diagonal:

$$(Q^T A)Q = \begin{bmatrix} 2.7071 & 3.8284 & 2.7071 \\ -2.8284 & 0.0000 & 2.8284 \\ -1.2929 & 1.8284 & -1.2929 \end{bmatrix} \times \begin{bmatrix} 0.5000 & -0.7071 & -0.5000 \\ 0.7071 & 0.0000 & 0.7071 \\ 0.5000 & 0.7071 & -0.5000 \end{bmatrix}$$

$$= \begin{bmatrix} 5.4142 & 0.0000 & 0.0000 \\ 0.0000 & 4.0000 & 0.0000 \\ 0.0000 & 0.0000 & 2.5858 \end{bmatrix}.$$

■ **EXAMPLE 8.2**

This time, let

$$A = \begin{bmatrix} 5 & 4 & 3 \\ -1 & 0 & -3 \\ 1 & -2 & 1 \end{bmatrix}.$$

The characteristic polynomial is now

$$p(\lambda) = \lambda^3 - 6\lambda^2 + 32,$$

and the eigenvalues are

$$\lambda_1 = 4, \lambda_2 = 4, \lambda_3 = -2.$$

Since $\lambda_1 = \lambda_2$ we are not guaranteed that the corresponding eigenvectors are independent, and in this case they are not. An $n \times n$ matrix with $k < n$ independent eigenvectors is called *defective*. Note that having repeated eigenvalues is a *necessary* condition for a matrix being defective, but it is not *sufficient*; there exist many nondefective matrices with repeated eigenvalues, the most obvious example being the identity matrix. In Problem 4 we explore a little bit more about defective matrices.

A certain amount of new terminology and notation will be useful. The collection of all eigenvalues of a matrix is typically called the *spectrum* of the matrix, which we denote by $\sigma(A)$:

$$\sigma(A) = \{\lambda \in \mathbb{C}, Ax = \lambda x, x \neq 0\}.$$

The vector space spanned by all the eigenvectors corresponding to a single eigenvalue λ is called the *eigenspace* and is denoted by $E_\lambda(A)$:

$$E_\lambda(A) = \{x \in \mathbb{C}^n, Ax = \lambda x\}.$$

Note that the eigenspaces, in order to be vector spaces, must include the zero vector, which is not an eigenvector. If there exists a nonsingular matrix P such that $A = PBP^{-1}$, then A and B are said to be *similar*, which we will denote $A \sim B$. Similar matrices have the same eigenvalues and related eigenvectors (Problem 2). Finally, we note that a commonplace term used to refer to the combination of an eigenvalue λ and a corresponding eigenvector is *eigenpair*.

An important tool in eigenvalue approximation is the ability to *localize* the eigenvalues, and the most important tool in eigenvalue localization is Gerschgorin's[2] Theorem.

---

[2]Semyon Aranovich Gerschgorin (1901–1933) was born in the Belarus region of Tsarist Russia. He studied mechanics at the Petrograd Technological Institute. He became a professor at the Institute of Mechanical

Theorem 8.2 (Gerschgorin's Theorem) *Let $A \in \mathbb{R}^{n \times n}$ be given, and define the quantities*

$$r_i = \sum_{\substack{j=1 \\ j \neq i}}^{n} |a_{ij}|,$$

$$D_i = \{z \in \mathbf{C} \mid |z - a_{ii}| \leq r_i\}.$$

Then every eigenvalue of A lies in the union of the disks D_i, that is,

$$\lambda_k \in \bigcup_{i=1}^{n} D_i$$

for all $k = 1, 2, \ldots, n$. Moreover, if any collection of p disks is disjoint from the other $n - p$ disks, then we know that exactly p eigenvalues are contained in the union of the set of p disks, and exactly $n - p$ eigenvalues are contained in the set of $n - p$ disks.

To illustrate the theorem, consider the following examples.

■ **EXAMPLE 8.3**

If we have

$$A = \begin{bmatrix} 2 & 1 & 0 \\ 1 & 2 & 1 \\ 0 & 1 & 2 \end{bmatrix},$$

then the disks are defined by the centers $a_{ii} = 2$, $1 \leq i \leq 3$, so the disks are concentric, and the radii are

$$r_1 = 1, \quad r_2 = 2, \quad r_3 = 1.$$

Figure 8.1 shows the circles in the complex plane along with the exact eigenvalues (shown as small circles) $\lambda_1 = 3.1414$, $\lambda_2 = 2.000$, and $\lambda_3 = 0.5859$. Since the matrix is symmetric, we expect all the eigenvalues to be real, thus they lie on the x-axis.

■ **EXAMPLE 8.4**

On the other hand, if

$$A = \begin{bmatrix} 2 & -1 & -2 & 1 & 2 \\ 2 & 2 & 1 & 0 & 0 \\ 0 & -2 & -1 & 2 & -1 \\ 2 & -2 & -2 & 0 & 1 \\ -2 & -1 & -2 & 0 & 2 \end{bmatrix},$$

Engineering in Leningrad in 1930, then at the Leningrad Mechanical Engineering Institute, as well as doing some teaching at Leningrad State University. His result bounding the eigenvalues of a matrix, "Über die Abgrenzung der Eigenwerte einer Matrix," appeared in 1931. Like many results in mathematics, it was first discovered by someone else: The French mathematician Lucien Lévy obtained an equivalent result (but only for real matrices) in 1881.

then we can no longer assume that the eigenvalues are real. The disks are defined by the centers

$$a_{11} = 2, \quad a_{22} = 2, \quad a_{33} = -1, \quad a_{44} = 0, \quad a_{55} = 2,$$

and the radii

$$r_1 = 6, \quad r_2 = 3, \quad r_3 = 5, \quad r_4 = 7, \quad r_5 = 5.$$

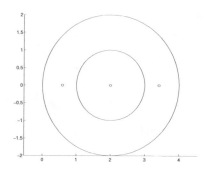

Figure 8.1 Illustration of Gerschgorin's Theorem (Example 8.3).

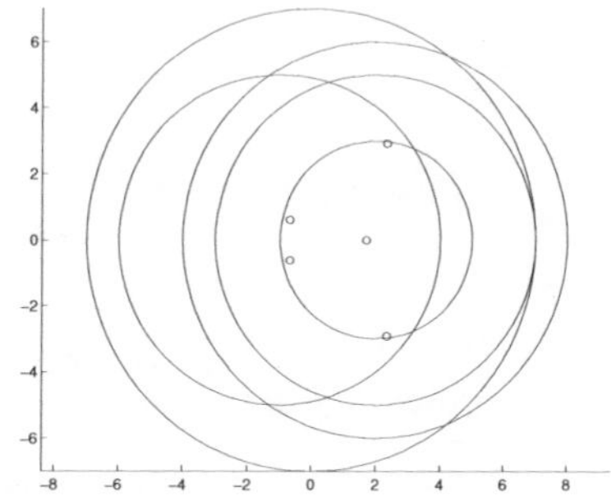

Figure 8.2 Illustration of Gerschgorin's Theorem (Example 8.4).

This situation is pictured in Figure 8.2. Note that all the eigenvalues are actually clustered within the smallest disk—they are all within all the disks.

Exercises:

1. For each matrix below, find the characteristic polynomial and the eigenvalues by a hand calculation. For some of the exercises, the correct eigenvalues are given, to four decimal places, so you can check your work. Feel free to use a root-finding method from Chapter 3 to do the computations.

   (a)
   $$A = \begin{bmatrix} 4 & 1 & 0 \\ 1 & 4 & 1 \\ 0 & 1 & 4 \end{bmatrix}$$
   for which $\sigma(A) = \{5.4142, 4.0000, 2.5858\}$;

   (b)
   $$A = \begin{bmatrix} 1 & 2 & 0 \\ 0 & 4 & 5 \\ 0 & 5 & 7 \end{bmatrix},$$
   for which $\sigma(A) = \{1.0000, 0.2798, 10.7202\}$;

   (c)
   $$A = \begin{bmatrix} 6 & -2 & 0 \\ 2 & 6 & -2 \\ 0 & 2 & 4 \end{bmatrix},$$

for which $\sigma(A) = \{5.5698 \pm 2.6143i, 4.8603\}$;

(d)

$$A = \begin{bmatrix} 4 & 2 & 0 \\ 1 & 4 & 2 \\ 0 & 1 & 4 \end{bmatrix},$$

for which $\sigma(A) = \{2, 4, 6\}$;

(e)

$$A = \begin{bmatrix} 6 & -1 & 0 \\ 2 & 6 & -1 \\ 0 & 2 & 4 \end{bmatrix};$$

(f)

$$A = \begin{bmatrix} 0 & 2 & 0 \\ 2 & 7 & 1 \\ 0 & 1 & 4 \end{bmatrix}.$$

2. Prove that similar matrices have identical eigenvalues and related eigenvectors.

3. Apply Gerschgorin's Theorem to the matrices in Problem 1 and determine the intervals or disks in which the eigenvalues must lie.

4. Show that the matrix

$$A = \begin{bmatrix} 3 & 2 & -1 \\ 0 & 2 & 2 \\ 0 & -2 & 7 \end{bmatrix},$$

has repeated eigenvalues and is defective, but that the matrix

$$A = \begin{bmatrix} -1 & -1 & 2 \\ -2 & 0 & 2 \\ -4 & -2 & 5 \end{bmatrix},$$

which also has repeated eigenvalues, is *not* defective.

5. Let A and B be similar matrices of size $n \times n$, with

$$B = \begin{bmatrix} \lambda & a^T \\ 0 & A_{22} \end{bmatrix}$$

where $\lambda \in \mathbb{R}$, $a \in \mathbb{R}^{n-1}$, and $A_{22} \in \mathbb{R}^{n-1 \times n-1}$.

(a) Prove that $\lambda \in \sigma(A)$. *Hint:* What is the product Be_1, where e_1 is the first standard basis vector?

(b) Prove that each eigenvalue of A_{22} is also an eigenvalue of A.

6. Generalize the above; let A and B be similar matrices of size $n \times n$, with

$$B = \begin{bmatrix} D & a^T \\ 0 & A_{22} \end{bmatrix},$$

where $D \in \mathbb{R}^{p \times p}$ is diagonal, $a \in \mathbb{R}^{n-p}$, and $A_{22} \in \mathbb{R}^{n-p \times n-p}$. Prove that each diagonal element of D is an eigenvalue of A: $d_{ii} \in \sigma(A), 1 \leq i \leq p$, and that each eigenvalue of A_{22} is also an eigenvalue of A.

7. Let A be given by

$$A = \begin{bmatrix} 2 & 0 \\ a & A_{22} \end{bmatrix}.$$

Prove that $2 \in \sigma(A)$.

8. Let $A_1 = BC$ and $A_2 = CB$ for given B, C. Show that any nonzero eigenvalue of A_1 is also an eigenvalue of A_2, and vice-versa.

◁ • • ▷

8.2 REDUCTION TO HESSENBERG FORM

Since the computation of eigenvalues is equivalent to finding the roots of a polynomial, it follows that for $n \geq 5$ there will be no general algorithm that works in a finite number of steps.[3] Accordingly, we expect an eigenvalue solver to be an inherently iterative process, that is, one that recursively computes better and better approximations. For this reason, it is usually more efficient to *pre-process* the matrix by making as much of a reduction as possible in a finite number of steps. This leads us to the notion of the *Hessenberg form* of a matrix.

Definition 8.1 (Hessenberg Form) *A matrix $A \in \mathbb{R}^{n \times n}$ is in Hessenberg form if $a_{ij} = 0$ for all i, j such that $i - j > 1$.*

Thus, the matrix

$$A = \begin{bmatrix} 1 & 2 & 3 & 4 \\ 5 & 6 & 7 & 8 \\ 0 & 9 & 8 & 7 \\ 0 & 0 & 6 & 5 \end{bmatrix}$$

is in Hessenberg form. Note that one way to characterize Hessenberg form is that it is "almost" triangular. This is important, since the eigenvalues of a triangular matrix are the diagonal elements. Note, also, that a symmetric Hessenberg matrix is tridiagonal.

The important result, then, is the following.

Theorem 8.3 *Let $A \in \mathbb{R}^{n \times n}$ be given. Then there exists $A_H \in \mathbb{R}^{n \times n}$, Hessenberg, which can be computed from A in a finite number of steps and that has the same eigenvalues as A.*

Proof: The proof is constructive, but somewhat involved, and depends heavily on a preliminary result.

Claim: For any vector $x \in \mathbb{R}^n$, there exists an orthogonal matrix Q such that $Qx = \|x\|_2 e_1$, where e_1 is the first standard basis vector.

We will prove this claim after the main proof of the theorem. What we will do to prove the theorem is use the claim to construct a matrix P such that $B = PAP^{-1}$ is in Hessenberg form. This means that A and B have the same eigenvalues (by virtue of similarity) and we will be done.

---

[3]The reason is because there can be no algorithm for finding the roots of a polynomial of degree $n \geq 5$ in a finite number of steps. This follows from the work of Niels Henrik Abel(1802–1829) and Evariste Galois (1811–1832).

We start by writing A in a partitioned form:

$$
A = \left[\begin{array}{c|c} a_{11} & b^T \\ \hline a & A_{22} \end{array} \right],
$$

where $a \in \mathbf{R}^{n-1}$, $b \in \mathbf{R}^{n-1}$, and $A_{22} \in \mathbf{R}^{n-1 \times n-1}$. Now, let Q_1 be an orthogonal matrix such that $Q_1 a = \|a\|_2 e_1$ and use this to construct the matrix

$$
P_1 = \left[\begin{array}{c|ccc} 1 & \cdots & 0 & \cdots \\ \hline \vdots & & & \\ 0 & & Q_1 & \\ \vdots & & & \end{array} \right].
$$

(Note that Q_1 is in $\mathbf{R}^{(n-1) \times (n-1)}$.) Then P_1 is also orthogonal (thus, $P_1^{-1} = P_1^T$) and

$$
A_1 = P_1 A P_1^T = \left[\begin{array}{c|c} a_{11} & b^T Q_1^T \\ \hline \|a\|_2 e_1 & Q_1 A_{22} Q_1^T \end{array} \right].
$$

Note that the first column of A_1 satisfies the criterion for being in Hessenberg form.

We can now continue the process in what is probably an obvious way. We write A_1 in the partitioned form

$$
A_1 = \left[\begin{array}{cc|c} a_{11} & a_{12}^H & b_1^T \\ \|a\|_2 & a_{22}^H & b_2^T \\ \hline 0 & a_1 & A_{22}' \end{array} \right]
$$

and define Q_2 to be the matrix such that $Q_2 a_1 = \|a_1\|_2 e_1$; define P_2 as

$$
P_2 = \left[\begin{array}{cc|ccc} 1 & 0 & \cdots & 0 & \cdots \\ 0 & 1 & \cdots & 0 & \cdots \\ \hline \vdots & \vdots & & & \\ 0 & 0 & & Q_2 & \\ \vdots & \vdots & & & \end{array} \right],
$$

so that $A_3 = P_2 A_1 P_2^T = P_2 P_1 A P_1^T P_2^T$ is in Hessenberg form. The process stops after $n - 2$ such steps.

It remains only to prove the claim, which we state as a separate lemma. ●

Lemma 8.1 (Claim) *For any vector $x \in \mathbb{R}^n$, there exists an orthogonal matrix Q such that $Qx = \|x\|_2 e_1$, where e_1 is the first standard basis vector.*

Proof: The proof is constructive, using what are known as *Householder transformations.*[4] Define Q as

$$
Q = I - \gamma w w^T,
$$

---

[4] Alston Householder (1904–1993) was born in Rockford, Illinois but raised in Alabama. He earned his Ph.D. in mathematics from the University of Chicago, in 1937, and spent the first part of his career working in mathematical

where we regard w as a vector to be determined, and $\gamma = 2/\|w\|_2^2$. It is easily verified that Q is orthogonal since we have

$$
\begin{aligned}
Q^T Q &= (I - \gamma w w^T)(I - \gamma w w^T), \\
&= I - 2\gamma w w^T + \gamma^2 \|w\|_2^2 w w^T, \\
&= I - 4\frac{w w^T}{\|w\|_2^2} + 4\frac{\|w\|_2^2 w w^T}{\|w\|_2^2}, \\
&= I.
\end{aligned}
$$

Then[5]

$$
Qx = x - \gamma(w^T x)w.
$$

Since we want Qx to be a constant times e_1, we are going to want most of these components to vanish, so let's look for w in the form $w = x + ce_1$, where c is to be determined later. This gives us

$$
\begin{aligned}
Qx &= x - \gamma(x + ce_1)^T x(x + ce_1), \\
&= x - \gamma(\|x\|_2^2 + cx_1)(x + ce_1), \\
&= (1 - \gamma(\|x\|_2^2 + cx_1))x - \gamma c(\|x\|_2^2 + cx_1)e_1.
\end{aligned}
$$

We want the coefficient of x to vanish, and the coefficient of e_1 to be $\|x\|_2$. Since $w = x + ce_1$, we know that

$$
\gamma = \frac{2}{\|x\|_2^2 + 2cx_1 + c^2},
$$

so we have (after a bit of manipulation) that $c = -\|x\|_2$. This completes the proof. •

Note: If we choose $c = \|x\|_2$, then we get that $Qx = -\|x\|_2 e_1$. (See Problem 1.) This is just as good a result for our purposes, and can sometimes be more stable, numerically. To minimize subtractive cancellation errors, it is best to use $c = \|x\|_2$ when $x_1 > 0$, and $c = -\|x\|_2$ when $x_1 < 0$.

The special form of the Householder matrices allows us to accomplish the reduction to Hessenberg form very easily, with an algorithm that is very closely akin to Gaussian elimination. If we are not interested in computing the eigenvectors of the original matrix, then very little additional storage is needed. On the other hand, if we do want to compute the eigenvectors, we will need to save the w vectors that are computed along the way so that we can reverse the transformation.

The operation count for the Hessenberg reduction is $\frac{5}{3}n^3 + \mathcal{O}(n^2)$ for the general case, but only $\frac{2}{3}n^3 + \mathcal{O}(n^2)$ if A is symmetric. In either case, this is inexpensive enough compared to the cost of operating with the full matrix that it is a worthwhile preliminary step for eigenvalue computations. For the sake of simplicity, we will consider only the symmetric case.

Assume that we have performed the necessary reductions on the first $i - 1$ columns of A. Then the partially reduced matrix can be partitioned as follows:

$$
P_{i-1}AP_{i-1} = \begin{bmatrix} A_{H,11} & B^T \\ B & A_{22} \end{bmatrix},
$$

---

biology. In 1946, he joined the Division of Mathematics at Oak Ridge National Laboratory, where he stayed until his retirement in 1969. It was at Oak Ridge that Householder took up numerical analysis as his area of study, in which he made a number of important contributions. What we know as "Householder transformations" were first known as "elementary hermitian matrices" and are used in a variety of places in numerical linear algebra. Householder's early text [4] remains a classic in the field of numerical linear algebra.

[5] Note how easy it is to operate with these matrices. Matrix–vector multiplication is accomplished with a single dot product and a subtraction.

where $A_{H,11} \in \mathbf{R}^{i \times i}$ is in Hessenberg (therefore, tridiagonal) form, all columns of B are zero except for the last one, and A_{22} is the remaining lower quadrant of the matrix. Let a be the nonzero column in B, and let w be the vector such that

$$Q = I - \gamma w w^T$$

is the Householder matrix for which $Qa = \pm \|a\|_2 e_1$. Then

$$P_i = \begin{bmatrix} I & 0 \\ 0 & Q \end{bmatrix},$$

so that the next stage of the reduction is

$$P_i P_{i-1} A P_{i-1} P_i = \begin{bmatrix} A_{H,11} & U^T \\ U & Q A_{22} Q \end{bmatrix},$$

where all columns of U are zero except for the last one, which is $\pm \|a\|_2 e_1$. Thus, we see that the work in each step of the reduction breaks down into three relatively simple steps:

1. Identify the vector a on which the Householder transformation is to operate.

2. Compute the vector w that defines the Householder transformation.

3. Update the lower quadrant of the matrix accordingly.

The first two steps are straightforward. The third step is equally so, using the equation

$$Q A_{22} Q = (I - \gamma w w^T) A_{22} (I - \gamma w w^T) = A_{22} - \gamma (w p^T + p w^T) + \gamma^2 \alpha w w^T,$$

where

$$p = A_{22} w, \quad \alpha = w^T p = w^T A_{22} w.$$

The assumed symmetry of A and A_{22} is used in simplifying some of these expressions. If A is not symmetric, then there will be some changes.

■ **EXAMPLE 8.5**

Consider the matrix

$$A = \begin{bmatrix} 6 & 2 & 1 & 1 \\ 2 & 6 & 2 & 1 \\ 1 & 2 & 6 & 2 \\ 1 & 1 & 2 & 6 \end{bmatrix}.$$

Since this is just 4×4, we only have to do two steps to accomplish the reduction. In the first step, the vector a is defined by

$$a = \begin{bmatrix} 2 \\ 1 \\ 1 \end{bmatrix},$$

so that

$$w = \begin{bmatrix} 4.4495 \\ 1 \\ 1 \end{bmatrix}$$

and $\gamma = 0.0918$. Thus,

$$Q_1 = I - \gamma w w^T = \begin{bmatrix} -0.8165 & -0.4082 & -0.4082 \\ -0.4082 & 0.9082 & -0.0918 \\ -0.4082 & -0.0918 & 0.9082 \end{bmatrix}$$

and

$$P_1 = \begin{bmatrix} 1 & 0 & 0 & 0 \\ 0 & -0.8165 & -0.4082 & -0.4082 \\ 0 & -0.4082 & 0.9082 & -0.0918 \\ 0 & -0.4082 & -0.0918 & 0.9082 \end{bmatrix}.$$

Therefore,

$$A_2 = P_1 A P_1 = \begin{bmatrix} 6.0000 & -2.4495 & -0.0000 & -0.0000 \\ -2.4495 & 8.6667 & -1.5749 & -0.7584 \\ -0.0000 & -1.5749 & 4.2584 & 0.6667 \\ -0.0000 & -0.7584 & 0.6667 & 5.0749 \end{bmatrix}.$$

Now, for the second (and final) step of the reduction, we have that the vector a is given by

$$a = \begin{bmatrix} -1.5749 \\ -0.7584 \end{bmatrix},$$

so that

$$w = \begin{bmatrix} -3.3229 \\ -0.7584 \end{bmatrix}$$

and $\gamma = 0.1722$. Thus,

$$Q_2 = I - \gamma w w^T = \begin{bmatrix} -0.9010 & -0.4339 \\ -0.4339 & 0.9010 \end{bmatrix}$$

and

$$P_2 = \begin{bmatrix} 1 & 0 & 0 & 0 \\ 0 & 1 & 0 & 0 \\ 0 & 0 & -0.9010 & -0.4339 \\ 0 & 0 & -0.4339 & 0.9010 \end{bmatrix}.$$

Therefore,

$$A_H = P_2 A_2 P_2 = \begin{bmatrix} 6.0000 & -2.4495 & 0.0000 & -0.0000 \\ -2.4495 & 8.6667 & 1.7480 & 0 \\ 0.0000 & 1.7480 & 4.9333 & -0.7348 \\ -0.0000 & -0.0000 & -0.7348 & 4.4000 \end{bmatrix}.$$

We leave it as an exercise (Problem 3) to finish the calculation to find P such that $A = P A_H P^T$.

Early editions of this book gave a lengthy pseudocode for performing a reduction to Hessenberg form; starting with the present edition, we rely entirely upon MATLAB's hess command. Many of the examples were done with the author's own code, which normalizes things differently than hess does, so the results are slightly different.[6] For example, hess applied to Example 8.5 produces

$$H = \begin{bmatrix} 4.4000 & -0.7348 & 0 & 0 \\ -0.7348 & 4.9333 & 1.7480 & 0 \\ 0 & 1.7480 & 8.6667 & -2.4495 \\ 0 & 0 & -2.4495 & 6.0000 \end{bmatrix}$$

and

$$P = \begin{bmatrix} 0.8581 & -0.3114 & -0.4082 & 0 \\ -0.4767 & -0.7785 & -0.4082 & 0 \\ -0.1907 & 0.5449 & -0.8165 & 0 \\ 0 & 0 & 0 & 1.0000 \end{bmatrix}$$

The reader should check that $A = PHP^T$.

Exercises:

1. Show that in the Householder construction, if we use $c = -\|x\|_2$, then we get $Qx = -\|x\|_2 e_1$, and that this will work just as well to construct the Hessenberg matrix.

2. Compute (by hand) the Hessenberg form of the matrix

$$A = \begin{bmatrix} 6 & 1 & 1 & 1 \\ 1 & 6 & 1 & 1 \\ 1 & 1 & 6 & 1 \\ 1 & 1 & 1 & 6 \end{bmatrix}.$$

You should get

$$A_H = \begin{bmatrix} 6.0000 & 1.7321 & 0 & 0 \\ 1.7321 & 8.0000 & -0.0000 & 0 \\ 0 & -0.0000 & 5.0000 & 0 \\ 0 & 0 & 0 & 5.0000 \end{bmatrix}.$$

3. Complete the computation in Example 8.5 by finding the matrix P such that $A = PA_H P^T$.

4. Use the hess command to compute the Hessenberg form for each of H_4, H_8, and H_{10}. Verify that the original matrix can be recovered from A_H.

◁ • • ▷

8.3 POWER METHODS

The simplest methods for approximating eigenvalues are based on the observation that the eigenvectors represent the directions along which the matrix operates, and the eigenvalues

---

[6]Note that this implies that the Hessenberg reduction is *not* unique: For a given matrix, A, we can have two pairs of matrices (P, H_1) and (Q, H_2) such that both $H_1 = P^{-1}AP$ and $H_2 = Q^{-1}AQ$ are in Hessenberg form.

represent the gain along those directions. Thus, the quantity $A^N x$ should eventually begin to "line up" in the direction of the eigenvector associated with the largest (in absolute value) eigenvalue. Although this is not an efficient approach for finding *all* the eigenvalues and eigenvectors of a matrix, it is useful for finding some of them, and much of the theory of more general methods is based on the essential ideas of the power methods.

We start with a formal theorem, from which we develop the first algorithm.

Theorem 8.4 *Let $A \in \mathbb{R}^{n \times n}$ be given, and assume that*

1. A has n linearly independent eigenvectors, x_k, $1 \le k \le n$.

2. The eigenvalues λ_k satisfy

$$|\lambda_1| > |\lambda_2| \ge |\lambda_3| \ge \ldots \ge |\lambda_n|.$$

3. The vector $z \in \mathbb{R}^n$ is such that

$$z = \sum_{k=1}^{n} \xi_k x_k$$

and $\xi_1 \neq 0$.

Then

$$\lim_{N \to \infty} \frac{A^N z}{\lambda_1^N} = c x_1 \tag{8.1}$$

for some $c \neq 0$, and

$$\lim_{N \to \infty} \frac{(z, A^N z)}{(z, A^{N-1} z)} = \lambda_1. \tag{8.2}$$

Proof: The key issue is that we can expand the vector z in terms of the eigenvectors, since the independence of the eigenvectors means they are a basis for $\mathbb{R}^n$. Thus, we have

$$z = \xi_1 x_1 + \xi_2 x_2 + \ldots + \xi_n x_n = \sum_{k=1}^{n} \xi_k x_k,$$

where $\xi_1 \neq 0$. Then

$$A^N z = \sum_{k=1}^{n} \xi_k \lambda_k^N x_k,$$

so that

$$\frac{A^N z}{\lambda_1^N} = \xi_1 x_1 + \sum_{k=2}^{n} \xi_k \theta_k^N x_k,$$

where

$$|\theta_k| = \left| \frac{\lambda_k}{\lambda_1} \right| < 1$$

for $k \ge 2$. Thus, as $N \to \infty$, the θ_k^N vanish and we are left with (8.1).

The proof of (8.2) is very similar. We can write

$$\frac{(z, A^N z)}{(z, A^{N-1} z)} = \frac{\lambda_1^N \eta_1 + \lambda_2^N \eta_2 + \cdots + \lambda_n^N \eta_n}{\lambda_1^{N-1} \eta_1 + \lambda_2^{N-1} \eta_2 + \cdots + \lambda_n^{N-1} \eta_n},$$

where $\eta_k = (z, x_k)$. Now factor out the λ_1 terms in both numerator and denominator to get

$$\frac{(z, A^N z)}{(z, A^{N-1} z)} = \lambda_1 \frac{\eta_1 + \theta_2^N \eta_2 + \cdots + \theta_n^N \eta_n}{\eta_1 + \theta_2^{N-1} \eta_2 + \cdots + \theta_n^{N-1} \eta_n}.$$

As $N \to \infty$ the fraction goes to 1 since each of the θ_k terms goes to zero, and we are done.

●

We can use the results of this theorem to construct an algorithm to find the *dominant* eigenvalue (i.e., the one that is largest in absolute value) and the corresponding eigenvector. We can't use the precise computations from this theorem, since in (8.1) we scaled the matrix powers by the exact eigenvalue, which, of course, we do not know. We also do not want to explicitly form the powers A^k. However, we can avoid forming A^k by using recursion, and it turns out (Problem 3) that the particular form of the scaling is not all that important.

Consider the following algorithm:

Algorithm 8.1 *Basic Power Method.*
  For k from 1 until convergence, do

  1. Compute $y^{(k)} = A z^{(k-1)}$;

  2. $\mu_k = y_i^{(k)}$, where $\|y^{(k)}\|_\infty = |y_i^{(k)}|$;

  3. Set $z^{(k)} = y^{(k)} / \mu_k$.

Concerning this, we can prove the following.

Theorem 8.5 (Basic Power Method) *Let $A \in \mathbb{R}^{n \times n}$ be given, and assume the following three conditions:*

  1. A has n linearly independent eigenvectors, x_k, $1 \le k \le n$.

  2. The eigenvalues λ_k satisfy

$$|\lambda_1| > |\lambda_2| \ge |\lambda_3| \ge \cdots \ge |\lambda_n|.$$

  3. The vector $z^{(0)} \in \mathbb{R}^n$ is such that

$$z^{(0)} = \sum_{k=1}^{n} \xi_k x_k$$

  and $\xi_1 \ne 0$.

Then the basic power method converges in the sense that

$$\lim_{k \to \infty} \mu_k = \lambda_1, \quad |\lambda_1 - \mu_k| = \mathcal{O}\left(\left(\frac{\lambda_2}{\lambda_1} \right)^k \right),$$

and there exists $c \ne 0$ such that

$$\lim_{k \to \infty} z^{(k)} = c x_1, \quad \|c x_1 - z^{(k)}\|_\infty = \mathcal{O}\left(\left(\frac{\lambda_2}{\lambda_1} \right)^k \right).$$

Proof: An inductive argument (see Problem 2) can be used to show that $z^{(k)} = c_k A^k z^{(0)}$, where c_k depends on all the scaling factors μ_k. We then expand in terms of the eigenvector basis, as we did before, to get

$$z^{(k)} = c_k \lambda_1^k \left(\xi_1 x_1 + \sum_{i=2}^{n} \xi_i \theta_i^k x_i \right),$$

where $\theta_i = \lambda_i / \lambda_1$ and $|\theta_i| < 1$ for all $i \geq 2$. Since each $z^{(k)}$ has infinity norm equal to 1, it follows that

$$\lim_{k \to \infty} c_k \lambda_1^k = \frac{1}{\xi_1 \|x_1\|_\infty} < \infty,$$

so that

$$\lim_{k \to \infty} z^{(k)} = \frac{1}{\|x_1\|_\infty} x_1.$$

Moreover, we clearly have that

$$\left\| z^{(k)} - \alpha_k x_1 \right\|_\infty = \left\| c_k \lambda_1^k \sum_{i=2}^{n} \xi_i \theta_i^k x_i \right\|_\infty \leq C \theta_2^k,$$

which establishes the error estimate for the eigenvector.

As for the eigenvalue, we have that

$$\mu_k z^{(k)} = y^{(k)} = A z^{(k-1)},$$

so that, taking dot products of both sides with $z^{(k)}$ we get

$$\mu_k = \frac{(z^{(k)}, A z^{(k-1)})}{(z^{(k)}, z^{(k)})} \to \frac{(x_1, A x_1)}{(x_1, x_1)} = \lambda_1.$$

The error estimate comes by manipulating with the ratio as was done in the proof of Theorem 8.4. See Problem 4. ●

The power method requires several hypotheses for the theory to apply, and it is worth asking if all of them are necessary. What happens if some of them are violated?

Recall that if A does not have n linearly independent eigenvectors, then it is said to be *defective*; eigenvalue computations with defective matrices are unavoidably more difficult. However, just as all singular matrices are close to a nonsingular matrix, it is the case that all defective matrices are arbitrarily close to a nondefective matrix. Thus, applying the power method to a defective matrix usually results in, at worst, slow convergence (possibly *very* slow, if the dependent eigenvectors correspond to the two largest eigenvalues, λ_1 and λ_2).

If the two largest (in absolute value) eigenvalues are not separated, i.e., if $|\lambda_1| = |\lambda_2|$, then the iteration will cycle in to a family of vectors in the subspace spanned by x_1 and x_2.

The last hypothesis is actually the least important. Except in very simple situations (involving integer components), the rounding error associated with the computations will introduce some component in the direction of x_1 into the initial guess, and the iteration will then begin to converge. Making an initial guess that has absolutely no component in the direction of x_1 could result in convergence to the wrong eigenpair, however. For this reason, random noninteger initial guesses are best.

Let's look at some examples.

■ **EXAMPLE 8.6**

Consider

$$A = \begin{bmatrix} 3 & 0 & 0 \\ -4 & 6 & 2 \\ 16 & -15 & -5 \end{bmatrix},$$

which has exact eigenvalues $\lambda_i = 3, 1, 0, i = 1, 2, 3$, and corresponding eigenvectors

$$x_1 = \begin{bmatrix} 1 \\ 0 \\ 2 \end{bmatrix}, \quad x_2 = \begin{bmatrix} 0 \\ 2 \\ -5 \end{bmatrix}, \quad x_3 = \begin{bmatrix} 0 \\ 1 \\ -3 \end{bmatrix}.$$

Note that the eigenvalues are all distinct; therefore, the eigenvectors are independent. We take the initial guess as (from a random number generator)

$$z^{(0)} = \begin{bmatrix} 0.21895918632809 \\ 0.04704461621449 \\ 0.67886471686832 \end{bmatrix}.$$

Then, the first two iterations go as follows:

$$y^{(1)} = Az^{(0)} = \begin{bmatrix} 0.65687755898427 \\ 0.76416038571122 \\ -0.59664584630951 \end{bmatrix}, \quad \mu_1 = 0.76416038571122,$$

$$z^{(1)} = y^{(1)}/\mu_1 = \begin{bmatrix} 0.85960692449779 \\ 1.00000000000000 \\ -0.78078615100441 \end{bmatrix};$$

$$y^{(2)} = Az^{(1)} = \begin{bmatrix} 2.57882077349338 \\ 1.00000000000000 \\ 2.65764154698677 \end{bmatrix}, \quad \mu_2 = 2.65764154698677,$$

$$z^{(2)} = y^{(2)}/\mu_2 = \begin{bmatrix} 0.97034183425423 \\ 0.37627346740338 \\ 1.00000000000000 \end{bmatrix}.$$

If we continue the computation (letting a computer do the work), then we get convergence (using the difference in consecutive eigenvector approximations as our convergence criterion, with a tolerance of 10^{-6}) to $\lambda_1 = 3$ in 14 iterations. The approximate eigenvector is found to be

$$x_1 = \begin{bmatrix} 0.50000045604132 \\ 0.00000036483306 \\ 1.00000000000000 \end{bmatrix}.$$

It should be noted that we did not bother to put A in Hessenberg form for this example.

■ **EXAMPLE 8.7**

We take as our second example the matrix

$$A = \begin{bmatrix} 4 & 1 & 0 \\ 1 & 4 & 1 \\ 0 & 1 & 4 \end{bmatrix},$$

whose eigenvalues are $\lambda_i = 5.414214, 4.0, 2.585786$, which are much closer together than was the case in Example 8.6, thus we expect slower convergence. With the initial guess

$$z^{(0)} = \begin{bmatrix} 0.67929640583661 \\ 0.93469289594083 \\ 0.38350207748986 \end{bmatrix}$$

(again, randomly generated), the first two iterations are

$$y^{(1)} = Az^{(0)} = \begin{bmatrix} 3.65187851928727 \\ 4.80157006708979 \\ 2.46870120590027 \end{bmatrix}, \quad \mu_1 = 4.80157006708979,$$

$$z^{(1)} = y^{(1)}/\mu_1 = \begin{bmatrix} 0.76055924796713 \\ 1.00000000000000 \\ 0.51414457592130 \end{bmatrix};$$

$$y^{(2)} = Az^{(1)} = \begin{bmatrix} 4.04223699186850 \\ 5.27470382388842 \\ 3.05657830368519 \end{bmatrix}, \quad \mu_2 = 5.27470382388842,$$

$$z^{(2)} = y^{(2)}/\mu_2 = \begin{bmatrix} 0.76634387954860 \\ 1.00000000000000 \\ 0.57947865998511 \end{bmatrix}.$$

If we continue the calculation we converge—using the same criterion as before—to $\lambda_1 = 5.414214$ in 37 iterations.

■ **EXAMPLE 8.8**

We now want to consider two defective matrices,

$$A = \begin{bmatrix} -2 & 4 & -1 \\ -3 & 5 & -1 \\ -1 & 1 & 1 \end{bmatrix},$$

which has eigenvalues $\lambda_i = 2, 1, 1$; and

$$B = \begin{bmatrix} 5 & 4 & 3 \\ -1 & 0 & -3 \\ 1 & -2 & 1 \end{bmatrix},$$

which has eigenvalues $\lambda_i = 4, 4, -2$. When we apply the power method to A, we find convergence occurs in about 23 iterations, using the initial guess

$$z^{(0)} = \begin{bmatrix} 0.51941637206795 \\ 0.83096534611237 \\ 0.03457211052746 \end{bmatrix}.$$

However, if we apply the power method to B, then we find that after 80 iterations the approximate eigenvalue is $\mu_k = 4.04888750871386$, which is not a good approximation to $\lambda_1 = 4$. The reason for the difference in performance is that for this matrix, it is the eigenvalue we are trying to find that is repeated, whereas for A the dominant eigenvalue $\lambda_1 = 2$ was not repeated. Note also that the iteration for the largest eigenvalue of B *was* converging, just very slowly. (The error in the eigenvector was going down by about 0.00001 each iteration.)

■ **EXAMPLE 8.9**

Finally, consider trying to find the largest eigenvalue of

$$A = \begin{bmatrix} 3 & 0 & 0 \\ -4 & 6 & 2 \\ 16 & -15 & -5 \end{bmatrix},$$

which has eigenvalues $\lambda = 3, 1, 0$, using the initial guess

$$z^{(0)} = \begin{bmatrix} 0 \\ -0.375 \\ 1 \end{bmatrix}.$$

The exact eigenvectors are

$$x_1 = \begin{bmatrix} 1 \\ 0 \\ 2 \end{bmatrix}, x_2 = \begin{bmatrix} 0 \\ -2 \\ 5 \end{bmatrix}, x_1 = \begin{bmatrix} 0 \\ -1 \\ 3 \end{bmatrix},$$

and $z^{(0)}$ has been chosen to have no component in the direction of the dominant eigenvector, $x_1 = (1, 0, 2)^T$. If we use the power method, we converge to a multiple of the eigenvector x_2 in six iterations. By making a random choice of the initial vector, say

$$z^{(0)} = \begin{bmatrix} 0.6789 \\ 0.6793 \\ 0.9347 \end{bmatrix},$$

we minimize the chance of having absolutely no component in the x_1 direction. For this choice of initial vector, the power method converges to the correct eigenvector in 15 iterations.

There are many variants to the basic power method. We first observe that if we use the power method on A^{-1}, then we get an algorithm that converges to the *smallest* eigenvalue of A.

Algorithm 8.2 *Inverse Power Method.*
For k from 1 until convergence, do

1. Solve $Ay^{(k)} = z^{(k-1)}$;

2. $\mu_k = y_i^{(k)}$, where $\|y^{(k)}\|_\infty = |y_i^{(k)}|$;

3. Set $z^{(k)} = y^{(k)}/\mu_k$.

The theorem is now the following.

Theorem 8.6 (Inverse Power Method) *Let $A \in \mathbb{R}^{n \times n}$ be given, nonsingular, and assume the following three conditions:*

1. A has n linearly independent eigenvectors, x_k, $1 \leq k \leq n$.

2. The eigenvalues λ_k satisfy

$$|\lambda_1| \geq |\lambda_2| \geq |\lambda_3| \geq \ldots \geq |\lambda_{n-1}| > |\lambda_n|.$$

3. The vector $z^{(0)} \in \mathbb{R}^n$ is such that

$$z^{(0)} = \sum_{k=1}^{n} \xi_k x_k$$

and $\xi_n \neq 0$.

Then the inverse power method converges in the sense that

$$\lim_{k \to \infty} \mu_k = \lambda_n^{-1}, \quad |\lambda_n^{-1} - \mu_k| = \mathcal{O}\left(\left(\frac{\lambda_n}{\lambda_{n-1}}\right)^k\right),$$

and there exists $c \neq 0$ such that

$$\lim_{k \to \infty} z^{(k)} = c x_n, \quad \|c x_n - z^{(k)}\|_\infty = \mathcal{O}\left(\left(\frac{\lambda_n}{\lambda_{n-1}}\right)^k\right).$$

Proof: This is equivalent to applying Theorem 8.5 to A^{-1}, so everything follows, with the largest (in absolute value) eigenvalue of A^{-1} being λ_n^{-1}. •

Of course, we do not implement the inverse power method by constructing A^{-1}; rather, we compute the LU factorization of A and use it to solve the linear system defined in Step 1 of the algorithm.

A single example at this point ought to illustrate.

■ **EXAMPLE 8.10**

Consider

$$A = \begin{bmatrix} 4 & 1 & 0 \\ 1 & 4 & 1 \\ 0 & 1 & 4 \end{bmatrix},$$

one of our previous example matrices. We factor this to obtain

$$A = LU = \begin{bmatrix} 1 & 0 & 0 \\ -0.25 & 1 & 0 \\ 0 & -0.26666... & 1 \end{bmatrix} \begin{bmatrix} 4 & 1 & 0 \\ 0 & 3.75 & 1 \\ 0 & 0 & 3.7333... \end{bmatrix}$$

and then use this factorization to solve the systems in Step 1 of Algorithm 8.2. Using the initial guess

$$z^{(0)} = \begin{bmatrix} 0.05346163504453 \\ 0.52970019333516 \\ 0.67114938407724 \end{bmatrix},$$

the first two iterations now look like this:

$$y^{(1)} = A^{-1}z^{(0)} = \begin{bmatrix} -0.01153083684992 \\ 0.09958498244420 \\ 0.14289110040826 \end{bmatrix}, \quad \mu_1 = 0.14289110040826,$$

$$z^{(1)} = y^{(1)}/\mu_1 = \begin{bmatrix} -0.08069667611890 \\ 0.69692921504333 \\ 1.00000000000000 \end{bmatrix};$$

$$y^{(2)} = A^{-1}z^{(1)} = \begin{bmatrix} -0.05353869646351 \\ 0.13345810973516 \\ 0.21663547256621 \end{bmatrix}, \quad \mu_2 = 0.21663547256621,$$

$$z^{(2)} = y^{(2)}/\mu_2 = \begin{bmatrix} -0.24713725702125 \\ 0.61604920078067 \\ 1.00000000000000 \end{bmatrix}.$$

If we continue, we get convergence to within 10^{-6} of x_3 in 39 iterations. The approximate eigenvalue is $\lambda_3 = 2.58578643767867$, correct to 11 places.

It is worth repeating that the inverse power method is *not* implemented by actually multiplying by the inverse matrix. Rather, the system $Ay^{(k+1)} = z^{(k)}$ is solved at each step, using, typically, an LU decomposition.

It is the inverse power method that opens up the possibility of efficient computation of more eigenvalues, because by introducing *shifts* we can converge to almost any eigenvalue we want. Moreover, we can increase the rate at which the method converges.

Algorithm 8.3 *Shifted Inverse Power Method.*

Let λ_* be given, with λ_* not an eigenvalue of A; define $A_* = A - \lambda_* I$. For k from 1 until convergence, do

1. Solve $A_* y^{(k)} = z^{(k-1)}$;

2. $\mu_k = y_i^{(k)}$, where $\|y^{(k)}\|_\infty = |y_i^{(k)}|$;

3. Set $z^{(k)} = y^{(k)}/\mu_k$.

The theorem now becomes the following.

Theorem 8.7 (Shifted Inverse Power Method) *Let $A \in \mathbb{R}^{n \times n}$ and a scalar λ_* be given, with λ_* not an eigenvalue of A. Define $A_* = A - \lambda_* I$, and assume the following three conditions:*

1. *A has n linearly independent eigenvectors, x_k, $1 \leq k \leq n$.*

2. *There exists an index J such that λ_J is the eigenvalue of A that is strictly closest to λ_*:*

$$|\lambda_J - \lambda_*| < |\lambda_k - \lambda_*|$$

for $k \neq J$.

3. *The vector $z^{(0)} \in \mathbb{R}^n$ is such that*

$$z^{(0)} = \sum_{k=1}^{n} \xi_k x_k$$

and $\xi_J \neq 0$.

Then the shifted inverse power method converges in the sense that

$$\lim_{k \to \infty} \mu_k = (\lambda_J - \lambda_*)^{-1}, \quad |(\lambda_J - \lambda_*)^{-1} - \mu_k| = \mathcal{O}\left(\left(\frac{\lambda_J - \lambda_*}{\lambda_K - \lambda_*}\right)^k\right),$$

where K is the index of the eigenvalue that is second *closest to λ_*; and, there exists $c \neq 0$ such that*

$$\lim_{k \to \infty} z^{(k)} = c x_J, \quad \|c x_J - z^{(k)}\|_\infty = \mathcal{O}\left(\left(\frac{\lambda_J - \lambda_*}{\lambda_K - \lambda_*}\right)^k\right).$$

Proof: This is Problem 5 •

The point of using shifts is that it allows us to control which eigenvector we converge to, as well as how fast we converge. Recall that the inverse power iteration converges to the eigenvector corresponding to the *smallest* eigenvalue. Using a shift λ_* means that, for some index J, $\lambda_J - \lambda_*$ is the smallest eigenvalue; thus, we converge to the eigenvector corresponding to the eigenvalue *closest to* the shift. Furthermore, the convergence rate is affected by how close to an actual eigenvalue the shift is—taking $\lambda_* \approx \lambda_J$ will generally result in much faster convergence than otherwise.

■ **EXAMPLE 8.11**

To illustrate the utility of shifts, we again look at

$$A = \begin{bmatrix} 4 & 1 & 0 \\ 1 & 4 & 1 \\ 0 & 1 & 4 \end{bmatrix}.$$

We now apply the shifted inverse power method, using the same initial guess as before, this time with a shift of $\lambda_* = 2.5$, chosen because we know (from Gerschgorin's Theorem) that the smallest eigenvalue is at least as large as 2. From Example 8.10 we have that the approximate eigenvalue, after two iterations, was $\frac{1}{\mu_2} = 4.61604920078069$, which is not a very good approximation to the correct value of 2.58578643762691. This time the first two iterations (using the same initial vector as for the unshifted iteration) are

$$y^{(1)} = (A - 2.5I)^{-1} z^{(0)} = \begin{bmatrix} 1.45298541901634 \\ -1.64486177807920 \\ 1.54400744143763 \end{bmatrix}, \quad \mu_1 = -1.64486177807920,$$

$$z^{(1)} = y^{(1)}/\mu_1 = \begin{bmatrix} -0.88334803469813 \\ 1.00000000000000 \\ -0.93868522085829 \end{bmatrix}$$

$$y^{(2)} = (A - 2.5I)^{-1} z^{(1)} = \begin{bmatrix} -9.44765403794921 \\ 13.28813302222568 \\ -9.48454549538931 \end{bmatrix}, \quad \mu_2 = 13.28813302222568,$$

$$z^{(2)} = y^{(2)}/\mu_2 = \begin{bmatrix} -0.71098430623377 \\ 1.00000000000000 \\ -0.71376057716502 \end{bmatrix}.$$

The approximate eigenvalue is now

$$\lambda \approx \lambda_* + \frac{1}{\mu_2} = 2.57525511660121,$$

which is substantially better than we achieved in two iterations with the unshifted method.

The performance of the power methods can be improved somewhat when A is symmetric by the use of the *Rayleigh quotient* as the approximate eigenvalue.

Definition 8.2 (Rayleigh Quotient) *Given $A \in \mathbb{R}^{n \times n}$ and $x \in \mathbb{R}^n$, $x \neq 0$, the Rayleigh quotient of A and x is defined by*

$$R(A, x) = \frac{(x, Ax)}{(x, x)}.$$

Note that if x is an eigenvector of A, then the Rayleigh quotient is the corresponding eigenvalue. If we define the approximate eigenvalue in the power methods via the Rayleigh quotient, in other words,

$$\mu_k = R(A, y^{(k)}),$$

then the iteration converges to the correct eigenvalue *faster* than before if A is symmetric. The reason for this is contained in the following theorem.

Theorem 8.8 *Let λ be an eigenvalue for the symmetric matrix A, with x the corresponding eigenvector. If $z \approx x$, $\mu = R(A, z)$, and $\|x\|_2 = \|z\|_2 = 1$, then*

$$|\lambda - \mu| \leq C_A \|x - z\|_2^2.$$

Proof: We start by defining the error $e = x - z$, and we note that

$$x - e = z \Rightarrow \|x - e\|_2^2 = 1 \Rightarrow 1 - 2(x, e) + \|e\|_2^2 = 1,$$

so that $2(x, e) = \|e\|_2^2$. Then we have

$$\mu = R(A, z) = (x-e, A(x-e)) = (x, Ax)-(x, Ae)-(e, Ax)+(e, Ae) = \lambda-2(e, Ax)+(e, Ae),$$

where we used the symmetry of A to write $(x, Ae) = (A^T x, e) = (Ax, e)$. Thus, we have

$$\mu = \lambda - 2\lambda(e, x) + (e, Ae) = \lambda - \lambda(e, e) + (e, Ae) = \lambda + (e, (A - \lambda I)e),$$

so that

$$|\mu - \lambda| = |(e, (A - \lambda I)e)| \leq \|A - \lambda I\|_2 \|e\|_2^2 = C_A \|e\|_2^2. \qquad \bullet$$

Thus, if we write a power iteration for a symmetric matrix that uses the Rayleigh quotient to compute the approximate eigenvalue (and which scales the iteration so that $\|z^{(k)}\|_2 = 1$), then the eigenvalue error will be the square of the eigenvector error.

■ **EXAMPLE 8.12**

In Example 8.7 we considered the matrix

$$A = \begin{bmatrix} 4 & 1 & 0 \\ 1 & 4 & 1 \\ 0 & 1 & 4 \end{bmatrix},$$

which has eigenvalues $\lambda_i = 5.414214, 4.0, 2.585786$. With the initial guess

$$z^{(0)} = \begin{bmatrix} 0.67929640583661 \\ 0.93469289594083 \\ 0.38350207748986 \end{bmatrix},$$

we got that the second approximate eigenvalue was $\mu_2 = 5.27470382388842$. If we use the same initial guess, but this time compute the approximate eigenvalue using the Rayleigh quotient while scaling the eigenvectors so that $\|z^{(k)}\|_2 = 1$, we get

$$y^{(1)} = Az^{(0)} = \begin{bmatrix} 3.65187851928727 \\ 4.80157006708979 \\ 2.46870120590027 \end{bmatrix},$$

so that the approximate eigenvalue is

$$r_1 = \frac{(z^{(0)}, Az^{(0)})}{(z^{(0)}, z^{(0)})} = \frac{(z^{(0)}, y^{(0)})}{(z^{(0)}, z^{(0)})} = 5.34045535705311.$$

The scaled approximate eigenvector is

$$z^{(1)} = \frac{y^{(1)}}{\|y^{(1)}\|_2} = \begin{bmatrix} 0.56026634732951 \\ 0.73665049610143 \\ 0.37874485692028 \end{bmatrix},$$

so for the second step we get

$$y^{(2)} = Az^{(1)} = \begin{bmatrix} 2.97771588541948 \\ 3.88561318865551 \\ 2.25162992378256 \end{bmatrix},$$

$$r_2 = \frac{(z^{(1)}, Az^{(1)})}{(z^{(1)}, z^{(1)})} = (z^{(1)}, y^{(2)}) = 5.38344613891082,$$

and

$$z^{(2)} = \frac{y^{(2)}}{\|y^{(2)}\|_2} = \begin{bmatrix} 0.55261820977325 \\ 0.72110996710610 \\ 0.41786783744055 \end{bmatrix}.$$

Note that the approximate eigenvalue, r_2, is substantially more accurate than the second approximate eigenvalue obtained by the ordinary power iteration.

The Rayleigh quotient can be combined with the shifted inverse power iteration to produce a useful, if somewhat erratic iteration, called the *Rayleigh quotient iteration*. The basic idea is to update the shift at each step using the most recently computed approximate eigenvalue.

Algorithm 8.4 *Rayleigh Quotient Iteration.*
   Given $A \in \mathbb{R}^{n \times n}$ and an initial guess $z^{(0)} \in \mathbb{R}^n$, compute as follows: For $k = 0$ to convergence, do

    1. Define $\mu_k = R(A, z^{(k)})$;

    2. Solve $(A - \mu_k I)y^{(k+1)} = z^{(k)}$;

    3. Set $z^{(k+1)} = y^{(k+1)}/\|y^{(k+1)}\|_2$.

Because we are updating the shift at each iteration, we have to do a new factorization of A, and this can be expensive. However, if A is symmetric, the Hessenberg reduction means we can assume that A is tridiagonal, and thus this is not a serious additional cost. Moreover, the rapid convergence that occurs because the shift keeps getting closer and closer to the correct eigenvalue makes the iteration very fast. The problem with the Rayleigh quotient iteration is that its performance is much akin to Newton's method; sometimes we have to start the iteration very close to the correct eigenvalue in order to converge to it. Note also that Step 2 involves solving a linear system for a matrix that is getting closer and closer to singular. (Why is this?) This requires that some care be taken in doing the factorization; typically, the shift is no longer updated once it is a sufficiently accurate approximation to the eigenvalue.

■ **EXAMPLE 8.13**

Let's continue with the simple 3×3 example matrix that we have so far been using,

$$A = \begin{bmatrix} 4 & 1 & 0 \\ 1 & 4 & 1 \\ 0 & 1 & 4 \end{bmatrix},$$

which we note is symmetric. Using the same initial guess as in Examples 8.10 and 8.11, the first two iterations of Rayleigh quotient iteration are

$$\mu_0 = R(A, z^{(0)}) = 5.04601728854858, \quad y^{(1)} = (A - \mu_0 I)^{-1} z^{(0)} = \begin{bmatrix} 1.29838130282511 \\ 1.41159092492783 \\ 0.70786740234284 \end{bmatrix},$$

$$z^{(1)} = y^{(1)}/\|y^{(1)}\|_2 = \begin{bmatrix} 0.63510013864028 \\ 0.69047635711817 \\ 0.34625166304280 \end{bmatrix};$$

$$\mu_1 = R(A, z^{(1)}) = 5.35520043415499, \quad y^{(2)} = (A - \mu_1 I)^{-1} z^{(1)} = \begin{bmatrix} 8.18704621259896 \\ 11.73018872040136 \\ 8.40018699112711 \end{bmatrix},$$

$$z^{(2)} = y^{(2)}/\|y^{(2)}\|_2 = \begin{bmatrix} 0.49352861955263 \\ 0.70711508106093 \\ 0.50637709645942 \end{bmatrix};$$

and the next approximate eigenvalue—the next Rayleigh quotient—is

$$\mu_2 = R(A, z^{(2)}) = 5.41409682286229,$$

which is very close to the exact largest eigenvalue $\lambda_1 = 5.41421356237310$, in only two iterations.

Deflation   Once we have found a single eigenvalue, how do we keep from "recomputing" it when we try to find other eigenvalues? The answer, in general, is a process called *deflation*, in which we create a new matrix from the old matrix by "deleting" (in some sense) the eigenvalue and eigenvector that we have already found.

Let λ and x be an eigenvalue and corresponding eigenvector for A, and let Q be an orthogonal matrix such that $Qx = e_1$. (Here we have assumed that $\|x\|_2 = 1$.) Then it is not difficult to show (Problem 10) that

$$B = QAQ^T = \begin{bmatrix} \lambda & a^T \\ 0 & A_2 \end{bmatrix},$$

where $A_2 \in \mathbf{R}^{(n-1)\times(n-1)}$. Similarity implies that B and A have the same eigenvalues; thus, A_2 has all the eigenvalues of A *except* for λ.

We can use this to compute eigenvalues in the obvious way. Once we have converged to λ and x, we define the vector w such that

$$Q = I - \gamma w w^T$$

satisfies $Qx = e_1$. This we already know how to do, from §8.2. Then compute B and pull out the lower submatrix to continue the computation. Although problems can develop due to the propagation of errors (since, in practice, neither λ nor x will be exact), this can be made an accurate procedure for eigenvalue computation. Note, however, that the deflation will usually destroy the Hessenberg form of the matrix, so that process will have to be repeated.

We close this discussion of the power methods by showing the details of a worked-out example.

■ **EXAMPLE 8.14**

Consider the matrix

$$A = \begin{bmatrix} 10 & -2 & 3 & 2 & 0 \\ -2 & 10 & -3 & 4 & 5 \\ 3 & -3 & 6 & 3 & 3 \\ 2 & 4 & 3 & 6 & 6 \\ 0 & 5 & 3 & 6 & 13 \end{bmatrix},$$

which has the spectrum

$$\sigma(A) = \{0.4000, 2.8156, 7.0186, 14.1179, 20.6479\}.$$

We will use the power methods to find *all* the eigenvalues of this matrix. We will do it two ways: First, using the ordinary power method with deflation; Second, using Rayleigh quotient iteration with deflation. The point of the exercise is to show that the power methods *can* be used to find all the eigenvalues (and eigenvectors, which we choose not to do here), and also to provide a basis for understanding that the power methods are not the most efficient way to do so.

The first step is to reduce A to Hessenberg form, using tha algorithm discussed earlier, or MATLAB'S hess command. We get

$$A_{H,0} = \begin{bmatrix} 10.0000 & -4.1231 & 0 & 0 & 0 \\ -4.1231 & 9.2941 & 3.0920 & 0 & 0 \\ 0 & 3.0920 & 6.4587 & -5.3189 & 0 \\ 0 & 0 & -5.3189 & 17.9989 & 3.1371 \\ 0 & 0 & 0 & 3.1371 & 1.2483 \end{bmatrix}.$$

We now use the power method to find a single eigenpair. This takes 76 iterations, using the random initial guess

$$z^{(0)} = \begin{bmatrix} 0.7703 \\ 1.0000 \\ 0.8373 \\ 0.2883 \\ 0.0521 \end{bmatrix}.$$

We get the pair

$$(\lambda_1, x_1) = (20.6479, (0.0494, -0.1276, -0.4027, 1.0000, 0.1617)^T),$$

and now deflate to get the smaller matrix

$$A_1 = \begin{bmatrix} 8.2460 & 1.2308 & 4.6223 & 0.7475 \\ 1.2308 & 5.1488 & -2.0659 & 0.5260 \\ 4.6223 & -2.0659 & 9.9202 & 1.8307 \\ 0.7475 & 0.5260 & 1.8307 & 1.0370 \end{bmatrix},$$

which we have to convert to Hessenberg form:

$$A_{H,1} = \begin{bmatrix} 8.2460 & 4.8414 & 0 & 0 \\ 4.8414 & 8.9781 & -3.0113 & 0 \\ 0 & -3.0113 & 5.9943 & 1.7071 \\ 0 & 0 & 1.7071 & 1.1336 \end{bmatrix}.$$

We do another power iteration, this time lasting 43 iterations, using the random initial vector

$$z^{(0)} = \begin{bmatrix} 0.9731 \\ 0.4339 \\ 0.8364 \\ 1.0000 \end{bmatrix}$$

and resulting in the eigenpair

$$(\lambda_2, x_2) = (14.1179, (0.8245, 1.0000, -0.3812, -0.0501)^T).$$

Note that this vector is in $\mathbb{R}^4$, not $\mathbb{R}^5$. This is because it is an eigenvector for the 4×4 matrix $A_{H,1}$, *not* our original A. Some additional work will have to be done to recover the correct eigenvector for the original matrix.

In any event, we deflate again to get

$$A_2 = \begin{bmatrix} 3.2896 & -1.6907 & 0.1736 \\ -1.6907 & 5.8141 & 1.6834 \\ 0.1736 & 1.6834 & 1.1305 \end{bmatrix},$$

which has the Hessenberg form

$$A_{H,2} = \begin{bmatrix} 3.2896 & -1.6996 & 0 \\ -1.6996 & 5.4231 & -2.1243 \\ 0 & -2.1243 & 1.5215 \end{bmatrix}.$$

Using the initial vector

$$z^{(0)} = \begin{bmatrix} 1.0000 \\ 0.3686 \\ 0.2493 \end{bmatrix}$$

we use the power method for 34 iterations, getting the eigenpair

$$(\lambda_3, x_3) = (7.0186, (-0.4558, 1.0000, -0.3864)^T).$$

Deflation yields the 2×2 matrix

$$A_3 = \begin{bmatrix} 1.9061 & -1.1704 \\ -1.1704 & 1.3095 \end{bmatrix},$$

whose eigenvalues are found directly from the quadratic formula.

Now let's look at the Rayleigh quotient iteration. To make a fair comparison, we will use the same initial vector,

$$z^{(0)} = \begin{bmatrix} 0.7703 \\ 1.0000 \\ 0.8373 \\ 0.2883 \\ 0.0521 \end{bmatrix}.$$

The Hessenberg form is, of course, also the same. We now converge in only *three* iterations, to the eigenpair

$$(\lambda_3, x_3) = (7.0186, (-0.6578, -0.4756, -0.5271, -0.2210, -0.1201)^T).$$

We can deflate to get the new matrix

$$A_1 = \begin{bmatrix} 3.5922 & 3.1235 & 0.0132 & 0.0072 \\ 3.1235 & 13.5312 & -2.3536 & 1.6121 \\ 0.0132 & -2.3536 & 19.2422 & 3.8130 \\ 0.0072 & 1.6121 & 3.8130 & 1.6158 \end{bmatrix},$$

which has to be reduced to Hessenberg form:

$$A_{H,1} = \begin{bmatrix} 3.5922 & 3.1235 & 0 & 0 \\ 3.1235 & 13.5188 & 2.8192 & 0 \\ 0 & 2.8192 & 10.0064 & -9.6050 \\ 0 & 0 & -9.6050 & 10.8640 \end{bmatrix}.$$

Then, with the initial vector

$$z^{(0)} = \begin{bmatrix} 0.3835 \\ 0.5194 \\ 0.8310 \\ 0.0346 \end{bmatrix},$$

we again take only three iterations to converge to the eigenpair

$$(\lambda_2, x_2) = (14.1179, (-0.2687, -0.9056, 0.1053, -0.3109)^T).$$

We again deflate and reconstruct the Hessenberg form, getting

$$A_2 = \begin{bmatrix} 5.1137 & 4.2466 & -4.2133 \\ 4.2466 & 9.7881 & -8.9606 \\ -4.2133 & -8.9606 & 8.9617 \end{bmatrix}$$

and

$$A_{H,2} = \begin{bmatrix} 5.1137 & 5.9821 & 0 \\ 5.9821 & 18.3385 & -0.3427 \\ 0 & -0.3427 & 0.4113 \end{bmatrix}.$$

Choose a random initial vector

$$z^{(0)} = \begin{bmatrix} 0.0535 \\ 0.5297 \\ 0.6711 \end{bmatrix}$$

and iterate, in this case five times, to get the eigenpair

$$(\lambda_4, x_4) = (2.8156, (-0.9323, 0.3581, -0.0511)^T).$$

We then deflate one last time to get

$$A_3 = \begin{bmatrix} 19.3308 & -4.9934 \\ -4.9934 & 1.7171 \end{bmatrix},$$

whose eigenvalues we can find from the quadratic formula.

The cost here is substantially less than for the ordinary power method, largely because the Rayleigh quotient iterations converged so much more quickly than the ordinary power iterations did.

The power methods are very useful for finding selected eigenvalues and corresponding eigenvectors, but to find all the eigenvalues and eigenvectors of a general matrix, it is more efficient to make use of methods that work on all of the eigenvalues at once. In the next section, we give a survey of the most commonly used such method, the QR iteration.

Exercises:

1. Explain, in your own words, why it is necessary to scale the power iterations. *Hint:* What happens if we don't scale the iteration?

2. Use an inductive argument to show that in the basic power method (Theorem 8.5) we have
$$z^{(k)} = c_k A^k z^{(0)}.$$

What is c_k?

3. Consider the iteration defined by

$$y^{(k+1)} = Az^{(k)},$$
$$z^{(k+1)} = y^{(k+1)}/\sigma_{k+1},$$

where σ_{k+1} is some scaling parameter. Assume that the iteration converges in the sense that $z^{(k)} \to z$; prove that σ_k must also converge, and that therefore (σ, z), where σ is the limit of the σ_k, must be an eigenpair of A.

4. Fill in the details of the eigenvalue error estimate in Theorem 8.5. In particular, show that there is a constant C such that

$$|\lambda_1 - \mu_k| \le C \left(\frac{\lambda_2}{\lambda_1} \right)^k.$$

5. Prove Theorem 8.7.

6. Write a program that does the basic power method. Use it to produce a plot of the largest eigenvalue of the T_n family of matrices, as a function of n, over the range $2 \le n \le 20$. Test your results against what MATLAB's `eig` command produces.

7. Write a program that does the inverse power method. Use it to produce a plot of the smallest eigenvalue of the H_n family of matrices, as a function of n, over the range $2 \le n \le 20$.

8. Write a program that does inverse power iteration on a symmetric matrix. Assume that the matrix is tridiagonal, and store only the necessary nonzero elements of the matrix. Test it on the T_n and K_n families, and on your results from finding the Hessenberg forms for the H_n matrices. Produce a plot of the smallest eigenvalues of each of these matrices, as a function of n.

9. Add shifts to your inverse power iteration program. Test it on the same examples, using a selection of shifts.

10. Let λ and x be an eigenvalue and corresponding eigenvector for $A \in \mathbb{R}^{n \times n}$, and let Q be an orthogonal matrix such that $Qx = e_1$. Assume that $\|x\|_2 = 1$. Show that

$$B = QAQ^T = \begin{bmatrix} \lambda & a^T \\ 0 & A_2 \end{bmatrix}.$$

Hint: Consider each of the inner products (e_i, Be_j), using the fact that $Qx = e_1$ and that Q is symmetric.

11. An alternate deflation for symmetric matrices is based on the relationship

$$A' = A - \lambda_1 x_1 x_1^T.$$

If $A \in \mathbb{R}^{n \times n}$ is symmetric, $\lambda_1 \in \sigma(A)$ with corresponding eigenvector x_1, and if $\|x_1\|_2 = 1$, show that A' has the same eigenvalues as A, except that λ_1 has been replaced by zero. In other words, if

$$\sigma(A) = \{\lambda_1, \lambda_2, \ldots, \lambda_n\},$$

then

$$\sigma(A') = \{0, \lambda_2, \ldots, \lambda_n\}.$$

12. Show how to implement the deflation from Problem 11 (known as *Hotelling's deflation*) without forming the product $x_1 x_1^T$. *Hint:* What is $A'z$ for any vector z? Can we write this in terms of the dot product $x_1^T z$? (According to Wilkinson [10], this deflation is prone to excessive rounding error and so is not of practical value.)

13. What is the operation count for one iteration of the basic power method applied to a symmetric matrix? Compute two values, one for the matrix in Hessenberg form, one for the full matrix.

14. What is the cost of the basic power method for a symmetric matrix if it runs for N iterations. Again, compute two values depending on whether or not a Hessenberg reduction is done. Is it always the case that a reduction to Hessenberg form is cost-effective?

15. Repeat Problems 13 and 14 for the inverse power method.

16. Consider the family of matrices T defined by

$$t_{ij} = \begin{cases} a, & i - j = 1, \\ b, & i = j, \\ c, & i - j = -1, \\ 0, & \text{otherwise,} \end{cases}$$

Thus, T is tridiagonal, with constant diagonal elements $a, b,$ and c. It is commonplace to abbreviate such a matrix definition by saying that

$$T = \text{tridiag}(a, b, c).$$

Write a computer program that uses the power method and inverse power method to find the largest and smallest (in absolute value) eigenvalues of such a matrix, for any user-specified values of a, b, and c. Test the program on the following examples; note that the exact values are given (to four places).

(a) $a = c = 4, b = 1, n = 6, \lambda_1 = -6.2708, \lambda_6 = -0.7802$;

(b) $a = b = c = 6, n = 7, \lambda_1 = 17.0866, \lambda_6 = 1.4078$;

(c) $a = 2, b = 3, c = 4, n = 5, \lambda_1 = 7.8990, \lambda_6 = 0.1716$;

(d) $a = c = 1, b = 10, n = 9, \lambda_1 = 11.9021, \lambda_6 = 8.0979$;

(e) $a = c = 1, b = 4, n = 8, \lambda_1 = 5.8794, \lambda_6 = 2.1206$.

17. Modify the program from the Problem 16 to do the inverse power method with shifts. Test it on the same examples by finding the eigenvalue closest to $\mu = 1$.

◁ ● ● ▷

8.4   AN OVERVIEW OF THE QR ITERATION

The power methods are very useful and efficient for finding single eigenvalues and eigenvectors; for finding many—or all—of the eigenvalues and vectors of a matrix, they are simply too costly. Since it was first published in the early 1960s, the standard algorithm for computing all the eigenvalues and eigenvectors of a general matrix has been the QR iteration, so-named because it is heavily based on the QR factorization of a matrix.

It is an unfortunate fact—and no less true for being unfortunate—that motivating and explaining the QR iteration for computing eigenvalues is *very* difficult, especially in comparison to the simplicity of the algorithm. To give a complete explanation of why the method works or how and why it might have evolved from earlier methods would require more background and more time than we want to devote to the subject. We will, however, make an attempt to provide some explanation.

One of the primary drawbacks to the power methods is that they work on a single vector at a time; if we want all the eigenpairs of a matrix, then we spend a lot of computation time in deflation and re-reducing the new matrices to Hessenberg form. Why not do the power method on several vectors at once? Let Z_0 be an initial guess matrix, and do the computation

$$Y_{k+1} = AZ_k, \quad Z_{k+1} = Y_{k+1}D_{k+1}, \tag{8.3}$$

where D_{k+1} is some properly chosen diagonal scaling matrix. The problem is that this won't work—the repeated multplications by A will simply cause *all* the columns of Y_{k+1} to line up in the direction of the dominant eigenvalue, so we will converge to n copies of the dominant eigenpair—hardly an improvement on the power methods.

This algorithm does have something to recommend it, though. If the Z_k matrices were all *orthogonal* matrices, then we get something that might work, because the orthogonality of the columns of Z_k would prevent the columns of the Y_{k+1} from all lining up in the same direction. But how can we guarantee this?

An important tool in computational linear algebra, used in solving a number of important applied problems, is the QR factorization, in which we factor an arbitrary square matrix[7] A into the product of an orthogonal matrix Q and an uppper triangular matrix R.

Theorem 8.9 *Given a matrix $A \in \mathbb{R}^{n \times n}$, there exists an orthogonal matrix Q and an upper triangular matrix R such that $A = QR$.*

Proof: Let a_1 be the first column of A, and let Q_1 be the orthogonal matrix such that $Qa_1 = \|a_1\|_2 e_1$. Then

$$A_2 = Q_1 A = \begin{bmatrix} \|a_1\|_2 & b^T \\ 0 & A_{22} \end{bmatrix}.$$

Now, let a_2 be the first column of A_{22}, and define Q_{22} as the orthogonal matrix such that $Q_{22}a_2 = \|a_2\|_2 e_1$. Define

$$Q_2 = \begin{bmatrix} 1 & 0 \\ 0 & Q_{22} \end{bmatrix}.$$

Then

$$A_3 = Q_2 A_2 = Q_2 Q_1 A$$

---

[7]In fact, A does not even have to be square, and in many important applications it isn't. We choose to look at only the square case for simplicity.

and we continue in this vein. After $n - 1$ steps, we have

$$A_{n-1} = Q_{n-1}Q_{n-2}\cdots Q_2 Q_1 A$$

and A_{n-1} is upper triangular. Thus,

$$R = Q_{n-1}Q_{n-2}\cdots Q_2 Q_1 A \Rightarrow A = QR$$

for

$$Q = Q_1^T Q_2^T \cdots Q_{n-1}^T. \qquad \bullet$$

Note: If A is nonsingular and we insist on the diagonal elements of R being positive, then the decomposition is unique. The QR factorization has a number of applications in other areas of computational mathematics, most notably in the solution of least squares problems and least-squares and ill-posed problems.

■ **EXAMPLE 8.15**

Let

$$A = \begin{bmatrix} 4 & 1 & 0 \\ 1 & 4 & 1 \\ 0 & 1 & 4 \end{bmatrix}.$$

Then $a_1 = (4, 1, 0)^T$ and we use what we know about Householder matrices (§8.2) to find that

$$Q_1 = \begin{bmatrix} -0.9701 & -0.2425 & 0 \\ -0.2425 & 0.9701 & 0 \\ 0 & 0 & 1.0000 \end{bmatrix},$$

so that

$$A_1 = Q_1 A = \begin{bmatrix} -4.1231 & -1.9403 & -0.2425 \\ 0 & 3.6380 & 0.9701 \\ 0 & 1.0000 & 4.0000 \end{bmatrix}.$$

To continue on, we note that $a_2 = (3.6380, 1.0000)^T$ from which we get that

$$Q_2 = \begin{bmatrix} 1.0000 & 0 & 0 \\ 0 & -0.9642 & -0.2650 \\ 0 & -0.2650 & 0.9642 \end{bmatrix}.$$

Thus,

$$R = A_2 = Q_2 A_1 = Q_2 Q_1 A = \begin{bmatrix} -4.1231 & -1.9403 & -0.2425 \\ 0 & -3.7730 & -1.9956 \\ 0 & 0 & 3.5998 \end{bmatrix}$$

and

$$Q = Q_1^T Q_2^T = \begin{bmatrix} -0.9701 & 0.2339 & 0.0643 \\ -0.2425 & -0.9354 & -0.2571 \\ 0 & -0.2650 & 0.9642 \end{bmatrix}.$$

We use the QR decomposition to modify (8.3) in a slight but important fashion.

Algorithm 8.5 *Orthogonal Iteration.*
Given an initial orthogonal matrix, Q_0, for k from 1 until convergence, do

1. *Set $Y_{k+1} = AQ_k$;*

2. *Define Q_{k+1} and R_{k+1} as the QR decomposition of Y_{k+1}: $Q_{k+1}R_{k+1} = Y_{k+1}$.*

One consequence of this algorithm is that we can write

$$R_{k+1} = Q_{k+1}^T AQ_k, \tag{8.4}$$

and it can be shown, under the proper conditions, that the diagonal elements of the triangular matrix R_{k+1} converge to the eigenvalues of A. Let's assume that the iteration does converge, and call the limiting upper triangular matrix R_∞ and the limiting orthogonal matrix Q_∞.

Let's look now at the slightly different iteration based on the computation

$$A_k = Q_k^T AQ_k, \tag{8.5}$$

where Q_k is exactly the same as in orthogonal iteration. Note that the A_k matrices are similar to A, thus they have the same eigenvalues. Moreover, if the orthogonal iteration converges then so does this modified iteration (see Problem 3), and it is not difficult to show that $A_k \to R_\infty$. In addition, this is not much different from orthogonal iteration since it comes from merely shifting the index by one in the iteration (8.4). So the question arises: Would it be possible to use the new iteration (8.5) instead of the one from orthogonal iteration (8.4)? If so, how could we efficiently organize the computation?

Note the following:

$$A_k = Q_k^T AQ_k = \left(Q_k^T AQ_{k-1}\right) Q_{k-1}^T Q_k = R_k \left(Q_{k-1}^T Q_k\right),$$

$$A_{k-1} = Q_{k-1}^T AQ_{k-1} = Q_{k-1}^T \left(AQ_{k-1}\right) = \left(Q_{k-1}^T Q_k\right) R_k.$$

Thus, since products of orthogonal matrices are themselves orthogonal (Problem 1), we have that there exists an orthogonal matrix $Q_k' = Q_{k-1}^T Q_k$ and an upper triangular matrix R_k such that

$$A_{k-1} = Q_k'R_k, \quad A_k = R_kQ_k'.$$

Thus, given A_{k-1}, we can get the next member of the sequence simply by doing a QR decomposition and then reverse-multiplying the two matrices. Now, suppose that we start the iteration with the special orthogonal matrix $Q_0 = I$; then we have

$$A_0 = A, \quad A_1 = R_1Q',$$

where $Q'R_1 = A$ is the QR decomposition of A. This is the final piece of the puzzle. We can do essentially the same computations as orthogonal iteration—which is an attempt to do the power methods on more than one vector at a time—simply by doing a QR decomposition on A and then reverse-multiplying the two factors. To be specific, we have the following algorithm.

Algorithm 8.6 *Basic QR iteration.*
Given $A \in \mathbb{R}^{n \times n}$, set $A_0 = A$ and compute as follows: for k from 1 until convergence, do

1. Define Q_k and R_k as the QR decomposition of A_{k-1}: $Q_k R_k = A_{k-1}$;

2. Define $A_k = R_k Q_k$.

The important theorem is the one that tells us that this is a sequence of similar matrices, thus, they all have the same eigenvalues.

Theorem 8.10 *Given $A \in \mathbb{R}^{n \times n}$, set $A_0 = A$ and compute the following sequence of matrices:*

$$A_{k+1} = R_k Q_k,$$

where $A_k = Q_k R_k$ is the QR factorization of A_k. Then the A_k matrices are all similar.

Proof: We have

$$A_{k+1} = R_k Q_k,$$

so that

$$Q_k A_{k+1} = Q_k R_k Q_k = A_k Q_k \Rightarrow A_{k+1} = Q_k^{-1} A_k Q_k.$$

We know that the inverse exists since Q_k is orthogonal and hence $Q_k^{-1} = Q_k^T$. •

At this point it might be worth looking at an example.

■ **EXAMPLE 8.16**

Let

$$A = \begin{bmatrix} 10 & -2 & 3 & 2 & 0 \\ -2 & 10 & -3 & 4 & 5 \\ 3 & -3 & 6 & 3 & 3 \\ 2 & 4 & 3 & 6 & 6 \\ 0 & 5 & 3 & 6 & 13 \end{bmatrix}.$$

which is the same matrix that we looked at, extensively, at the end of §8.3. We will perform what might be called "naive" QR on this matrix.[8] The Hessenberg form is

$$A_{H,0} = A_0 = \begin{bmatrix} 10.0000 & -4.1231 & 0 & 0 & 0 \\ -4.1231 & 9.2941 & 3.0920 & 0 & 0 \\ 0 & 3.0920 & 6.4587 & -5.3189 & 0 \\ 0 & 0 & -5.3189 & 17.9989 & 3.1371 \\ 0 & 0 & 0 & 3.1371 & 1.2483 \end{bmatrix}.$$

Since our discussion of the QR iteration is not going to get into implementation details, we will simply illustrate how the computation proceeds. The QR decomposition of A_0 gives us

$$Q = \begin{bmatrix} -0.9245 & -0.3488 & 0.1024 & 0.1072 & 0.0402 \\ 0.3812 & -0.8461 & 0.2485 & 0.2601 & 0.0974 \\ 0 & -0.4031 & -0.6102 & -0.6387 & -0.2393 \\ 0 & 0 & 0.7453 & -0.6244 & -0.2339 \\ 0 & 0 & 0 & -0.3508 & 0.9364 \end{bmatrix},$$

---

[8]Actual QR codes do not explicitly form either factor—the process is organized in an almost elimination-like format, progressing down the diagonal.

$$R = \begin{bmatrix} -10.8167 & 7.3546 & 1.1786 & 0 & 0 \\ 0 & -7.6715 & -5.2193 & 2.1438 & 0 \\ 0 & 0 & -7.1370 & 16.6596 & 2.3380 \\ 0 & 0 & 0 & -8.9417 & -2.3967 \\ 0 & 0 & 0 & 0 & 0.4351 \end{bmatrix},$$

so that the next matrix in the sequence is

$$A_1 = RQ = \begin{bmatrix} 12.8034 & -2.9242 & -0.0000 & -0.0000 & -0.0000 \\ -2.9242 & 8.5943 & 2.8766 & -0.0000 & -0.0000 \\ 0 & 2.8766 & 16.7709 & -6.6639 & 0 \\ 0 & 0 & -6.6639 & 6.4239 & -0.1526 \\ 0 & 0 & 0 & -0.1526 & 0.4074 \end{bmatrix}.$$

Several things are worth noting at this point.

1. A_1 is tridiagonal—that is, the QR step and reverse multiplication preserve the Hessenberg form.

2. The $(5, 4)$ element of A_1 is substantially smaller than it was in A_0. This is significant, because when this element reaches zero, it means that the $(5, 5)$ element is an eigenvalue. (See Problem 2.)

We now continue the computation, without showing all the details. After six iterations we have

$$A_6 = \begin{bmatrix} 16.0164 & -2.9678 & -0.0000 & -0.0000 & -0.0000 \\ -2.9678 & 18.7347 & 0.3964 & -0.0000 & 0.0000 \\ 0 & 0.3964 & 7.0332 & -0.0167 & -0.0000 \\ 0 & 0 & -0.0167 & 2.8157 & 0.0000 \\ 0 & 0 & 0 & 0.0000 & 0.4000 \end{bmatrix}.$$

We again make several observations:

1. Since the $(5, 4)$ element is zero (to four digits) we can take $\lambda_5 \approx 0.4000$ as our first approximate eigenvalue.

2. Note that the iteration found the smallest eigenvalue first.

3. Note that the other off-diagonal elements are generally smaller than in A_0. This is significant; one reason that the QR iteration is superior to the power methods is that each step contributes something toward finding *all* the eigenvalues, not just a single one.

4. We can deflate by simply extracting the upper left 4×4 submatrix; there is no need to do any special computation, and the Hessenberg form is preserved. This is one of the key cost-saving features of the QR algorithm over the power methods and orthogonal iteration.

Although this example perhaps suggests that the QR iteration would outperform the power methods, the method can still be improved. Just as shifts accelerated the convergence of the inverse power methods, they can also be used to improve the performance of the QR iteration. The algorithm becomes the following:

Algorithm 8.7 *Shifted QR Iteration.*
   Given $A \in \mathbb{R}^{n \times n}$, set $A_0 = A$ and compute as follows: for k from 1 until convergence, do

   1. Define Q_k and R_k as the QR decomposition of $A_{k-1} - \mu_{k-1}I$: $Q_k R_k = A_{k-1} - \mu_{k-1}I$;

   2. Define $A_k = R_k Q_k + \mu_{k-1}I$.

It can be shown (Problem 4) that the A_k matrices are all similar, and it is possible to arrange the computation so that the shifted matrices are never explicitly formed (this is known in the literature as *implicit QR*).

What do we use as shifts? The general philosophy is the same as for the power methods: Choose shifts that are close to the eigenvalue you are trying to compute. Two common shift strategies are then the following:

Rayleigh Shift: Set $\mu_k =$ the lower right diagonal element of A_k.

Wilkinson Shift: Let

$$S = \begin{bmatrix} \alpha_{11} & \alpha_{12} \\ \alpha_{21} & \alpha_{22} \end{bmatrix}$$

denote the lower right 2×2 submatrix in A_k. Then the Wilkinson shift is to set $\mu_k =$ the eigenvalue of S that is closest to α_{22}.

■ **EXAMPLE 8.17**

To illustrate the value of shifts in the QR iteration, we return to the same example matrix as we have used all along. The Hessenberg form is

$$A_{H,0} = A_0 = \begin{bmatrix} 10.0000 & -4.1231 & 0 & 0 & 0 \\ -4.1231 & 9.2941 & 3.0920 & 0 & 0 \\ 0 & 3.0920 & 6.4587 & -5.3189 & 0 \\ 0 & 0 & -5.3189 & 17.9989 & 3.1371 \\ 0 & 0 & 0 & 3.1371 & 1.2483 \end{bmatrix}.$$

If we use the Rayleigh shift, then $\mu_0 = 1.2483$ and the initial QR factorization gives us

$$Q = \begin{bmatrix} -0.9046 & -0.3719 & 0.1068 & 0.1478 & 0.1005 \\ 0.4262 & -0.7893 & 0.2268 & 0.3137 & 0.2134 \\ 0 & -0.4886 & -0.4477 & -0.6192 & -0.4213 \\ 0 & 0 & 0.8583 & -0.4242 & -0.2886 \\ 0 & 0 & 0 & -0.5625 & 0.8268 \end{bmatrix}$$

and

$$R = \begin{bmatrix} -9.6743 & 7.1590 & 1.3178 & 0 & 0 \\ 0 & -6.3281 & -4.9864 & 2.5989 & 0 \\ 0 & 0 & -6.1967 & 16.7589 & 2.6927 \\ 0 & 0 & 0 & -5.5769 & -1.3307 \\ 0 & 0 & 0 & 0 & -0.9054 \end{bmatrix},$$

so that

$$A_1 = RQ + \mu_0 I = \begin{bmatrix} 13.0511 & -2.6970 & 0.0000 & 0.0000 & 0 \\ -2.6970 & 8.6795 & 3.0278 & 0.0000 & 0.0000 \\ 0 & 3.0278 & 18.4072 & -4.7869 & -0.0000 \\ 0 & 0 & -4.7869 & 4.3626 & 0.5093 \\ 0 & 0 & 0 & 0.5093 & 0.4997 \end{bmatrix}.$$

The next shift is then $\mu_1 = 0.4997$. If we continue the iteration, we find that

$$A_3 = \begin{bmatrix} 14.3388 & -1.5148 & -0.0000 & -0.0000 & -0.0000 \\ -1.5148 & 17.5527 & 5.4211 & 0.0000 & 0.0000 \\ 0 & 5.4211 & 9.8879 & -0.1601 & -0.0000 \\ 0 & 0 & -0.1601 & 2.8207 & 0.0000 \\ 0 & 0 & 0 & 0.0000 & 0.4000 \end{bmatrix}.$$

Thus, we have found the first eigenvalue in only three iterations, whereas the unshifted method took six. We can now deflate, by defining

$$A_3' = \begin{bmatrix} 14.3388 & -1.5148 & -0.0000 & -0.0000 \\ -1.5148 & 17.5527 & 5.4211 & 0.0000 \\ 0 & 5.4211 & 9.8879 & -0.1601 \\ 0 & 0 & -0.1601 & 2.8207 \end{bmatrix}$$

and continue the iteration. Thus, $\mu_3 = 2.8207$ and A_4 is defined by

$$A_4 = \begin{bmatrix} 14.7851 & -2.0073 & -0.0000 & 0.0000 \\ -2.0073 & 19.7542 & 1.6637 & -0.0000 \\ 0 & 1.6637 & 7.2452 & -0.0002 \\ 0 & 0 & -0.0002 & 2.8156 \end{bmatrix}$$

and

$$A_5 = \begin{bmatrix} 15.5758 & -2.7220 & -0.0000 & -0.0000 \\ -2.7220 & 19.1737 & 0.4287 & -0.0000 \\ 0 & 0.4287 & 7.0349 & -0.0000 \\ 0 & 0 & -0.0000 & 2.8156 \end{bmatrix},$$

showing that we have found another eigenvalue. Hence, in fewer steps than it took to find a single eigenvalue in the unshifted algorithm, we have found two eigenvalues by using the very simple Rayleigh shift.[9] Using the Wilkinson shift on this example yields very similar results.

Which shift do we use? We might expect that the simpler Rayleigh shift is less reliable than the more involved Wilkinson shift, and we would be right. While the Rayleigh shift can produce rapid convergence of the off-diagonal elements to zero, there exist examples for which it will not converge at all. On the other hand, it can be shown that QR using the Wilkinson shift will always converge [6]. Problem 5 illustrates some of this, drawing on an example from Trefethan and Bau [6].

Why does QR work? A complete discussion is beyond the intended level of this text. Good explanations are in the works of Watkins ([7] and [8]) and Trefethen and Bau [6]; a

---

[9]To within four decimals, that is. For greater accuracy we would need to take more iterations before deflating, and perhaps some more after deflating.

brief answer is that the QR iteration is equivalent to a properly scaled and shifted inverse power iteration being done on an orthonormal set of vectors. Thus, we are working on *all* the eigenvectors at the same time. Some of this is suggested by the motivation for the QR iteration that we did at the beginning of this section, using orthogonal iteration.

Since the underlying theory for the QR iteration is similar to that of the power methods, it should have the same weak points, and this in fact is the case. Defective matrices, or matrices that are nearly defective (meaning that two or more eigenvectors are nearly parallel) will slow down the convergence of the iteration; fundamentally, this occurs because the eigenvalue problem is a polynomial root-finding problem, and defective matrices are caused by multiple roots for the polynomial (recall §3.11.4). But software such as the codes in LAPACK or the `eig` command in MATLAB is robust and reliable, and can be counted on to deliver accurate results for almost all problems.

Exercises:

1. Show that the product of two orthogonal matrices, Q_1 and Q_2, is also orthogonal.

2. Let $A \in \mathbb{R}^{n \times n}$ have the partioned form

$$A \begin{bmatrix} A_{11} & a \\ 0 & \lambda \end{bmatrix},$$

   where $A_{11} \in \mathbb{R}^{(n-1) \times (n-1)}$ and $\lambda \in \mathbb{R}$. Show that λ must be an eigenvalue of A. How are the eigenvalues of A and A_{11} related?

3. Assume that the iteration in Algorithm 8.5 converges, in the sense that $R_k \to R_\infty$, which is upper triangular, and $Q_k \to Q_\infty$, which is orthogonal.

   (a) Prove that the matrices A_k, defined in (8.5), must also converge to R_∞.

   (b) Prove then that the eigenvalues of A can be recovered from the diagonal elements of A_k, in the limit as $k \to \infty$.

4. Show that the matrices in the shifted QR iteration all have the same eigenvalues.

5. Consider the matrix

$$A = \begin{bmatrix} 0 & 1 \\ 1 & 0 \end{bmatrix}.$$

   (a) Find the exact eigenvalues of A.

   (b) Show that

$$Q = \begin{bmatrix} 0 & 1 \\ 1 & 0 \end{bmatrix}, \quad R = \begin{bmatrix} 1 & 0 \\ 0 & 1 \end{bmatrix}$$

   is a valid QR decomposition of A.

   (c) Use this to show that the shifted QR iteration for A, using the Rayleigh shift, will not converge.

6. The QR factorization is usually carried out for the QR iteration by means of *Givens transformations*, defined in part (a), below. In this exercise we will introduce the basic ideas of this kind of matrix operation.

(a) Show that the matrix

$$G(\theta) = \begin{bmatrix} \cos\theta & -\sin\theta \\ \sin\theta & \cos\theta \end{bmatrix}$$

is an orthogonal matrix for any choice of θ.

(b) Show that we can always choose θ so that

$$G(\theta)A = \begin{bmatrix} \cos\theta & -\sin\theta \\ \sin\theta & \cos\theta \end{bmatrix}\begin{bmatrix} a_{11} & a_{12} \\ a_{21} & a_{22} \end{bmatrix} = \begin{bmatrix} r_{11} & r_{12} \\ 0 & r_{22} \end{bmatrix} = R.$$

(c) Use a Givens transformation to perform one step of the basic QR iteration for the matrix

$$\begin{bmatrix} 4 & 1 \\ 1 & 4 \end{bmatrix}.$$

7. (a) If $G(\theta)$ is a Givens transformation, show that

$$\hat{G}(\theta) = \begin{bmatrix} 1 & 0 \\ 0 & G(\theta) \end{bmatrix}$$

is orthogonal.

(b) Show how to use a sequence of *two* Givens transformations to do a single QR step for the matrix

$$\begin{bmatrix} 4 & 1 & 0 \\ 1 & 4 & 1 \\ 0 & 1 & 4 \end{bmatrix}.$$

8. Write a MATLAB code that automates the kind of computation we did in Examples 8.16 and 8.17: Take an input matrix (square), and perform "naive" QR by executing the commands

$$[q,r] = qr(A);$$
$$A = r*q;$$

Stop when A is triangular. Use no shifts. Test your code by forming a random symmetric matrix[10] and using your routine to find its eigenvalues. Remember to reduce the matrix to Hessenberg form! Use the flops command to measure the cost of your routine, and compare this to what MATLAB's eig command costs.

9. Repeat Problem 8, this time using the Rayleigh shift. You should save the random matrix from Problem 8 so that a flops comparison is meaningful.

10. Repeat Problem 8, this time using the Wilkinson shift. Again, use the same matrix and do a flops cost comparison.

◁ ● ● ● ▷

---

[10]Recall that $S = A + A^T$ is always symmetric.

8.5 APPLICATION: ROOTS OF POLYNOMIALS, PART II

We know that the eigenvalues of a matrix are the roots of a polynomial, the characteristic polynomial of the matrix. This raises an obvious question: Can we solve the polynomial root-finding problem by finding the eigenvalues of a properly defined matrix? The answer is "yes," as we outline, next.

Definition 8.3 (Companion Matrix) *Given a (monic) polynomial*

$$p(x) = x^n + a_{n-1}x^{n-1} + \cdots + a_2x^2 + a_1x + a_0,$$

the companion matrix *is defined as*

$$C = \begin{bmatrix} 0 & 0 & 0 & \cdots & 0 & -a_0 \\ 1 & 0 & 0 & \cdots & 0 & -a_1 \\ 0 & 1 & 0 & \cdots & 0 & -a_2 \\ 0 & 0 & 1 & \cdots & 0 & -a_3 \\ \vdots & \vdots & & \ddots & \vdots & \vdots \\ 0 & 0 & \cdots & 0 & 1 & -a_{n-1} \end{bmatrix}.$$

(Some authors use C^T, instead.)

It is a simple exercise to use a cofactor expansion to show that the characteristic polynomial of C is, indeed, $p(\lambda)$.

Because of the special (sparse) structure of C, it is possible to develop specialized iterations for this purpose, and several have been outlined in the literature, most notably based on the several papers of Jenkins and Traub. MATLAB simply forms the companion matrix and applies the `eig` command to it.

One advantage of using the eigenvalue approach to polynomial root-finding is that we can use shifting strategies to accelerate convergence. This is central to the work of Jenkins and Traub.

If we apply Gerschgorin's Theorem to the companion matrix, we find that all the eigenvalues (hence, all the roots) are in the union of the regions

$$\begin{aligned} R_0 &= \{z \in \mathbb{C} \mid |z| \le |a_0|\}, \\ R_j &= \{z \in \mathbb{C} \mid |z| \le (1 + |a_{j-1}|)\}, \quad 2 \le j \le n-1, \\ R_n &= \{z \in \mathbb{C} \mid |z + a_{n-1}| \le 1\}. \end{aligned}$$

In other words,

$$z_k \in R_{\text{roots}} = \cup_{j=1}^{n} R_j$$

for all k, $1 \le k \le n$.

This is where the bounds in Theorem 3.8 come from. Other localizations can be derived by manipulating the matrix.

Exercises:

1. For the polynomial $p(x) = x^3 - 2x^2 + 5x + 1$, construct the companion matrix, then use a cofactor expansion to confirm that the characteristic polynomial is, indeed, $p(\lambda)$.

2. For each polynomial in Problem 2 of §3.10, construct the companion matrix and use MATLAB's `eig` to find the roots. Use `flops` to compare the costs with the Durand–Kerner method from §3.10.

3. Consider the polynomial $p(x) = x^4 - 10x^3 + 35x^2 - 50x + 24$ and form its companion matrix. Using MATLAB's `qr` command, do several iterations of the unshifted QR iteration. Are the roots being isolated? Is the structure of the companion matrix being maintained? Comment.

◁ ● ● ▷

8.6 LITERATURE AND SOFTWARE DISCUSSION

Most of the works cited at the end of Chapter 7 contain good discussions of the numerical solution of the algebraic eigenvalue problem. The classic work is by James Wilkinson [10]. An outstanding explanation of the QR algorithm is in the SIAM Review article by Watkins [8] (the texts by Watkins [7] and Trefethen and Bau [6] contain some of the same material). Parlett's book [5] is a good discussion of the symmetric eigenvalue problem, which is substantially easier than the general case, as we have seen. Two excellant articles on the history of the QR algorithm are by Golub and Frank [3] and Watkins [9]. Watkins is of the opinion [7] that we should be calling the "QR algorithm" the "Francis algorithm;" not only does this recognize the author of the original algorithm, but the methodology of modern computer codes is much closer to what Francis wrote about in [1] and [2].

There exist other eigenvalue approximation ideas beyond what was discussed here, and Watkins [7] has a brief discussion of many of these, some of which are of mostly historic interest. Of particular interest is the "divide and conquer" algorithm due to Cuppen. The rise of parallel computing has meant that some of these older methods, such as the Jacobi method, are of more interest because they can be programmed in a manner that takes greater advantage of the parallel computing architecture.

REFERENCES

1. Francis, J. G. F., "The QR transformation, part I," *Computer J.* vol. 4, pp. 265–272, 1961.

2. Francis, J. G. F., "The QR transformation, part II," *Computer J.* vol. 4 pp. 332–345, 1961.

3. Golub, Gene; and Uhlig, Frank, "The QR algorithm: 50 years later its genesis by John Francis and Vera Kublanovskaya and subsequent developments," *IMA J. Numer. Anal.*, vol. 29 (2009), no. 3, 467-485, 2009.

4. Householder, Alston, *The Theory of Matrices in Numerical Analysis*, Dover, New York, 1975 (originally by Blaisdell Publishing Co., 1964).

5. Parlett, Beresford, *The Symmetric Eigenvalue Problem*, SIAM Classics in Applied Mathematics, SIAM, Philadelphia, 1998; originally published by Prentice-Hall, 1980.

6. Trefethen, L. N., and Bau, David, *Numerical Linear Algebra*, SIAM, Philadelphia, 1997.

7. Watkins, David, *Fundamentals of Matrix Computations*, John Wiley & Sons, Inc., New York, 3rd Edition, 2010.

8. Watkins, David, "Understanding the QR algorithm," *SIAM Rev.*, vol. 24, pp. 427–440, 1982.

9. Watkins, David, "Francis's algorithm," *Amer. Math. Monthly*, vol. 118, no. 5, 387–403, 2011.

10. Wilkinson, J. H., *The Algebraic Eigenvalue Problem*, Oxford University Press, New York, 1965.

CHAPTER 9

A SURVEY OF NUMERICAL METHODS FOR PARTIAL DIFFERENTIAL EQUATIONS

Partial differential equations (PDEs) are differential equations involving functions of more than one independent variable, such as the temperature at each point (x, y, z) in an iron bar. They are necessarily more complicated in some respects than ordinary differential equations, and it is beyond our intended scope here to provide an in-depth treatment of all the methods used for their approximate solution. Indeed, this is such an active research topic that it is best, in an introductory text, to confine ourselves to a brief survey of major ideas.

9.1 DIFFERENCE METHODS FOR THE DIFFUSION EQUATION

9.1.1 The Basic Problem

Perhaps the simplest PDE is the *diffusion equation*, so-called because it can be used to model a number of processes that are driven by diffusion, such as heat transfer and some types of slow mass transfer.

We seek the unknown function $u(x, t)$ such that

$$u_t = au_{xx} + f(x, t), \quad t > 0, 0 < x < 1; \tag{9.1}$$
$$u(0, t) = g_0(t); \tag{9.2}$$
$$u(1, t) = g_1(t); \tag{9.3}$$
$$u(x, 0) = u_0(x). \tag{9.4}$$

An Introduction to Numerical Methods and Analysis, Second Edition. By James F. Epperson
Copyright © 2013 John Wiley & Sons, Inc.

Here g_0 and g_1 are the (known) *boundary data*, f is a known source term, and u_0 is the (known) *initial data*. More general and involved boundary conditions are possible, and we can generalize the equation somewhat, but this is the usual standard form. For simplicity we assume *homogeneous boundary data* , meaning that $g_0 = g_1 = 0$ for all time t. In the most commonplace examples, the unknown u gives the temperature at each point x along a thin rod, for each time t. The boundary conditions (9.3) and (9.4) give the temperature at which each end of the rod is maintained, and the initial condition (9.4) gives the temperature distribution at the start of the process. See a text, such as Humi and Miller [8], for background material on the diffusion equation.

The parameter a, which must be positive, is the *thermal diffusivity* of the material being studied. If a is large, it means that diffusion is rapid; if a is small, then diffusion is slow.

We will study two basic methods for the diffusion equation in this section: the explicit method, and the Crank–Nicolson method. In the next section we discuss the finite element method for problems of this type.

Before proceeding with the derivations and analysis, we need to deal with some notational issues. Our approximations will be constructed, as was the case for ODE problems in Chapter 6, only at discrete points on a grid defined over space and time. There will be times when we will want to consider the approximation as a discretely defined function, and there will be times when we want to consider it as a vector in a standard Euclidean space. At times we will also want or need to consider the collection of exact solution values defined at the same grid points. To accomplish all this with a minimum of excessive notation, we will use the following conventions.

If the dependent variable in a problem is u, then $u_{h,\Delta t}$ will denote the approximation, defined on a grid using h as the spatial step and Δt as the time step. Specific values can be denoted using the standard functional notation, $u_{h,\Delta t}(x_i, t_n)$, which we occasionally abbreviate as u_i^n. If we wish to consider the Euclidean vector of $u_{h,\Delta t}$ values at a specific point in time, t_n, then we will write $u_{h,\Delta t}^n$. The similar notation u^n (note the lack of subscript) will be used for the vector of exact solution values at time t_n, using the same grid points. Some of this notation is abusive of standard conventions, but it enables us to work with a minimum of confusing new symbols. The need for most of this will not occur until §9.1.3, where we begin to discuss implicit methods and the Crank–Nicolson method.

9.1.2   The Explicit Method and Stability

The explicit method is constructed by using the most straight-forward difference approximations for the derivatives in the PDE. We start by defining a grid in the spatial variable x:

$$0 = x_0 < x_1 < x_2 < \cdots < x_{N-1} < x_N = 1;$$

and we assume, for simplicity, that the mesh spacing is uniform: $x_k - x_{k-1} = h$ for all k. Define a time step $\Delta t > 0$, and a time grid by $t_n = n\Delta t$ for $n \geq 0$. Then, at each point in our space–time grid we can define difference approximations to the derivatives, thus:

$$\frac{\partial u}{\partial t}(x,t) = \frac{u(x, t+\Delta t) - u(x,t)}{\Delta t} - \frac{1}{2}\Delta t \frac{\partial^2 u}{\partial t^2}(x, \theta),$$

$$\frac{\partial^2 u}{\partial x^2}(x,t) = \frac{u(x-h,t) - 2u(x,t) + u(x+h,t)}{h^2} + \frac{1}{12}h^2 \frac{\partial^4 u}{\partial x^4}(\eta, t),$$

where $t \leq \theta \leq t + \Delta t$ and $x - h \leq \eta \leq x + h$. We use these in the PDE to replace $u_t(x, t)$ and $u_{xx}(x, t)$ and get

$$\frac{u(x, t + \Delta t) - u(x, t)}{\Delta t} = a \left(\frac{u(x - h, t) - 2u(x, t) + u(x + h, t)}{h^2} \right) + f(x, t) + T_E(x, t), \tag{9.5}$$

where the truncation error $T_E(x, t)$ is given by

$$T_E(x, t) = \frac{1}{12} h^2 \frac{\partial^4 u}{\partial x^4}(\eta, t) + \frac{1}{2} \Delta t \frac{\partial^2 u}{\partial t^2}(x, \theta). \tag{9.6}$$

We now define our approximate solution $u_{h, \Delta t}(x, t)$ at the grid points by dropping the truncation error from (9.5) and replacing $u(x_i, t_n)$ with $u_i^n = u_{h, \Delta t}(x_i, t_n)$. We get

$$\frac{u_i^{n+1} - u_i^n}{\Delta t} = a \left(\frac{u_{i-1}^n - 2u_i^n + u_{i+1}^n}{h^2} \right) + f(x_i, t_n), \tag{9.7}$$

which can be simplified to get

$$u_i^{n+1} = u_i^n + \frac{a \Delta t}{h^2} \left(u_{i-1}^n - 2u_i^n + u_{i+1}^n \right) + \Delta t f(x_i, t_n). \tag{9.8}$$

This is called the *explicit method* because it yields the value of the approximation at time t_{n+1} explicitly in terms of the approximate solution at time t_n.

Coding this is very easy, so it is worth going directly to some examples.

■ **EXAMPLE 9.1**

Consider the problem (9.1)–(9.4) with $a = \pi^{-2}$, $u_0(x) = \sin \pi x + \sin 2\pi x$ and $g_0 = g_1 = 0$, for which the exact solution is $u(x, t) = e^{-t} \sin \pi x + e^{-4t} \sin 2\pi x$. (As always, the student should check this.) Figure 9.1 shows the solution profile for the approximate solution computed using $h = 1/16$, $\Delta t = 1/64$, and Figure 9.2 shows the error profile, both at time $t = 1$. The approximation appears to be doing a decent job. However, if we next look at the case defined by $h = 1/32$, $\Delta t = 1/128$, then Table 9.1 shows how the values in the middle of the interval evolve for the first 45 time steps. Note that the approximate solution appears to be evolving normally for the first 40 or so steps, then suddenly begins to "blow up." If we continue the computation, the solution values eventually get as large as 10^{29}. In fact, they will grow without bound.

What is going on here?

The problem, in a nutshell, is that the time step is too large for the speed of diffusion and the spatial step. The explicit method requires that the time step be sufficiently small in order that the computation be stable and accurate. A formal result is the following.

Theorem 9.1 *Let $u \in C^{4,2}(Q)$ for $Q = \{(x, t) \mid 0 \leq x \leq 1, 0 < t \leq T\}$ for some $T > 0$. If h and Δt are such that*

$$\Delta t \leq \frac{h^2}{2a}, \tag{9.9}$$

then there exists a constant $C > 0$, independent of h and Δt, such that

$$\max_{(x_i, t_n) \in Q} |u(x_i, t_n) - u_{h, \Delta t}(x_i, t_n)| \leq \max_{0 \leq i \leq N} |u_0(x_i) - u_{h, \Delta t}(x_i, 0)| + Ch^2 M, \tag{9.10}$$

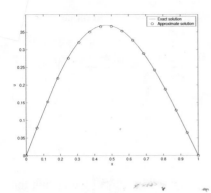

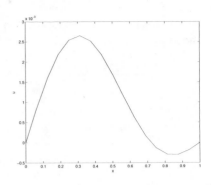

Figure 9.1 Solution profile at $t = 1$ for Example 9.1; explicit method, $h = 1/16, \Delta t = 1/64$.

Figure 9.2 Error profile at $t = 1$ for Example 9.1; explicit method, $h = 1/16, \Delta t = 1/64$.

where

$$M = \max_{(x,t) \in Q} \left\{ |u_{xxxx}(x,t)|, |u_{tt}(x,t)| \right\}.$$

Proof: To construct an error estimate, we subtract (9.5) and (9.7); letting

$$e_i^n = u(x_i, t_n) - u_i^n,$$

we then have

$$e_i^{n+1} = e_i^n + \lambda \left(e_{i-1}^n - 2e_i^n + e_{i+1}^n \right) + \left(\frac{1}{12} h^2 \Delta t \frac{\partial^4 u}{\partial x^4}(\eta_i, t_n) + \frac{1}{2} \Delta t^2 \frac{\partial^2 u}{\partial t^2}(x_i, \theta_n) \right),$$

where $x_{i-1} \leq \eta_i \leq x_{i+1}$ and $t_n \leq \theta_n \leq t_{n+1}$, and we have used $\lambda = \frac{a \Delta t}{h^2}$. Taking absolute values and using the definition of M, plus some simplification, gives us

$$|e_i^{n+1}| \leq |1 - 2\lambda||e_i^n| + \lambda \left(|e_{i-1}^n| + |e_{i+1}^n| \right) + \left(\frac{1}{12} h^2 \Delta t + \frac{1}{2} \Delta t^2 \right) M, \quad 1 \leq i \leq N - 1.$$

Now, let

$$E_n = \max_{1 \leq i \leq N-1} |e_i^n|,$$

so that we have

$$|e_i^{n+1}| \leq (|1 - 2\lambda| + 2\lambda) E_n + \left(\frac{1}{12} h^2 \Delta t + \frac{1}{2} \Delta t^2 \right) M.$$

Note that if $1 - 2\lambda \geq 0$, then we can remove the absolute value symbol and write

$$|1 - 2\lambda| + 2\lambda = 1 - 2\lambda + 2\lambda = 1.$$

Therefore (if $1 - 2\lambda \geq 0$),

$$|e_i^{n+1}| \leq E_n + \left(\frac{1}{12} h^2 \Delta t + \frac{1}{2} \Delta t^2 \right) M,$$

Table 9.1   Illustration of instability for Example 9.1; $h = 1/32$, $\Delta t = 1/128$, $a = 1/\pi^2$.

k	u_{14}^k	u_{15}^k	u_{16}^k	u_{17}^k	u_{18}^k
1	0.134389E+01	0.117643E+01	0.992194E+00	0.798403E+00	0.602366E+00
$\vdots$	$\vdots$	$\vdots$	$\vdots$	$\vdots$	$\vdots$
5	0.126978E+01	0.112348E+01	0.961573E+00	0.790403E+00	0.616417E+00
$\vdots$	$\vdots$	$\vdots$	$\vdots$	$\vdots$	$\vdots$
10	0.118573E+01	0.106234E+01	0.924624E+00	0.778003E+00	0.627984E+00
$\vdots$	$\vdots$	$\vdots$	$\vdots$	$\vdots$	$\vdots$
15	0.111007E+01	0.100618E+01	0.889094E+00	0.763449E+00	0.633948E+00
$\vdots$	$\vdots$	$\vdots$	$\vdots$	$\vdots$	$\vdots$
20	0.104172E+01	0.954414E+00	0.854929E+00	0.747210E+00	0.635278E+00
$\vdots$	$\vdots$	$\vdots$	$\vdots$	$\vdots$	$\vdots$
25	0.979764E+00	0.906558E+00	0.822080E+00	0.729670E+00	0.632816E+00
$\vdots$	$\vdots$	$\vdots$	$\vdots$	$\vdots$	$\vdots$
30	0.923295E+00	0.862416E+00	0.789955E+00	0.712294E+00	0.625040E+00
31	0.913011E+00	0.853028E+00	0.785740E+00	0.704519E+00	0.631205E+00
32	0.901453E+00	0.847107E+00	0.774446E+00	0.710929E+00	0.610682E+00
33	0.894185E+00	0.832261E+00	0.781858E+00	0.681157E+00	0.656032E+00
34	0.875812E+00	0.841599E+00	0.741088E+00	0.742416E+00	0.539034E+00
35	0.888149E+00	0.787860E+00	0.823636E+00	0.576484E+00	0.818067E+00
36	0.819241E+00	0.898150E+00	0.594303E+00	0.972637E+00	0.136538E+00
37	0.964831E+00	0.587900E+00	0.114726E+01	−0.117450E−01	0.176970E+01
38	0.552823E+00	0.134683E+01	−0.245593E+00	0.237169E+01	−0.214521E+01
39	0.157751E+01	−0.587537E+00	0.316667E+01	−0.341106E+01	0.717104E+01
40	−0.106303E+01	0.421043E+01	−0.520808E+01	0.104982E+02	−0.149249E+02
41	0.556995E+01	−0.769844E+01	0.151573E+02	−0.228400E+02	0.372474E+02
42	−0.110780E+02	0.215827E+02	−0.341683E+02	0.566645E+02	−0.855143E+02
43	0.303114E+02	−0.500811E+02	0.846480E+02	−0.132208E+03	0.202351E+03
44	−0.720490E+02	0.124290E+03	−0.200336E+03	0.314752E+03	−0.470614E+03
45	0.179537E+03	−0.297988E+03	0.480310E+03	−0.739356E+03	0.109814E+04

and this holds for all i; hence,

$$E_{n+1} \le E_n + \left(\frac{1}{12} h^2 \Delta t + \frac{1}{2} \Delta t^2 \right) M. \tag{9.11}$$

This is a simple recursive inequality, very much akin to those we dealt with in Chapter 6, and applying the solution procedures from those sections gives us

$$
\begin{aligned}
E_n &\le E_0 + n \left(\frac{1}{12} h^2 \Delta t + \frac{1}{2} \Delta t^2 \right) M = E_0 + \left(\frac{1}{12} h^2 + \frac{1}{2} \Delta t \right) (M t_n) \\
&\le E_0 + \left(\frac{1}{12} h^2 + \frac{1}{2} \Delta t \right) (M T)
\end{aligned}
$$

from which the conclusion (9.10) follows. The condition $1 - 2\lambda \geq 0$ is equivalent to (9.9), and thus we are done. •

If (9.9) does not hold, then (9.11) becomes

$$E_{n+1} \leq cE_n + \left(\frac{1}{12}h^2\Delta t + \frac{1}{2}\Delta t^2 \right) M, \qquad (9.12)$$

for $c = 4\lambda - 1 > 1$. We then have

$$E_n \leq c^n E_0 + \left(\frac{1}{12}h^2\Delta t + \frac{1}{2}\Delta t^2 \right) M \sum_{k=0}^{n-1} c^k.$$

This is important for a number of reasons. First, it shows that if the *stability condition* (9.9) does not hold, then any perturbation in the initial condition (which amounts to having $E_0 \neq 0$) might be amplified by a factor of c^n, making the error eventually very large. In addition, the presence of the c^k factors in the summation means that, even in the total absence of rounding error, the mathematical error is also amplified by exponentially growing factors.

The attentive reader might be concerned that (9.12), being an inequality, does not lead to the conclusion that E_n *must* grow as c^n does; it simply *allows* for such growth to occur. However, it is possible to establish a strict inequality which shows that E_n will grow if the stability condition is not satisfied, and this growth is observed in practice, as our simple example showed.

The condition (9.9) is called a *stability condition* because it is both necessary and sufficient to ensure that the approximate solutions do not "blow up" during the course of the computation. It can be shown that if a numerical method for the approximate solution of the diffusion equation is stable and has a truncation error that goes to zero as the mesh parameters do, then the approximate solutions will converge to the exact solution. Note that (9.9) requires us to use a time step size that goes like the square of the spatial step; thus, we will often be forced to use very tiny time steps. This is usually considered too much of a restriction for practical use, although the simplicity of the explicit method makes it attractive for nonlinear problems and problems involving more than one space dimension.

■ EXAMPLE 9.2

To demonstrate that the explicit method works fine when condition (9.9) is satisfied, we return to the same example, taking Δt to be 90% of the maximum possible time step (for $h = 1/32$), and computing all the way out to $t = 1$. Fig. 9.3 shows a plot of the maximum absolute error as a function of t. The rapid decay is due to the nature of the diffusion equation, so it might be better to look at Fig. 9.4, which is almost the same plot, except that we have scaled the error by the maximum absolute value of the exact solution. Finally, note that it took 231 time steps to reach $t = 1$.

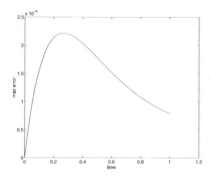

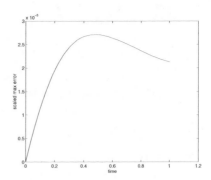

Figure 9.3   Maximum absolute error (unscaled) for $0 \le t \le 1$ for the explicit method (Example 9.2), $h = 1/32$.

Figure 9.4   Maximum absolute error (scaled) for $0 \le t \le 1$ for the explicit method (Example 9.2), $h = 1/32$.

9.1.3   Implicit Methods and the Crank–Nicolson Method

The time-step restriction in the explicit method can be removed by using a very slightly different set of difference approximations in constructing the numerical method. If we use

$$
\frac{\partial u}{\partial t}(x, t + \Delta t) = \frac{u(x, t + \Delta t) - u(x, t)}{\Delta t} + \frac{1}{2} \Delta t \frac{\partial^2 u}{\partial t^2}(x, \theta)
$$

$$
\frac{\partial^2 u}{\partial x^2}(x, t + \Delta t) = \frac{u(x - h, t + \Delta t) - 2u(x, t + \Delta t) + u(x + h, t + \Delta t)}{h^2}
$$

$$
+ \frac{1}{12} h^2 \frac{\partial^4 u}{\partial x^4}(\eta, t + \Delta t)
$$

where $t \le \theta \le t + \Delta t$ and $x_{i-1} \le \eta \le x_{i+1}$, then the numerical method becomes

$$
u_i^{n+1} - \frac{a\Delta t}{h^2} \left(u_{i-1}^{n+1} - 2u_i^{n+1} + u_{i+1}^{n+1} \right) = u_i^n + f(x_i, t_{n+1}). \tag{9.13}
$$

This is a tridiagonal system of $N-1$ equations in the $N-1$ unknowns $u_1^{n+1}, u_2^{n+1}, \ldots, u_{N-1}^{n+1}$, and we can solve it using Algorithm 2.6. Moreover, the method is unconditionally stable; i.e., it is stable for any value of Δt. The problem is that it is only first-order accurate in the time step.

Theorem 9.2 *For the implicit method (9.13) there exists a constant $C > 0$, independent of h or Δt, such that*

$$
\max_{(x_i, t_n) \in Q} |u(x_i, t_n) - u_{h, \Delta t}(x_i, t_n)| \le \max_{0 \le i \le N} |u_0(x_i) - u_{h, \Delta t}(x_i, 0)|
$$
$$
+ C(h^2 + \Delta t)M \tag{9.14}
$$

where

$$
M = \max_{(x, t) \in Q} \{ |u_{xxxx}(x, t)|, |u_{tt}(x, t)| \}.
$$

Proof: Using the same argument as for the explicit method, it is easy to show that the pointwise error $e_i^n = u(x_i, t_n) - u_i^n$ satisfies

$$
e_i^{n+1} - \lambda \left(e_{i-1}^{n+1} - 2e_i^{n+1} + e_{i+1}^{n+1} \right) = e_i^n + \Delta t T_E(x_i, t_{n+1}), \tag{9.15}
$$

where $\lambda = a\Delta t h^{-2}$ and $T_E(x_i, t_{n+1})$ is the truncation error:

$$T_E(x_i, t_{n+1}) = -\frac{1}{2}\Delta t \frac{\partial^2 u}{\partial t^2}(x_i, \theta_{n+1}) + \frac{1}{12}h^2 \frac{\partial^4 u}{\partial x^4}(\eta_i, t_{n+1}).$$

Now, let J be the index such that

$$|e_J^{n+1}| = \max_{1 \le i \le n-1} |e_i^{n+1}|.$$

Then (9.15) implies that

$$(1 + \lambda)|e_J^{n+1}| \le \lambda \left(|e_{J-1}^{n+1}| + |e_{J+1}^{n+1}|\right) + |e_J^n| + \Delta t \tau, \tag{9.16}$$

where τ is the maximum (in absolute value) truncation error:

$$\tau = \left|\frac{1}{2}\Delta t + \frac{1}{12}h^2\right| M.$$

Since $|e_J^{n+1}| \ge |e_j^{n+1}|$ for any j, we can write

$$(1 + 2\lambda)|e_J^{n+1}| \le 2\lambda|e_J^{n+1}| + |e_J^n| + \Delta t \tau, \tag{9.17}$$

from which

$$E_{n+1} = \max_{1 \le i \le n-1} |e_i^{n+1}| = |e_J^{n+1}| \le |e_J^n| + \Delta t \tau \le \max_{1 \le i \le n-1} |e_i^n| + \Delta t \tau = E_n + \Delta t \tau \tag{9.18}$$

follows. Therefore, as in the proof of Theorem 9.1, we solve the recursion $E_{n+1} \le E_n + \Delta t \tau$ to get

$$E_n \le E_0 + t_n \tau,$$

from which the conclusion follows. $\bullet$

This method is called *unconditionally stable* because there is no condition on the step size for the (approximate) solution to be bounded and accurate. The problem with this method, however, is that it is only $\mathcal{O}(\Delta t + h^2)$ accurate; the explicit method was $\mathcal{O}(h^2)$ accurate, since $\Delta t = \mathcal{O}(h^2)$ was required for stability. For comparison's sake, consider the same example that we looked at for the explicit method. If we use the same grid sizes ($h = 1/16, \Delta t = 1/64$), then we get the error plot shown in Figure 9.5; moreover, we can take much larger time steps and the solution remains stable. If we use $h = 1/32, \Delta t = 1/128$, then we get the error profile shown in Figure 9.6; recall that the explicit approximation blew up for this case.

To regain $\mathcal{O}(h^2)$ accuracy, we use an idea developed in the late 1940s by the British mathematicians Crank and Nicolson.[1]

---

[1]John Crank (1916–2006) was born in Lancashire, England and received all of his education at Manchester University. From 1957 until his retirement in 1981, he was professor of mathematics at Brunel University in Acton.

Phyllis (Lockett) Nicolson (1917–1968) was born in Macclesfield, England and received her early education (bachelor's and master's degrees) at Manchester University. In the immediate postwar period she was a research fellow at Girton College, Cambridge. In 1952 she was appointed lecturer in physics at the University of Leeds.

The Crank–Nicolson method resulted from a collaboration that took place in the immediate postwar period and was first presented in a 1947 paper in the *Proceedings of the Cambridge Philosophical Society*, titled "A practical method for numerical evaluation of solutions to partial differential equations of the heat-conduction type."

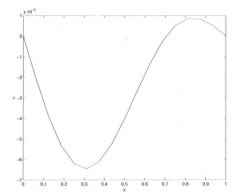

Figure 9.5 Error profile at $t = 1$ for the heat equation; implicit method, $h = 1/16$, $\Delta t = 1/64$.

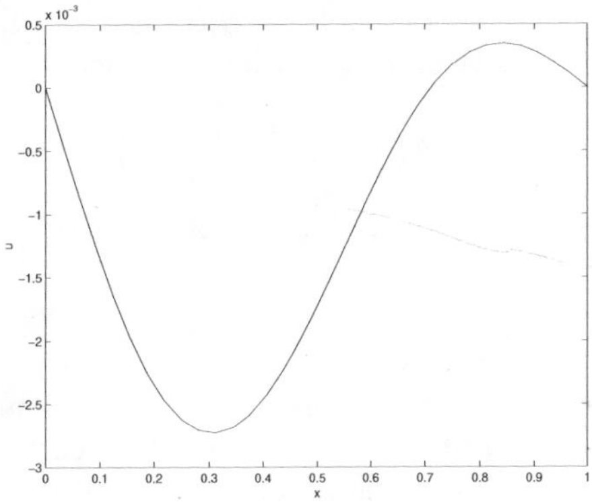

Figure 9.6 Error profile at $t = 1$ for the heat equation; implicit method, $h = 1/32$, $\Delta t = 1/128$.

There are a number of ways to derive the Crank–Nicolson scheme. Perhaps the best way is to first "solve" the PDE by integrating in time on both sides of the equation:

$$u(x, t + \Delta t) - u(x, t) = a \int_{t}^{t+\Delta t} (a u_{xx}(x, s) + f(x, s)) \, ds$$

and then use the trapezoid rule to evaluate the integral on the right side:

$$
\begin{aligned}
u(x, t + \Delta t) - u(x, t) =\ & \frac{1}{2} \Delta t a \left(u_{xx}(x, t) + u_{xx}(x, t + \Delta t) \right) \\
& + \frac{1}{2} \Delta t \left(f(x, t) + f(x, t + \Delta t) \right) \\
& - \frac{1}{12} \Delta t^3 u_{xxtt}(x, \theta_t) \\
& - \frac{1}{12} \Delta t^3 f_{tt}(x, \eta_t).
\end{aligned}
$$

Then, replacing the spatial derivatives with difference quotients, we get the lengthy expression

$$
\begin{aligned}
u(x, t + \Delta t) - u(x, t) =\ & \frac{1}{2} \Delta t a \left[\frac{u(x + h, t + \Delta t) - 2u(x, t + \Delta t) + u(x - h, t + \Delta t)}{h^2} \right] \\
& + \frac{1}{2} \Delta t a \left[\frac{u(x + h, t + \Delta t) - 2u(x, t + \Delta t) + u(x - h, t + \Delta t)}{h^2} \right] \\
& + \frac{1}{2} \Delta t \left(f(x, t + \Delta t) + f(x, t) \right) \\
& + \frac{1}{24} h^2 \Delta t a \left(\frac{\partial^4 u}{\partial x^4}(\eta_1, t + \Delta t) + \frac{\partial^4 u}{\partial x^4}(\eta_0, t) \right) - \frac{1}{12} \Delta t^3 u_{xxtt}(x, \theta_t) \\
& - \frac{1}{12} \Delta t^3 f_{tt}(x, \eta_t),
\end{aligned}
$$

where $\eta_t \in [t, t+h]$, $\theta_t \in [t, t+h]$, $\eta_1 \in [x-h, x+h]$, and $\eta_0 \in [x-h, x+h]$. Finally, we drop the remainders to define the numerical method:

$$u_j^{n+1} - \mu(u_{j+1}^{n+1} - 2u_j^{n+1} + u_{j-1}^{n+1}) = u_j^n + \mu(u_{j+1}^n - 2u_j^n + u_{j-1}^n)$$
$$+ \frac{1}{2}\Delta t(f(x_j, t_{n+1}) + f(x_j, t_n)) \quad (9.19)$$

where $u_j^n = u_{h,\Delta t}(t_i, t_n) \approx u(x_j, t_n)$ is the approximate solution and $\mu = \frac{1}{2}a\Delta t h^{-2}$. This is again a system of $N-1$ equations in $N-1$ unknowns, but it is slightly more involved than is the system for the implicit method (9.13). However, it is still unconditionally stable and it retains second-order accuracy in both space and time, as we shall soon see.

We can write the system (9.19) in matrix–vector form as

$$(I + \mu K)u_{h,\Delta t}^{n+1} = (I - \mu K)u_{h,\Delta t}^n + \frac{1}{2}\Delta t F^n,$$

where

$$F^n = (f(x_1, t_{n+1}) + f(x_1, t_n), \ldots, f(x_{N-1}, t_{n+1}) + f(x_{N-1}, t_n))^T$$

is the vector of source function values, and

$$K = \begin{bmatrix} 2 & -1 & 0 & \cdots & & 0 & 0 \\ -1 & 2 & -1 & 0 & \cdots & & 0 \\ \vdots & & \ddots & & & & \vdots \\ 0 & \cdots & 0 & -1 & 2 & -1 \\ 0 & 0 & 0 & \cdots & & -1 & 2 \end{bmatrix}.$$

Theorem 9.3 *Crank–Nicolson is unconditionally stable and second-order accurate in space and time, in the sense that*

$$\left(\sum_{i=1}^{N-1} (u(x_i, t_n) - u_{h,\Delta t}(x_i, t_n))^2\right)^{\frac{1}{2}} \leq C_0^m \left(\sum_{i=1}^{N-1} (u(x_i, 0) - u_{h,\Delta t}(x_i, 0))^2\right)^{\frac{1}{2}} + C_1(h^2 + \Delta t^2)M,$$

where C_0 and C_1 are positive and independent of h and Δt, $C_0 < 1$, and M depends on the derivatives u_{xxxx}, u_{xxtt}, and f_{tt}.

Proof: The proof involves a linear algebra argument that leans heavily on the fact that K is a symmetric, positive definite matrix. This means that the eigenvalues of K are real and positive, and that there is an orthogonal matrix Q which diagonalizes K (see Theorem 8.1).

We note that the vector of errors

$$e_{h,\Delta t}^n = (u(x_1, t_n) - u_1^n, \ldots, u(x_{N-1}, t_n) - u_{N-1}^n)^T$$

satisfies the equation

$$(I + \mu K)e_{h,\Delta t}^{n+1} = (I - \mu K)e_{h,\Delta t}^n + \Delta t T^n, \quad (9.20)$$

where T^n is the truncation error vector at time t_n, given by

$$T_i^n = \frac{1}{24}h^2 a\left(\frac{\partial^4 u}{\partial x^4}(\eta_{1,i}, t_{n+1}) + \frac{\partial^4 u}{\partial x^4}(\eta_{0,i}, t_n)\right) - \frac{1}{12}\Delta t^2 u_{xxtt}(x_i, \theta_t) - \frac{1}{12}\Delta t^2 f_{tt}(x_i, \eta_t).$$

Since K is symmetric, there exists an orthogonal matrix Q such that

$$K = QDQ^T,$$

where D is the diagonal matrix of the eigenvalues of K. If we substitute this into (9.20) and multiply by $Q^T = Q^{-1}$, we get

$$(I + \mu D)Q^T e_{h,\Delta t}^{n+1} = (I - \mu D)Q^T e_{h,\Delta t}^n + \Delta t Q^T T^n$$

so that

$$Q^T e_{h,\Delta t}^{n+1} = (I + \mu D)^{-1}(I - \mu D)(Q^T e_{h,\Delta t}^n) + \Delta t(I + \mu D)^{-1} Q^T T^n$$

or,

$$Q^T e_{h,\Delta t}^{n+1} = D_1(Q^T e_{h,\Delta t}^n) + \Delta t D_2 Q^T T^n$$

for the diagonal matrices

$$D_1 = (I + \mu D)^{-1}(I - \mu D)$$

and

$$D_2 = (I + \mu D)^{-1}.$$

This recursion is easily solved to get

$$Q^T e_{h,\Delta t}^n = D_1^n (Q^T e_{h,\Delta t}^0) + \Delta t \sum_{k=0}^{n-1} D_1^k D_2 Q^T T^{n-1-k},$$

so that

$$e_{h,\Delta t}^n = QD_1^n (Q^T e_{h,\Delta t}^0) + \Delta t \sum_{k=0}^{n-1} QD_1^k D_2 Q^T T^{n-1-k}.$$

Now take the 2-norm of both sides to get

$$\|e_{h,\Delta t}^n\|_2 \leq \|Q\|_2 \|D_1^n\|_2 \|Q^T\|_2 \|e_{h,\Delta t}^0\|_2 + \Delta t \|Q\|_2 \sum_{k=0}^{n-1} \|D_1^k D_2 Q^T T^{n-1-k}\|_2.$$

But $\|Q\|_2 = 1$ for any orthogonal matrix Q (Problem 7) and it is easily shown (Problem 6) that

$$\|D\|_2 = \max_{1 \leq k \leq m} |d_{kk}|$$

for any $m \times m$ diagonal matrix. Thus, we have

$$\|e_{h,\Delta t}^n\|_2 \leq \|D_1^n\|_2 \|e_{h,\Delta t}^0\|_2 + t_n \|D_1 D_2\|_2 \max_{0 \leq k \leq n} \|T^k\|_2,$$

from which the conclusion follows. ●

To illustrate the additional accuracy, consider the same simple example that we used to test the explicit and implicit methods. If we use Crank–Nicolson we get a better error for both the $h = 1/16$, $\Delta t = 1/64$ case as well as the $h = 1/32$, $\Delta t = 1/128$ case, as Figures

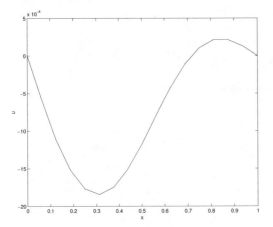

Figure 9.7   Error profile at $t = 1$ for heat equation; Crank–Nicolson method, $h = 1/16, \Delta t = 1/64$.

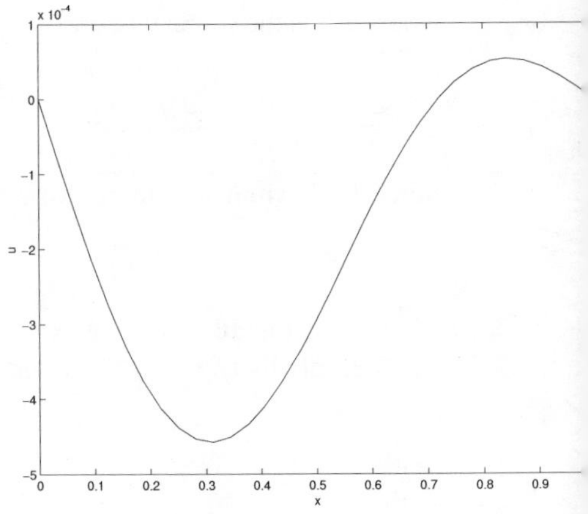

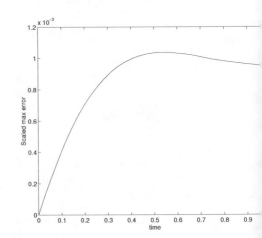

Figure 9.8   Error profile at $t = 1$ for heat equation; Crank–Nicolson method, $h = 1/32, \Delta t = 1/128$.

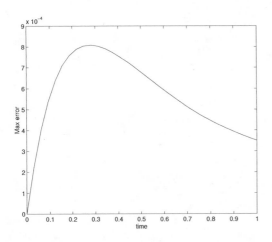

Figure 9.9   Maximum absolute error (unscaled) for $0 \leq t \leq 1$ for Crank–Nicolson, $h = 1/32$.

Figure 9.10   Maximum absolute error (scaled) for $0 \leq t \leq 1$ for Crank–Nicolson, $h = 1/32$.

9.7 and 9.8 show. Of more importance is the fact that we can get the same accuracy that the explicit method had with a lot fewer time steps, as Figure 9.9 shows. The peak error for Crank–Nicolson is about 9×10^{-4}, whereas for the explicit method it was 2.25×10^{-3}, and Crank–Nicolson required only 32 time steps to achieve this.

Because the values of the vector $u_{h,\Delta t}^n$ are supposed to represent function values, it is often preferable to scale the 2-norm by the mesh size h. We will call this the *discrete integral 2-norm*, because it can be considered a trapezoid rule approximation to the integral of (the square of) $u_{h,\Delta t}(\cdot, t_n)$ at the grid points. We will write $\|u_{h,\Delta t}^n\|_{2,h}$ for this norm, which is formally defined as

$$\|u_{h,\Delta t}(\cdot, t_n)\|_{2,h} = \left(h \sum_{i=1}^{N-1} u_{h,\Delta t}^2 (x_i, t_n) \right)^{1/2}, \tag{9.21}$$

and we note that in this notation the Crank–Nicolson error estimate can be written as

$$\|u^n - u_{h,\Delta t}^n\|_{2,h} \leq C_0^n \|u_0 - u_{h,\Delta t}^0\|_{2,h} + C_1 (h^2 + \Delta t^2)\|M_D\|_{2,h},$$

where M_D is a grid function dependent on the derivatives of the exact solution.

Note, however, that while the results for both the explicit method and the implicit method were in the infinity norm, the results for Crank–Nicolson are in the 2-norm. What does this mean, and what difference does it make?

Potentially, it can make quite a difference. The infinity norm gives us "pointwise" results, but the 2-norm does not; it gives average values. This allows the error in Crank–Nicolson to exhibit more "wiggles" than for the other methods; it also allows Crank–Nicolson to generate a negative approximation to a positive value, which will not happen with the other methods.

■ **EXAMPLE 9.3**

To illustrate what the difference between the two norms gives us, consider the new example defined by the initial data

$$u_0(x) = \begin{cases} 0, & 0 \leq x \leq \frac{1}{3}; \\ 1, & \frac{1}{3} < x < \frac{2}{3}; \\ 0, & \frac{2}{3} \leq x \leq 1. \end{cases}$$

This function is piecewise constant, discontinuous at $x = \frac{1}{3}$ and $x = \frac{2}{3}$, but it can be shown that the exact solution will be very smooth for any $t > 0$.

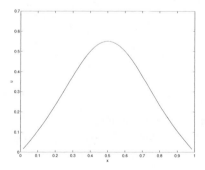

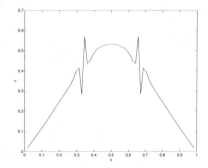

Figure 9.11   Solution profile at $t = \frac{1}{4}$ for discontinuous initial data; implicit method.

Figure 9.12   Solution profile at $t = \frac{1}{4}$ for discontinuous initial data; Crank–Nicolson method.

We approximated the solution to the heat equation using this initial condition, with $a = \pi^{-2}, h = 1/64$, and $\Delta t = 1/32$, and computed out to $t = 1/4$ (thus using eight time steps) with both the implicit and Crank–Nicolson methods. As Figures 9.11

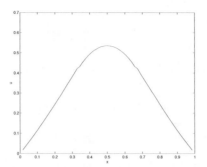

Figure 9.13   Solution profile at $t = \frac{1}{4}$ for discontinuous initial data; Crank–Nicolson method with implicit first step.

and 9.12 show, the implicit method handles the discontinuity in the initial data much better than Crank–Nicolson does. To remove the oscillations, we have to do one of two things—either take Δt very small in Crank–Nicolson—small enough to make the explicit method stable—or we can take the first step using the implicit method, and then proceed with Crank–Nicolson (this smooths out the discontinuity enough to avoid the oscillations we see in Crank–Nicolson, without introducing too much error). Figures 9.11 to 9.13 illustrate this.

Exercises:

1. Write a program to use the explicit method to solve the diffusion equation

$$
\begin{aligned}
u_t &= u_{xx}, \quad t > 0, 0 < x < 1; \\
u(0,t) &= 0; \\
u(1,t) &= 0; \\
u(x,0) &= \sin \pi x.
\end{aligned}
$$

which has exact solution

$$
u(x,t) = e^{-\pi^2 t} \sin \pi x.
$$

(The student should check that this is indeed the exact solution.)   Use $h^{-1} = 4, 8, \ldots, \frac{1}{1024}$, and take Δt as large as possible to maintain stability. Confirm that the approximate solution is as accurate as the theory predicts. Compute out to $t = 1$.

2. Repeat Problem 1, but this time take Δt 10% too large to satisfy the stability condition, and attempt to compute solutions out to $t = 1$. Comment on what happens.

3. Apply Crank–Nicolson to the same PDE as in Problem 1. For each value of h, adjust the choice of Δt to obtain comparable accuracy in the pointwise norm to what was achieved above. Comment on your results. Try to estimate the number of operations needed for each computation.

4. Write a program to use the explicit method to solve the diffusion equation

$$
\begin{aligned}
u_t &= u_{xx}, \quad t > 0, 0 < x < 1; \\
u(0,t) &= 0; \\
u(1,t) &= e^{-\pi^2 t/4}; \\
u(x,0) &= \sin \pi x/2.
\end{aligned}
$$

which has the exact solution

$$
u(x,t) = e^{-\pi^2 t/4} \sin \pi x/2.
$$

(The student should check that this is indeed the exact solution.) Use $h^{-1} = 4, 8, \ldots, \frac{1}{1024}$, and take Δt as large as possible to maintain stability. Confirm that the approximate solution is as accurate as the theory predicts. Compute out to $t = 1$.

5. Modify the three algorithms for the diffusion equation to handle nonhomogeneous boundary conditions, i.e., to handle problems of the form

$$
\begin{aligned}
u_t &= u_{xx} + f(x,t), \quad t > 0, 0 < x < 1; \\
u(0,t) &= g_0(t); \\
u(1,t) &= g_1(t); \\
u(x,0) &= u_0(x);
\end{aligned}
$$

where $g_0(t)$ and $g_1(t)$ are not identically zero. Test your work by applying it to the problem defined by

$$
f(x,t) = -2e^{x-t}; \quad g_0(t) = e^{-t}; \quad g_1(t) = e^{1-t}; \quad u_0(x) = e^x,
$$

for which the exact solution is $u(x,t) = e^{x-t}$.

6. Recall the definition of the matrix 2-norm:

$$
\|A\|_2 = \max_{u \neq 0} \frac{\|Au\|_2}{\|u\|_2},
$$

where $\| \cdot \|_2$ is the usual vector 2-norm. If D is an $n \times n$ diagonal matrix, use this definition to show that

$$
\|D\|_2 = \max_{1 \leq k \leq n} |d_{ii}|,
$$

where d_{ii} are the diagonal elements of D.

7. Let Q be an arbitrary $n \times n$ orthogonal matrix; show that $\|Q\|_2 = 1$.

8. Compute the number of operations needed to compute out to $t = T$ using the explicit method, as a function of n, the number of points in the spatial grid. Assume that the time step is chosen to be as large as possible to satisfy the stability condition.

9. Compute the number of operations needed to compute out to $t = T$ using Crank–Nicolson, taking $\Delta t = ch$ for some constant $c > 0$, again as a function of n, the number of points in the spatial grid.

10. Consider the *nonlinear* equation

$$u_t = u_{xx} + V u u_x.$$

Take $u_0(x) = \sin \pi x$ and homogeneous boundary data, and solve this using the explicit method, taking Δt to be 90% of the maximum value allowed for stability. Compute out to $t = 1$, and plot your solution, for $V = 1, 5, 10$.

11. Write down the nonlinear system that results from applying the implicit method to Problem 10, using $h = \frac{1}{8}$. How might you try to solve this system?

12. One way to attack this kind of nonlinear system would be to treat the u_{xx} term as usual in the implicit method, but treat the nonlinear term *explicitly*, i.e., use $(uu_x)(x_i, t_n)$ in the discretization. Write a program to do this approximation and compare your results to the fully explicit method in Problem 10

9.2   FINITE ELEMENT METHODS FOR THE DIFFUSION EQUATION

In §6.10.3 we outlined the finite element method for two-point boundary value problems. Can we do something like that for the diffusion equation? Of course we can. Consider the following simple example:

$$
\begin{aligned}
u_t &= a u_{xx} + f(x, t) \quad 0 \le x \le 1, \quad t > 0, & (9.22)\\
u(0, t) &= 0, & (9.23)\\
u(1, t) &= 0, & (9.24)\\
u(x, 0) &= u_0(x). & (9.25)
\end{aligned}
$$

It is relatively straight-forward to apply a finite element discretization to this problem. We assume a grid on $[0, 1]$ so that

$$0 = x_0 < x_1 < \cdots < x_{N-1} < x_N = 1$$

and then look for an approximate solution in the form

$$u_h(x, t) = \sum_{j=1}^{N-1} v_j(t) \phi_j^h(x),$$

where the ϕ_j^h are the tent functions from §6.10.3. Note that the coefficients $\{v_j\}$ are now taken to be functions of time, instead of constants. We proceed generally as we do in §6.10.3: First, multiply the PDE (9.22) by a smooth function v which also satisfies the boundary conditions, and integrate by parts over the interval. We get

$$\int_0^1 u_t(x, t) v(x) dx + a \int_0^1 u_x(x, t) v_x(x) dx = \int_0^1 f(x, t) v(x) dx.$$

This is analogous to (6.105). We now replace u with u_h and v with ϕ_j^h to get

$$\int_0^1 \left\{ (\phi_i^h(x)) \left(\sum_{j=1}^{N-1} v_j'(t) \phi_j^h(x) \right) + a \left(\phi_i^h(x) \right)' \left(\sum_{j=1}^{N-1} v_j(t) \left(\phi_j^h(x) \right)' \right) \right\} dx = \int_0^1 \phi_i^h(x) f(x) dx.$$

As in (6.106), this looks very imposing, but it can quickly be reduced to the following system of ordinary differential equations for the coefficients $v_j(t)$:

$$Mv'_h(t) + Kv_h(t) = F_h, \qquad (9.26)$$

where

$$k_{ij} = a \int_0^1 \left(\phi_i^h(x)\right)' \left(\phi_j^h(x)\right)' dx,$$

$$m_{ij} = \int_0^1 \phi_i^h(x)\phi_j^h(x)dx,$$

and

$$F_j^h(t) = \int_0^1 f(x,t)\phi_j^h(x)dx.$$

The matrix K is very similar to the coefficient matrix from §6.10.3, and is usually called the *stiffness matrix*, a terminology that comes out of the field of structural dynamics; M is similarly often called the *mass matrix*. Because of the local nature of the tent functions, both are tridiagonal. If the grid is uniform, with

$$x_{i+1} - x_i = h$$

for all i, then we have

$$K = \frac{1}{h}\operatorname{tridiag}(-1, 2, -1)$$

and

$$M = \frac{1}{6}h\operatorname{tridiag}(1, 4, 1).$$

If the grid is not uniform, then the matrices are of course not this simple.

We can apply this to the same example that we used in §9.1, but first we have to figure out how to solve the ODE system (9.26).

We can use almost any of the methods from Chapter 6, but we choose the trapezoid rule method, which is essentially the basis for Crank–Nicolson. This leads to the implicit time-stepping

$$\left(M + \frac{1}{2}a\Delta tK\right)v^{n+1} = \left(M - \frac{1}{2}a\Delta tK\right)v^n + \frac{1}{2}\Delta t(F_j^h(t_{n+1}) + F_j^h(t_n)).$$

Let's look at an example.

■ **EXAMPLE 9.4**

We take as our problem the same example we have been looking at in this section:

$$\begin{aligned}
u_t &= \frac{1}{\pi^2}u_{xx}, \quad t > 0, 0 < x < 1; \\
u(0,t) &= 0; \\
u(1,t) &= 0; \\
u(x,0) &= \sin \pi x + \sin 2\pi x.
\end{aligned}$$

Since Crank–Nicolson time-stepping is unconditionally stable, we take $\Delta t = h$ and compute out to $t = 1$. Our plots of the maximum absolute error, akin to the several

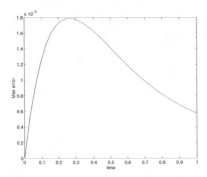

Figure 9.14 Error (unscaled) in finite element approximation of diffusion equation, $h = \Delta t = 1/32$.

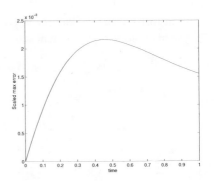

Figure 9.15 Error (scaled) in finite element approximation of diffusion equation, $h = \Delta t = 1/32$.

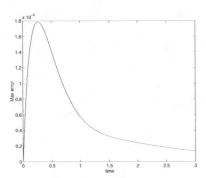

Figure 9.16 Error (unscaled) in finite element approximation of diffusion equation, $h = \Delta t = 1/32$, $T_{\max} = 3$.

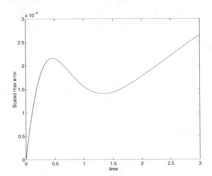

Figure 9.17 Error (scaled) in finite element approximation of diffusion equation, $h = \Delta t = 1/32$, $T_{\max} = 3$.

plots produced earlier in the chapter, are given in Fig. 9.14 to Fig. 9.17. Note that the error here is less than for Crank–Nicolson differencing.

Note that the scaled error is beginning to rise. Is this the result of some instability in the computation? No, it is not. The fact that it is a smooth rise suggests that it is not. In addition, this is the *scaled* error; the unscaled error is not rising. This is simply a reflection of the fact that the error is declining more slowly with time than is the solution.

Exercises:

1. Use a finite element approach to solve the problem

$$u_t = au_{xx}, \quad 0 \le x \le 1, \quad t > 0, \tag{9.27}$$
$$u(0,t) = 0, \tag{9.28}$$
$$u(1,t) = 0, \tag{9.29}$$
$$u(x,0) = u_0(x) = \sin \pi x + \sin 4\pi x. \tag{9.30}$$

For $a = 1/\pi^2$, this has the exact solution $u(x,t) = e^{-t}\sin\pi x + e^{-16t}\sin 4\pi x$ (Confirm this.[2])

2. Use a finite element approach to solve the problem

$$u_t = au_{xx} + 1, \quad 0 \le x \le 1, \quad t > 0 \tag{9.31}$$
$$u(0,t) = 0 \tag{9.32}$$
$$u(1,t) = 0 \tag{9.33}$$
$$u(x,0) = u_0(x) = \phi_M^h(x), \tag{9.34}$$

where $\phi_M^h(x)$ is the "hat function" centered nearest the middle of the interval. Solve the problem for a range of (positive) values of a, and compute out to $t = 1$. How does the value of a affect the results?

3. Discuss how you know that your code is working in Problem 2, given that we have no exact solution.

4. Use a finite element approach to solve the problem

$$u_t = au_{xx}, \quad 0 \le x \le 1, \quad t > 0 \tag{9.35}$$
$$u(0,t) = 1 \tag{9.36}$$
$$u(1,t) = 0 \tag{9.37}$$
$$u(x,0) = u_0(x) = \sin\pi x \tag{9.38}$$

For various values of h, solve this using the finite element method, for $a = 1$. Plot your solutions for several values of t.

5. Repeat Problem 4, this time using different (positive) values of a, the diffusion coefficient. How does varying a affect the solution?

9.3 DIFFERENCE METHODS FOR POISSON EQUATIONS

9.3.1 Discretization

In this section we discuss methods for the approximate solution of PDEs of the form

$$-\Delta u = f, \quad (x,y) \in D; u = g, \quad (x,y) \in \Gamma = \partial D,$$

where Δ is the Laplace operator,

$$\Delta u = \frac{\partial^2 u}{\partial x^2} + \frac{\partial^2 u}{\partial y^2};$$

---

[2]The author is chagrined to admit that several exercises in the Revised Edition had very "inexact exact solutions." He would like to try to claim that it was all done deliberately, to catch unwary students who failed to check these things, but that would be a claim of dubious honesty. So, be advised!

D is an open, connected subset of $\mathbb{R}^2$; and Γ is the boundary of D. The unknown function u therefore depends on two independent variables x and y. Equations of this type are known as Poisson equations, after the French mathematician Siméon-Denis Poisson.[3]

For simplicity's sake we will assume that D is the unit square $(0,1) \times (0,1)$, although much more general domains and, of course, much more general equations can be considered. It is also possible to consider problems posed in three dimensions.

In terms of the general theory, problems of this form have much in common with the boundary value problems studied in §6.10.1. The approximation is constructed by replacing the derivatives in the PDE with difference quotients, resulting in a system of linear equations. While we can apply Gaussian elimination and similar ideas to the solution of these systems, the size of the problem often mandates that we look at different methods.

Since the difference methods are fairly straightforward to apply, we will spend most of our time discussing efficient means for solving the large systems of linear equations that result from discretizing such problems. Thus, although the chapter is about approximating PDEs, this section has more to do with linear algebra than anything else.

We first define grids in both the x and y directions, in the usual way:

$$0 = x_0 < x_1 < x_2 < \cdots < x_{N-1} < x_N = 1$$

and

$$0 = y_0 < y_1 < y_2 < \cdots < y_{M-1} < y_M = 1.$$

For simplicity's sake, we assume that $N = M$, which implies an equivalent grid, with the same spacing h in each direction. We denote the approximate solution as the array of values $u_{i,j} \approx u(x_i, y_j)$, where x_i and y_j are points on a grid over D. Second, we use these in the usual difference approximation to the second derivative to write the approximate form of the PDE as

$$-\left[\frac{u_{i+1,j} - 2u_{i,j} + u_{i-1,j}}{h^2}\right] - \left[\frac{u_{i,j+1} - 2u_{i,j} + u_{i,j-1}}{h^2}\right] = f(x_i, y_j) \qquad (9.39)$$

which can be simplified somewhat to yield

$$-u_{i+1,j} - u_{i-1,j} + 4u_{i,j} - u_{i,j+1} - u_{i,j-1} = h^2 f_{i,j}, \quad 1 \le i, j \le N-1,$$

where we have written $f_{i,j} = f(x_i, y_j)$. This is a system of $(N-1)^2$ equations in $(N-1)^2$ unknowns. To write it in a matrix–vector format, we define the vector u_h as a one-dimensional arrangement of the $u_{i,j}$ values:

$$u_h = (u_{1,1}, u_{1,2}, u_{1,3}, \ldots, u_{1,N-1}, u_{2,1}, \ldots, u_{N-1,N-1})^T.$$

We could just as easily have ordered the $u_{i,j}$ in another way (by columns, instead of rows, for example). The ordering of the unknowns can make a difference, especially in more complicated settings.

---

[3]Siméon-Denis Poisson (1781–1840) was born in Pithiviers, France, about 50 miles south of Paris. He was sent by his father to become a doctor, but the young Poisson had little interest in medicine, and lacked the manual coordination necessary to be a surgeon. He was then enrolled in the École Centrale, where his mathematical talent was first recognized. He was encouraged to apply to the École Polytechnique, which he entered in 1798. (Two of his teachers were Laplace and Lagrange.) He published over 300 articles in a wide range of areas. His name is attached to this differential equation because of his study of potential theory, published in the journal *Bulletin de la Société Philomatique* in 1813.

The resulting linear system can be quite large but has a very interesting structure. For $N = 4$, we have the following 9×9 system:

$$
\left[
\begin{array}{ccc|ccc|ccc}
4 & -1 & 0 & -1 & 0 & 0 & 0 & 0 & 0 \\
-1 & 4 & -1 & 0 & -1 & 0 & 0 & 0 & 0 \\
0 & -1 & 4 & 0 & 0 & -1 & 0 & 0 & 0 \\
\hline
-1 & 0 & 0 & 4 & -1 & 0 & -1 & 0 & 0 \\
0 & -1 & 0 & -1 & 4 & -1 & 0 & -1 & 0 \\
0 & 0 & -1 & 0 & -1 & 4 & 0 & 0 & -1 \\
\hline
0 & 0 & 0 & -1 & 0 & 0 & 4 & -1 & 0 \\
0 & 0 & 0 & 0 & -1 & 0 & -1 & 4 & -1 \\
0 & 0 & 0 & 0 & 0 & -1 & 0 & -1 & 4
\end{array}
\right]
\left[
\begin{array}{c}
u_{1,1} \\ u_{1,2} \\ u_{1,3} \\ u_{2,1} \\ u_{2,2} \\ u_{2,3} \\ u_{3,1} \\ u_{3,2} \\ u_{3,3}
\end{array}
\right]
= h^2
\left[
\begin{array}{c}
f_{1,1} \\ f_{1,2} \\ f_{1,3} \\ f_{2,1} \\ f_{2,2} \\ f_{2,3} \\ f_{3,1} \\ f_{3,2} \\ f_{3,3}
\end{array}
\right] . \quad (9.40)
$$

We can take advantage of the structure in this system by defining some notation. Let

$$
K = \left[
\begin{array}{ccc}
4 & -1 & 0 \\
-1 & 4 & -1 \\
0 & -1 & 4
\end{array}
\right]
$$

and denote by u_i the column subvector of elements $u_{i,j}$, $1 \le j \le N - 1$ (and similarly for f_i). Then we can write the large system more compactly as

$$
\left[
\begin{array}{ccc}
K & -I & 0 \\
-I & K & -I \\
0 & -I & K
\end{array}
\right]
\left[
\begin{array}{c}
u_1 \\ u_2 \\ u_3
\end{array}
\right]
= h^2
\left[
\begin{array}{c}
f_1 \\ f_2 \\ f_3
\end{array}
\right].
$$

There are a number of things that can be observed about this system of equations. The most important one, for our purposes, is that it contains a lot of zeros. In fact, the coefficient matrix is "mostly" zeros. Of $\mathcal{O}(N^4)$ elements in the matrix, only $\mathcal{O}(5N^2)$ will be nonzero. Some of the other properties—the symmetry, the very nice and tidy block structure, the simplicity of the numbers—will be drastically affected by changes in the original PDE or the computational parameters or the shape of the domain D. But the *sparseness* of the matrix will remain.

We could solve this system by applying the standard techniques from Chapter 7, and for moderate-sized problems, this is perhaps the best choice. The coefficient matrix is SPD, so the Cholesky algorithm is the natural first choice. However, for very large problems, so-called direct solution algorithms are not the best choice. They take too much computer time, and use too much computer storage.

An alternative approach is to use an iterative technique like those studied in §7.7 to solve the system. This seems counter-intuitive, perhaps, since we have available direct methods for solving linear systems exactly. Why would we use an iterative method that is only going to approximate the solution? The answer lies in the size of the problem. For large values of N—in other words, for small values of the mesh parameters—performing Gaussian elimination on a system such as (9.40), but for larger N, will involve an enormous amount of computer storage, and the overhead in managing that storage can make the computation excessively expensive (even on modern computers). On the other hand, the iterative techniques that are used in practice converge very quickly, and the individual iterations are so cheap that the overall algorithms are cost-competitive with Gaussian elimination.

9.3.2 Banded Cholesky Solvers

Note that the coefficient matrix in our example above is entirely zero outside of three sub- or super-diagonals of the main diagonal. This is the essential feature of a *banded* matrix, that all the elements are zero if we are far enough away from the diagonal. Formally, we say a that matrix A is banded, with *half-bandwidth* p, if

$$a_{i,j} = 0$$

for all i, j such that $|i - j| > p$. Thus, our example is banded with half-bandwidth $p = 3$.

The important feature of banded systems is that the elimination algorithm can be carried out "within the band." That is, we don't need to access, loop through, or even store those elements outside the band. This allows us to write elimination and solution routines that not only run faster but use much less storage.

To illustrate, consider a Cholesky factorization routine applied to systems such as (9.40). A naive approach, ignoring the band structure, would waste a lot of time looping through indices of elements that are always zero, in addition to the storage costs of saving all those zeros. To save storage, we have to decide upon a scheme by which we put the band of the original matrix A into a smaller array, and then we have to modify the indexing of our factorization algorithm accordingly.

Perhaps the simplest storage scheme is to store each column below (and including) the main diagonal as a column in a rectangular array. Thus we store the band of our example A as

$$
B_A = \begin{bmatrix}
4 & 4 & 4 & 4 & 4 & 4 & 4 & 4 & 4 \\
-1 & -1 & 0 & -1 & -1 & 0 & -1 & -1 & 0 \\
0 & 0 & 0 & 0 & 0 & 0 & 0 & 0 & 0 \\
-1 & -1 & -1 & -1 & -1 & 0 & 0 & 0 & 0
\end{bmatrix}.
$$

Instead of being a full 9×9, this is only 4×9. In terms of the discretization parameters, instead of storing a matrix of size $(N - 1)^2 \times (N - 1)^2$, we can get by with one that is $N \times (N - 1)^2$. In large problems, this can be a substantial savings.

To finish the banded algorithm, we need an *index map* that takes us from the indices for the original matrix A and returns the appropriate indices for the band-only array B_A. One advantage of the band storage system that we have chosen to use is that all of the second (i.e., column) indices remain the same. Thus, all we need to do is determine the single-variable map that takes us from $a_{i,j}$ to $b^a_{k(i,j),j}$. A little experimentation with the indices ought to convince the reader that the map we want is

$$k(i, j) = i - j + 1,$$

so that we get our band algorithm by replacing every occurence of $a_{i,j}$ in our Cholesky factorization with $b^a_{k(i,j),j}$. This gives us Algorithm 9.1.

The scheme we have used here to handle the banded matrix is somewhat crude, but easy to understand. It is possible to construct much more sophisticated procedures that allow us to efficiently handle matrices that have "variable bandwidths" as we move down the diagonal. Some discussion of this is in the text by Watkins [16]; a complete treatment is in Duff et al. [5].

This is the type of algorithm that is used in the next section in the comparison tests with the conjugate gradient and other iterative methods. Although this is a very reasonable approach to take, for problems where a fine grid is required to obtain the accuracy needed in the PDE approximation, the storage requirements will still be excessive on any but the largest computers; note, also, that even with the band storage system, as n gets larger we have to store a lot of elements that begin as zeros, but which are overwritten with nonzeros during the elimination process. (This is known as "fill-in.") This is why we are led to the use of iterative methods.

Algorithm 9.1 *Band-limited Cholesky Algorithm (with Band Storage).*

Here the matrix is stored in the variable ba *with only that part within the band being stored. The matrix thus has* n *rows and* n2 *columns. The function* kk *is the index map and is defined by*

$$kk(i,j) = i - j + 1.$$

```
ba(kk(1,1),1) = sqrt(ba(kk(1,1),1))
for i=2,n do
    ba(kk(i,1),1) = ba(kk(i,1),1)/ba(kk(1,1),1)
end
for j=2,n2-1 do
    sum = 0.0d0
    kmin = max(1,j-n+1)
    for k=kmin,j-1 do
        sum = sum + ba(kk(j,k),k)*ba(kk(j,k),k)
    end
    ba(kk(j,j),j) = sqrt(ba(kk(j,j),j) - sum)
    imax = min(n2,j+n-1)
    for i=j+1,imax do
        sum = 0.0d0
        kmin = max(1,j-n+1)
        for k=kmin,j-1 do
            sum = sum + ba(kk(i,k),k)*ba(kk(j,k),k)
        end
        ba(kk(i,j),j) = (ba(kk(i,j),j) - sum)/ba(kk(j,j),j)
    end
end
sum = 0.0d0
for k=n2-n+1,n2-1 do
    sum = sum + ba(kk(n2,k),k)*ba(kk(n2,k),k)
end
ba(kk(n2,n2),n2) = sqrt(ba(kk(n2,n2),n2) - sum)
```

9.3.3   Iteration and the Method of Conjugate Gradients

In §7.7 we looked at some very simple iterative methods for solving large sparse linear systems. In this section we will show how to implement them when applied to the type of systems generated by discretizing Poisson equations. We will also introduce a very powerful iterative technique, the method of conjugate gradients.

Recall the equation for the approximate value at (x_i, y_j):

$$-u_{i+1,j} - u_{i-1,j} + 4u_{i,j} - u_{i,j+1} - u_{i,j-1} = h^2 f_{i,j}. \tag{9.41}$$

We can solve this for $u_{i,j}$ as follows:

$$u_{i,j} = \frac{1}{4}\left(u_{i+1,j} + u_{i-1,j} + u_{i,j+1} + u_{i,j-1} + h^2 f_{i,j}\right). \tag{9.42}$$

Suppose now that we take $u_{i,j}^{(0)}$ to be an initial guess, for $1 \le i \le N-1$ and similarly for j, and convert (9.42) into an iteration:

$$u_{i,j}^{(k+1)} = \frac{1}{4}\left(u_{i+1,j}^{(k)} + u_{i-1,j}^{(k)} + u_{i,j+1}^{(k)} + u_{i,j-1}^{(k)} + h^2 f_{i,j}\right). \tag{9.43}$$

This is the same as the Jacobi iteration applied to the original system defined by (9.41), as can be seen by a modest manipulation (Problem 1). Thus, the Jacobi iteration can be implemented by a very simple set of loops.

The Gauss–Seidel iteration is not much more complicated than Jacobi. We have (Problem 2)

$$u_{i,j}^{(k+1)} = \frac{1}{4}\left(u_{i+1,j}^{(k)} + u_{i-1,j}^{(k+1)} + u_{i,j+1}^{(k)} + u_{i,j-1}^{(k+1)} + h^2 f_{i,j}\right). \tag{9.44}$$

Note that some of the u values on the right side are at iteration k, and some are at iteration $k+1$. This is how Gauss–Seidel differs from Jacobi: We are using the most current value of the approximate solution for those elements for which it has been computed. This leads to an algorithm that is actually more compact than for Jacobi.

The most complicated of the three splitting methods is SOR. The best way to organize the SOR iteration is to first take a Gauss–Seidel step:

$$\bar{u} = \frac{1}{4}\left(u_{i+1,j}^{(k)} + u_{i-1,j}^{(k+1)} + u_{i,j+1}^{(k)} + u_{i,j-1}^{(k+1)} + h^2 f_{i,j}\right),$$

and then average this value with the previous iterate value, using the relaxation parameter:

$$u_{i,j}^{(k+1)} = \omega\bar{u} + (1-\omega)u_{i,j}^{(k)}. \tag{9.45}$$

Problem 3 asks the student to show that this is equivalent to the matrix formulation in Chapter 7. Note that if $\omega = 1$, then Gauss–Seidel and SOR are equivalent.

Of course, the convergence behavior of the SOR iteration will be heavily dependent on the choice of ω. A very complete theory for the choice of the "best" value of ω is available [17] for the types of SPD systems that come from discretizing Poisson equations. The precise details are more than we want to get into in this text, but a summary of the important results is the following.

Theorem 9.4 *If the system of equations $Au = f$ comes from discretizing a Poisson equation of the form*

$$-a_1 u_{xx} - a_2 u_{yy} + a_3 u = f,$$

where $a_1 > 0$, $a_2 > 0$, and $a_3 \ge 0$, and the unknowns are ordered within the vector u either row-wise or column-wise, then:

1. *The spectral radius of the Jacobi iteration matrix, ρ_J, and the spectral radius of the Gauss–Seidel iteration matrix, ρ_{GS}, are related according to*

$$\rho_{GS} = \rho_J^2;$$

2. *The "best" value of ω for the SOR iteration ("best" in the sense that it leads to the smallest spectral radius for the SOR iteration matrix) is given by*

$$\omega_* = \frac{2}{1 + \sqrt{1 - \rho_J^2}}.$$

Note that to use the optimal value of ω we apparently have to solve an eigenvalue problem first. In practice, an SOR iteration is usually begun with $\omega = 1$, which is Gauss–Seidel, and from the way that the iterates change it is possible to estimate ρ_{GS} and therefore ρ_J, thus allowing us to use an estimate of the optimal ω (see Problem 6). However, there is no denying that the need to come up with a value for ω makes effective use of SOR difficult, and is one reason that it is no longer as widely used as it once was.

The most important modern technique for the solution of large, sparse symmetric linear systems is known as the *method of conjugate gradients*, or "CG." The basic ideas go back to the mid-1950s, but the utility of the method was not completely recognized until the late 1980s. A full treatment of the theory underlying the method requires more background than we have assumed here, so we will simply present the algorithm and outline some of the major ideas. Several more omplete treatments are in the references [4, 9, 12].

Theorem 9.5 *Let $A \in \mathbb{R}^{n \times n}$ be given, SPD. For any $f \in \mathbb{R}^n$, the vector $u \in \mathbb{R}^n$ satisfies $u = A^{-1}f$ if and only if*

$$(u, Au) - 2(u, f) < (v, Av) - 2(v, f)$$

for all $v \in \mathbb{R}^n$, $u \neq v$.

The point of the theorem is that we can find the solution of $Ax = b$ by finding the vector that minimizes
$$\phi(u) = (u, Au) - 2(u, f).$$

Many techniques exist for the solution of minimization problems, and many have the form

$$u^{(k+1)} = u^{(k)} + \alpha_k p^{(k)},$$

where α_k is a scalar (called the *step length*) and $p^{(k)}$ is a vector (called the *search direction*). Thus, we go from the current approximate minimizer to the next approximate minimizer by determining a direction in which to move, and a distance to go along that direction. Often, α_k is chosen to minimize ϕ along the search direction. Obviously, a lot will depend on how the search direction is chosen.

The CG method works by choosing the search direction so that the residual vectors $r^{(k)} = b - Ax^{(k)}$ are all mutually orthogonal:

$$(r^{(k)}, r^{(j)}) = 0, \quad k \neq j.$$

It therefore follows that if A is $m \times m$, then $r^{(m)} = 0$, because otherwise we would have $m + 1$ mutually orthogonal vectors in an m-dimensional space. (See Problem 5) Thus, CG is not really an iterative method, since it finds the theoretically exact solution in a finite number of steps. However, as a practical matter, CG converges much faster than this would suggest, as some of the exercises demonstrate.

The complete algorithm is as follows:

Algorithm 9.2 *Conjugate Gradient Iteration.*
Assume $A \in SPD$ and compute as follows.

1. *Given $u^{(0)}$, the initial guess, compute $r^{(0)} = f - Au^{(0)}$ and $p^{(0)} = r^{(0)}$. Let ϵ be the convergence tolerance.*

2. *For $k = 1$ to convergence do*

   (a) $w = Ap^{(k-1)}$;

   (b) $\alpha = (r^{(k-1)}, r^{(k-1)})/(p^{(k-1)}, w)$;

   (c) $u^{(k)} = u^{(k-1)} + \alpha p^{(k-1)}$;

   (d) $r^{(k)} = r^{(k-1)} - \alpha w$;

   (e) *If $\|r^{(k)}\|_2 \le \epsilon$ then stop;*

   (f) $\beta = (r^{(k)}, r^{(k)})/(r^{(k-1)}, r^{(k-1)})$;

   (g) $p^{(k)} = r^{(k)} + \beta p^{(k-1)}$.

Note that we can execute this algorithm using only the storage for the matrix plus four additional vectors (u, w, r, and p), since we can over-write the values from one iteration onto the next. In addition, we do not really need to store the entire matrix A; all we have to do is have a routine that can compute the product $w = Ap$. For a matrix as structured for the discretization of Poisson equations, this can be done very efficiently and with a minimum of storage.

■ **EXAMPLE 9.5**

To compare the performance of all four of these methods, we will try them each out on the linear system that we get from the PDE

$$-\Delta u = f(x,y), \ (x,y) = (0,1) \times (0,1) \in R; \quad (9.46)$$
$$u = 0, \quad (x,y) \in \partial R, \quad (9.47)$$

for $f(x,y) = (2 + \pi^2 y(1-y))\sin \pi x + (2 + \pi^2 x(1-x))\sin \pi y$, which has the exact solution

$$u(x,y) = y(1-y)\sin \pi x + x(1-x)\sin \pi y.$$

The discretization we are using is second-order accurate, so taking $h = 1/100$ gives us an expected error on the order of 10^{-4}; it also means that $N = 99^2 = 9801$. Doing a direct solution, even taking advantage of the symmetry and band structure of the matrix, would require working with a 99×9801 rectangular matrix, thus, we would have to store 970,299 elements. Table 9.2 shows the results for each of the four iterations for a sequence of values of $h^{-1} = 4, 8, 16, \ldots, 128$; we show the number of iterations to convergence, an estimate of the number of "flops," or floating-point operations (additions, subtractions, multiplications, divisions, square roots) performed in doing the computation, and the error between the exact solution and the converged iteration, measured in the infinity norm. For SOR the optimal ω was precomputed and used, which is entirely unrealistic but allows us to see the best possible performance for SOR.

Table 9.2   Iterative solutions of (9.46)-(9.47).

Jacobi Iteration			
h^{-1}	Iterations	Error	Flops
4	15	1.1E–02	1908
8	82	2.5E–03	56,350
16	406	6.3E–04	1,279,350
32	1917	1.6E–04	25,793,240
64	8827	3.9E–05	490,489,020
128	39921	9.8E–06	9,014,433,584

Gauss–Seidel Iteration			
h^{-1}	Iterations	Error	Flops
4	9	1.2E–02	1152
8	42	2.6E–03	28,910
16	204	6.3E–04	643,050
32	960	1.6E–04	12,917,762
64	4415	3.9E–05	245,331,828
128	19962	9.8E–06	4,507,571,630

SOR Iteration with optimal ω			
h^{-1}	Iterations	Error	Flops
4	6	1.3E–02	990
8	15	3.3E–03	13,328
16	33	8.0E–04	134,100
32	72	2.0E–04	1,247,378
64	158	4.9E–05	11,295,774
128	354	1.3E–05	102,806,246
256	831	3.2E–06	972,774,000

Conjugate Gradient Iteration			
h^{-1}	Iterations	Error	Flops
4	3	6.8E–03	291
8	5	1.7E–03	3,045
16	9	4.2E–04	27,465
32	17	1.0E–04	232,593
64	39	2.6E–05	2,270,331
128	86	6.5E–06	20,596,860
256	178	1.6E–06	172,771,680

The iterations were all stopped when the residual $f - Au$ was smaller than h^4; Jacobi, Gauss–Seidel, and SOR measured this in the infinity norm, but the 2-norm is much more natural for CG. This makes the comparison something of an "apples and oranges" comparison, but it allows us to use the most natural norms for each method. Making the convergence tolerance dependent on h was done to ensure that the iteration error was much less than the $\mathcal{O}(h^2)$ mathematical error in the approximation technique. For comparison's sake, the solution of the same system on the same computer, using a Cholesky factorization that took advantage of the band structure of the matrix, is summarized in Table 9.3. Note that CG was just as accurate as the Cholesky decomposition, but much cheaper in terms of the operation count. CG was also faster than the three splitting methods. Moreover, CG does not depend on a parameter like SOR does. Note that SOR and CG were the only methods for which we could run the $n = 256$ case. This last point deserves emphasizing, since one of the drawbacks to a direct solution technique such as the Cholesky decomposition is the memory requirement for storing even a band-limited form of the matrix. Figs. 9.18 and 9.19 show the solution and error surfaces, respectively.[4]

Table 9.3 Cholesky solution of (9.46).

h^{-1}	Error	Flops
4	6.8298357260145E–03	322
8	1.6705855964484E–03	6,974
16	4.1526959889103E–04	122,230
32	1.0366786255403E–04	2,029,286
64	2.5907605235998E–05	33,021,382
128	6.4763166429495E–06	532,642,694

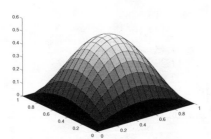

Figure 9.18 Surface plot of approximate solution to (9.46)-(9.47), using $h = 1/16$.

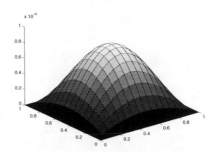

Figure 9.19 Plot of the error in Fig. 9.18.

The method of conjugate gradients requires that the matrix $A \in$ SPD, thus it will not work if the Poisson equation has a term such as $b(x, y)u_x$ in it (see Problem 8). CG can be applied to asymmetric problems by multiplying by A^T; i.e., if we want to solve the linear system

$$Au = f$$

---

[4]Color versions of these plots should be on the text website. See the preface for details on how to access the online material.

for $A \neq A^T$, we apply CG to the new system

$$A^T A u = A^T f.$$

If A is nonsingular, then $A^T A$ is SPD, and this can be implemented with only modest additional work in the algorithm (see Problem 9). However, it can be shown that the condition number of $A^T A$ is the square of the condition number of A, and this can have drastic effects on the convergence of the CG algorithm. The development of an efficient version of CG that can be applied to asymmetric problems is a very active area of numerical analysis research.

The fundamental error and convergence theorem for CG can be stated once we define two notations, the first of which we used in Chapter 8.

Definition 9.1 (Spectrum of a Matrix) *If A is a square matrix, the set $\sigma(A)$ is the set of all eigenvalues of A.*

Definition 9.2 *For any k, the set Q_k will denote all polynomials p of degree k with the additional property that $p(0) = 1$.*

Then the basic theorem is as follows.

Theorem 9.6 *Let the CG algorithm be applied to the solution of the linear system $Au = f$, with $A \in SPD$. Let $u^{(k)}$ denote the kth iterate. Then the error satisfies*

$$\|u - u^{(k)}\|_2 \leq \sqrt{\kappa_2(A)} \left(\max_{\lambda \in \sigma(A)} \min_{p \in Q_k} |p(\lambda)| \right) \|u - u^{(0)}\|_2.$$

Stating the error in terms of a minimum over a set of polynomials allows us to use some advanced properties of Chebyschev polynomials to obtain the following very conservative result.

Theorem 9.7 *The error in CG satisfies the conservative upper bound*

$$\|u - u^{(k)}\|_2 \leq 2\sqrt{\kappa_2(A)} \left(\frac{\sqrt{\kappa_2(A)} - 1}{\sqrt{\kappa_2(A)} + 1} \right)^k \|u - u^{(0)}\|_2.$$

From this we see why the condition number of A is so important. For a poorly conditioned matrix, the method can converge very slowly, and this leads to the idea of *preconditioning*. If we want to solve the system

$$Au = f, \tag{9.48}$$

this is equivalent (assuming that S is nonsingular) to solving

$$SAS^T S^{-T} u = Sf$$

or,

$$SAS^T v = Sf \tag{9.49}$$

for $u = S^T v$. Note that the product SAS^T is always symmetric; moreover, if $(S^T S)^{-1} = M \approx A$, then it can be shown that

$$\kappa_2(SAS^T) \approx 1;$$

thus, the CG iteration for solving (9.49) will converge more quickly than the CG iteration for solving the original system, and we recover u from v via $u = S^T v$. We can organize the CG iteration applied to the preconditioned system in such a way that we only need to do a few extra computations.

Algorithm 9.3 *Preconditioned Conjugate Gradient Algorithm.*

   1. Given $u^{(0)}$, the initial guess, compute $r^{(0)} = f - Au^{(0)}$ and $p^{(0)} = z^{(0)} = M^{-1}r^{(0)}$. Let ϵ be the convergence tolerance.

   2. For $k = 1$ to convergence do

      (a) $w = Ap^{(k-1)}$;

      (b) $\alpha = (z^{(k-1)}, r^{(k-1)})/(p^{(k-1)}, w)$;

      (c) $u^{(k)} = u^{(k-1)} + \alpha p^{(k-1)}$;

      (d) $r^{(k)} = r^{(k-1)} - \alpha w$;

      (e) If $\|r^{(k)}\|_2 \le \epsilon$ then stop;

      (f) $z^{(k)} = M^{-1}r^{(k)}$;

      (g) $\beta = (z^{(k)}, r^{(k)})/(z^{(k-1)}, r^{(k-1)})$;

      (h) $p^{(k)} = z^{(k)} + \beta p^{(k-1)}$.

Note that we do not actually compute the inverse of M; rather, we solve the system $Mz^{(k)} = r^{(k)}$. This means that the preconditioner must be a matrix for which the linear systems problem is *easier* to solve than it is for A.

The choice of preconditioners is an active area of research. Common choices amount to some kind of partial factorization of the original matrix A, or some direct approximation to A^{-1}.

One choice is the so-called *incomplete Cholesky factorization*; in this method, we compute only those elements of the Cholesky factorization of A that correspond to nonzero elements of A. (This avoids the massive fill-in that slows down the Cholesky factorization.) If L is the incomplete Cholesky factor, then the preconditioner is $M = LL^T$. The solution of the preconditioning equation is simple because M is already factored, so the solution for $z^{(k)}$ is accomplished by a couple of triangular solution steps.

Table 9.4 shows the results of using a simple incomplete Cholesky preconditioner to solve our model problem; note that the number of iterations to convergence decreased markedly, but the operation counts are initially actually greater, and it is only for the larger problems that the preconditioning pays off. This is because the preconditioning step involves a nontrivial number of additional operations.

See Chapter 10 of Saad [12] for a good introduction to preconditioning ideas. MATLAB contains software to implement some preconditioners.

Time-Dependent Problems and Finite Elements

Two questions appear to be unanswered here:

   1. Can we treat higher-dimensional problems for the diffusion equation, such as problems governed by the PDE $u_t = a(u_{xx} + u_{yy})$?

   2. Could we apply the finite element method to the Poisson equation?

Table 9.4   Incomplete Cholesky preconditioned conjugate gradient solution of (9.46).

h^{-1}	Iterations	Error	Flops
4	2	6.8E–03	799
8	3	1.7E–03	6,027
16	5	4.6E–04	42,667
32	9	1.6E–04	309,387
64	16	5.9E–05	2,195,175
128	30	1.7E–05	16,373,343
256	59	4.7E–06	128,241,523

The answer to both equations is an emphatic *yes*, and it was simply an editorial judgment on the author's part to skip these issues. We have probably already stretched the boundaries of what should be in an introductory text!

Exercises:

1. Show that the Jacobi iteration (9.43) is equivalent to the matrix iteration (7.19) from Chapter 7.

2. Show that the Gauss–Seidel iteration (9.44) is equivalent to the matrix iteration (7.20) from Chapter 7.

3. Show that the SOR iteration (9.45) is equivalent to the matrix iteration (7.22) from Chapter 7.

4. What is the truncation error in the approximation defined by (9.39)?

5. Let $x_1, x_2, \ldots, x_{n-1}, x_n$ be orthogonal vectors in a vector space V of dimension n. Show that if $z \in V$ is orthogonal to each one of the x_k, then $z = 0$.

6. For an iteration of the form

$$u^{(k+1)} = Tu^{(k)} + c,$$

   show that

$$\frac{\|u^{(k+1)} - u^{(k)}\|_\infty}{\|u^{(k)} - u^{(k-1)}\|_\infty} \leq \|T\|_\infty.$$

   Can we use this to estimate ρ_J and therefore ω_*? *Hint:* Recall Problem 8 from §7.7.

7. Write programs to do

   (a) Banded Cholesky;

   (b) CG;

   Test your program on the example problem defined by

$$-\Delta u = \pi^2 \sin \pi x \sin \pi y, (x, y) \in (0, 1) \times (0, 1),$$

   with $u = 0$ on the boundary, using $h = \frac{1}{8}$.

8. Discretize the Poisson equation

$$-u_{xx} - u_{yy} + bu_x = f; (x, y) \in (0, 1) \times (0, 1);$$
$$u(x, 0) = u(x, 1) = 0; x \in (0, 1);$$
$$u(0, y) = u(1, y) = 0; y \in (0, 1);$$

in the case $h = \frac{1}{3}$, $b \neq 0$. Is the resulting system symmetric?

9. Consider the linear system problem

$$Au = f,$$

where A is *not* symmetric. Modify the CG algorithm to solve the symmetrized system

$$A^T Au = A^T f$$

without explicitly forming the matrix $A^T A$.

10. Write a code to implement the algorithm you wrote in Problem 9. Test it on the system obtained by discretizing the PDE

$$-u_{xx} - u_{yy} + u_x = \pi \cos \pi x \cos \pi y + 2\pi^2 \sin \pi x \sin \pi y; (x, y) \in (0, 1) \times (0, 1);$$
$$u(x, 0) = u(x, 1) = 0; x \in (0, 1);$$
$$u(0, y) = u(1, y) = 0; y \in (0, 1);$$

for $h = \frac{1}{4}$ and $h = \frac{1}{8}$. The exact solution is $u(x, y) = \sin \pi x \sin \pi y$; use this to ensure your algorithm is working properly.

11. How does the discretization change when the boundary data is nonhomogeneous (i.e., nonzero)? Demonstrate by writing down the discrete system for the PDE

$$-u_{xx} - u_{yy} = -2e^{x-y}; (x, y) \in (0, 1) \times (0, 1);$$
$$u(x, 0) = e^x; \quad u(x, 1) = e^{x-1}; \quad x \in (0, 1);$$
$$u(0, y) = e^{-y}; \quad u(1, y) = e^{1-y}; \quad y \in (0, 1);$$

for $h = \frac{1}{4}$. *Hint:* It will help to write the values of the approximate solution at the grid points in two vectors, one for the interior grid points where the approximation is unknown, and one at the boundary grid points where the solution is known.

12. Apply the following solution techniques to the system in Problem 11, this time using $h = \frac{1}{8}$. Use the exact solution of $u(x, y) = e^{x-y}$ to verify that the code is working properly.

   (a) Jacobi iteration;

   (b) Gauss–Seidel iteration;

   (c) SOR iteration using $\omega = 1.4465$;

   (d) Conjugate gradient iteration.

◁ ● ● ▷

9.4 LITERATURE AND SOFTWARE DISCUSSION

The numerical solution of partial differential equations is one of the most active areas of numerical analysis research, and is usually not treated extensively in undergraduate texts. Good sources for the basic difference methods that we outlined here are the books by Ames [1], Smith [13], Morton and Mayers [11], and Strikwerda [15]. A classic, though by now somewhat dated source is the book by Forsythe and Wasow [6]. The question of numerical methods for the resulting linear systems gets extensive treatment. The best references for the CG method are Datta [4], Kelley [9], and Saad [12]. More classical iterative methods appear in the books by Young [17], Hageman and Young [7], and Birkhoff and Lynch [2].

One class of equations not treated here are those first-order equations often styled as "conservation laws." The general form of the equation in one space dimension is

$$u_t + (f(x, u))_x = 0.$$

These are important equations in areas such as fluid flow and gas dynamics; a good reference is the book by LeVeque [10].

Perhaps the most important technique for approximating solutions to PDEs is the finite element method, which is sufficiently complicated that it deserves its own course. The classic reference is the book by Strang and Fix [14]; a more modern treatment is the book by Brenner and Scott [3].

Easy-to-use software for approximating the solution of PDEs is difficult to write and therefore difficult to find. The generality of the problem, the need to specify solution techniques, data functions, problem geometries, boundary conditions, etc., all make it a challenging project. MATLAB contains a number of tools to assist in this.

REFERENCES

1. Ames, W. F., *Numerical Methods for Partial Differential Equations*, Academic Press, New York, 1977.

2. Birkhoff, Garrett, and Lynch, Robert, *Numerical Solution of Elliptic Problems*, SIAM, Philadelphia, 1984.

3. Brenner, Suzanne, and Scott, Ridgway, *The Mathematical Theory of Finite Element Methods*, Springer-Verlag, New York, 1994.

4. Datta, Biswa Nath, *Numerical Linear Algebra and Applications*, Brooks/Cole, Pacific Grove, California, 1995.

5. Duff, Ian, Erisman, A. M., and Reid, J. K., *Direct Methods for Sparse Matrices*, Oxford University Press, Oxford, UK, 1989.

6. Forsythe, George, and Wasow, Wolfgang, *Finite-Difference Methods for Partial Differential Equations*, John Wiley & Sons, Inc., New York, 1960.

7. Hageman, L. A., and Young, David, *Applied Iterative Methods*, Academic Press, New York 1981.

8. Humi, Mayer, and Miller, William B., *Boundary Value Problems and Partial Differential Equations*, PWS-Kent, Boston, 1992.

9. Kelley, C. T., *Iterative Methods for Linear and Nonlinear Equations*, SIAM, Philadelphia, 1995.

10. LeVeque, Randall, *Numerical Methods for Conservation Laws*, Birkhäuser, Basel, Switzerland, 1990.

11. Morton, K. W., and Mayers, D. F. *Numerical Solution of Partial Differential Equations*, Cambridge University Press, Cambridge, UK, 1994.

12. Saad, Yousef, *Iterative Methods for Sparse Linear Systems*, PWS Publishing, Boston, 1996.

13. Smith, G. D., *Numerical Solution of Partial Differential Equations: Finite Difference Methods*, The Clarendon Press, Oxford University Press, New York, 1985.

14. Strang, Gilbert, and Fix, George; *An Analysis of the Finite Element Method*, Wellesley-Cambridge, Cambridge, MA, 2nd edition, 2008.

15. Strikwerda, John, *Finite Difference Schemes and Partial Differential Equations*, Wadsworth & Brooks/Cole, Pacific Grove, CA, 1989.

16. Watkins, David, *Fundamentals of Matrix Computations*, John Wiley & Sons, New York, 3rd Edition, 2010.

17. Young, David, *Iterative Solution of Large Linear Systems*, Academic Press, New York, 1971.

CHAPTER 10

AN INTRODUCTION TO SPECTRAL METHODS

In Chapters 6 and 9 we discussed using finite difference and finite element methods to get approximate solutions to ordinary and partial differential equations. Both approaches can produce very accurate approximations, and a lot of important computations are being done these days using both of these techniques.

But in §4.9 we did something different—we defined our approximation as a linear combination of *smooth* functions, and imposed the differential equation at a discrete set of points, which we called *collocation points*. Also in Chapter 4, we observed that the Chebyshev polynomials can produce very accurate approximations to functions. So we have an obvious question to ask and explore: Can we use the Chebyshev polynomials to produce approximate solutions to differential equations that are as accurate as the approximations in §4.11.2 might suggest? The answer is "yes" (we wouldn't have posed the question if it couldn't be done), and the resulting body of work is generally known, collectively, as *spectral methods*, although a variety of terms are used.[1] The presentation here relies heavily on the work of Boyd [2] and Trefethen [9].[2]

It should be said that the methods as presented here are in some ways suboptimal. "True" spectral methods rely on some techniques not covered in this text, in order to achieve their full efficiency, but it was thought worthwhile to expose students (and perhaps some instructors) to these techniques in a way that was not cluttered with lengthy digressions into

---

[1] Some authors call what we do here the *pseudospectral method*, while others call it *spectral collocation*.

[2] The author would like to express his appreciation to both Prof. Boyd and Prof. Trefethen for their valuable assistance in learning this material well enough to write this chapter. Any errors are of course the author's responsibility

An Introduction to Numerical Methods and Analysis, Second Edition. By James F. Epperson
Copyright © 2013 John Wiley & Sons, Inc.

other material. The presentation in this chapter is also long on illustration and exposition (by example) and very short on theory. This is deliberate. The mathematical foundations of spectral methods are quite deep, and getting lost in those details simply for the sake of theoretical completeness would, in the author's opinion, get in the way of demonstrating the power of these methods, which is the main point. Finally, I should mention the chebfun project[3], which enables MATLAB users to explore these types of approximations with a very easy-to-use interface.

To outline the basic ideas, we will begin with an introduction to spectral methods for ODEs.

10.1 SPECTRAL METHODS FOR TWO-POINT BOUNDARY VALUE PROBLEMS

We begin by summarizing a number of results about Chebyshev polynomials from Chapter 4:

Definition:

$$T_n(x) = \cos(n \arccos x). \tag{10.1}$$

Orthogonality relation:

$$\int_{-1}^{1} \frac{T_i(x)T_j(x)}{\sqrt{1-x^2}} dx = 0, \quad i \neq j,$$

Explicit form of the first five polynomials:

$$
\begin{aligned}
T_0(x) &= 1, \\
T_1(x) &= x, \\
T_2(x) &= 2x^2 - 1, \\
T_3(x) &= 4x^3 - 3x, \\
T_4(x) &= 8x^4 - 8x^2 + 1.
\end{aligned}
$$

Three-term recurrence relation:

$$T_{n+1}(x) = 2xT_n(x) - T_{n-1}(x). \tag{10.2}$$

We consider almost the same type of problem that we studied in §§2.7 and 6.10.3, namely,

$$
\begin{aligned}
-u'' + u &= f(x), \quad -1 \leq x \leq 1, \tag{10.3} \\
u(-1) &= 0, \tag{10.4} \\
u(1) &= 0. \tag{10.5}
\end{aligned}
$$

The change from the interval $(0,1)$ to $(-1,1)$ is to accommodate the special properties of the Chebyshev polynomials, which are defined on $(-1,1)$.

We look for an approximate solution in the form

$$u_N(x) = \sum_{i=1}^{N} v_i T_{i-1}(x). \tag{10.6}$$

---

[3]http://www2.maths.ox.ac.uk/chebfun/.

(The shift in the index for the Chebyshev polynomial is to ensure that we include $T_0(x) = 1$ in our basis.) The simplest way to proceed is to impose that u_N satisfies the differential equation (10.3) at $N-2$ collocation points $\{\zeta_i\}$, and at the same time force $u_N(x)$ to satisy the boundary conditions

$$u_N(-1) = \sum_{i=1}^{N} v_i T_{i-1}(-1) = 0$$

and

$$u_N(1) = \sum_{i=1}^{N} v_i T_{i-1}(1) = 0.$$

This will lead to a system of N equations in N unknowns.

Let's illustrate this by working through two specific cases in some detail.

■ EXAMPLE 10.1

First we will take perhaps the smallest possible case, $N = 3$. We therefore are looking for an approximation in the form

$$u_3(x) = v_1 T_0(x) + v_2 T_1(x) + v_3 T_2(x).$$

Imposing the boundary conditions gives us two equations:

$$\begin{aligned} v_1 T_0(-1) + v_2 T_1(-1) + v_3 T_2(-1) &= 0, \\ v_1 T_0(1) + v_2 T_1(1) + v_3 T_2(1) &= 0. \end{aligned}$$

We get a third equation by imposing the differential equation at an interior point. The obvious choice seems to be $x = 0$, because that is the middle of the interval. We thus have

$$v_1 \left(-\frac{d^2 T_0}{dx^2} + T_0 \right)(0) + v_2 \left(-\frac{d^2 T_1}{dx^2} + T_1 \right)(0) + v_3 \left(-\frac{d^2 T_2}{dx^2} + T_2 \right)(0) = 1.$$

Because of the simplicity of the Chebyshev polynomials, it is easy to work out this system to get the following 3×3 matrix problem:

$$\begin{bmatrix} 1 & -1 & 1 \\ 1 & 0 & -5 \\ 1 & 1 & 1 \end{bmatrix} \begin{bmatrix} v_1 \\ v_2 \\ v_3 \end{bmatrix} = \begin{bmatrix} 0 \\ 1 \\ 0 \end{bmatrix}.$$

Note that we have organized this with the left endpoint ($x = -1$) as the first row, the right endpoint ($x = 1$) as the bottom row, and the interior point ($x = 0$) as the middle row. This system can be easily solved to get

$$v_1 = \frac{1}{6}, \quad v_2 = 0, \quad v_3 = -\frac{1}{6}.$$

The exact solution is

$$u(x) = 1 - \beta e^x - \beta e^{-x}, \quad \beta = \frac{e}{e^2 + 1}.$$

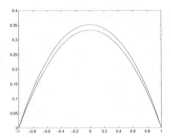

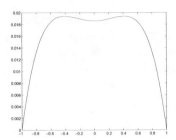

Figure 10.1   Spectral solution for $N = 3$.

Figure 10.2   Spectral error for $N = 3$.

(This is slightly different from what we had in §§2.7 and 6.10.3, again because the interval is different.) Figs. 10.1 and 10.2 plot the approximation, the exact solution, and the error, which seems to compare favorably to what we got for finite differences in §2.7 and finite elements in §6.10.3.

Even though we only used three functions to compute our approximation, we still got decent accuracy compared to the finite difference and finite element cases. What happens if we take more terms in our approximation?

■ **EXAMPLE 10.2**

Now let's consider the same problem, but this time we will take $N = 5$ terms in our approximation. We are therefore looking for the coefficients in the expansion

$$u_5(x) = v_1 T_0(x) + v_2 T_1(x) + v_3 T_2(x) + +v_4 T_3(x) + v_5 T_4(x).$$

Our general process is exactly as before. Imposing the boundary conditions leads to the two equations

$$v_1 T_0(-1) + v_2 T_1(-1) + v_3 T_2(-1) + v_4 T_3(-1) + v_5 T_4(-1) = 0,$$
$$v_1 T_0(1) + v_2 T_1(1) + v_3 T_2(1) + v_4 T_3(1) + v_5 T_4(1) = 0,$$

which become

$$v_1 - v_2 + v_3 - v_4 + v_5 = 0,$$
$$v_1 + v_2 + v_3 + v_4 + v_5 = 0.$$

We now need to impose our approximation at three interior collocation points. How are these defined? It might seem obvious to take the middle three from the equidistant set of points

$$\zeta_k^{(5)} = \left\{ -1, -\frac{1}{2}, 0, \frac{1}{2}, 1 \right\},$$

but it is in fact better to use the so-called *Gauss–Lobatto Chebyshev points* defined by[4]

$$\zeta_k^{(5)} = \cos \frac{\pi k}{N - 1}. \tag{10.7}$$

---

[4]Note that these are not the same as the *Chebyshev nodes* defined in §4.12.3.

We will work this example with both sets of collocation points. Regardless of our choice of points, we need to construct and solve the 5×5 system

$$
\begin{bmatrix}
1 & -1 & 1 & -1 & 1 \\
1 & a_{22} & a_{23} & a_{24} & a_{25} \\
1 & a_{32} & a_{33} & a_{34} & a_{35} \\
1 & a_{42} & a_{43} & a_{44} & a_{45} \\
1 & 1 & 1 & 1 & 1
\end{bmatrix},
\begin{bmatrix}
v_1 \\ v_2 \\ v_3 \\ v_4 \\ v_5
\end{bmatrix}
=
\begin{bmatrix}
0 \\ 1 \\ 1 \\ 1 \\ 0
\end{bmatrix}
\tag{10.8}
$$

where the "interior" coeficients are defined by

$$
a_{kj} = \left(-\frac{d^2 T_{j-1}}{dx^2} + T_{j-1} \right) (\zeta_k^{(5)}),
\tag{10.9}
$$

for $k = 1, 2, 3, 4$.

Case 1: Equidistant points. Here $\zeta_k^{(5)} = \left\{ -1, -\frac{1}{2}, 0, \frac{1}{2}, 1 \right\}$, and a little work with (10.8) and (10.9) shows that the matrix is

$$
A =
\begin{bmatrix}
1 & -1 & 1 & -1 & 1 \\
1 & -0.5 & -4.5 & 13 & -8.5 \\
1 & 0 & -5 & 0 & 17 \\
1 & 0.5 & -4.5 & -13 & -8.5 \\
1 & 1 & 1 & 1 & 1
\end{bmatrix}.
$$

The resulting solution and error are plotted in Figs. 10.3 and 10.4. Note that the approximate and exact solutions are indistinguishable on the graph. The maximum absolute error is 3.0949×10^{-4}.

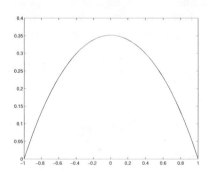

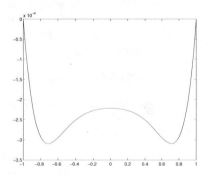

Figure 10.3 Spectral solution for a uniform grid, $N = 5$.

Figure 10.4 Spectral error for a uniform grid, $N = 5$.

Case 2: Chebyshev points. We have

$$
\zeta_k^{(5)} = \{ -1, -0.7071, 0, 0.7071, 1 \},
$$

and, again, a little work with (10.8) and (10.9) shows that the matrix now is

$$
A =
\begin{bmatrix}
1 & -1 & 1 & -1 & 1 \\
1 & -0.7071 & -4 & 17.6777 & -33.0000 \\
1 & 0 & -5 & 0 & 17.0000 \\
1 & 0.7071 & -4 & -17.6777 & -33.0000 \\
1 & 1 & 1 & 1 & 1
\end{bmatrix}
$$

The solution coefficients are

$$v = (0.1796, \, 0.0000, \, -0.1761, \, 0.0000, \, -0.0035)^T.$$

The resulting solution and error are plotted in Figs. 10.5 and 10.6, and the approximate and exact solutions are again indistinguishable on the graph. The maximum absolute error is 1.6849×10^{-4}, slightly better than the uniform grid. (Note that the coefficients of the odd Chebyshev polynomials are zero; we will revisit this issue later.)

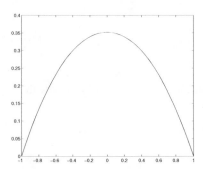

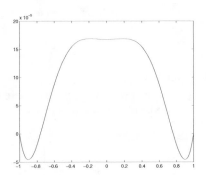

Figure 10.5   Spectral solution for Chebyshev points, $N = 5$.

Figure 10.6   Spectral error for Chebyshev points, $N = 5$.

So, what do we take away from this pair of examples? The following list shows a comparison of the error in our two spectral approximations for $N = 5$ with the finite difference (FD) and finite element (FEM) approximations for $h = 0.125$ (which implies taking seven points in the approximation).

- Finite Difference method (§2.7): 1.3314×10^{-4};

- Finite Element method (§6.10.3): 1.3366×10^{-3};

- Spectral method using uniform grid: 3.0949×10^{-4};

- Spectral method using Chebyshev grid: 1.6849×10^{-4}.

There's a sense in which this is an apples-and-oranges comparison, because the problems are posed on different intervals and are using different discretization parameters, but the point we want to emphasize is that we got *comparable* accuracy (actually, slightly better) with fewer computational elements. Taking more terms in the spectral approximations would show almost exponentially fast convergence (as we shall eventually demonstrate). However, there are some issues to be addressed. Note that the matrices for the spectral methods are *full*, whereas we should recall that the matrices for the FD and FEM methods were tridiagonal. In fact, looking at the two 5×5 matrices in Example 10.2, it is difficult to discern much structure at all. In addition, our construction of the matrices was, to be frank, rather "clunky." Nonetheless, there is quite a bit of structure to these matrices (although we won't use much of it in our discussion), and we will soon develop a more robust and orderly means of constructing them. The fact that they are full matrices is compensated for by the fact that, to get high accuracy, we do not need to take nearly as large a matrix as we did for the FD or FEM approximations.

To develop a more coherent construction of the matrices, let's now consider the more general BVP

$$
\begin{aligned}
Lu &= f(x), \quad -1 \le x \le 1, \\
u(-1) &= a, \\
u(1) &= b,
\end{aligned}
$$

where L is the differential operator defined by

$$
Lv = -p(x)v'' + q(x)v' + r(x)v.
$$

(Note the minus sign on the second derivative term.) If we wish to write this operator applied to a function w and then evaluated at a specific point x, we will write $(Lw)(x)$. So, constructing our interior matrix elements amounts to evaluating

$$
a_{kj} = (LT_j)(\zeta_k^{(N)}) \quad 1 \le j \le N - 1, \tag{10.10}
$$

where k ranges over the "interior" points of our grid. The challenge now is to develop an efficient means of computing these values.

The most efficient means involves the concept of *differentiation matrices*. The idea is to develop a matrix D such that, for a set of points arranged in a vector v, Dv returns the derivative at each point on the collocation grid. We will use the approach favored by Boyd [2], which relies on the properties of the Chebyshev polynomials, especially their definition in terms of cosines:

$$
T_n(x) = \cos(n \arccos x).
$$

An obvious consequence of this is that if $x = \cos t$, then $T_n(x) = \cos nt$. We need to develop corresponding formulas for derivatives of T_n.

Theorem 10.1 *If $x = \cos t$, then*

$$
\frac{dT_n}{dx}(x) = \frac{n \sin nt}{\sin t} \tag{10.11}
$$

and

$$
\frac{d^2 T_n}{dx^2}(x) = -\left(\frac{n^2 \cos nt}{\sin^2 t} \right) + \left(\frac{n \sin(nt) \cos t}{\sin^3 t} \right). \tag{10.12}
$$

Proof: This is a simple, if lengthy, exercise in the chain rule plus some trigonometry. From (10.1) we have

$$
\begin{aligned}
\frac{dT_n}{dx}(x) &= -n \sin(n \arccos x) \frac{d}{dx} \arccos x = -n \sin(n \arccos x) \left(\frac{-1}{\sqrt{1 - x^2}} \right) \\
&= \frac{n \sin nt}{\sin t}.
\end{aligned}
$$

To get the second derivative result, we compute from the above:

$$
\begin{aligned}
\frac{d^2 T_n}{dx^2}(x) &= \frac{d}{dx} \left(n \sin(n \arccos x) \frac{1}{\sqrt{1 - x^2}} \right) \\
&= n^2 \cos(n \arccos x) \left(\frac{-1}{1 - x^2} \right) - \frac{1}{2} n \sin(n \arccos x) \left(\frac{-2x}{(1 - x^2)^{3/2}} \right) \\
&= -n^2 \left(\frac{\cos nt}{\sin^2 t} \right) + \left(\frac{n \sin(nt) \cos t}{\sin^3 t} \right),
\end{aligned}
$$

and we are done.●

(We can, of course, compute higher-order derivatives in a similar fashion. See Problem 4.)

So how do we use these to implement our spectral methods? Given a set of nodes $\zeta_k^{(N)}$, find the values $t_k^{(N)}$ such that $\zeta_k^{(N)} = \cos t_k^{(N)}$; i.e., $t_k^{(N)} = \arccos \zeta_k^{(N)}$. For the Chebyshev nodes this is especially simple, since the nodes are defined as cosines. For the uniform grid—which our brief experiments suggest is suboptimal—we will have to do a modest amount of computation. Then use the $t_k^{(N)}$ values to compute the matrix values as defined in (10.10).

To illustrate this approach, we return to our simple example, this time running a sequence of tests for $5 \le N \le 12$. Table 10.1 shows the maximum error for each case, for both grid choices. Note that both node sets achieved comparable accuracy, but the main point is that we got 10^{-9} accuracy using only nine terms in the expansion (solving a 9×9 full matrix problem). The finite difference method, using $h = 1/1024$, got only 10^{-8} accuracy, but it was solving a 1024×1024 tridiagonal matrix problem.

Table 10.1   Errors in spectral example: $-u'' + u = 1$

N	Uniform grid	Chebyshev grid
5	3.094864761792604e–4	1.684850110583724e–4
6	1.854730677157135e–4	7.071177963646758e–5
7	2.905885602449976e–6	6.223496339663459e–7
8	1.837634214518768e–6	3.265847972055980e–7
9	1.735935828417467e–8	1.287307857533726e–9
10	1.133209905079813e–8	8.136106788292351e–10
11	7.135687873915942e–11	1.900909984975385e–12
12	4.752973015165196e–11	1.564748330906696e–12

An imperfect estimate of the operations counts would have the finite difference method costing about $3,000$ flops (for $h = 1/1024$), and the spectral method about 243 flops (for the $N = 9$ case). This does ignore the cost of forming the spectral matrix, which involves significant computations with sines and cosines. (MATLAB's flops command says that the author's code used 2111 flops for $N = 9$, but the author makes no claims about the ultimate efficiency of his programming.) But there is no question that the spectral collocation method delivers high accuracy for modest effort.

Even though we are not going to present a lot of analysis underlying these methods, we can learn a lot by looking at some data. Consider Fig. 10.7, which shows a semilog plot of the (absolute value of) the spectral coefficients for an $N = 24$ expansion for this same example. There are two things to note about this plot: (1) The oscillations in the size of the coefficients; and, (2) The "flattening" of the graph toward the high end. Both of these are easily explainable:

- The oscillations are due to the example. Note that the exact solution is symmetric about the line $x = 0$. Another way to say this is that u is an *even* function. This means a polynomial expansion—including a Chebyshev expansion—has only even coefficients; thus, the odd terms of our approximation have coefficients that are essentially zero, which is where the oscillations come from. Recall that this was noted in Example 10.2.

- The flattening for the high-index coefficients is because the computation is becoming so accurate that we have hit the rounding-error limit in the computation.

Fig. 10.8 looks very similar to Fig. 10.7, but it is a different plot which reveals different information. Here we have done a semilog plot of the *last* coefficient in the approximation; the oscillations and plateau are caused by the same phenomena as in Fig. 10.7. However, it can be shown that a useful "rule of thumb" for spectral approximations is

$$\|u - u_N\| \sim \mathcal{O}(|v_N|),$$

where $|v_N|$ is the absolute value of the last coefficient used in the approximation. We have to be careful when using this kind of informal result—recall that the odd coefficients are going to be essentially zero since the solution is an even function. Fig. 10.9 shows the last (even) coefficient and the actual error for $5 \leq N \leq 32$, from which we see that this is a very approximate estimation, but still a useful one.

(Could we take advantage of the known symmetry to compute an approximation using only even-order polynomials? Certainly—see Problem 2.)

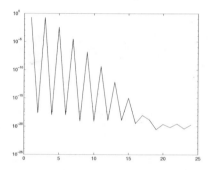

Figure 10.7 Semilog plot of coefficients for $N = 24$ spectral expansion of (10.3)–(10.5).

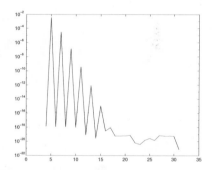

Figure 10.8 Semilog plot of the last coefficient for (10.3)–(10.5) as a function of N.

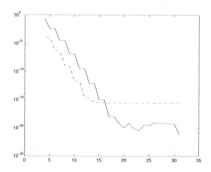

Figure 10.9 Plot of actual error (solid) and last even coefficient (dashed).

So why use anything else? Spectral methods are very accurate for problems with *smooth* solutions, such as our initial example. There are many applications in which the solution is not very smooth. Spectral methods applied to those examples will not work as well. Perhaps the simplest example is the problem

$$-u'' + Vu' + u = f(x), \quad -1 \le x \le 1, \tag{10.13}$$

$$u(-1) = 0, \quad u(1) = 1, \tag{10.14}$$

for V large and positive. Now this is simply a challenging problem—the finite difference and finite element methods would also have trouble with it—but it serves to illustrate a situation for which spectral methods do not work so well. The exact solution is graphed in Fig. 10.10 for $V = 50$. The difficulty is the sharp gradient in the solution near the right endpoint. Fig. 10.11 shows a spectral solution on the Chebyshev grid, using $N = 8$ terms (and $V = 50$), and it is evident that the approximation is not doing well. Taking more terms does address the issue, as Fig. 10.12 ($N = 20$) indicates, although we still have not converged, as the approximation is still slightly oscillating.

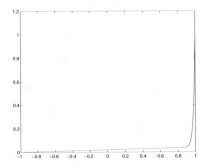

Figure 10.10   Exact solution for (10.13)–(10.14), $V = 50$.

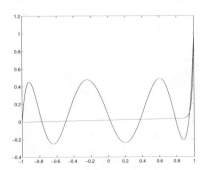

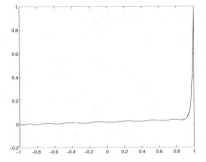

Figure 10.11 Approximation to (10.13)–(10.14), $V = 50$, on a Chebyshev grid, using eight terms.

Figure 10.12 Approximation to (10.13)–(10.14), $V = 50$, on a Chebyshev grid, using 20 terms.

We close this section by discussing an alternative means of handling the boundary conditions. What we have done so far is called "boundary bordering" in the literature, and for most ODE problems it works well. But there are circumstances—the treatment of time-dependent PDEs (next section) is one example—in which a different approach is better. Suppose that we wish to solve the BVP:

$$\begin{aligned}
-u'' + u &= 1, \quad -1 \le x \le 1, \\
u(-1) &= 1, \\
u(1) &= 0,
\end{aligned}$$

Instead of imposing the boundary conditions directly, we can modify our basis of Chebyshev polynomials so that each basis function satisfies homogeneous boundary conditions, thus:

$$\begin{aligned}
C_{2n}(x) &= T_{2n}(x) - T_0(x) & (10.15) \\
C_{2n+1}(x) &= T_{2n+1}(x) - T_1(x) & (10.16)
\end{aligned}$$

This is called "basis recombination"[5]. If the problem in question has inhomogeneous boundary conditions, then we modify our approximation by adding a simple function that satisfies those conditions. Let's illustrate all this with one more example.

■ **EXAMPLE 10.3**

Consider the BVP (with inhomogneous boundary conditions):

$$\begin{aligned}
-u'' + u &= 1, \quad -1 \le x \le 1, \\
u(-1) &= 1, \\
u(1) &= 0,
\end{aligned}$$

We will look for a solution in the form

$$u_N(x) = \frac{1}{2}(1 - x) + \sum_{k=1}^{N} c_k C_{k+1}(x),$$

where the C_k are defined as above. The exact solution is (the reader should confirm this) $u(x) = 1 + Ae^{-x} + Be^x$, for $A = \frac{e}{e^4-1}$, $B = \frac{-e^3}{e^4-1}$. We need to compute

$$a_{jk} = (LC_{j+1})(\zeta_k^{(N)}) \quad 1 \le j, k \le N,$$

and

$$f_j = 1 - (L\phi)(\zeta_k^{(N)}), \quad \phi(x) = \frac{1}{2}(1 - x), \quad 1 \le j \le N,$$

where L is the differential operator

$$L = -\frac{\partial^2}{\partial x^2} + I,$$

and then solve the matrix problem $Av = f$ to get the approximation

$$u_N = \sum_{j=1}^{N-2} v_j C_{j+1};$$

the index offset for C is because our shifted basis has no C_0 or C_1. Table 10.2 shows

---

[5] Both terms are from Boyd [2].

Table 10.2 Errors in Example 10.3.

N	Error on Chebyshev grid
5	6.823559855666517e−4
6	4.209627083873180e−5
7	3.047557035618098e−6
8	1.632923987138213e−7
9	1.098231416918338e−8
10	4.289586463812611e−10
11	2.306366209126054e−11
12	7.824851877558103e−13

the maximum absolute error for $5 \le N \le 12$, and Fig. 10.13 shows the error for $N = 10$, both using the Chebyshev nodes.[6] Note the oscillatory nature of the error over the interval. This is typical of spectral methods, and is indicative of a very accurate approximation.

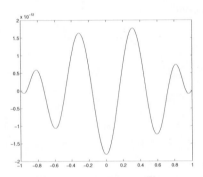

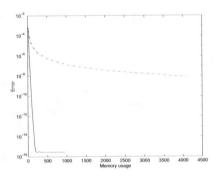

Figure 10.13 Error in Example 10.3, $N = 10$.

Figure 10.14 Error vs. memory use, finite difference (dashed) and spectral (solid).

Basis recombination does require some careful adjustment of the computation, however. By subtracting out two of our basis elements, we are essentially removing them from our approximation set, so a five-term approximation (using T_0, T_1, T_2, T_3, T_4) becomes a three-term approximation (using C_2, C_3, C_4). We use the same collocation points as before, but ignore the boundary points since the (homogeneous) boundary conditions are satisfied automatically; thus, we still have a square system, but it is 3×3 instead of 5×5.

It should be evident from our examples that spectral methods can be *very* accurate. But we have done essentially no analysis of the method—we haven't even explained where the particular collocation points (10.7) came from. So, did we simply choose "nice" examples? *No!* Spectral methods, for problems with smooth solutions, deliver very high accuracy for the resources used. They tend to be *memory minimizing*, in the sense that they use a small

---

[6]From now on we will use only the Chebyshev nodes.

amount of memory compared to finite difference or finite element methods. This might seem counterintuitive, since the matrices produced for spectral methods are full, and those for difference or element methods are tridiagonal. But let's look at some numbers; Fig. 10.14 plots the errors for Examples 2.8 and 10.3 against the memory requirements for the associated linear algebra problems (again, using a semilog scale). Note that the spectral method yields a *lot* more accuracy for a lot less memory use than did the finite difference method.

This is an imperfect comparison, because the problems are posed on different domains ($[-1, 1]$ vs. $[0, 1]$), but it still shows that spectral methods deliver much more accuracy for the memory required than do finite difference methods. A comparison with a finite element approximation would have given a very similar graph.

One problem with spectral approximations is that the matrices can be very poorly conditioned as N increases. Since we do not usually need to take large values of N, this is not often an issue, but it can become one. In Fig 10.15 we plot the condition number as a function of N (again, using a semilog plot) for the matrices generated in Example 10.3.

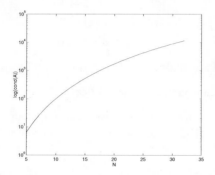

Figure 10.15   Condition number of spectral matrix as a function of N.

Exercises:

1. Write a program to solve the BVP

$$
\begin{aligned}
-u'' + 5u' + u &= 1, \quad -1 \le x \le 1 \\
u(-1) &= 0 \\
u(1) &= 1
\end{aligned}
$$

using the spectral method, with either "boundary bordering" or "basis recombination." The exact solution is

$$u(x) = Ae^{r_1 x} + Be^{r_2 x} + 1$$

for $r_1 = 5.19258$, $r_2 = -0.19258$, and

$$A = 0.003781 \quad B = -0.824844$$

(confirm this); plot the solution and the error, and produce a table of maximum absolute errors, for $4 \le N \le 32$.

2. Solve the BVP in Example 10.2, for $4 \leq N \leq 32$, but take advantage of the fact that the solution is even by looking for an approximation that uses only the even Chebyshev polynomials.

3. It has been suggested that a better way to do "basis recombination" would be as follows:

$$
\begin{aligned}
C_{2N}(x) &= C_{2N}(x) - C_{2N-2}(x) \\
C_{2N-1}(x) &= C_{2N-1}(x) - C_{2N-3}(x)
\end{aligned}
$$

Repeat Problem 1 using this basis. In addition to the same plots as requested in Problem 1, plot the condition number of the matrix A as a function of N for both methods.

4. Extend the work in Theorem 10.1 to include third and fourth derivatives.

5. Use the formulas from Problem 4 to approximate the solution to

$$
u'''' = 1,
$$
$$
u(-1) = u(1) = 0,
$$
$$
u'(-1) = u'(1) = 0.
$$

Compute a spectral approximation for $4 \leq N \leq 32$. Plot your solution for $N = 16$.

6. Compute the condition number of the matrices in Problem 5, as a function of N.

7. As an alternative to the trigonometric formulas from Theorem 10.1, we could use the three-term recursion (10.2) as a basis for constructing the spectral coefficient matrix. Show that

$$
T'_{n+1}(x) = 2T_n(x) + 2xT'_n(x) - T'_{n-1}(x), \quad T'_0(x) = 0, \quad T'_1(x) = 1,
$$

and similarly for the second derivative. Write a program to solve Problem 1 in this way. Use MATLAB's flops command to compare the costs of forming the spectral coefficient matrix this way, compared to the procedure outlined in the text.

10.2   SPECTRAL METHODS FOR TIME-DEPENDENT PROBLEMS

The standard way to apply spectral methods to time-dependent problems is to use a spectral approximation for the spatial variables, which reduces the original problem to an ODE system. This "method of lines" approach allows us to then apply any ODE method to obtain the final approximation. We retain spectral accuracy in the spatial variables, but only the accuracy of the ODE method for the variation in time. This is considered acceptable because it is generally less costly to take a smaller time step than to refine the spatial approximation.

We will make all this more specific by considering the following example problem:

$$
\begin{aligned}
u_t &= u_{xx}, \quad -1 \leq x \leq 1 & (10.17) \\
u(-1, t) &= u(1, t) = 0, \quad t > 0, & (10.18) \\
u(x, 0) &= \cos \pi x / 2 - \sin \pi x, & (10.19)
\end{aligned}
$$

which has the exact solution (verify this) $u(x,t) = e^{-(\pi/2)^2 t}\cos \pi x/2 - e^{-\pi^2 t}\sin \pi x$. (This is essentially the same example we looked at in §§9.1 and 9.2, adjusted for the interval $(-1,1)$.) We look for an approximation in the form

$$u_N(x,t) = \sum_{k=1}^{N-1} v_k(t) C_{k+1}(x),$$

where the C_i are defined as in (10.15)–(10.16). If we substitute this into the PDE (10.17) and evaluate at the collocation points $\{\zeta_j^{(N)}\}_{j=1}^{j=N-2}$, we quickly get the equation

$$\sum_{k=1}^{N-1} v_k'(t) C_{k+1}(\zeta_j^{(N)}) = \left(\sum_{k=1}^{N-1} v_k(t) C_{k+1}''(\zeta_j^{(N)})\right), \quad 1\le k \le N-1, \quad 1 \le j \le N-2.$$

The boundary conditions are automatically satisfied, and thus we have the following ODE system for the coefficients $v_k(t)$:

$$Mv'(t) = Kv(t), \quad v(0) = v_0, \tag{10.20}$$

where the matrices M and K are defined by

$$M = \begin{bmatrix} C_2(\zeta_1^{(N)}) & C_3(\zeta_1^{(N)}) & \cdots & \cdots & C_{N+1}(\zeta_1^{(N)}) \\ C_2(\zeta_2^{(N)}) & \vdots & \vdots & \vdots & C_{N+1}(\zeta_2^{(N)}) \\ \vdots & \vdots & \vdots & \vdots & \vdots \\ C_2(\zeta_{N-2}^{(N)}) & \cdots & \cdots & \cdots & C_{N+1}(\zeta_{N-2}^{(N)}) \end{bmatrix},$$

and

$$K = \begin{bmatrix} C_2''(\zeta_1^{(N)}) & C_3''(\zeta_1^{(N)}) & \cdots & \cdots & C_{N+1}''(\zeta_1^{(N)}) \\ C_2''(\zeta_2^{(N)}) & \vdots & \vdots & \vdots & C_{N+1}''(\zeta_2^{(N)}) \\ \vdots & \vdots & \vdots & \vdots & \vdots \\ C_2''(\zeta_{N-2}^{(N)}) & \cdots & \cdots & \cdots & C_{N+1}''(\zeta_{N-2}^{(N)}) \end{bmatrix}.$$

The vector $v(t)$ is the vector of coefficients in our spectral expansion:

$$v(t) = (v_1(t), v_2(t), \ldots, v_{N-2}(t))^T,$$

and the initial condition v_0 is defined by a collocation approximation to the initial function, $u_0(x)$; i.e., we solve the linear system

$$Mw = (u_0(\zeta_1^{(N)}), u_0(\zeta_2^{(N)}), \ldots, u_0(\zeta_{N-2}^{(N)})^T$$

and set

$$v_0 = \sum_{k=1}^{N} w_k C_{k+1}(x).$$

(Note that in all of this we have adjusted the indices on the basis functions and collocation points to reflect the shift of basis, as discussed at the end of §10.1.)

To finish the approximation, we need only decide on a method to solve the ODE system (10.20). Almost any of the methods from Chapter 6 would work—although there can be stability issues, as we will see—and we will initially use the trapezoid rule method from §6.4, which is essentially the Crank–Nicolson idea from §9.1. We therefore have

$$Mv_{n+1} = Mv_n + \frac{1}{2}\Delta t(Kv_n + Kv_{n+1}),$$

or,

$$\left(M - \frac{1}{2}\Delta t K\right)v_{n+1} = \left(M + \frac{1}{2}\Delta t K\right)v_n, \quad n = 1, 2, \ldots.$$

Fig. 10.16 is a plot of the maximum absolute value of the error (as a function of time) for $N = 4, 8, 16, 32$, using $\Delta t = 1/N$ and computing out to $T_{\max} = 2.5$. In order to see things better, we have plotted the $N = 8, 16, 32$ cases separately in Fig. 10.17. Table 10.3 shows the maximum error over the spatial interval $(-1, 1)$ and the time interval $0 < t \leq 2.5$, for the indicated cases.

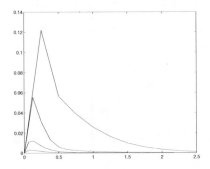

Figure 10.16 Maximum error as a function of time for trapezoidal time-stepping, $N = 4, 8, 16, 32$, $\Delta t = 1/N$.

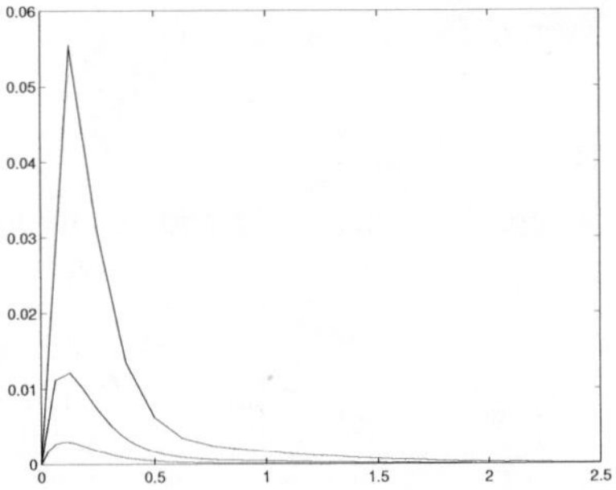

Figure 10.17 Maximum error as a function of time for trapezoidal time-stepping, $N = 8, 16, 32$, $\Delta t = 1/N$.

Table 10.3 Spectral method applied to heat equation, trapezoid rule time-stepping.

N	Maximum error	Ratio
4	0.06823696502527	N/A
8	0.00294684103558	23.1560
16	7.288544676949393e–4	4.0431
32	1.823820974664048e–4	3.9963

There are several points to take away from the plots and the table:

- The exact solution decays to zero exponentially, so the decay in the error is expected.

- For $N = 4$ we are using a very crude approximation (even for spectral methods) as well as a crude time step ($\Delta t = 0.25$), so the coarseness of the error is not unexpected for that case.

- On the other hand, for the next case, $N = 8 \Rightarrow \Delta t = 0.125$, the error is much smaller and decays smoothly to zero.

- Based on the last two entries (which is admittedly not a lot of data on which to base a conclusion) the total error appears to be going down by a factor of 4, suggesting (no surprise) that the $\mathcal{O}(\Delta t^2)$ error for the trapezoid time-stepping method is what is dominating the computation.

To get better error performance in time, we need to use a smaller time step or a more accurate ODE method. Let's look at some of our possible choices for the second option, based on material in Chapter 6:

- Fourth-order Runge–Kutta (§6.5);

- Higher-order Adams–Bashforth methods (§6.6.1);

- Higher-order Adams–Moulton methods (§6.6.1);

- Higher-order BDF methods (§6.6.2).

All of these choices are potentially effective, with different advantages and disadvantages. However, because of the issues with the stability of the explicit finite difference method for the heat equation (§9.1), we are leery of the first two options, because they are also explicit methods. So we will look at the last two options, which are both implicit and therefore (more) stable than the explicit methods.

■ **EXAMPLE 10.4   Adams–Moulton**

Since we already have a second-order method—trapezoid rule/Crank–Nicolson, which is also second-order Adams–Moulton—we will look at the third-order Adams–Moulton method, usually written as

$$y_{n+1} = y_n + \frac{h}{12} \left(5f_{n+1} + 8f_n - f_{n-1} \right),$$

where $y' = f(t, y)$ is the differential equation. Our differential equation is given by (10.20), and is a system of equations in which $f(t, y) \leftrightarrow M^{-1}Kv$. So, AM3 applied to this equation initially looks like this:

$$v_{n+1} = v_n + \frac{\Delta t}{12} \left(5M^{-1}Kv_{n+1} + 8M^{-1}Kv_n - M^{-1}Kv_{n-1} \right).$$

We can simplify this to get

$$\left(M - \frac{5\Delta t}{12} K \right) v_{n+1} = \left(M + \frac{8\Delta t}{12} K \right) v_n - Kv_{n-1}. \tag{10.21}$$

We need a single starting value, v_1, which we can get from using our trapezoid rule method for a single step; then it is a straightforward recursion to produce new solution values from (10.21). If we apply this to our example problem, using $N = 8$, $\Delta t = 1/N$, $T_{\max} = 2.5$, we get the plot in Fig. 10.18. Note that the maximum error over time begins to *increase* for $t > 1$ (roughly). This does not seem to be a good thing. Why is it happening?

Recall that multistep methods—of which Adams–Moulton is an example—have stability issues. We must take our step size small enough so that the "computation remains within the stability region" given in Fig. 6.8. What does this mean? It means

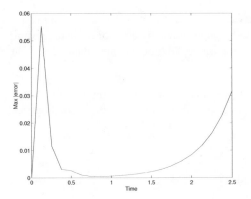

Figure 10.18 Spectral solution to (10.17) using third-order Adams–Moulton time-stepping.

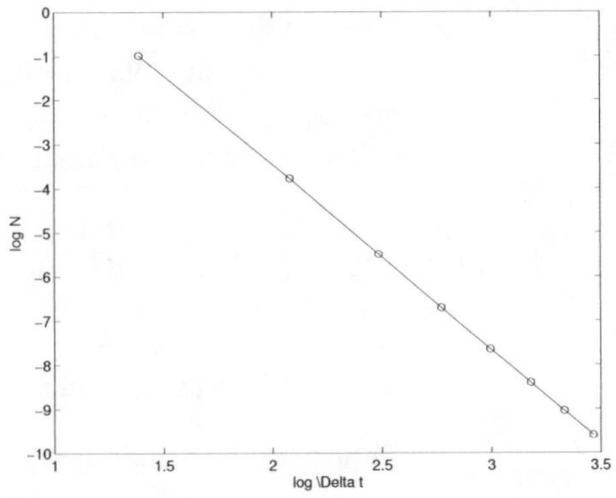

Figure 10.19 Plot (log-log) of largest stable Δt vs. N.

we must choose Δt so that $\Delta t \lambda_k$ is within the stability region for all eigenvalues λ_k of the matrix $G = M^{-1}K$. For our particular case here, we have that all the eigenvalues are real, with $\max_k |\lambda_k| = 131$ and $\Delta t = \frac{1}{8}$, so the product $\Delta t \lambda_k = 16.375$ is clearly not inside the stability region in Fig. 6.8. We need to have $\Delta t \leq 6/\lambda_{\max}$; thus, $\Delta t = 0.458 = c/N$ for $c = 0.3664$ and now $N = 8$ will suffice. If we now try $\Delta t = 0.3664/N$ for the $N = 12$ case, we find that the computation blows up catastrophically. The maximum eigenvalue for $M^{-1}K$ is now over 700—the eigenvalues are growing too fast for our choice of time steps. Fig 10.19 shows a log-log plot of the largest possible Δt as a function of N, and a least squares analysis (see §4.11) suggests that a relationship of the form

$$\Delta t = CN^{-4}$$

is present. Obviously this means that the best (largest) time step we can use is going to be rapidly decreasing. What can we do?

Frankly, we have two choices: We can live with *very* small time steps, or we can consider a more stable time-stepping method, such as the BDF family.

■ EXAMPLE 10.5 BDF

The BDF methods are stable along the entire negative real axis, so they appear to be the appropriate methods to use. The third-order BDF formula is

$$y_{n+1} - \frac{6h}{11}f(t_{n+1}, y_{n+1}) = \frac{18}{11}y_n - \frac{9}{11}y_{n-1} + \frac{2}{11}y_{n-2}$$

and the fourth-order formula is

$$y_{n+1} - \frac{12h}{25}f(t_{n+1}, y_{n+1}) = \frac{48}{25}y_n - \frac{36}{25}y_{n-1} + \frac{16}{25}y_{n-2} - \frac{3}{25}y_{n-3}.$$

Applied to our particular problem (recall $f \leftrightarrow M^{-1}K$) we have

$$Mv_{n+1} - \frac{6h}{11}Kv_{n+1} = M\left(\frac{18}{11}v_n - \frac{9}{11}v_{n-1} + \frac{2}{11}v_{n-2}\right),$$

or

$$Mv_{n+1} - \frac{12h}{25}Kv_{n+1} = M\left(\frac{48}{25}v_n - \frac{36}{25}v_{n-1} + \frac{16}{25}v_{n-2} - \frac{3}{25}v_{n-3}\right).$$

The problem is that we need accurate starting values.

We will start with the third-order BDF method, using our second-order Crank–Nicolson code to generate the two necessary starting values. (Recall from §6.4.3 that we can often get away with using a one-order-lower method to generate the starting values without incurring any significant extra error.) Figs. 10.20 and 10.21 show the evolution of the maximum absolute error over time, for $N = 4, 8, 16, 32$ points, and $N = 8, 16, 32$ points, respectively; and Table 10.4 shows the maximum absolute error over time. Two things are of concern or interest here: (1) The error does not appear to be going down by the expected factor of 8 that we should expect from an $\mathcal{O}(\Delta t^3)$ method (in fact, a least squares fit to the logarithms of the data suggests that the order is around 2); (2) In all four cases, the maximum absolute error occurred at the second or third step, i.e., at one of the starting values. Perhaps using an $\mathcal{O}(\Delta t^2)$ method to generate the starting values is not such a good idea in this case. So what do we do?

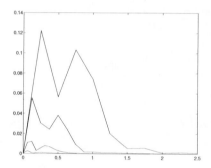

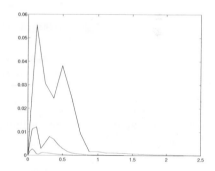

Figure 10.20   **Figure 10.21**

Table 10.4   Maximum error for BDF3 time-stepping using Crank–Nicolson starting values.

N	$\max_{t \leq 2.5} |u - u_N|$
4	0.12205771461309
8	0.05536342126983
16	0.01216091356321
32	0.00271680076022

If we had a convenient implicit third-order single-step method to use, we would do that. There exist implicit Runge–Kutta methods, but they are very difficult to derive and costly to use. Instead, we will continue using the Crank–Nicolson method for the starting values, but with shorter time steps.

The trick is to take a *series* of short time steps of length Δt, with the property that ΔT (the step we want to take with the high-order method) is an integer multiple of Δt; i.e., $\Delta T = M \Delta t$ for some M. Then our first starting value is the (approximate) solution at $M\Delta t$, and the second one is the (approximate) solution at $2M\Delta t$. We implemented this for the BDF3 scheme, using $\Delta T = 0.25/N$ and $\Delta t = \Delta T/32$. We computed out to $t = 2.5$ and tracked two different errors

$$E_{\max} = \max_{0<t\leq 2.5} \|u(\cdot,t) - u_N(\cdot,t)\|_\infty$$

(this is the maximum pointwise error in the entire computation), and

$$E_1 = \|u(\cdot,1) - u_N(\cdot,1)\|_\infty$$

(this is simply the pointwise error at the arbitrary value $t = 1$). We got the results in Table 10.5. (Note that we are no longer computing the $N = 4$ case; the results are simply too crude.)

Table 10.5 Errors for BDF3

N	$E_{\max}$	E_1
8	0.00196351553612	0.00002593734915
12	0.00062352684952	0.00000761067383
16	0.00027970078950	0.00000318853246
20	0.00014831615102	0.00000162602736
24	0.00008808570623	0.00000093852185
28	0.00005651598554	0.00000058995674
32	0.00003836696856	0.00000039469439

If we do the same kind of thing using the BDF4 formula, we get the results in Table 10.6

Table 10.6 Errors for BDF4

N	$E_{\max}$	E_1
8	0.00065906557204	0.00000351957149
12	0.00009256524086	0.00000047008815
16	0.00003163313807	0.00000014416519
20	0.00001375731857	0.00000005804163
24	0.00000691388232	0.00000002770166
28	0.00000383871479	0.00000001485340
32	0.00000230094409	0.00000000866881

We did a least squares fit on the logarithm of all this data, and produced the two plots in Fig. 10.22 and 10.23. The lines with circles are the errors $E_{\max}$, and the lines with asterisks are E_1. The unmarked line is a reference line at the slope -3 (for BDF3) and -4 (for BDF4). It is apparent that both errors are decreasing at the expected rate.

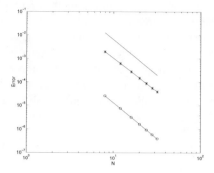

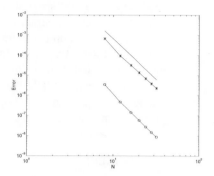

Figure 10.22   Log-log plot of errors for BDF3.

Figure 10.23   Log-log plot of errors for BDF4.

The Costs of Time-Stepping   It must be confessed that we have overlooked a number of important issues in our discussion of time-stepping. Let's consider a cost comparison of a spectral method and a finite difference method, both using Crank–Nicolson time-stepping, for an N-point approximation applied to a simple heat equation such as the one we treated here. In either case we will get a computation in the following form:

$$G_1 v_{n+1} = G_2 v_n + F$$

for matrices G_1 and G_2. What are the costs associated with these different methods?

Finite difference method: A finite difference method would require roughly $3N$ flops to form the right-hand side $G_2 v_n + F$, and a similar amount to do the solution steps of the matrix factorization, as well as the initial factorization of the matrix. So, to compute m steps would cost (very roughly)

$$C_{\text{FD}} = 6Nm + 3N$$

flops.

Spectral method: Because the spectral method uses full matrices, the costs are higher. Forming the right-hand side now costs N^2 flops (because it is a full matrix–vector multiplication), and the solution steps cost another N^2, plus we have the $\frac{1}{3}N^3$ cost of the factorization. Our total is now

$$C_{\text{spectral}} = 2N^2 m + \frac{1}{3}N^3$$

flops. Even though we get much more accuracy out of a given value of N for the spectral method, $2N^2$ will grow much faster than $6N$. Eventually, the cost of solving the problem over lengthy time intervals will become an issue.

Additionally, we have to consider that this is a *very* simple, one-dimensional linear example. If the problem is nonlinear, then there will be additional costs associated with forming the coefficient matrix on the left, and the matrix factorization will have to be repeated at each time step, perhaps multiple times, as a nonlinear system solver of some kind is employed. If the problem is in higher dimensions, the costs grow even more. All these manipulations will continue to involve full matrices for the spectral method, and only tridiagonal matrices (or banded matrices in higher dimensions) for the finite difference method. The cost of the spectral method will escalate, rapidly. At some point the user will be thinking, "Spectral accuracy is a good thing, but so is a timely solution!"

For this reason, a number of transformations are typically employed to reduce the cost of spectral methods. We do not discuss these here, because the intent of this chapter was to introduce spectral methods to the reader in a style that avoided some of the messy—but not unimportant—details. Students with access to recent editions of MATLAB are encouraged to inquire about adding the chebfun system[7] to their installation; chebfun is a very powerful graphical user interface that automates almost all of the messy details in spectral collocation.

Exercises:

1. Use spectral collocation with Crank–Nicolson time-stepping to solve the following PDE:

$$
\begin{aligned}
u_t &= u_{xx}, \\
u(-1, t) &= 0, \\
u(1, t) &= 0, \\
u(x, 0) &= \cos \pi x / 2 - \sin 4\pi x.
\end{aligned}
$$

The exact solution is $u(x, t) = e^{-\pi^2 t/4} \cos \pi x / 2 - e^{-16\pi^2 t} \sin 4\pi x$. Compute out to $t = 1$; use a sequence of values of N; plot your approximation and the error for one of them at $t = 1$.

2. Use your spectral code to solve the problem

$$
\begin{aligned}
u_t &= a u_{xx}, \\
u(-1, t) &= 1, \\
u(1, t) &= 1, \\
u(x, 0) &= (x^2 - 1)^8.
\end{aligned}
$$

Assume that $a = 1$ and compute out to $t = 1$ using a sequence of values of N. Plot the solution profile as the computation advances. Now vary a (you must keep it positive, of course) and investigate how this affects the solution.

3. Now change the initial condition to $u(x, 0) = (x^4 - 1)^8$ and repeat Problem 2.

4. Consider how to implement spectral collocation with variable coefficients. Construct the general linear system that would result from solving the problem

$$
\begin{aligned}
u_t &= a(x) u_{xx}, \\
u(-1, t) &= 0, \\
u(1, t) &= 0, \\
u(x, 0) &= u_0(x).
\end{aligned}
$$

[7] http://www2.maths.ox.ac.uk/chebfun/.

5. Apply your results from Problem 4 to approximate solutions to

$$
\begin{aligned}
u_t &= a(x)u_{xx}, \\
u(-1,t) &= 0, \\
u(1,t) &= 0, \\
u(x,0) &= \cos \pi x/2,
\end{aligned}
$$

for the following choices of a:

(a) $a(x) = (1 + x^2)$;

(b) $a(x) = e^{-x^2}$;

(c) $a(x) = (1 - x^2)$. (Because a vanishes at the boundary, this problem is known as *degenerate*, but you should be able to compute solutions.)

6. Consider the nonlinear problem

$$
\begin{aligned}
u_t &= u_{xx} + V u u_x \\
u(-1,t) &= 0, \\
u(1,t) &= 0, \\
u_0(x) &= \sin \pi x.
\end{aligned}
$$

Use spectral collocation to attack this problem, as suggested in Problem 12 of §9.1, by treating the nonlinearity explicitly (at time $t = t_n$) and the differential equation implicitly (at time $t = t_{n+1}$). Compare your spectral solution to the explicit solution computed in Problem 10 of §9.1. Comment on your results.

10.3 CLENSHAW–CURTIS QUADRATURE

A skeptical reader might wonder why a section on quadrature appears at the end of a chapter on spectral methods for ODEs and PDEs. The answer is very simple: Our spectral methods for differential equations are based on using Chebyshev expansions, and that is also the basis for Clenshaw-Curtis quadrature [4].

Consider the integration problem

$$
I(f) = \int_{-1}^{1} f(x)dx.
$$

In Gaussian quadrature (§5.6) we found N weights w_i and N abscissas ξ_i, so that the quadrature

$$
\int_{-1}^{1} f(x)dx \approx \sum_{i=1}^{N} w_i f(x_i)
$$

is exact for all polynomials of degree $\leq 2N - 1$. This produced some remarkably accurate quadrature rules, but the weights and abscissas are not easily computed.[8] The Clenshaw-Curtis idea, which leads to simpler weights and abscissas, is simply to expand f in a series

---

[8]It is possible to compute Gaussian quadrature weights and abscissas by solving a tridiagonal eigenvalue problem; see the article by Golub and Welch [7]. Trefethen gives a very tidy MATLAB code for doing this in [10].

of Chebyshev polynomials, and integrate this exactly. We have

$$\int_{-1}^{1} f(x)dx \approx \sum_{k=1}^{N} w_k^{(N)} f(\xi_k^{(N)}),$$

which is very similar in form to Gaussian quadrature. The abscissas are the same as our collocation points,

$$\xi_k^{(N)} = \cos\left(\frac{k\pi}{N+1}\right).$$

The weights can be computed very easily—much more easily than for Gaussian quadrature. Following Boyd [1], we have

$$w_k^{(N)} = \frac{2\sin\left(\frac{k\pi}{N+1}\right)}{N+1} \sum_{j=1}^{N} \sin\left(\frac{jk\pi}{N+1}\right)\left(\frac{1-\cos j\pi}{j}\right)$$

for $1 \le k \le N$.

For comparison purposes, we tested Clenshaw-Curtis quadrature against Gaussian quadrature on a variety of integrals over the range $N = 4, 8, 16, 32, \ldots, 512$. Fig. 10.24 shows log-log plots of the error vs. the number of quadrature points. The plots show that Clenshaw-Curtis (marked with an "x") is almost always slightly less accurate than Gaussian quadrature (marked with an "o"), for a given value of N. However, it is a lot easier to construct the quadrature for a given value of N, and it is similarly much easier to create an adaptive version, although we will not do that here.

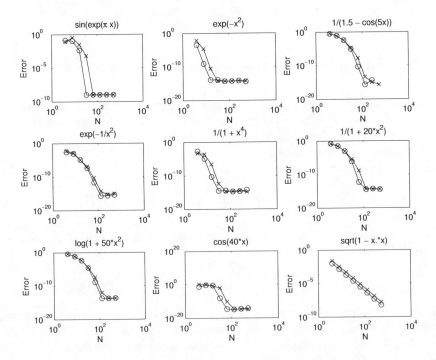

Figure 10.24   Comparison of Clenshaw-Curtis quadrature (x) with Gaussian quadrature (o).

Exercises:

1. Use the appropriate change of variable to show how to apply Clenshaw-Curtis quadrature to an integral over an arbitrary interval $[a, b]$.

2. Write a program to do Clenshaw-Curtis quadrature on each of the integrals in Problem 4 of §5.6. Compare your results to those obtained with Gaussian quadrature. Produce a log-log plot of the error as a function of N for each integral.

3. Looking at the plots in Fig. 10.24, we see that most of them show a very rapid decrease of the error, and that a "rounding error plateau" is reached for most of the examples. The exception is the last one, where the integrand is given by $f(x) = \sqrt{1 - x^2}$. Explain why this example is the one that displays this kind of suboptimal performance.

4. Let $C_N = \sum_{k=1}^{N} w_k^{(N)} f(\xi_k^{(N)}) \approx \int_{-1}^{1} f(x)dx$ be the Clenshaw–Curtis quadrature operator. Show that C_{2N} uses some of the same function values as C_N. Why is this important?

◁ ● ● ● ▷

10.4   LITERATURE AND SOFTWARE DISCUSSION

The classic reference for the spectral method is the book by Gottlieb and Orszag [8]. Trefethen's book [9], the small volume by Gheorghiu [6], and Fornberg's book [5] were all useful in preparing this chapter. There is also the book by Canuto et al. [3]. Boyd's book [2] has an unbelievably extensive bibliography, and a very unacademic style (that comment is intended as a compliment—the book is a pleasure to read). The author has to repeat, however, that there is much to the presentation here that can be criticized in terms of omissions. My goal was to present an introduction to spectral methods that could be understood and appreciated by students in their first numerical analysis course.

REFERENCES

1. Boyd, John, "Exponentially convergent Fourier–Chebyshev quadrature schemes on bounded and infinite intervals," *J. Sci. Comput.*, vol. 2, pp. 99–109, 1987.

2. Boyd, John, *Chebyshev and Fourier Spectral Methods*, Dover, New York, 2001, Second Edition; originally published by Springer-Verlag, 1989.

3. Canuto, C., Quarteroni, A., Hussaini, M. Y., and Zang, T. A., *Spectral Methods: Fundamentals in Single Domains*, Springer-Verlag, Berlin, 2006.

4. Clenshaw, C. W., and Curtis, A. R., "A method for numerical integration on an automatic computer," *Numer. Math.*, vol. 2, pp. 197-205, 1960.

5. Fornberg, Bengt, *A Practical Guide to Pseudospectral Methods*, Cambridge University Press, Cambridge, UK, 1995.

6. Gheorghiu, C. I., *Spectral Methods for Differential Problems*, Casa Cartii de Stiinta, Cluj-Napoca, Romania, 2007. (Out of print, available online at:
http://www.ictp.acad.ro/gheorghiu/spectral.pdf.

7. Golub, G. H., and Welsch, J. H., "Calculation of Gauss quadrature rules," *Math. Comp.*, vol. 23, pp. 221-230, 1969.

8. Gottleib, David, and Orszag, Steven A., *Numerical Analysis of Spectral Methods: Theory and Applications*, SIAM, Philadelphia, 1977.

9. Trefethen, L. N., *Spectral Methods in MATLAB*, SIAM, Philadelphia, 2000.

10. Trefethen, L. N., "Is Gauss quadrature better than Clenshaw-Curtis?" *SIAM Rev.*, vol. 50, no. 1, pp. 67-87, 2008.

APPENDIX A

PROOFS OF SELECTED THEOREMS, AND ADDITIONAL MATERIAL

A.1 PROOFS OF THE INTERPOLATION ERROR THEOREMS

The interpolation error theorems all depend on applying a generalization of Rolle's Theorem to a very carefully constructed function. First, we will state and prove the general version of Rolle's Theorem.

Theorem A.1 (Generalized Rolle's Theorem) *Let $f \in C^n([a,b])$ be given, and assume that there are n points, $z_k, 1 \leq k \leq n$ in $[a,b]$ such that $f(z_k) = 0$. Then there exists at least one point $\xi \in [a,b]$ such that $f^{(n-1)}(\xi) = 0$.*

Proof: By Rolle's Theorem, there exists at least one point η_k between each z_k and z_{k+1} such that $f'(\eta_k) = 0$. Thus there are $n-1$ points where $g_1(x) = f'(x)$ is zero. By the same argument, then, there are $n-2$ points where $g_2(x) = f''(x)$ is zero. Continuing onward, then, we end up with a single point where $g_{n-1}(x) = f^{(n-1)}(x)$ is zero. ●

This result allows us to prove, rather easily, both of Theorems 4.3 and 4.5.

Theorem A.2 (Lagrange Interpolation Error Theorem) *Let $f \in C^{n+1}([a,b])$ and let the nodes $x_k \in [a,b]$ for $0 \leq k \leq n$. Then, for each $x \in [a,b]$, there is a $\xi_x \in [a,b]$ such that*

$$f(x) - p_n(x) = \frac{w_n(x)}{(n+1)!} f^{(n+1)}(\xi_x).$$

An Introduction to Numerical Methods and Analysis, Second Edition. By James F. Epperson
Copyright © 2013 John Wiley & Sons, Inc.

Proof: For any $t \in [a, b], t \neq x_k$, define the function

$$G(x) = E(x) - \frac{w(x)}{w(t)} E(t),$$

where $E(x) = f(x) - p_n(x)$ and

$$w(x) = \prod_{k=0}^{n} (x - x_k).$$

Then $G(x_k) = 0$, for $0 \leq k \leq n$ and $G(t) = 0$; thus, there are $n + 2$ points where G is zero. Therefore, by the Generalized Rolle's Theorem, there is a point $\xi \in [a, b]$ such that $G^{n+1}(\xi) = 0$. But

$$
\begin{aligned}
G^{(n+1)}(x) &= E^{(n+1)}(x) - \frac{w^{(n+1)}(x)}{w(t)} E(t), \\
&= f^{(n+1)} - \frac{(n+1)!}{w(t)} E(t).
\end{aligned}
$$

Therefore, $G^{n+1}(\xi) = 0$ implies that

$$f^{(n+1)}(\xi) - \frac{(n+1)!}{w(t)} E(t) = 0 \Rightarrow f(x) - p_n(x) = \frac{w_n(x)}{(n+1)!} f^{(n+1)}(\xi_x),$$

and we are done. ●

Theorem A.3 (Hermite Interpolation Error Theorem) *Let $f \in C^{2n}([a, b])$ and let the nodes $x_k \in [a, b]$ for all $k, \quad 1 \leq k \leq n$. Then, for each $x \in [a, b]$, there is a $\xi_x \in [a, b]$ such that*

$$f(x) - H_n(x) = \frac{\psi_n(x)}{(2n)!} f^{(2n)}(\xi_x), \tag{A.1}$$

where

$$\psi_n(x) = \prod_{k=1}^{n} (x - x_k)^2.$$

Proof: Essentially the same as for the Lagrange theorem. This time we define the auxiliary function as

$$G(x) = E(x) - \frac{\psi_n(x)}{\psi_n(t)} E(t),$$

where $E(x) = f(x) - H_n(x)$. Then the Generalized Rolle's Theorem gives us a point $\xi \in [a, b]$ such that

$$0 = G^{(2n)}(\xi) = f^{(2n)} - \frac{(2n)!}{\psi_n(t)} E(t),$$

from which the result follows. ●

A.2   PROOF OF THE STABILITY RESULT FOR ODES

The theorem as stated in the text (Theorem 6.2) is the following.

Theorem A.4 *Let $f(t, y)$ be continuous for all (t, y) in a rectangle R, and smooth and uniformly monotone decreasing in y. Then for all $(t_0, y_0) \in R$ there exists a unique solution to the initial value problem*

$$y' = f(t, y), \quad y(t_0) = y_0.$$

Moreover, if $z(t)$ is the solution to the same problem with initial data $z(t_0) = z_0$, then

$$|y(t) - z(t)| \leq e^{-m(t-t_0)}|y_0 - z_0|. \tag{A.2}$$

Proof: The existence and uniqueness follows from Theorem 6.1 and was demonstrated in §6.1; all we need to do here is demonstrate the stability result (A.2). We have

$$y' = f(t, y), \quad y(t_0) = y_0,$$

and

$$z' = f(t, z), \quad z(t_0) = z_0.$$

Subtract and use the Mean Value Theorem on f to get

$$(y - z)' = f_y(t, \eta)(y - z), \quad y(t_0) - z(t_0) = y_0 - z_0.$$

Here η will be a function of y, z, and t, with value between $y(t)$ and $z(t)$. To simplify things, let's introduce some notation:

$$p(t) = f_y(t, \eta), \quad \phi(t) = y(t) - z(t), \quad \phi_0 = y_0 - z_0.$$

We therefore have

$$\phi' - p(t)\phi(t) = 0, \quad \phi(t_0) = \phi_0,$$

which is a simple first-order linear differential equation. We can solve it by introducing the function

$$P(t) = -\int_{t_0}^{t} p(s)ds$$

and the integrating factor $e^{P(t)}$. We then have

$$\left(e^{P(t)}\phi(t)\right)' = 0,$$

so we integrate both sides (from t_0 to t) to get

$$e^{P(t)}\phi(t) = \phi_0.$$

Hence,

$$\phi(t) = \phi_0 e^{-P(t)},$$

so we have

$$y(t) - z(t) = (y_0 - z_0)e^{-\int_{t_0}^{t} p(s)ds},$$

and therefore,

$$|y(t) - z(t)| \leq |y_0 - z_0| \max_{t_0 \leq s \leq t} e^{-P(t)}.$$

Maximizing the exponential on the right means that we must maximize the exponent. But

$$-P(t) = \int_{t_0}^{t} p(s)ds \leq -m(t - t_0),$$

using the smooth and uniformly monotone decreasing condition to bound p (which is f_y, remember). •

A.3   STIFF SYSTEMS OF DIFFERENTIAL EQUATIONS AND EIGENVALUES

In this part of the appendix, we provide some more details on the relationship between the stiffness of a linear system of differential equations and the eigenvalues of the coefficient matrix. As in the text, we take as our example the IVP

$$
\begin{align}
y_1' &= 198y_1 + 199y_2, \quad y_1(0) = 1; & \text{(A.3)}\\
y_2' &= -398y_1 - 399y_2, \quad y_2(0) = -1; & \text{(A.4)}\\
& & \text{(A.5)}
\end{align}
$$

which we write in matrix–vector form as

$$y' = Ay, \quad y(0) = y_0,$$

and we approximate its solution using the second-order Adams–Bashforth scheme. Thus, we look at the recursion

$$y_{n+1} = y_n + \frac{1}{2}h\left(3Ay_n - \frac{1}{2}Ay_{n-1}\right) = \left(I + \frac{3}{2}hA\right)y_n - \frac{1}{2}hAy_{n-1}. \qquad \text{(A.6)}$$

If A were a scalar instead of a matrix, it would be a simple matter to attack this recursion by looking for the solution in the form $y_n = r^n$ for some r, which we could determine by factoring a polynomial. But since A is a matrix, this approach appears fruitless.

If A were a *diagonal* matrix, we could use the old approach. In that case, each row of (A.6) would represent a scalar recursion for that component of the vector y_{n+1}. In general, A will not be a diagonal matrix, but it almost always is *diagonalizable*, and that is what we need to do.

Recall from linear algebra (see Theorem 8.1) that a matrix is diagonalizable if there exists an invertible matrix P such that $P^{-1}AP = D$, where D is diagonal. Recall further, that if the eigenvectors of A are independent, then A is diagonalizable, with P being the matrix of eigenvectors, and D the matrix of eigenvalues. We therefore have (if A is diagonalizable)

$$y_{n+1} = \left(I + \frac{3}{2}hP^{-1}DP\right)y_n - \frac{1}{2}hP^{-1}DPy_{n-1},$$

from which we get

$$Py_{n+1} = \left(I + \frac{3}{2}hD\right)Py_n - \frac{1}{2}hDPy_{n-1},$$

or, setting $z_n = Py_n$,

$$z_{n+1} = \left(I + \frac{3}{2}hD \right) z_n - \frac{1}{2}hD z_{n-1}.$$

Since D is diagonal, we can now break this apart into each separate scalar recursion, thus:

$$z_{n+1,i} = \left(1 + \frac{3}{2}hd_i \right) z_{n,i} - \frac{1}{2}hd_i z_{n-1,i},$$

where d_i are the diagonal elements of D (and, therefore, the eigenvalues of A). For simplicity, let $hd_i = \eta_i$, and solve this using our usual procedure. We get that

$$z_{n,i} = C_{1,i} r_{1,i}^n + C_{2,i} r_{2,i}^n$$

where the $r_{j,i}$ values are the roots of the polynomial

$$q(r) = r^2 - \left(1 + \frac{3}{2}\eta_i \right) r + \frac{1}{2}\eta_i.$$

If we compute the roots of this polynomial as a function of the parameter η_i (for η_i real and negative), we find that the absolute value of both roots is not less than 1 until we have $|\eta_i| \leq 1$. Recall that $\eta_i = hd_i$, where d_i are the eigenvalues (-1 and -200) of the coefficient matrix A for the ODE. Thus, in order to be within the stability region for second-order Adams–Bashforth, we need to have $h \leq 1/200$. Note that this is the case even when the e^{-200t} component of the solution is not present, and even though this component would be negligibly small after the first time step.

In the text we saw that the stiffness of this system could be "defeated" by using an implicit method to compute the numerical solution. Let's see why that worked.

If we apply the trapezoid rule method to our example, we get the recursion

$$y_{n+1} = y_n + \frac{1}{2}h(Ay_n + Ay_{n+1}) = \left(I + \frac{1}{2}hA \right) y_n + \frac{1}{2}hAy_{n+1},$$

which we can solve to get

$$\left(I - \frac{1}{2}hA \right) y_{n+1} = \left(I + \frac{1}{2}hA \right) y_n,$$

or,

$$y_{n+1} = \left(I - \frac{1}{2}hA \right)^{-1} \left(I + \frac{1}{2}hA \right) y_n.$$

When we diagonalize the recursion, we get

$$z_{n+1} = \left(I - \frac{1}{2}hD \right)^{-1} \left(I + \frac{1}{2}hD \right) z_n,$$

so the scalar recursion of interest is

$$z_{n+1,i} = \frac{(1 + \frac{1}{2}\eta_i)}{(1 - \frac{1}{2}\eta_i)} z_{n,i}.$$

It can be shown that the fraction is always less than 1 in absolute value, as long as the real part of η_i is negative. Thus, we do not require that h be sufficiently small to avoid the solution growing without bound as n increases.

A.4 THE MATRIX PERTURBATION THEOREM

Our concern here is the proof of the following theorem from Chapter 7.

Theorem A.5 *Let $A \in \mathbb{R}^{n \times n}$ be given, nonsingular, and let $E \in \mathbb{R}^{n \times n}$ be a perturbation of A. Let $x \in \mathbb{R}^n$ be the unique solution of $Ax = b$. If $\kappa(A)\|E\| < \|A\|$, then the perturbed system $(A + E)x_c = b$ has a unique solution and*

$$\frac{\|x - x_c\|}{\|x\|} \leq \frac{\kappa(A)\frac{\|E\|}{\|A\|}}{1 - \kappa(A)\frac{\|E\|}{\|A\|}}$$

Proof: We start by assuming that the perturbed matrix $A + E$ is nonsingular, and therefore that the perturbed system has a unique solution; we will then prove this as a lemma.

We have, then, the two systems

$$Ax = b$$

and

$$(A + E)x_c = b.$$

We can subtract and rearrange to get

$$A(x - x_c) = -Ex_c,$$

so that

$$
\begin{aligned}
\frac{\|x - x_c\|}{\|x\|} &\leq \frac{\|A^{-1}\|\|E\|\|x_c\|}{\|x\|} \\
&\leq \|A^{-1}\|\|E\| \left(\frac{\|x\| + \|x - x_c\|}{\|x\|} \right) \\
&= \kappa(A)\frac{\|E\|}{\|A\|} \left(1 + \frac{\|x - x_c\|}{\|x\|} \right);
\end{aligned}
$$

therefore,

$$\frac{\|x - x_c\|}{\|x\|} \leq \frac{\kappa(A)\frac{\|E\|}{\|A\|}}{1 - \kappa(A)\frac{\|E\|}{\|A\|}}.$$

The positivity of the denominator—which is required to preserve the direction of the inequality—is guaranteed by the assumptions of the theorem. •

Lemma A.1 *If $A \in \mathbb{R}^{n \times n}$ is nonsingular and $\kappa(A)\|E\| < \|A\|$, then $A + E$ is also nonsingular.*

Proof: Note that $A + E = A(I + A^{-1}E)$, so that $A + E$ is nonsingular if and only if $I + A^{-1}E$ is nonsingular. Note also that our hypothesis implies that $\|A^{-1}\|\|E\| < 1$. For simplicity, we write

$$F = -A^{-1}E.$$

Assume, then, that $I + A^{-1}E = I - F$ is singular. Then there exists $z \neq 0$ such that $(I - F)z = 0$; thus, $z = Fz$. Taking norms, then, we have that $\|z\| = \|Fz\| \leq \|F\|\|z\|$ so that $1 \leq \|F\|$, which is a contradiction. Thus, $I - F$ is nonsingular; therefore, $A + E$ is nonsingular. •

INDEX